40th Annual Conference - Meeting of the Departments of Fluid Mechanics and Thermomechanics in connection with XXIII International Scientific Conference - The Application of Experimental and Numerical Methods in Fluid Mechanics and Energy (40th MDFMT & XXIII AEaNMiFMaE-2022)

MATEC Web of Conferences Volume 369 (2022)

Piestany, Slovakia
12-14 September 2022

Editors:

Richard Lenhard
Katarina Kaduchová

ISBN: 978-1-7138-6277-2

Printed from e-media with permission by:

Curran Associates, Inc.
57 Morehouse Lane
Red Hook, NY 12571

proceedings.com

Some format issues inherent in the e-media version may also appear in this print version.

For additional information, please contact EDP Sciences – Web of Conferences at the address below.

EDP Sciences – Web of Conferences
17, Avenue du Hoggar
Parc d'Activité de Courtabœuf
BP 112
F-91944 Les Ulis Cedex A
France

Phone: +33 (0) 1 69 18 75 75
Fax: +33 (0) 1 69 28 84 91

contact-edps@webofconferences.org

Additional copies of this publication are available from:

Curran Associates, Inc.
57 Morehouse Lane
Red Hook, NY 12571 USA
Phone: 845-758-0400
Fax: 845-758-2633
Email: curran@proceedings.com
Web: www.proceedings.com

TABLE OF CONTENTS

Department of Power Engineering
Faculty of Mechanical Engineering
University of Žilina

40th annual conference

"Meeting of the Departments of Fluid Mechanics and Thermomechanics"

in the connection with

XXIII. international scientific conference

"The Application of Experimental and Numerical Methods in Fluid Mechanics and Energy"

12.09. - 14.09.2022

Piešťany – Slovakia

Organizing Committee - Editors and technical reviewers

assoc. prof. Ing. Richard Lenhard, PhD. (richard.lenhard@fstroj.uniza.sk)
Ing. Katarína Kaduchová, PhD. (katarina.kaduchova@fstroj.uniza.sk)

Guarantors of the conference

Head of department
doc. Ing. Michal Holubčík, PhD. ŽU Žilina, SR

Rector of the University of Žilina
prof. Ing. Jozef Jandačka, PhD., ŽU Žilina, SR

Scientific committee

prof. Andrzej Szlek, PhD., DSc., UT, Gliwice, PL

prof. Dr. Ing. René Pyszko, VŠB-TU Ostrava, CZ

prof. Ing. Augustín Varga, CSc., TU v Košiciach, SK

prof. Ing. František Kavička, CSc., VUT v Brně, CZ

prof. Ing. František Urban, PhD., STU v Bratislave, SK

prof. Ing. Jan Mareček, DrSc., MENDELU v Brně, CZ

prof. Ing. Jaromír Příhoda, CSc., ÚT AV ČR Praha, CZ

prof. Ing. Josef Soukup, CSc., UJEP Ústi nad Labem, CZ

prof. Ing. Josef Štětina, Ph.D., VUT v Brně, CZ

prof. Ing. Jozef Jandačka, PhD., ŽU v Žiline, SR

prof. Ing. Jozef Vlček, Ph.D., VŠB-TU Ostrava, CZ

prof. Ing. Ladislav Dzurenda, PhD., TU vo Zvolene, SK

prof. Ing. Mária Čarnogurská, CSc., TU v Košiciach, SK

prof. Ing. Miroslav Příhoda, CSc., VŠB-TU Ostrava, CZ

prof. Ing. Miroslav Rimár, CSc., TU v Košiciach, SK

prof. Ing. Pavel Šafařík, CSc., ČVUT v Praze, CZ

prof. Ing. Peter Tauš, PhD., TU v Košiciach, SK

prof. Ing. Tomáš Brestovič, PhD., TU v Košiciach, SK

prof. Ing. Václav Kopecký, CSc., TU v Liberci, CZ

prof. Ing. Václav Uruba, CSc., ZČU v Plzni, CZ

prof. Ing. Róbert Olšiak, PhD., STU Bratislava, SR

Scientific Reviewers

Ing. Marek Patsch, PhD. ŽU Žilina, SR

Ing. Patrik Nemec, PhD. ŽU Žilina, SR

The conference was supported:

Preface:
Proceedings of the 40th annual conference "Meeting of the Departments of Fluid Mechanics and Thermomechanics"

in the connection with

XXIII. international scientific conference "The Application of experimental and Numerical Methods in Fluid Mechanics and Energy"

The International Scientific Conference 40th. MDFMT & XXIII. AENMMTE has been scheduled to take place at Hotel Magnólia – Piešťany, Slovakia, between 12-14 September 2022.

The aim of the international conference is to bring together researchers, designers, experimenters, modeller's and numerical analysts from laboratories, industry and academia working in the field of two-phase and multiphase flow and experimental methods used in the field of fluid mechanics, thermodynamics and energy.

The international conference covers all topics in Measurement and calculation of state variables in the fluid flow, modelling and simulation in fluid mechanics and energy, optimization of flow parameters in thermal and hydraulic machines and systems, the application of the latest knowledge in the field of fluid mechanics and energy in technical practice, flow visualization, measurement of the energy systems and systems, special experimental methods in fluid mechanics and energy, renewable energy, current problems in the energy sector.

The conference should contribute to the establishment closer contacts, presentations on the latest knowledge, exchange of experience in the application of different experimental methods, preparation and implementation experiments, results processing and numerical simulations of experiments.

The scientific committee determined the following conference topics:

- Measurement and calculation of state variables in the fluid flow.
- Modelling and simulation in fluid mechanics and energy.
- Optimization of flow parameters in the thermal and hydraulic machines and systems.
- Application of latest developments in the field of fluid mechanics and energy in technical practice.
- Visualization of flow.
- Measurement of energy systems and systems.
- Special experimental methods in fluid mechanics and energy.
- Renewable energy.
- Current problems in the energy.

We would like thanks:

- The Scientific Committee and Reviewers of 40th. MDFMT & XXIII. AENMMTE for their help and support.
- The Session Chairman's for their considerable effort the managing of individual sections.
- The Guarantors of the conference:

  prof. Ing. Jozef Jandačka, PhD., Rector, University of Zilina, Slovakia,

  doc. Ing. Michal Holubčík, PhD., Head of department power engineering, University of Zilina, Slovakia.

Statement of Peer review

In submitting conference proceedings to *Web of Conferences*, the editors of the proceedings certify to the Publisher that

1. They adhere to its **Policy on Publishing Integrity** in order to safeguard good scientific practice in publishing.
2. All articles have been subjected to peer review administered by the proceedings editors.
3. Reviews have been conducted by expert referees, who have been requested to provide unbiased and constructive comments aimed, whenever possible, at improving the work.
4. Proceedings editors have taken all reasonable steps to ensure the quality of the materials they publish and their decision to accept or reject a paper for publication has been based only on the merits of the work and the relevance to the journal.

Title, date and place of the conference

40th annual conference "Meeting of the Departments of Fluid Mechanics and Thermomechanics"
in the connection with
XXIII. international scientific conference "The Application of experimental and Numerical Methods in Fluid Mechanics and Energy"
12-14 September 2022
Hotel Magnólia – Piešťany, Slovakia

Proceedings editor(s):

assoc. prof. Ing. Richard Lenhard, PhD. (richard.lenhard@fstroj.uniza.sk)
Ing. Katarína Kaduchová, PhD. (katarina.kaduchova@fstroj.uniza.sk)

Date and editor's signature

25. 10. 2022

Correlation Coefficients in Lee´s Model of Multiphase Flows

Natália Holešová[1,*], *Richard* Lenhard[1], *Katarína* Kaduchová[1] and *Milan* Malcho[1]

[1]Department of Power Engineering, Faculty of Mechanical Engineering, University of Žilina, Univerzitná 1, 010 26 Žilina, Slovakia

Abstract. Multiple technical and biological systems exhibit multiphase flow phenomena. The demands for accurate calculations of the physical phenomena that occur in engineering technologies have increased along with their rapid advancement. However, it is very hard to identify two- and multi-phase flow through experimental measurements, as a result, in addition to experimental measurements, numerical simulations are also performed, which can help improve our physical understanding of the complex phenomenon of phase transformations. The goal of numerical simulations of phase changes is to precisely simulate the real progress and experiment. These simulations operate on the basis of physical principles and correlation coefficients of phase changes. These correlation coefficients have a different range of values. The examination of the Lee model's correlation coefficients from the ANSYS Fluent environment, which is now the most popular for multiphase flow simulations, is the subject of the article that is being given. In this article is also described and tested the numerical simulation of interphase mass transport in a closed space.

Research background: The article is focused on the problematics of multiphase flows. In ANSYS Fluent, there are many types of models, which are used for the numerical simulations of this phenomena. In models are included correlation parameters, which are specific for every single situation and are within the given ranges. This paper is about the Lee model, which correlation coefficients are in the range from 10^{-3} to 10^{2}.
Purpose of the article: A detailed description of Lee's model with its testing on a heat pipe in the ANSYS Fluent program with determined correlation coefficients.
Methods: The use of CFD simulation of multiphase flow to determine correlation coefficients.
Findings & Value added: Testing the correlation coefficients of Lee's model and finding their appropriate values for the given situation.

Keywords: *Lee model, correlation coefficient, CFD simulation, multiphase flow*

* Corresponding author: natalia.holesova@fstroj.uniza.sk

1 Introduction

A unique characteristic of two-phase or two immiscible mixtures is the presence of one or more interfaces separating the phases or components. Examples of such flow systems can be found in a large number of engineering systems as well as in a wide variety of natural phenomena. Understanding the flow and heat transfer processes in two-phase systems is becoming increasingly important in nuclear, mechanical, and chemical engineering, as well as in the environmental and medical fields.

The design of engineering systems and the ability to predict their performance depend on the availability of experimental data and conceptual mathematical models that can be used to describe physical processes with the required degree of accuracy. It is imperative that the various characteristics and physics of two-phase flow be modelled and formulated on a rational basis and supported by detailed scientific experiments.

The evaporation and condensation are also included, which are crucial components of many important industrial processes, among the two-phase flow.

Evaporation is the process of turning a liquid into vapor at its saturation temperature, by the action of heat. Condensation, which is the reverse of evaporation, occurs when heat is removed from or absorbed by a vapor, turning it into a liquid. To understand the evaporation and condensation phenomena, several experimental analyses have been conducted. However, it is very difficult to detect these phenomena with experimental measurements, and therefore numerical simulations are usually performed with experimental measurements.

Numerical simulations can contribute to a better physical understanding of the complex phenomena of phase transformations - evaporation and condensation. The ANSYS Fluent program is currently the most widely used CFD analysis software, which enables multiphase flow modelling with heat and mass sharing for liquid-gas, liquid-solid phase combinations, and fluid-solid interaction.

The aim of the article is to describe the correlation coefficient of the Lee model, which is used for numerical simulations of interphase heat transport in a closed space, taking into account condensation and evaporation. This coefficient must be correlate to match the experimental data as closely as possible.

1.1 Lee model

Nowadays, there are already many different types of phase change models [10]. Using empirical expressions to quantify interfacial heat and mass transfer appears to be a common way to model phase change phenomena. This approach is indeed useful for certain simple geometries but limited for any other more complex geometry.

The phase change model proposed by Lee is the most commonly used. Mass transfers are given by the following equations [6], [8], [11], [13]:

If $T_l > T_{sat}$ (evaporation)

$$\dot{m}_{lv} = \lambda_c \alpha_l \rho_l \frac{T_l - T_{sat}}{T_{sat}} \tag{1}$$

If $T_v < T_{sat}$ (condensation)

$$\dot{m}_{vl} = \lambda_c \alpha_v \rho_v \frac{T_{sat} - T_v}{T_{sat}} \tag{2}$$

where $\dot{m}_{lv}, \dot{m}_{vl}$ are intensities of mass transfer due to evaporation and condensation, respectively (kg.s^{-1}.m^{-3}); α_l, α_v are phase volumes fractions (–); ρ_l, ρ_v are densities of liquid and vapor (kg.m^{-3}); T_l is the temperature of liquid (K); T_{sat} is the temperature of saturation (K) and T_v is temperature of vapor (K). λ_c denotes the mass transfer intensity factor with unit s^{-1}. It is recommended that the value of λ_c be such that it keeps the interfacial temperature reasonably close to the saturation temperature and avoids divergence problems [14].

As an empirical coefficient, λ_c has different values for different situations [9]. In numerical studies by Wu et al. [1], De Schepper et al. [2] and Alizadehdakhel et al. [3], λ_c was set to 0.1 s^{-1} and 1 s^{-1} to numerically keep the interface temperature close to the saturation temperature. However, in Yang [4] and Fang [5], λ_c was specified at a different value, 100 s^{-1}.

λ_c, which is inverse to relaxing time (s^{-1}) is defined:

$$\lambda_c = \frac{6}{d_b}\beta\sqrt{\frac{M}{2\pi R T_{sat}}}L\left(\frac{\alpha_v \rho_v}{\rho_l - \rho_v}\right) \tag{3}$$

where d_b is diameter of bubble (m); β is adaptation coefficient, which shows a part of the vapor molecules moving to the surface of the liquid and subsequently being absorbed by this surface (-); M is molar mass (kg.mol^{-1}); T_{sat} is saturation temperature (K); R is universal gas constant (J.K^{-1}.mol^{-1}); L is latent heat (J.kg^{-1}); α_v is phase volume fraction (-); ρ_v, ρ_l are densities of vapor and liquid (kg.m^{-3}).

This leads to the final expression for the evaporation process defined by equation (1). It can be considered implicitly as the source term in the conservation phase of the equation. A similar term can be obtained by condensation. In this case, even small droplets in the continuous gas phase regime are considered, even if the primary phase is liquid.

The coefficient λ_c should theoretically be different for condensation and evaporation. In addition, the theoretical expression is based on several assumptions:

- desktop interface,
- diffuse mode with constant diameter,
- known β.

The bubble diameter and the fitting coefficient are usually not known, which is why the coefficient λ_c must be tuned to match the experimental data. By default, the coefficient for evaporation and condensation is 0.1 s^{-1}. However, λ_c in practical cases can reach a value of the order of 10^2 s^{-1}.

2 Numerical simulation of multiphase flow in an enclosed space

The 2D model of the numerical non-stationary simulation of interphase mass transport in a closed space in the ANSYS Fluent program is a gravity heat pipe, which consists of an evaporative, adiabatic, and condensing part. The heat pipe contains water, steam, and a non-condensing component - air. The detail of the 2D model and the main dimensions are described in Figure 1 and Table 1.

Table 1. Dimensions of heat pipe.

V11	400 mm	V15	120 mm
V12	80 mm	V16	52 mm
H8	32 mm		

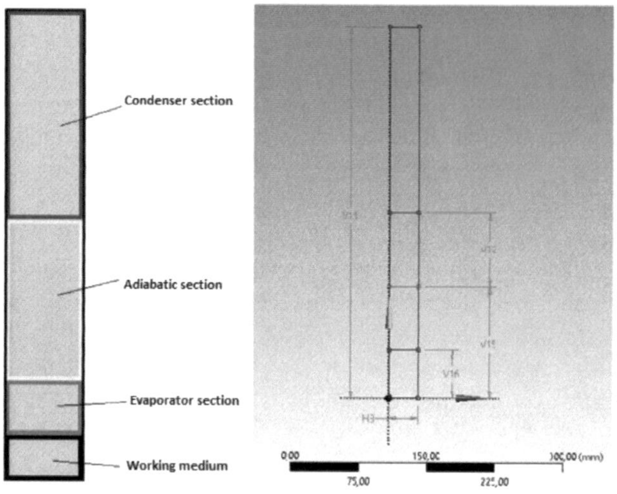

Fig. 1. Geometric parameters of heat pipe.

In multiphase mass transport (condensation and evaporation), it is important to pay attention to the modelled phenomena and adapt the computational mesh to capture these types of heat transport. During condensation and evaporation, it is necessary to thicken the mesh near the walls and just above the surfaces of the working substances, where phase changes occur. A square mesh with 57 256 cells was used to create the 2D model of the heat pipe. The Evaporation – Condensation model was used to describe the multiphase heat transport of mass, the settings of which are contained in the following tables with a description of the parameters of boundary, operational, material, and other key conditions. Data are obtained from experimental measurement and used as input values for numerical simulation. The values of the correlation coefficients are set to 0.1 s^{-1} and 10 s^{-1}, which describe the best our experiment.

Table 2. Boundary conditions.

Name	**Type of condition**	**Set parameters**		
		Velocity [m.s^{-1}]	**Temperature [°C]**	**Temperature flow [W.m^{-2}]**
chladenesteny	Wall	0	10	–
adiabatickacast	Wall	0	–	0
ohrievanesteny	Wall	0	200	–

Table 3. Operational conditions.

Pressure [Pa]	**Boiling temperature [°C]**	**Gravitation [m.s^{-2}]**
20 000	60	- 9,81

Table 4. Material conditions.

Material	Density [kg.m^3]	Thermal capacity [J.kg^{-1}.K^{-1}]	Thermal conductivity [W.m^{-1}.K^{-1}]	Dynamic viscosity [kg. m^{-1}.s^{-1}]
Air	1.225	1 006.43	0.0224	1.7894×10^{-5}
Water	998.2	4182	0.6	1003
Mixture	0.13	*	0.0261	1.34×10^{-5}

*$c_p(T) = 1563.077 + 1.603755T - 0.002332784T^2 + 3.216101\times10^{-6}T^3 - 1.156527\times10^{-9}T^4$

Table 5. Material conditions.

Material	**Molar mass [kg. kmol^{-1}]**	**Enthalpy [J. kmol^{-1}]**	**Reference temperature [°C]**
Air	28.966	0	25
Water	18.0152	0	25
Mixture	18.01534	4.399×10^{7}	25

Table 6. General conditions.

Space dimension	**Precision**	**Flow**	**Setting of the model**
2D	double	Non-stationary	evaporation - condensation: T_{var} = 60 °C, λ_l = **0.1**, λ_v = **10**
		Non-compressible, laminar	Multiphase model: mixture Liquid phase = water, Gaseous phase (mixture) = water vapor + air

Table 7. Setting of the solution.

Type	**Coupling**	**Discretization**	**Discretization of pressure**	**Transient formulation**
Pressure-based	Simple	1st order upwind	Standard	1st order implicit

Table 8. Under-relaxation factors.

Under-relaxation factors	
Density	0.9
Body Forces	0.9
Pressure	0.3
Momentum	0.7
Species	0.9
Energy	0.7
Volume Fraction	0.5
Slip Velocity	0.1
Vaporization Mass	0.9

Table 9. Setting of the solution.

Time stepping method	**Time step size**	**Number of time steps**	**Max iterations**	**Reporting interval**
Fixed	0.001	10 000	20	1

3 Results of numerical simulation

Even before starting the numerical calculation, it was necessary to correctly patch the phases where water and air are located. In Figure 2 individual phases are plotted through the Volume Fraction contour.

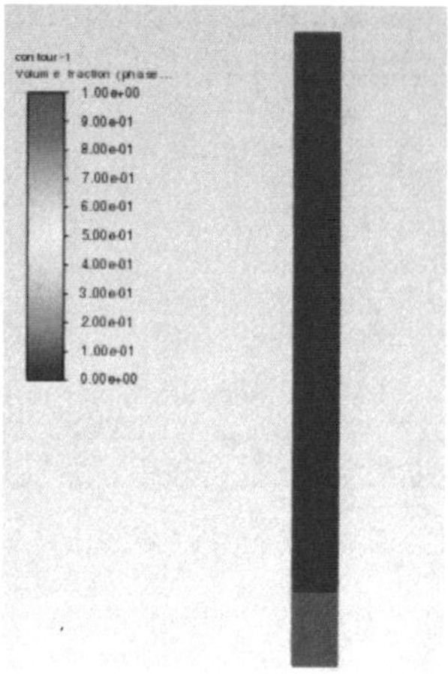

Fig. 2. Initial state of Volume Fraction from stationary simulation; phase-1 is water.

The simulation was set to calculate 10 000 numbers of time steps with 0.001 time step size, which means that the simulation captures 10 s of heat pipe heating. In Figure 3, it is possible to see the formation of a steam bubble at water in 2 s.

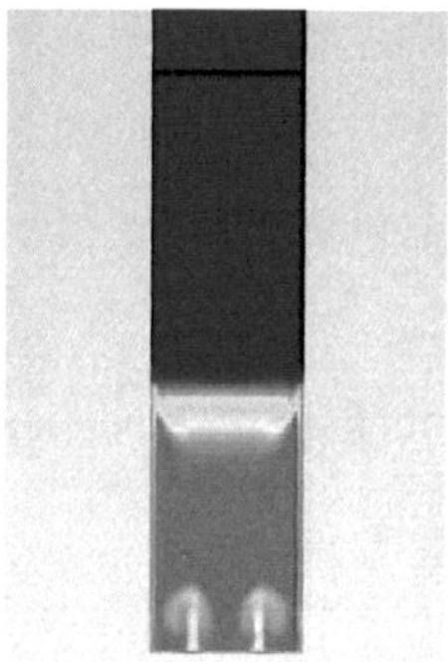

Fig. 3. The presence of steam bubble in the bottom of heat pipe.

Other figures show the time courses of state changes in a heat pipe. Figure 4 shows the time course of the density at times 0.5 s, 1 s, 2 s, 5 s and 10 s. At the same times, speed (Figure 5.) and temperature (Figure 6.) are also shown.

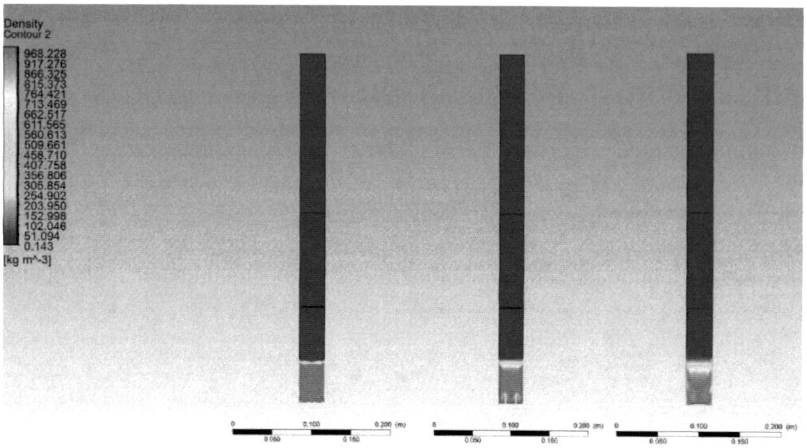

Fig. 4. Contours of densities at times 0.5 s, 1 s, 2 s.

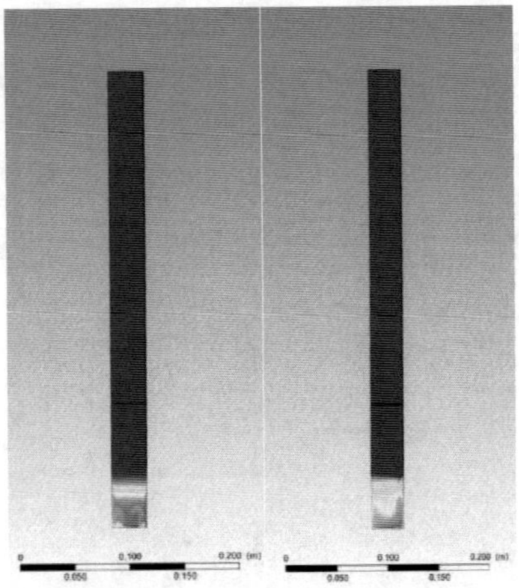

Fig. 5. Contour of density in 5 s and 10 s.

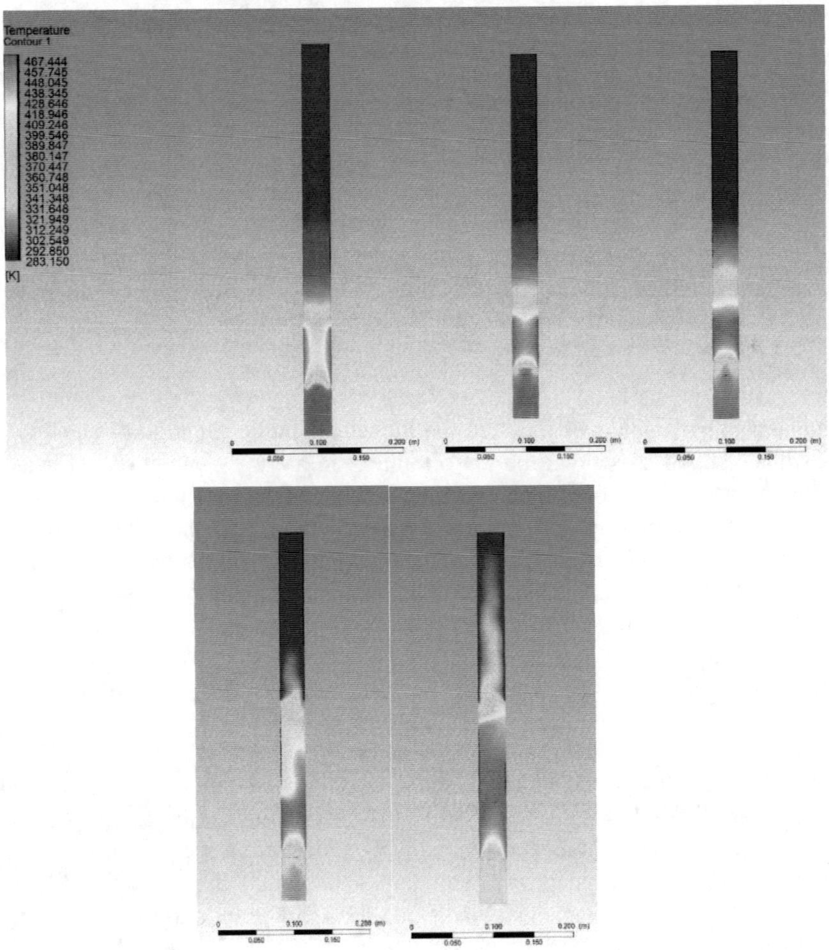

Fig. 6. Contours of temperatures at the times 0.5 s, 1 s, 2 s, 5 s, 10 s.

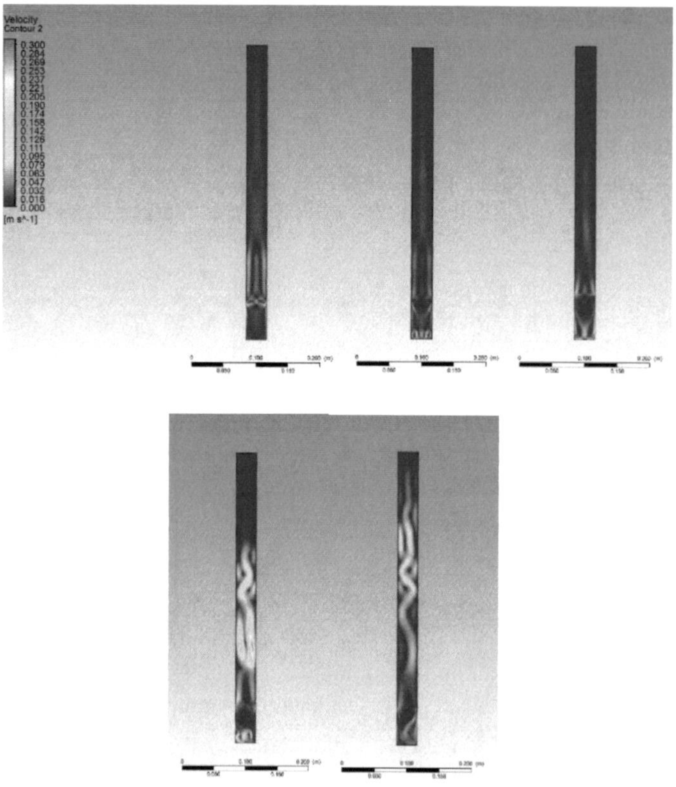

Fig. 7. Countours of velocities at the time 0.5s, 1s, 2s, 5s, 10s

4 Conclusion

In general, this article analyses the mathematical model of multiphase flow - the Lee model, which uses the ANSYS Fluent program for numerical simulations of heat transfer in multiphase flows. Lee's model includes certain correlation coefficients in its mass transfer relations. These correlation coefficients must be fine-tuned that they describe the real experimental measurement as accurately as possible. They depend on the amount of condensate formed and reach values from 10^{-3} to 10^{2} s^{-1}.

The output of the article is numerical simulation of interphase mass transport in a closed space - heat pipe capturing physical phenomena – evaporation with an optimized correlation coefficient of the Lee model - λ_c were 0.1s^{-1} and 10 s^{-1}. These coefficients are valid only for the given situation when pressure is 20 000 Pa and boiling temperature is 333.15 K.

In the future, by carrying out the experiment to obtain experimental data on the generated condensate and subsequent back-calculation of the correlation coefficients in program ANSYS Fluent, the dependence of the correlation coefficient on the amount of generated condensate would be created, thanks to which it would be able to create a correlation graph that would be used to refine the CFD simulation to a real experimental measurement.

This research is supported by the project KEGA 047ŽU-4/2022 Innovative approaches to fluid dynamics modelling in energy systems and KEGA 046ŽU-4/2021 Innovative methods of increasing the thermal efficiency of small heat sources using heat recovery through phase transformations.

References

1. Wu H. L., Peng X. F., Ye P., Eric Gong Y. (2007): Simulation of Refrigerant Flow Boiling in Serpentine Tubes, *Int. J. Heat Mass Transfer*, vol. 50, pp. 1186–1195.
2. De Schepper S. C. K., Heynderichx G. J., Marin G. B. (2009): Modeling the Evaporation of a Hydrocarbon Feedstock in the Convection Section of a Steam Cracker, *Comput. Chem. Eng.*, vol. 33, pp. 122–132.
3. Alizadehdakhel A., Rahimi M., Alsairafi A. A. (2010): CFD Modeling of Flow and Heat Transfer in a Thermosyphon, Int. Commun. *Heat Mass Transfer*, vol. 37, pp. 312–318.
4. Yang Z., Peng X. F., Ye P. (2008): Numerical and Experimental Investigation of Two Phase Flow during Boiling in a Coiled Tube, *Int. J. Heat Mass Transfer*, vol. 51, pp. 1003–1016.
5. Fang C., David M., Rogacs A., Goodson K. (2010): Volume of Fluid Simulation of Boiling Two-Phase Flow in a Vapor-Venting Microchannel, *Frontiers Heat Mass Transfer*, vol. 1, pp. 1–11.
6. Cao, Z.; Zhang, H.; Mei, H.; Yan, G.; Chu, W.; Wang, Q. (2022): Numerical Study on R32 Flow Condensation in Horizontally Oriented Tubes with U-Bends. *Energies*, 15, 4799.
7. Carey, V.P. (2020). *Liquid-Vapor Phase-Change Phenomena: An Introduction to the Thermophysics of Vaporization and Condensation Processes in Heat Transfer Equipment (3rd ed.).* CRC Press.
8. Chen G., Taotao N., Xiaohong Y. (2020): An explicit expression of the empirical factor in a widely used phase change model, *International Journal of Heat and Mass Transfer*, vol. 150.
9. Haibo L., Maocheng T., Min W., Jingzhi Z. (2022): Numerical investigation on interfacial characteristics of steam bubble condensation in subcooled water, *International Communications in Heat and Mass Transfer,* vol. 132.
10. Hongli L., Jiguo T., Licheng S., Zhengyu M., Guo X. (2020): An assessment and analysis of phase change models for the simulation of vapor bubble condensation, *International Journal of Heat and Mass Transfer,* vol. 157.
11. Cao, Z.; Zhang, H.; Mei, H.; Yan, G.; Chu, W.; Wang, Q. (2022): Numerical Study on R32 Flow Condensation in Horizontally Oriented Tubes with U-Bends. *Energies*, 15, 4799.
12. Čajová Kantová, N., Čaja, A., Patsch, M., Holubčík, M., Ďurčanský, P. (2021): Dependence of the Flue Gas Flow on the Setting of the Separation Baffle in the Flue Gas Tract, In: *Applied sciences-basel*, Volume 11, Issue 7, Article Number 2961.
13. Błasiak, P.; Opalski, M.; Parmar, P.; Czajkowski, C.; Pietrowicz, S. (2021): The Thermal—Flow Processes and Flow Pattern in a Pulsating Heat Pipe—Numerical Modelling and Experimental Validation. *Energies*, 14, 5952.
14. Yunfeng Z., Hanzhong T., Jingjing Z., Wei L., Jianjie Ch., Yannan L. (2022): Numerical simulation of flooding characteristics in steam-water countercurrent inside an inclined pipe under condensation conditions, *International Communications in Heat and Mass Transfer*, vol. 132, 105912, ISSN 0735-1933.
15. Myeongjin K., Joo Hyun M. (2021): Numerical investigation and modeling of thermal resistance and effective thermal conductivity for two-phase thermosyphon, *Case Studies in Thermal Engineering*, vol. 27, 101358, ISSN 2214-157X.

Study on Losses in the Intermediate-Pressure Steam Turbine Inlet Chamber

Vaclav Slama[1,*], *David* Simurda[2], and *Richard* Lenhard[3]

[1]University of West Bohemia, Power System Engineering Department, Univerzitni 8, 306 14, Pilsen, Czech Republic
[2]Institute of Thermomechanics of the Czech Academy of Sciences, Dolejskova 1402/5, 182 00 Praha 8, Czech Republic
[3]Department of Power Engineering, Faculty of Mechanical Engineering, University of Žilina, Univerzitna 1, 010 26 Žilina, Slovakia

Abstract. Know-how about energy and pressure losses in all steam turbine parts is crucial to guarantee enhanced operational reliability and efficiency. This paper focuses on studying pressure losses in the intermediate-pressure turbine inlet chamber. Measurements were performed on a complex model consisting of the turbine inlet chamber, a compact valve assembly situated upstream, and nozzles located downstream. These intermediate-pressure turbine parts are larger due to the greater volumetric mass flow than the high-pressure turbine parts. However, their inner parts are proportionately smaller, which causes greater pressure losses. Measurements were carried out in the Aerodynamic laboratory of the Institute of Thermomechanics of the Czech Academy of Sciences, where the model was installed in an in-draft wind tunnel. The results were complemented by numerical simulations performed in the Doosan Skoda Power company using ANSYS software tools. Pressure losses were evaluated using the total pressure loss coefficient and, as a result, can be predicted in similar turbine inlet chambers with the required accuracy.

1 Introduction

To some extent, each steam turbine part should be investigated in detail to enhance the know-how and predict the efficiency to create a highly competitive turbine bid. Therefore, extensive studies of losses in the most common valve assemblies have been conducted recently. There was a high-pressure valve assembly for the nozzle [1, 2] and throttle [3, 4] governing along with a compact valve assembly for the intermediate-pressure turbine [5, 6]. A valve assembly may consist of many parts, such as a stop valve and a control valve. In these parts, pressure losses and flow fields have been investigated using measurements and numerical simulations. As a result, a complex valve flow characteristic representing the dependence between the mass flow ratio, the pressure ratio, and the relative valve cone lift was created [6]. Except for these parts, there is the turbine inlet chamber between the valve

* Corresponding author: vaclav.slama@doosan.com

assembly and the first turbine stage. Measurement of pressure losses in such a chamber had never been performed. This paper presents results for the first time.

2 Experimental model

The whole model for the current measurement is in Fig. 1. It consists of the valves already measured and numerically simulated [5, 6] and the turbine inlet chamber, the subject of this new study. Downstream, there are a nozzle part and outlet pipelines, which do not represent any actual turbine part but connect the model with the vacuum storage.

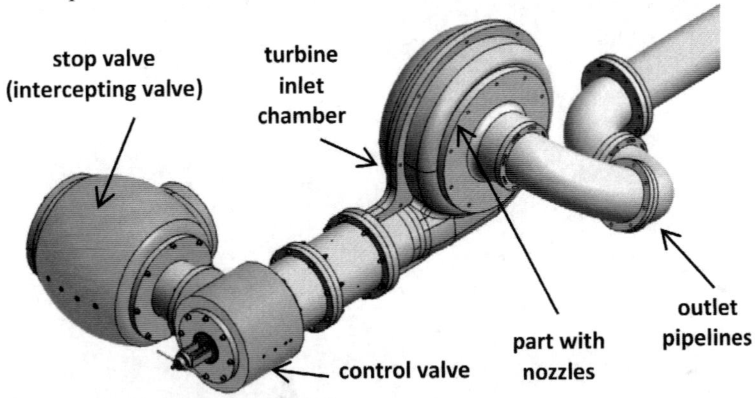

Fig. 1. Whole assembly of the measured model.

The aerodynamic suction-type modular tunnel is in Fig. 2. The air is sucked down from ambient conditions at the inlet through the tested model to the outlet pipelines, where a control nozzle sets conditions according to the required pressure ratio. The vacuum in the chamber can reach an absolute pressure of 10 kPa.

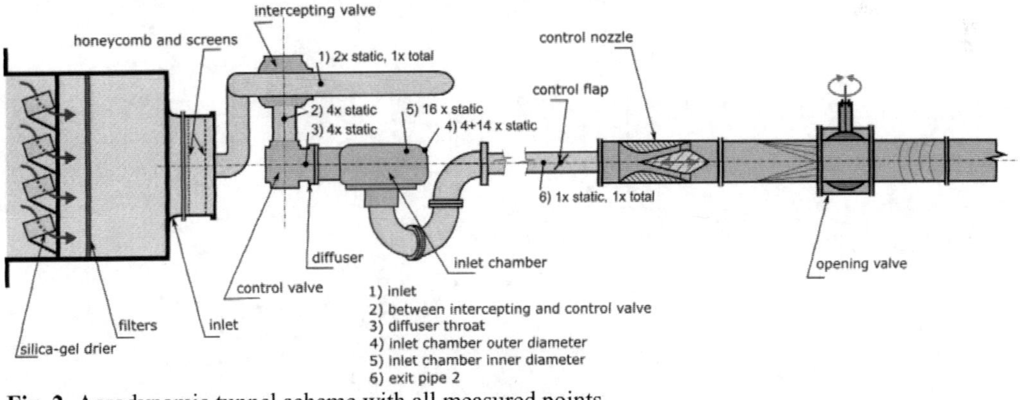

Fig. 2. Aerodynamic tunnel scheme with all measured points.

In order to study the flow field in the turbine inlet chamber, four pressures around the circumference were measured upstream. It is shown in Fig. 2 and Fig. 3 in position p_A. Other pressures p_B and p_C were measured on the outer chamber surface. Pressure p_D was measured upstream of the nozzles. Positions are shown in Fig. 4. Four pressure extractions were used for p_B, see Fig. 5, due to the transducer's limited capacity. Pressures p_C and p_D were measured at more positions. p_C at 14 positions, see Fig. 5, p_D was similar but at 16 positions because there is no inlet pipeline.

A transducer of Pressure Systems 9116 was used to evaluate all pressures. Besides, the total pressure upstream of the valves were measured by a Prandtl probe. The mass flow rate was measured in the outlet pipeline using the Annubar Deltatop DP 62D probe and

Thermometers Pt100. The relative uncertainty of pressure measurement was evaluated to be about 4 % with a confidence interval of 95 %. Because experiments were carried out with dry air at low pressure, Reynolds numbers are approximately two orders of magnitude lower than the real valve assembly with flowing steam. This difference was analysed in [7, 8], There is concluded that measurements at lower Reynolds numbers still provide accurate valve characteristics. It is based on the theory of models applied to valve assemblies [9]. There is also concluded that differences are minor and show slightly greater pressure losses. It is a conservative result to be applied to real valves.

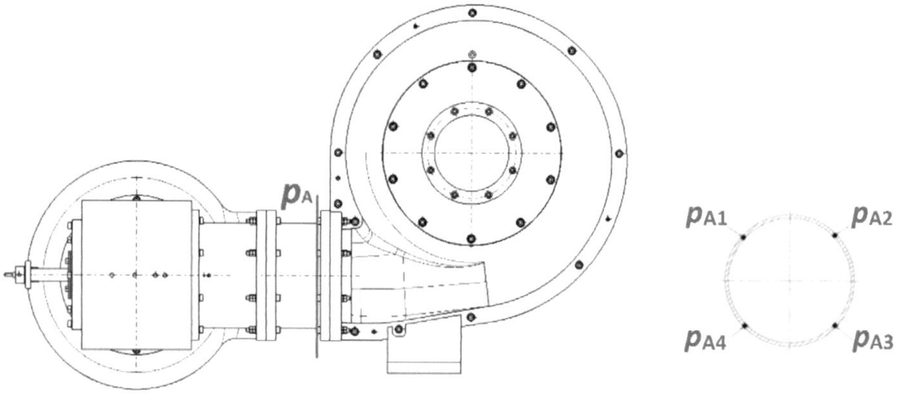

Fig. 3. Pressure measurement position at the chamber inlet.

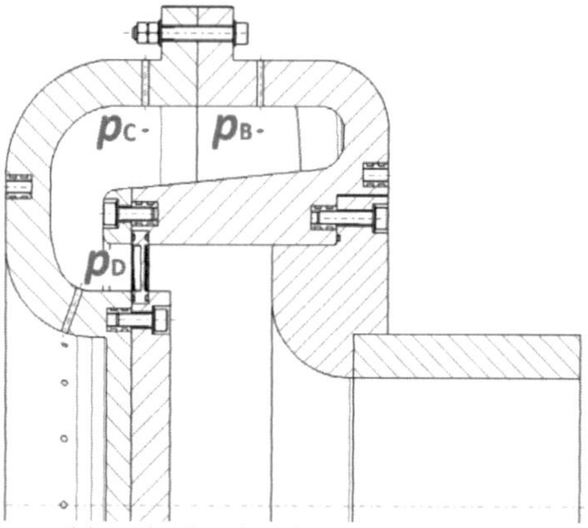

Fig. 4. Pressure measurement positions in the chamber.

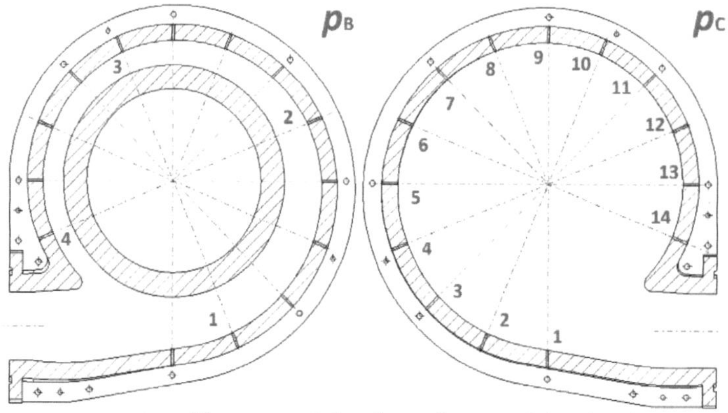

Fig. 5. Pressure measurement positions around the circumference of the chamber.

3 Numerical model

Numerical simulations were used especially to evaluate total pressures since only static pressures can be measured due to the model limited space. The numerical model is shown in Fig. 6 on the right.

Fig. 6. Numerical models – the former model (left) [6], the model with the inlet chamber (right).

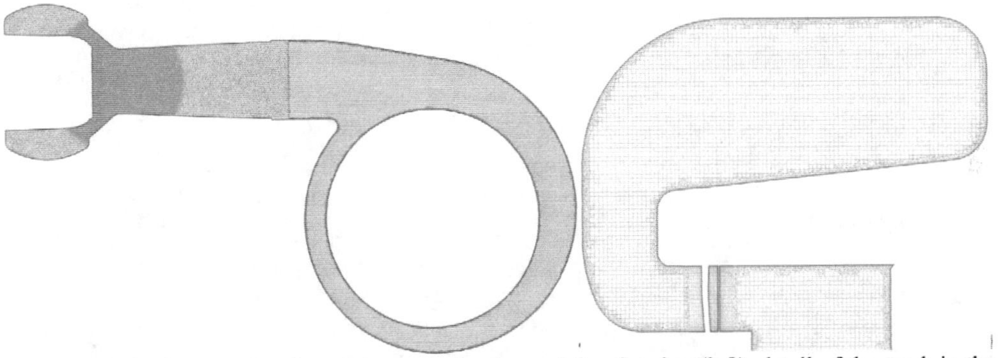

Fig. 7. Mesh in the cross-section of the control valve and the chamber (left), detail of the mesh in the cross-section of the chamber (right).

Hexahedral and tetrahedral meshes, which satisfy typical quality criteria, were created. It is shown in Fig. 7. It was also kept a recommendation for the compressible turbulence model k-ω SST to have $y+<5$ [10, 11]. It was based on similar projects [12], where mesh sensitivity was tested. The medium was defined as dry air following the experiment where the silica-gel drier guarantees dryness, see Fig. 2. To compute the steady-state RANS solution, a high-resolution scheme was used. Boundary conditions were defined according to the measured data. 10^{-4} residual target values judged simulation convergence criteria and imbalances of pressure losses in the chamber were monitored to be less than 0.1 %.

4 Results and discussion

At first, a valve flow characteristic representing the dependence of the mass flow ratio m/m_{cr} on the pressure ratio ε_2, and the relative valve cone lift z/D, was evaluated. The pressure ratio ε_2 can be calculated from outlet static p_2/p_{1t} or total pressures p_{2t}/p_{1t}. Index 1 means the inlet and index 2 means the outlet of the valve, respectively. m is the mass flow and m_{cr} is the critical mass flow in the valve calculated from the inlet flow parameters and the diffuser throat diameter D. z is the valve cone lift. Fig. 8 compares the measurement results on the valve assembly alone (the dashed lines) with the new model with the turbine inlet chamber (the full lines). The resulting differences are not significant. As a result, it can be concluded that the valve characteristics, widely used for pressure loss prediction, are not dependent on the downstream geometry. It also shows that the measurement is consistent with the previous one.

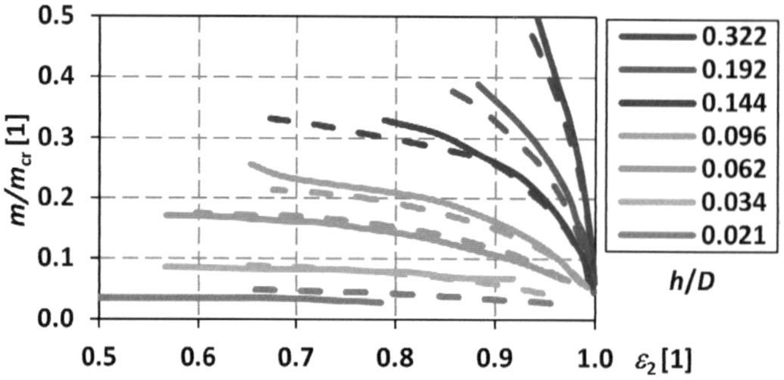

Fig. 8. Measured valve flow characteristic comparison.

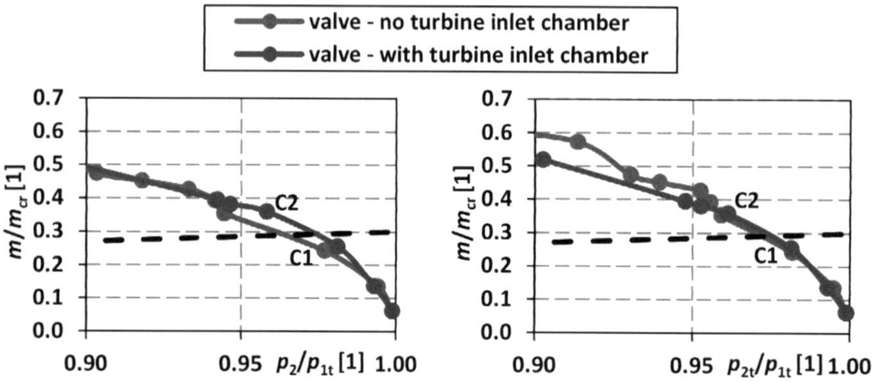

Fig. 9. CFD valve flow characteristics comparison.

The valve flow characteristic comparison was also evaluated from CFD results for $z/D = 0.322$, where the valve is fully open. For such a case, a pressure loss prediction is required. Using the CFD, static and total pressure ratios could be evaluated. Results are in Fig. 9. Some differences can be observed in the case of the total pressure ratios $p_{2t}/p_{1t} < 0.96$. Fortunately, this region is far away from the area of practical use, which is under the dashed line in Fig. 9.

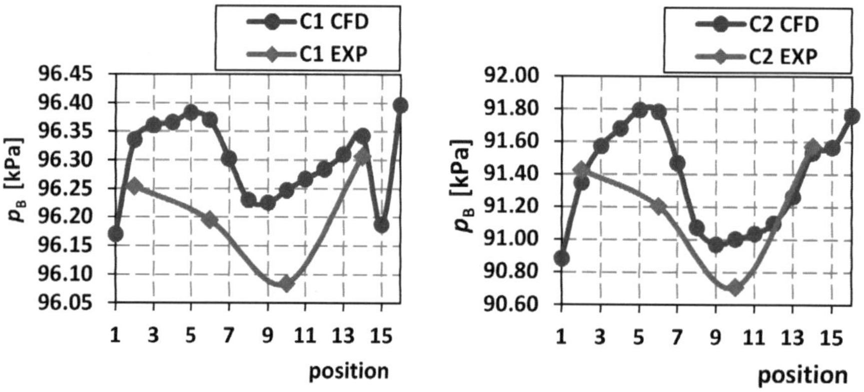

Fig. 10. Distribution of pressure p_B for cases C1 (left) and C2 (right).

After proving that the results of the control valve flow characteristics are consistent and with reasonable differences, the turbine inlet chamber can be evaluated. Comparison of measured and calculated static pressures p_B, p_C and p_D are shown in Figs. 10 to 12. Two cases were chosen: C1 and C2. Their positions in the valve flow characteristic are shown in Fig. 9. Differences between numerical (CFD) and experimental results (EXP) can be observed in both cases. The trends are similar. For case C1, the maximum difference is about 0.40 kPa and for case C2 it is about 1.00 kPa. It is a good match, given the complexity of the geometry.

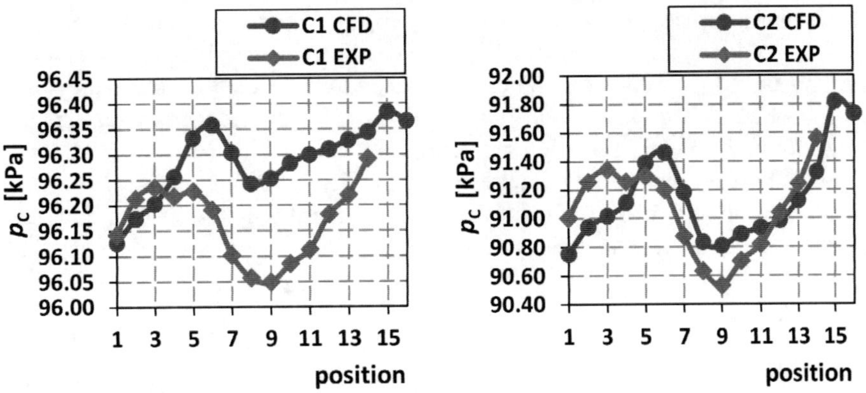

Fig. 11. Distribution of pressure p_C for cases C1 (left) and C2 (right).

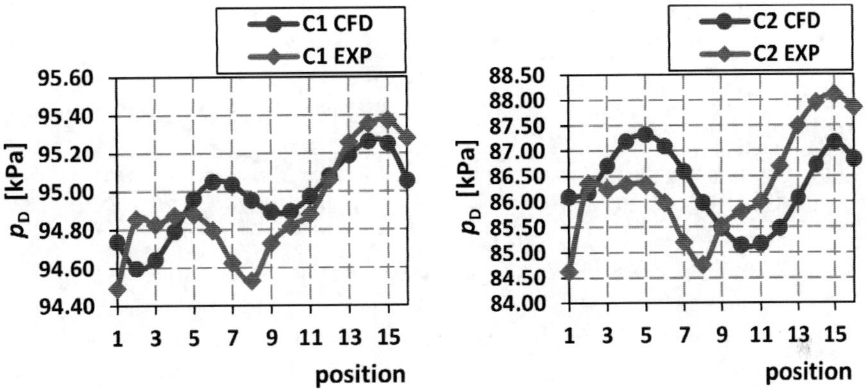

Fig. 12. Distribution of pressure p_D for cases C1 (left) and C2 (right).

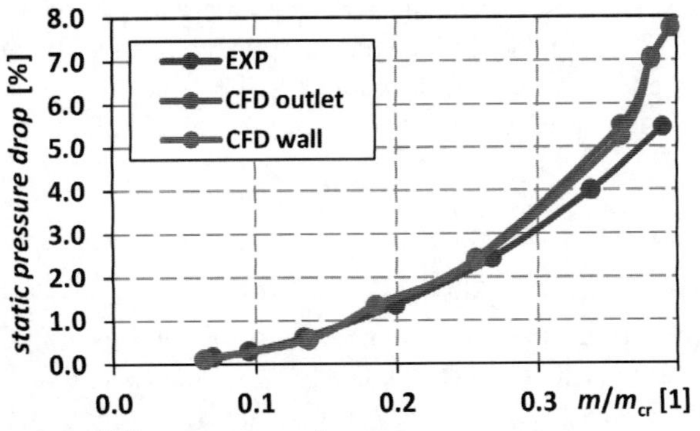

Fig. 13. Static pressure drop differences comparison.

In order to generalize the results, the pressure drop from the inlet to the outlet of the turbine inlet chamber should be evaluated not only from a few measuring points but from average values at the inlet and outlet plane. Furthermore, for practical use, the pressure drop between locations p_A and p_D should be better evaluated from the CFD results since the positions for measurement are limited, see Figs. 3 to 5. Results for the required case with $z/D = 0.322$ are shown in Fig. 13. It can be noticed that the static pressure drop is almost the same no matter whether it is calculated from the pressure at the whole outlet (before the nozzles in Fig. 4) or the measurement points (p_D). Comparing experimental and numerical results, it could be seen that the differences are reasonable up to $m/m_{cr} < 0.3$. For greater m/m_{cr}, the differences are significant. The reason is that in the region of $m/m_{cr} > 0.3$, strongly unsteady phenomena such as flow detachments appear, but the numerical simulations are performed as steady-state. It was described and shown in detail in [6]. Fortunately, for a practical pressure loss prediction, the valves and the turbine inlet chambers are designed to be used only for conditions where $m/m_{cr} < 0.3$ and where the differences are reasonably small.

If the differences in the static pressure drop are minor (for $m/m_{cr} < 0.3$), there is an assumption that differences in the total pressure drop are also small in this range. As a result, the total pressure losses should be evaluated. The standard way to evaluate and generalize the loss in the inlet turbine chamber is to use the total pressure loss coefficient Y [13]. It is defined as:

$$Y = \frac{p_{1c} - p_{2c}}{p_{2c} - p_2} \quad (1)$$

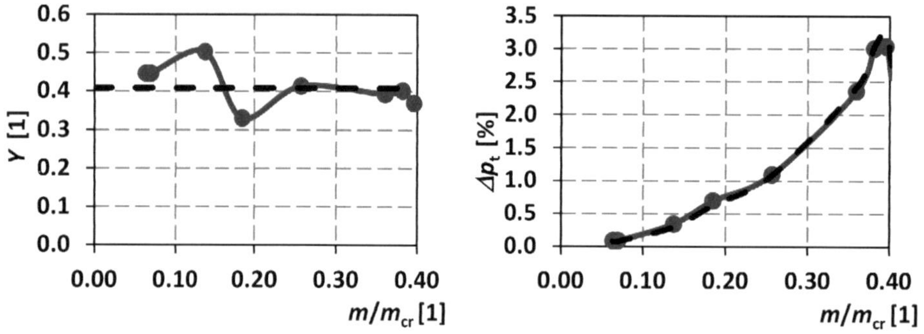

Fig. 14. Total pressure loss coefficient (left), total pressure loss dependence (right).

The results of Y for all calculated cases are in Fig. 14 on the left. It is visible that the total pressure loss coefficient is not constant. At lower relative mass flow rates, its value oscillates. The reasons for these oscillations are going to be analyzed in further studies. However, the average value of Y, shown by the dashed line, can be used very well for practical use. Results of calculated total pressure losses using the average value Y are shown in Fig. 14 on the right by the dashed line and using the exact values of Y by the solid line, respectively. For $m/m_{cr} < 0.3$, the differences in the total pressure loss between the exact and average values are lower than 0.10 %. It is a reasonable difference for a practical pressure loss estimation.

5 Conclusions

Measurement and numerical simulations were performed on the complex model of the intermediate-pressure turbine inlet chamber. As a result, it was found that the presented measurement is consistent with the previous one [5, 6], and the valve flow characteristic is not dependent on the downstream geometry. It can be observed that differences in pressure drop in the inlet turbine chamber between experimental and numerical results are reasonable for the often-used turbine operating range, where $m/m_{cr} < 0.3$. For greater mass flow ratios, the differences are significant. For practical use, total pressure loss prediction for inlet chambers of the intermediate-pressure turbine was generalized in the form of the average value of the total pressure loss coefficient.

This work was supported by the grants TN01000007/05 and supported by the project KEGA 047ŽU-4/2022 Innovative approaches to fluid dynamics modelling in energy systems

References

1. Slama, V., Mrozek, L., Rudas, B., Simurda, D., Hala, J., Luxa, M. (2020). Experimental and numerical study on pressure losses and flow fluctuations in a high-pressure valve assembly of steam turbine governing system. *ASME Turbo Expo 2020*, GT2020-14474.
2. Slama, V., Mrozek, L., Tajc, L., Klimko, M., Zitek, P. (2021). Flow analysis in a steam turbine control valve with through-flow valve chamber. *ASME J of Nuclear Rad Sci.*, NERS-20-1141.
3. Slama, V., Mrozek, L., Tajc, L., Simurda, D., Hala, J. (2019). Creation and validation of pressure loss model for steam turbine control valves. *AIP Conference Proceedings 2189.*
4. Mrozek, L., Slama, V., Sedlak, K., Tajc, L. (2021). Analysis of losses in steam turbine control valves. *AIP Conference Proceedings 2323.*
5. Slama, V., Mrozek, L., Rudas, B., Simurda, D., Hala, J., Radnic, T. (2021). Investigation of pressure losses and flow fluctuations in an intercept valve assembly of an intermediate-pressure turbine part. *ASME IMECE*, IMECE2020-23359.
6. Slama, V., Simurda, D., Mrozek, L., Tajc, L., Hala, J., Radnic, T. (2021). Pressure losses and oscillations in a compact valve of a steam turbine. *MATEC Web of Conferences 345*, 00027.
7. Brüggemann, P., Vinnemeier, P., Balkowski, I., Büscher, C., Stapper, P., Zimmer, G., Wirsum, M. (2014). A new emergency stop and control valves design – part 1: experimental verification with scaled models. *ASME Turbo Expo*, GT2014-25116.
8. Brüggemann, P., Vinnemeier, P., Balkowski, I., Büscher, C., Stapper, P., Zimmer, G., Wirsum, M. (2014). A new emergency stop and control valves design – part 2: validation of numerical model and transient flow physics. *ASME Turbo Expo*, GT2014-25117.
9. Zaryankin, A., Simonov, B. (2005). *Control and stop-control valves of steam turbines (Reguliruyuschie i stoporno-reguliruyuschie klapanyi parovyih turbin).* Moscow Energetic Institute, Moscow, ISBN 5-7046-1143-5.
10. Knopp, T. (2006). Model-consistent universal wall-functions for RANS turbulence modelling. *BAIL 2006 International Conference on Boundary and Interior Layers.*
11. Denton, J., D. (2010). Some limitations of turbomachinery CFD, *ASME Turbo Expo, Power for Land, Sea and Air.* Volume 7: Turbomachinery.
12. Kalista, R., Mrozek, L., Slama, V., Sedlak, K. (2019). Steam turbine axial exhaust diffuser investigation using a test rig and its numerical model. *ASME IMECE*, IMECE2019-10267.
13. Denton, J., D. (1993). Loss Mechanisms in Turbomachines. *Journal of Turbomachinery*, 115(4).

https://doi.org/10.1051/matecconf/202236901003

Impact of a Natural Gas Cooler Design on the Cooling Performance

Mária Čarnogurská[1,*], *Miroslav* Příhoda[2], and *Branislav* Žatkovič[1]

[1]Technical University of Košice, Faculty of Mechanical Engineering, Department of Power Engineering, 042 00 Košice, Slovak Republic
[2]VŠB - Technical University of Ostrava, Faculty of Materials Science and Technology, Department of Thermal Engineering, 17. listopadu 2172/15, 708 33 Ostrava-Poruba, Czech Republic

Abstract. This article describes a methodology for the identification of cooling performance of a natural gas cooler relative to the shape of its heat-transfer surface and presents the outputs of numerical solutions for four different shapes of heat-transfer surfaces in coolers designated as C_A, C_B, C_C and C_D. Calculations were carried out for a cooler with a single row of tubes, and for coolers with two through six rows of tubes that were positioned above one another with an alternating arrangement. In all of the surface shapes, the boundary conditions were respected in order to facilitate the identification of the shape of the heat-transfer surface which is the most appropriate for achieving maximum cooling performance. Out of these four shapes, the best results were observed with the heat-transfer surface of the cooler designated as C_A. The cooling performance of a 1 m long tube with such a surface was 1,650 W.

Keywords: *cooler; finned heat-transfer surface; natural gas; performance*

1 Introduction

Determination of cooling performance of natural gas coolers by applying an analytical method is a rather difficult and time-consuming task. The complexity consists primarily in the determination of individual parameters in the equations that characterise heat transmission or transfer through heat-transfer surfaces of various shapes. An optimal way how to identify the cooling performance is to apply numerical methods with clearly defined boundary conditions.

This article presents results of the investigation into intensity of natural gas cooling with four different types of heat-transfer surfaces of the coolers used at a compressor station (CS) of a gas transit pipeline. The output of the investigation was the evaluation of individual cooling surfaces in terms of cooling intensity with forced convection of the cooling air.

* Corresponding author: maria.carnogurska@tuke.sk

2 Problem analysis

A shape of an external heat-transfer surface of a natural gas cooler is determined by the finned tubes, i.e., the fins with circular cross-sections that are installed along the entire length of the cooler tubes. These fins may be of various thicknesses s_f (Fig. 1), various outer diameters d_f, and various pitches b. The total cooling performance of such a device greatly depends on the overall heat transfer coefficient (k) that expresses the heat transfer from the gas through the cooler tube wall into the cooling air. If this overall heat transfer coefficient may be regarded as a constant parameter across the entire heat-transfer surface, the equation for calculating the cooling performance P is as follows [1–3]:

$$P = k \cdot S \cdot \overline{\Delta t} \quad \text{(W)} \tag{1}$$

wherein k is the overall heat transfer coefficient ($W \cdot m^{-2} \cdot K^{-1}$); S is the external surface area of the tubes, which was identified as a sum of the surface areas of the fins S_f and of the smooth sections of the tubes between the fins S_b (m^2); and $\overline{\Delta t}$ is the mean temperature difference between the heat-transfer media (K).

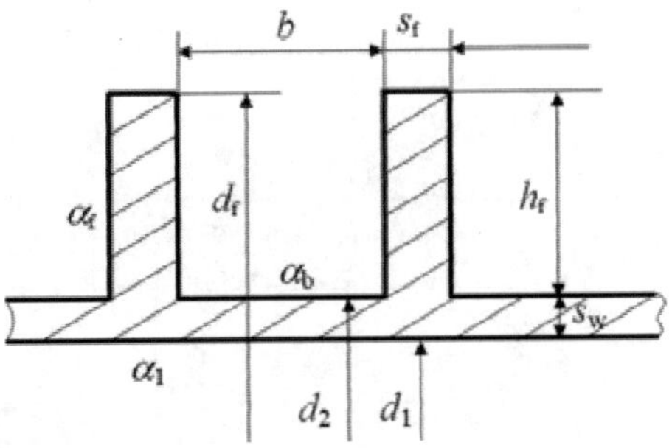

Fig. 1. Geometry of a tube and a fin.

Overall heat transfer coefficient k for a clean (without any sediments) heat-transfer surface of a finned tube is calculated using the following equation [2]:

$$\frac{1}{k} = \left(\frac{1}{\alpha_1} + \frac{s_w}{\lambda_w} \right) \cdot \frac{S}{S_1} + \frac{1}{\alpha_2 \cdot f(\eta_f)} \quad (W \cdot m^{-2} \cdot K^{-1}) \tag{2}$$

wherein s_w is the tube wall thickness (m); λ_w is the thermal conductivity coefficient of the tube material ($W \cdot m^{-1} \cdot K^{-1}$); S_1 is the internal surface area of the tubes that corresponds to a diameter d_1 (m^2); α_1 is the heat transfer coefficient for the inner side of the tubes ($W \cdot m^{-2} \cdot K^{-1}$); α_2 is the heat transfer coefficient for the heat transfer from the external finned surface into the surrounding environment ($W \cdot m^{-2} \cdot K^{-1}$); and the $f(\eta_f)$ formula is a function that depends on, inter alia, the fin efficiency η_f.

3 Thermal balance of the cooler

In the gas cooling process, a certain portion of the heat is not conducted directly into the cooling air but into the cooler structure itself. These losses are expressed by the loss coefficient η_z. The losses identified in the investigated coolers represented 0.5 – 3 %; i.e., the value of the η_z coefficient ranged from 0.995 to 0.97. The real cooling performance P_2 may then be calculated as follows [1]:

$$P_2 = P_1 \cdot \eta_z \quad \text{(W)} \tag{3}$$

or

$$Q_{m1} \cdot c_1 \cdot (t_{1,1} - t_{1,2}) \cdot \eta_z = Q_{m2} \cdot c_2 \cdot (t_{2,2} - t_{2,1}) \quad \text{(W)} \tag{4}$$

The $Q_{m1} \cdot c_1$ product contained in equation (4) represents the performance capacity of the gas C_1, while the $Q_{m2} \cdot c_2$ product represents the performance capacity of the air C_2.

The mean temperature difference $\overline{\Delta t}$ contained in equation (1) for cross-flows of heat-transfer media was calculated using the following equation [1]:

$$\overline{\Delta t} = \psi \cdot \frac{\Delta t' - \Delta t''}{\ln \dfrac{\Delta t'}{\Delta t''}} \quad \text{(K)} \tag{5}$$

wherein $\Delta t' = t_{1,1} - t_{2,2}$ and $\Delta t'' = t_{1,2} - t_{2,1}$.

The first figure in the temperature index represents the heat transfer medium (1 – gas; 2 – air) while the second figure after a comma means either the entry (1) or the exit (2). The correction coefficient ψ is expressed using the P and R criteria as follows:

$$\psi = \frac{\ln A}{n \cdot (1-R) \cdot \ln\left(1 + \dfrac{1}{R} \cdot \ln \dfrac{R-1}{R \cdot A^{1/n} - 1}\right)} \quad (1) \tag{6}$$

wherein $A = \dfrac{1-P}{1-R \cdot P}$ (1), $P = \dfrac{t_{1,2} - t_{1,1}}{t_{2,1} - t_{1,1}}$ (1), $R = \dfrac{t_{2,1} - t_{2,2}}{t_{1,2} - t_{1,1}}$ (1), n is the number of heat exchanger runs, as shown in Fig. 2.

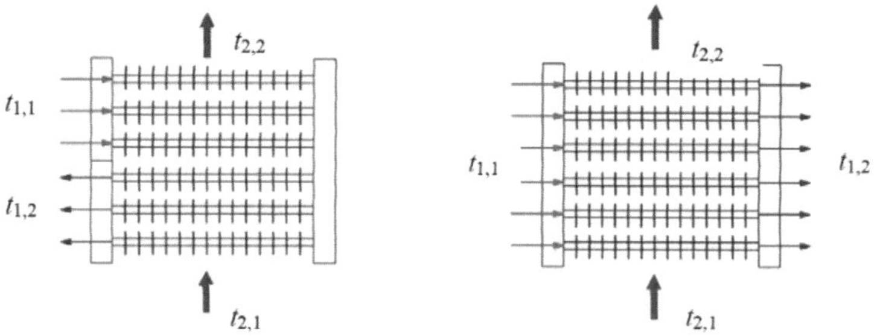

Fig. 2. A scheme of natural gas coolers with the media cross-flow.

Heat transfer coefficient inside a tube α_1 was calculated using the following equation:

$$Nu = 0.021 \cdot Re^{0.8} \cdot Pr^{0.4} \cdot \left(1.27 - 0.27 \cdot \frac{T_w}{T_{1,mean}}\right) \quad (1) \tag{7}$$

wherein T_w is the temperature of a tube wall (K) and $T_{1,mean}$ is the mean value of the gas temperature (K). A characteristic length in Nu and Re criteria in equation (7) is the inner diameter of the tube d_1.

The calculation of heat transfer coefficient on the external side of a bundle of finned tubes was made using the following equation:

$$Nu = C \cdot Re^{0.6} \cdot K_{\mathrm{f}}^{-0.15} \cdot \mathrm{Pr}^{1/3} \quad (1) \qquad (8)$$

The value of the constant parameter *C,* relative to the tube arrangement and a number of tube rows in the bundle, ranged from 0.20 to 0.38.

Coefficient K_f was identified using the following equation:

$$K_{\mathrm{f}} = \frac{S}{S_2} \quad (1) \qquad (9)$$

wherein S_2 is the external surface area of the smooth tubes (m^2).

Circular fins are subject to the following equation:

$$K_{\mathrm{f}} = 1 + 2 \cdot \frac{h_{\mathrm{f}} \cdot (h_{\mathrm{f}} + d_2 + s_{\mathrm{f}})}{(b + s_{\mathrm{f}}) \cdot d_2} \quad (1) \qquad (10)$$

A characteristic length in *Nu* and *Re* criteria in equation (8) is the outer diameter of the tube d_2.

4 Numerical calculations

Identification of a cooler's performance by applying an analytical procedure does not facilitate an adequate comparison of heat-transfer surfaces in terms of heat removal intensity since different coolers have different designs of their heat-transfer surface (tube diameter, fin diameter, fin thickness and fin pitch). The key parameters that affect the cooling performance of a cooler during its utilisation are as follows:

- Gas flow rate (an amount of the cooled natural gas);
- Gas temperature at the entry into the cooler;
- Air flow rate (determined by a number of fans and their performance);
- Shape and size of the heat-transfer surface;
- Tube arrangement in the cooler's block (above one another, alternating);
- Cooler blocks arrangement (cross-flow, multi-cross flow).

4.1 Characteristics of the heat-transfer surface of the examined coolers

The heat-transfer surface of the investigated natural gas coolers was determined by the parameters listed in Table 1 and by the geometry presented in Fig. 1. Numerical calculations were made separately for the case of forced convection with a single row of tubes in the cooler, and then for two through six rows. In a simulation, natural gas with a mean temperature of 65 °C was flowing through the cooler tubes. At this temperature, a numerical solution was made for the cooling intensity of finned tubes in all types of cooling surfaces.

The numerical solution was made with a symmetrical section of one tube fin in each row of tubes. The input boundary conditions were a gas pressure of 7.45 MPa and a gas temperature of 65 °C, while the heat-transfer surface was assumed to be clean.

Table 1. Typical dimensions of natural gas coolers at the CS.

Parameter		**C_A**	**C_B**	**C_C**	**C_D**	**Unit**
d_1	Inner diameter of the tube	0.025	0.02058	0.02	0.025	m
d_2	Outer diameter of the tube	0.03	0.0254	0.025	0.03	m
s_1	Tube spacing	0.065	0.065	0.065	0.065	m
s_w	Tube wall thickness	0.0025	0.00241	0.0025	0.0025	m
d_f	Fin diameter	0.058	0.0568	0.055	0.057	m
s_f	Fin thickness	0.00065	0.0002	0.0007	0.0007	m
b	Fin pitch	0.0025	0.0026	0.0023	0.0026	m
n	Number of tubes in the cooler	1,323	4,704	2,115	1,134	ks

4.2 Physical properties of the used media

The numerical solution was made with two different gases—natural gas and air. The properties of the gas with a temperature of 65°C and the air with a temperature of 20°C are listed in Table 2. The air flow was in the transit section; that is why the Shear-Stress Transport (SST) turbulence model was selected for the numerical calculation.

Table 2. Physical properties of the gas and air.

Parameter		**Value**	**Unit**
$t_{1,1}$	Gas temperature at the entry	75	°C
$t_{1,2}$	Gas temperature at the exit	55	°C
T_{mean}	Mean gas temperature	65	°C
Q_{Vg}	Volume flow rate of the gas	18,982	$m^3 \cdot h^{-1}$
p_g	Gas pressure	7.45	MPa
ρ_g	Gas density	45.96	$kg \cdot m^{-3}$
$c_{p,g}$	Gas heat capacity	2.7053	$kJ \cdot kg^{-1} \cdot K^{-1}$
λ_g	Gas thermal conductivity	0.046075	$W \cdot m^{-1} \cdot K^{-1}$
μ_g	Dynamic viscosity of the gas	$13.424 \cdot 10^{-6}$	$Pa \cdot s$
λ_a	Air thermal conductivity	0.02515	$W \cdot m^{-1} \cdot K^{-1}$
ν_a	Kinematic viscosity of the air	$1.54 \cdot 10^{-5}$	$m^2 \cdot s^{-1}$
$c_{p,a}$	Air heat capacity	1.005	$kJ \cdot kg^{-1} K^{-1}$
ρ_a	Air density	1.205	$kg \cdot m^{-3}$
υ_0	Air velocity at the outlet from the fan	4.7	$m \cdot s^{-1}$

4.3 Calculation area

The modelled area was investigated at a mean temperature of natural gas on the inner side of the tube. The calculation area represented the point on the heat-transfer surface located in the middle of the tube length and the width of the cooling surface. The model of the symmetrical part of one fin and one tube for the cooling surface with a single row of tubes is shown in Fig. 3. For this model of the heat-transfer surface, the cooling performance of all of the coolers was plotted by applying the following command *areaInt(Heat Flux)@Domain Interface1 Side2*. The results are shown in Fig. 4.

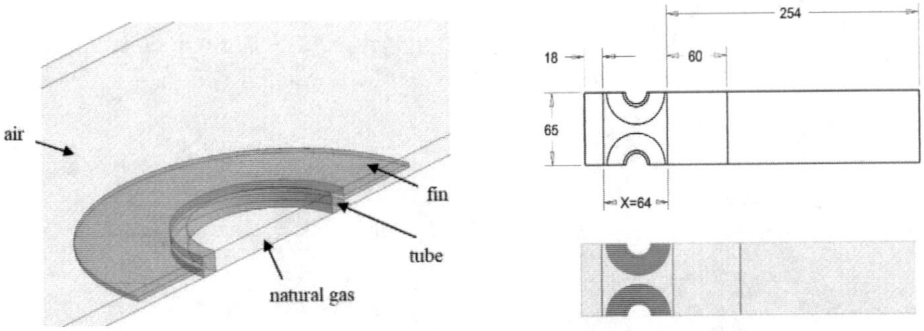

Fig. 3. Model of the symmetry of one tube and one fin – in one-row arrangement of the heat-transfer surface.

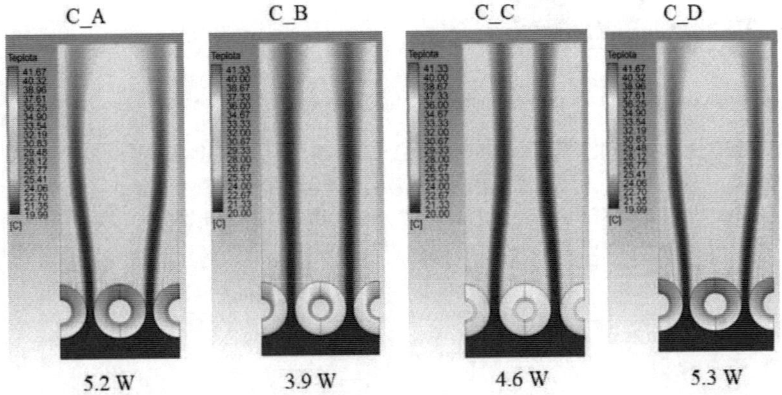

Fig. 4. Cooling performance with one row of tubes (P_{f1}).

The cooling performance for the six-row arrangement of the heat-transfer surface is shown in Fig. 5.

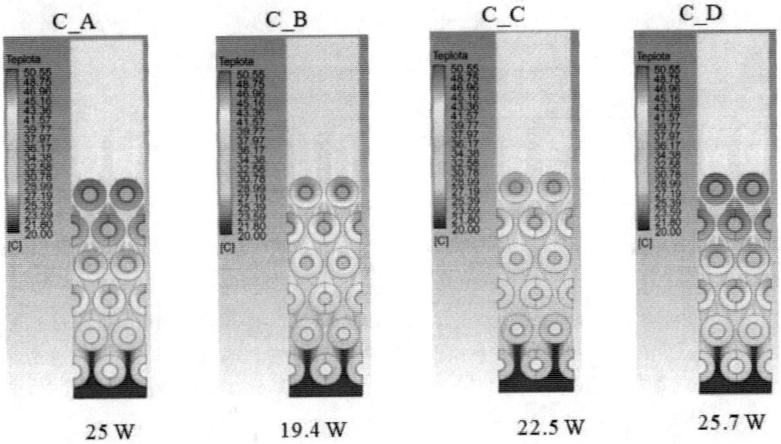

Fig. 5. Cooling performance with six rows of tubes (P_{f6}).

The evaluation of changes in the temperature along the height of the cooling surface that was formed of a single row of a finned tube, up to the six-row arrangement, was made using a section where several fins were located in the middle of the tube length. Thermal fields were plotted using the following command: *areaAve(Temperature)@Domain Interface1 Side2*. For a single row of fins, a 3D image of the thermal field is shown in (Fig. 6). The temperatures on the fins in individual coolers ranged from approximately 30 to 41°C.

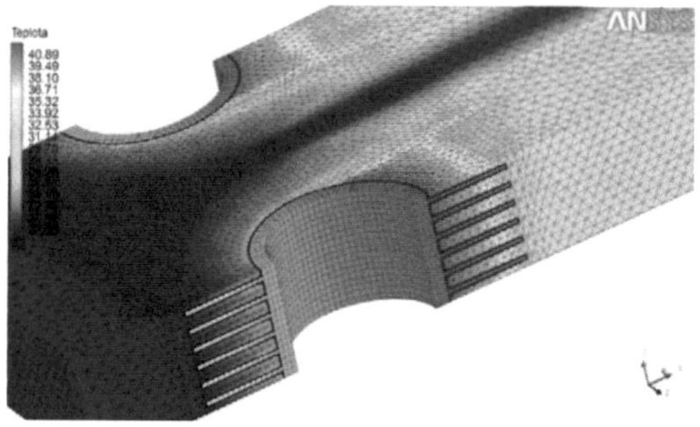

Fig. 6. Thermal field on the heat-transfer surface with a single row of tubes.

The correlations between the cooling performance and the number of rows of tubes in forced convection, for 1 row of tubes (P_f) and for 1 through 6 rows of tubes arranged above one another with an alternating arrangement, are graphically represented in Fig. 7.

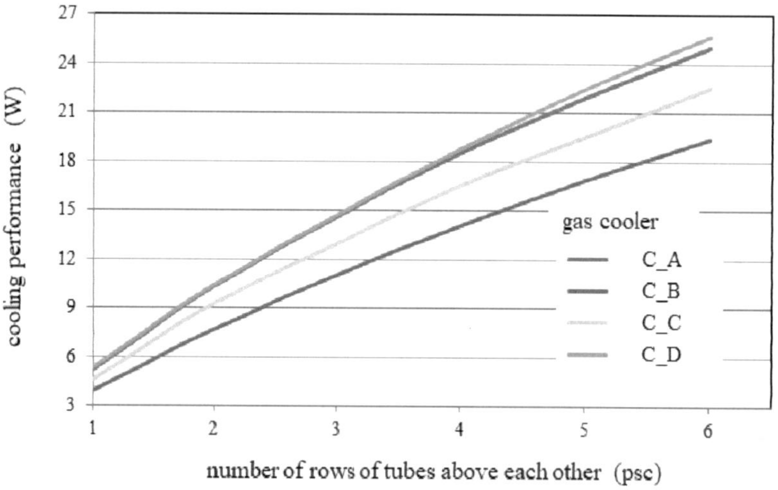

Fig. 7. Cooling performance of individual heat-transfer surfaces (1 fin and 1 through 6 rows of tubes).

The cooling performance of a 1 meter long finned tube for the individual types of the examined heat-transfer surfaces was identified for 1 through 6 rows of tubes in the heat-transfer surface based on the identified cooling performance. This required knowing the fin pitch, which was calculated as follows:

$$t_f = b + s_f \quad \text{(m)} \tag{11}$$

The t_f values for all four heat-transfer surfaces are listed in Table. 3.

Table 3. Fin pitches for individual coolers.

Fin pitch	C_A	C_B	C_C	C_D	Unit
t_f	$3.15 \cdot 10^{-3}$	$2.8 \cdot 10^{-3}$	$3 \cdot 10^{-3}$	$3.3 \cdot 10^{-3}$	m

For the C_A cooler with the single-row arrangement of the cooling surface (Fig. 4), the cooling performance of a 1 m long tube (P_{l1}) was calculated as follows:

$$P_{l1} = \frac{P_{f1}}{t_f} = \frac{5.2}{3.15 \cdot 10^{-3}} = 1,650 \quad (\mathrm{W \cdot m^{-1}}) \qquad (12)$$

Calculated values of cooling performance P_{l1} for the individual types of heat-transfer surfaces applicable to 1 meter of the finned tube are listed in Table 4.

Table 4. Cooling performance of 1m long heat-transfer surface with a single row of tubes.

Parameter	C_A	C_B	C_C	C_D	Unit
P_{l1}	1.650	1.393	1.533	1.606	$\mathrm{W \cdot m^{-1}}$

5 Conclusion

An analysis of the results of numerical solutions that were made for all four types of heat-transfer surfaces with 1 through 6 rows of tubes has brought the following conclusions:

- Cooling performance increased with a higher number of rows of tubes positioned above one another with an alternating arrangement. With forced convection, the best cooling performance was observed with the heat-transfer surfaces of C_A and C_D coolers. This may be attributed to the dimensions of the tubes and fins and the consequent value of K_f coefficient. The K_f value for the C_A cooler was 14.2 while for the C_D cooler the K_f value amounted to 13.1. With a higher value of this coefficient, a calculation based on equation (8) revealed a lower value of the *Nu* criterion and subsequently a lower value of the heat transfer coefficient α_2. As for C_B and C_C coolers, the values of K_f coefficient represented 19.2 and 17.3, accordingly.
- Dimensions of the tubes and fins affected not only the K_f coefficient but also the fin efficiency, and hence also the value of the $f(\eta_f)$ function.

This paper has been written with the financial support from the VEGA granting agency within the project no. 1/0626/20, VEGA 1/0532/22 and from FMT VŠB-TUO within the project no. SP 2022/13.

References

1. M. Příhoda, M. Čarnogurská, J. Molínek, Tepelný výkon chladiča zemného plynu s čistou a zanesenou teplovýmennou plochou. SLOVGAS **3** (2008).
2. D. B. Spalding, J. Taborek, *Heat Exchanger Theory, Heat Exchanger Design Handbook 1*. Hemisphere Publishing Corporation (1983).
3. *VDI-Wärmeatlas. 10. Auflage*. Berlin, Heidelberg, Springer Verlag (2006).
4. S. Kutateladze, V. M. Borišanskyj, *Příručka sdílení tepla*. Praha, SNTL (1972).
5. M. Rédr, M. Příhoda, *Základy tepelné techniky*. Praha, SNTL (1991).
6. ANSYS, Inc.: ANSYS CFX Tutorials. [online]. [cit.23-04-2014]. Available online: http://www.mecheng.osu.edu/documentation/Fluent14.5/145/cfx_tutr.pdf

CFD Simulation of Transonic Flow Through the Tip-Section Turbine Blade Cascade Intended for the Long Turbine Blade

Pavlo Kosiak[1,*], *Jindřich* Hála[1], *Martin* Luxa[1], and *Jaromír* Příhoda[1]

[1]Institute of Thermomechanics of the Czech Academy of Sciences, Dolejškova 5, 182 00 Prague 6, Czech Republic

Abstract. The paper deals with numerical simulations of transonic flow through the turbine blade cascade consisting of flat profiles. The cascade is one of variants of the tip section of ultra-long blades, which were designed for the last stage of the steam turbine. CFD simulations were realized by means of the ANSYS CFX commercial software using the γ-Re_θ bypass transition model completed by the two-equation SST turbulence model. Some simulations were made only by the SST turbulence model for comparison. Numerical results were compared with experimental data. Calculations performed for two nominal regimes and two computational domains. In addition to the standard computational domain, the calculation was performed for a domain with an extended output part for the suppression of reflected shock waves. The interaction of the inner branch of the exit shock wave with the boundary layer on the blade suction side leads in the both flow regimes to the flow separation followed by the transition to turbulence. The flow structure in the blade cascade obtained for the extended domain corresponds well to experimental results.

Keywords: *CFD simulation, tip-section blade cascade, flat profile, shock-wave/boundary-layer interaction*

1 Introduction

The last stages of the steam turbine have a significant impact on the total efficiency of the turbine, see Luxa [1]. Therefore, the considerable attention is paid especially to the flow through the tip sections of the last stage. The compressible flow through the tip section of the blade cascade has a transonic character with shock-waves in the flow field. The form of the tip/section blades is designed not only from viewpoint of aerodynamics but also from viewpoint of dynamics and vibration. These conditions have resulted to the design of flat blade profiles which are relatively sensitive to the angle of attack. This leads to the design of flat blades profiles that are relatively sensitive to the angle of attack. As follows from simulations of Musil et al. [2], the inlet angle has a strong influence on the inlet Mach number and for some regimes even a small change of the inlet angle changes the flow from the transonic to the supersonic character.

* Corresponding author: kosiak@it.cas.cz

The flow field in the cascade is then significantly influenced by the interaction of shock waves with the boundary layer, especially by the interaction of the inner branch of the exit shock wave with the boundary layer on the suction side of the blade (see Luxa et al. [3]). The shock-wave/boundary-layer interaction is important for the flow structure including the losses in the blade cascade, see Ligrani et al. [4]. A long-term research is devoted to this problem, but mainly on the interaction of the shock wave with the turbulent boundary layer. Ganapathisubramani et al. [5] dealt with the effect of the upstream boundary layer on the shock-induced separation. Giepman et al. [6] realized experimental investigation of the laminar and transitional oblique shock wave reflections for a range of Mach numbers.

This paper deals with the numerical simulation of the 2D compressible flow through the TR-U-9 blade cascade which presents one of variants of last-stage tip section blades. The blade cascade has been designed for two nominal regimes with the subsonic inlet Mach number M_1 = 0.8 and the outlet isentropic Mach numbers M_{2is} = 1.6 and 1.8 respectively. The CFD simulation was performed by means of the ANSYS CFX commercial software for the both regimes using two different computational domains. Numerical results are compared with experimental data of Luxa et al. [7].

2 Numerical simulations

Simulations of compressible flows were completed by the conditionally-averaged Navier-Stokes equations completed by constitutional relations for the ideal gas, the two-equation SST turbulence model of Menter [8] and by the bypass-transition γ-$Re_{\theta t}$ model of Langtry and Menter [9].

The transition model is based on transport equations for the intermittency coefficient and the local transition momentum Reynolds number. The intermittency coefficient is given by the equation

$$\frac{\partial(\rho\gamma)}{\partial t} + \frac{\partial(\rho \overline{U_j}\gamma)}{\partial x_j} = P_\gamma - E_\gamma + \frac{\partial}{\partial x_j}\left[\left(\mu + \frac{\mu_t}{\sigma_\gamma}\right)\frac{\partial \gamma}{\partial x_j}\right] \tag{1}$$

The production term

$$P_\gamma = F_{length} C_{a1} \rho S (\gamma F_{onset})^{\frac{1}{2}} (1 - C_{e1}\gamma) \tag{2}$$

contains empirical correlations F_{onset} for the transition onset and F_{length} for the transition length. The destruction term E_γ allows the prediction of the relaminarization of the boundary layer during a strong acceleration of the flow. The empirical correlation for the transition onset F_{onset} depends on the critical Reynolds number $Re_{\theta c}$ giving the position where the turbulent energy begins to grow up. The onset and length of the transition region, i.e. parameters $Re_{\theta c}$ and L_{length} are expressed by means of the local Reynolds number $\widetilde{Re_{\theta t}}$ The application of local variables is allowed by the relation between the Reynolds number related to the momentum thickness and the maximum of the vorticity Reynolds number used in the form $Re_\theta = Re_{vmax}/2.193$ valid for the Blasius boundary layer. The parameters $Re_{\theta c}$ and L_{length} are expressed by means of the local Reynolds number $\widetilde{Re_{\theta t}}$ given by the transport equation

$$\frac{\partial(\rho\widetilde{Re_{\theta t}})}{\partial t} + \frac{\partial(\rho U_j \widetilde{Re_{\theta t}})}{\partial x_j} = P_{\theta t} + \frac{\partial}{\partial x_j}\left[\sigma_{\theta t}(\mu + \mu_t)\frac{\partial \widetilde{Re_{\theta t}}}{\partial x_j}\right] \tag{3}$$

The model is switched over to a simple algebraic transition model in the case of the transition in separated flow. The effective value of the intermittency coefficient is given by the relation

$$\gamma_{eff} = max(\gamma, \gamma_{sep}) \tag{4}$$

Due to local variables the transition model can be used for the simulation of complex flows using unstructured computational grids. Calculations were realized by means of the ANSYS CFX commercial software.

Numerical simulation of flow through blade cascades are usually conducted using the computational domain corresponding to one pitch of the cascade with the inlet and the outlet in the distance about one half of the pitch from the leading and trailing edges respectively, see e. g. Cutrone et al. [10] and/or Louda et al. [11]. The outlet isentropic Mach number corresponds to the mean value of the static pressure in the traversing plane situated in a relatively short distance downstream the cascade. The other possibility is the application of the domain corresponding to the experimental arrangement of the test section in the wind tunnel (see Louda and Příhoda [12]). The computational domain covers a part of the inlet channel, the test section with the blade cascade and the prismatic section of settling chamber. However, this treatment puts greater demands on the computational technique.

The acceptable results can be as well achieved by the application of the extended outlet part of the computational domain with the prescription of the outlet static pressure corresponding to the pressure in the settling chamber behind the test section. The computational domain and mesh were created in two variants using GMSH free software. The scheme of the domain is shown in Fig. 1. The width is one pitch of the blade cascade. The both domains differ is the length of the outlet region. The short domain has the outlet length 1.78s while the extended one one has the outlet length 4s. Two computational domains with different outlet parts were used because of the study the effect of non-physical parasitic shock waves, which originates as reflection waves at the outlet boundary of the computational domain and might influence the flow field in the inter-blade channel.

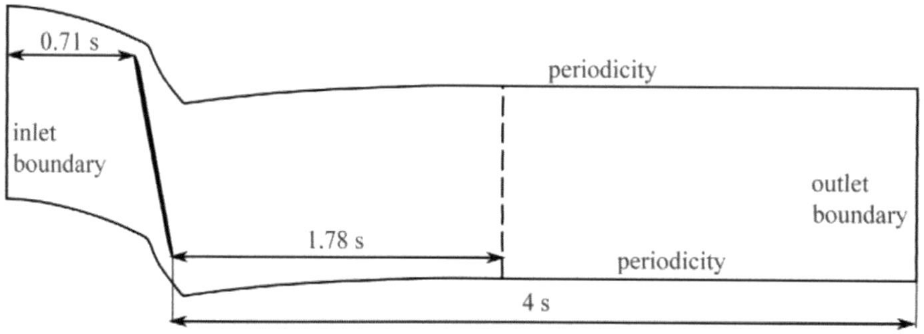

Fig. 1. Computational domain.

The computational domain consists of two grids - the structured O grid with quadrilateral cells around the blade and the triangular grid in the rest of the domain. The detail of the grid near to the trailing edge is shown in Fig. 2. The short domain contains about 167000 triangular cells and 54500 quadrilateral cells while the extended domain has 379000 triangular cells and 63000 quadrilateral cells. The grid has 665 nodes on the pressure side and 650 nodes on the suction side. The grid is refined near the wall with the first node in the distance $y^+ \approx 0.5$ from the wall. Due to used software the domain has one cell in the z direction.

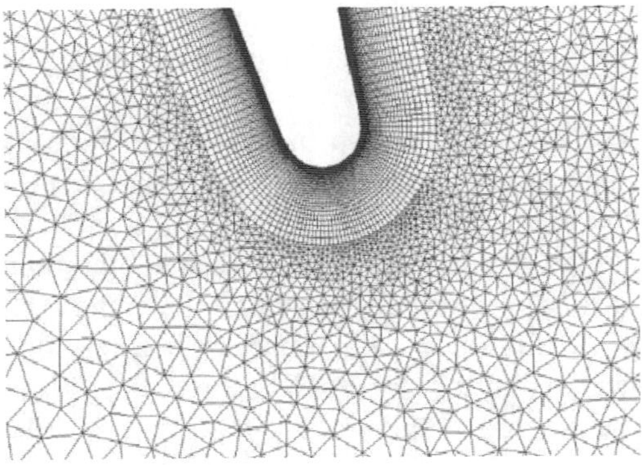

Fig. 2. Detail of the grid close to the trailing edges.

Numerical simulations were carried out for two nominal regimes characterized by the outlet isentropic Mach number M_{2is} = 1.6 and 1.8. Boundary conditions are the same for the both computational domain. The inlet boundary conditions are prescribed by the total pressure, the flow angle and by turbulence parameters, i.e. the turbulence level Tu = 10 % and the ratio of turbulent and molecular viscosity μ_t/μ = 250. The outlet boundary condition is given by the static pressure in accordance with the Mach number in the settling chamber behind the test section with the blade cascade. The turbulence intensity upstream the blade cascade $Tu \approx 1.5$ % corresponds to measured values at Mach number $M_1 \approx 0.8$. The periodic boundary condition is used on side boundaries of the computation domain.

3 Results

The TR-U-9 blade cascade consists of flat profiles with shaped leading and trailing parts. The cascade has the profile chord c = 0.15 m, the relative pitch s/c = 0.904, the maximum relative thickness $t/c \approx 0.02$ and the stagger angle γ= 79.23°measured from the blade suction side. Similarly, as for other tip-section blade cascades pressure and suction sides are on the inverse blade sides than usually. All simulations were realized according to experiments with the inlet angle α_1 = 79.23°. Experiments covering optical (interferometry and schlieren technique) and pneumatic measurements were accomplished in the in the suction-type high-speed wind tunnel in the Aerodynamic laboratory of the Institute of Thermomechanics at Nový Knín. The Mach-Zehnder interferometer was used for optical measurements. Pneumatic measurements were realized by the Prandtl probe upstream the blade cascade and by the five-hole conical probe in the traversing plane behind the blade cascade. A more detailed description is given e. g. by Luxa et al. [7].

The structure of the flow through the tip-section turbine blade cascade considerably depends on the character of the shock-wave/boundary-layer interaction. This interaction is usually related to the flow separation followed by the transition in separated flow in the case of the laminar boundary layer. The interferometric and schlieren pictures of the flow field for the Mach number M_{2is} = 1.8 are shown in Fig. 3.

a) Interferometric picture b) Schlieren picture

Fig. 3. Flow field for the Mach number M_{2is} = 1.8.

The effect of the interaction of the inner branch of the exit shock wave of the neighboring blade with the boundary layer on the suction side of the blade is well apparent in the interferometric and schlieren pictures.

The corresponding calculated field of Mach isolines is shown in Fig. 4 for the both computational domains. The flow field in the inter-blade channel is considerably influenced by the interaction of the inner branch of the exit shock wave of the neighboring blade with the boundary layer on the suction side of the blade. This is evident both from interferometric pictures and from the computed field of the Mach isolines. As can be seen in Fig. 4a), the field of Mach isolines obtained for the short domain contains a parasitic shock wave reflected from the domain boundary. Due to the extension of the outlet part of the domain the effect of parasitic shock wave is negligible and so only the results for the extended domain are presented.

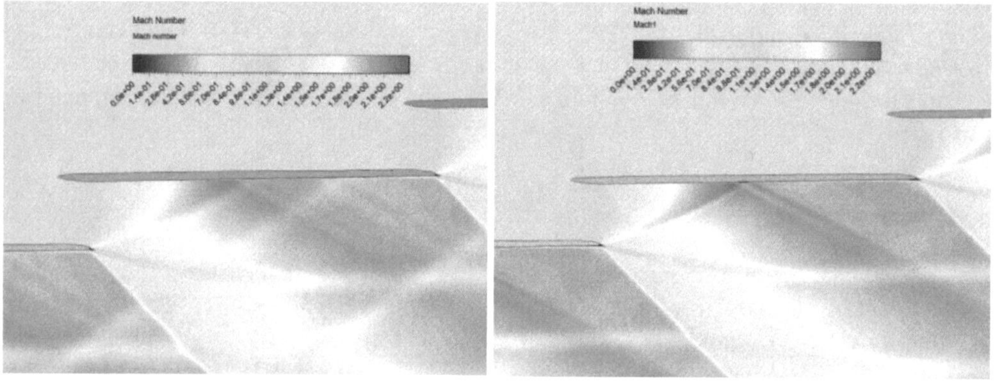

a) Short computational domain b) Long computational domain

Fig. 4. Field of the Mach number isolines for M_{2is} = 1.8.

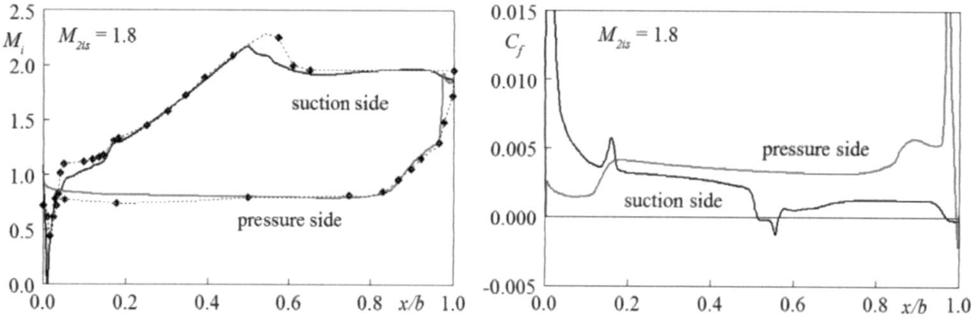

a) Isentropic Mach number b) Skin friction coefficient

Fig. 5. The distribution of the isentropic Mach number and the skin friction coefficient for M_{2is} = 1.8.

The distribution of the isentropic Mach number M_i on the blade profile for the Mach number M_{2is} = 1.8 is shown in Fig. 5a). Experimental results were evaluated from interferogram results. The so-called pressure and suction sides are for the tip-section cascades ordinarily on the inverse blade sides due to the stagger angles. The skin friction distribution is expressed in the Fig. 5b) by the coefficient C_f given by the ratio of the wall friction and the inlet dynamics pressure.

The structure of the flow field in the inter-blade channel depends largely on the interaction of the inner branch of the exit shock wave of the neighboring blade with the boundary layer on the suction side of the blade. It follows from interferometric pictures and from the computed field of the Mach isolines as well. The flow on the suction side accelerates from the leading edge up to the shock-wave interaction. The effect of the sudden change of the blade surface curvature can be seen at $x/b \approx 0.15$ where the curved leading part passes into the flat-plate profile. On the pressure side the velocity is mostly constant corresponding to the flow around the flat plate.

It can be seen from the interferometric picture and the isentropic Mach number distribution that the shock-wave interaction with the boundary layer takes place at the distance $x/b \approx 0.55$ where b is the axial chord of the blade. According to numerical simulations the position of the interaction is shifted somewhat upstream in the comparison with experimental results to the distance about $x/b \approx 0.51$. The distribution of the skin friction coefficient corresponds to the isentropic Mach number on the blade. On the suction side, there is obvious the effect of the sudden change in the surface curvature corresponding to changes in the Mach number. The interaction with the shock wave leads to the flow separation followed by the transition in the separated flow and the reattachment. The detail of the velocity field in the position of the shock-wave interaction for the $M_{2is} = 1.8$ is shown in Fig. 6. The reverse flow, forming a short separated region, can be seen near the wall.

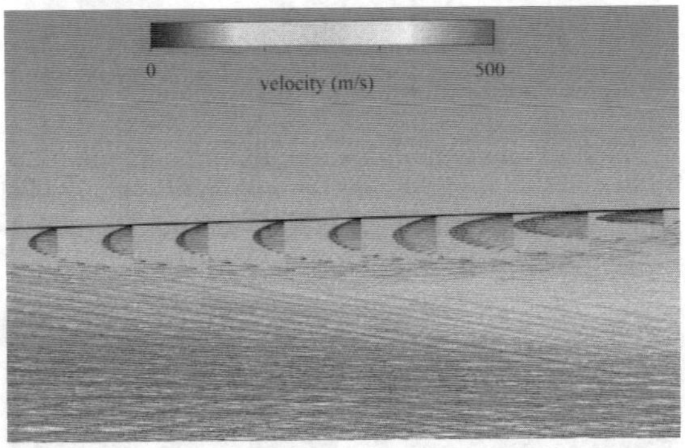

Fig. 6. Velocity field in the position of the shock-wave interaction.

However, it follows from the interferometric and hot-film measurements (see Luxa et al. [7]) that the transition on the suction side could occur in the attached flow upstream the interaction of the boundary layer with the shock wave. The velocity peak near the leading edge induces a very short laminar separation bubble on the pressure side. According to the simulation with the transition model the transition in attached flow occurs on the pressure side at the distance $x/b = 0.1$ from the leading edge. A great increase of C_f followed by the separation just upstream the trailing edge.

The distribution of the skin friction coefficient for the Mach number $M_{2is} = 1.8$ and for the numerical simulation by means of the turbulence model without the transition model is shown in Fig. 7. The course of the skin friction coefficient is similar as for the calculation with the transition model. On the pressure side, the coefficient C_f gradually decreases from the leading edge and for $x/b >$ 0.3 corresponds to the calculation with the transition model. Also on the suction side, the course of the C_f is similar. However, there is no separation at the interaction of the shock wave with the turbulent boundary layer. As well, values of the skin friction coefficient on the suction side are

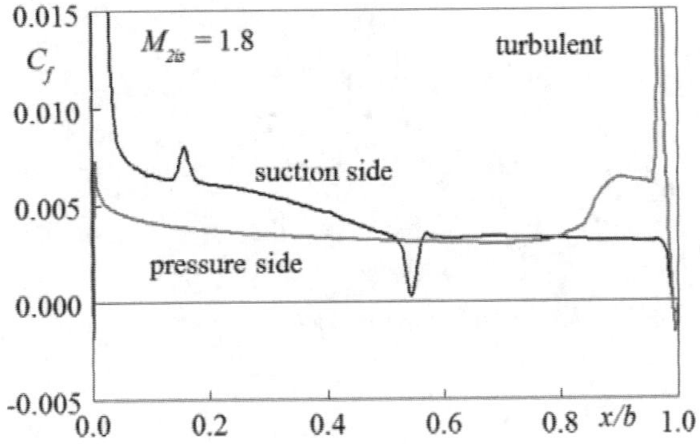

Fig. 7. Distribution of the skin friction coefficient for $M_{2is} = 1.8$ (turbulent model only).

higher. The supersonic expansion arises on the pressure side upstream of the asymmetric trailing edge. The near wake is shifted in the suction side direction and the inlet part of exit shock wave is very thick. The significant supersonic flow deviation takes place downstream the trailing edge, see Fig. 8. It leads to a short flow separation on the both blade sides.

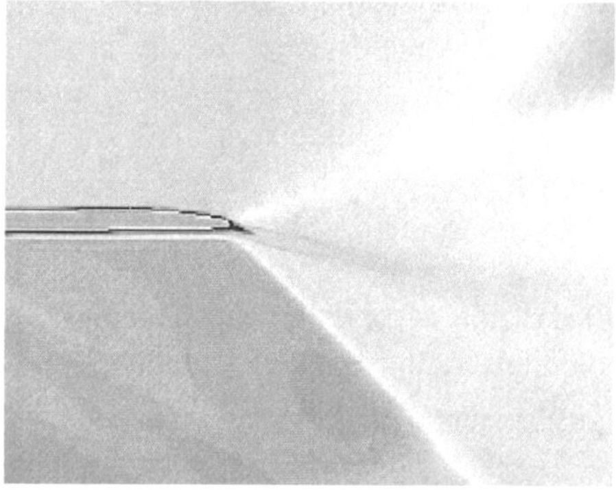

Fig. 8. Detail of the Mach number isolines near the trailing edge for $M_{2is} = 1.8$.

The investigation of the flow through the blade cascade at the Mach number $M_{2is} = 1.6$ was accomplished in the same experimental configuration. The interferometric and schlieren pictures of the flow field are shown in Fig. 9. The calculated field of Mach isolines for $M_{2is} = 1.6$ and the extended computational domain is shown in Fig. 10.

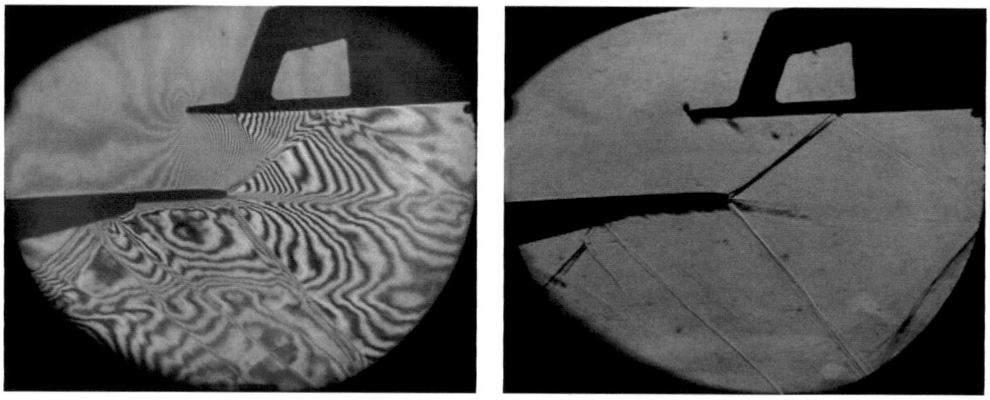

a) Interferometric picture b) Schlieren picture

Fig. 9. Flow field for the Mach number $M_{2is} = 1.6$.

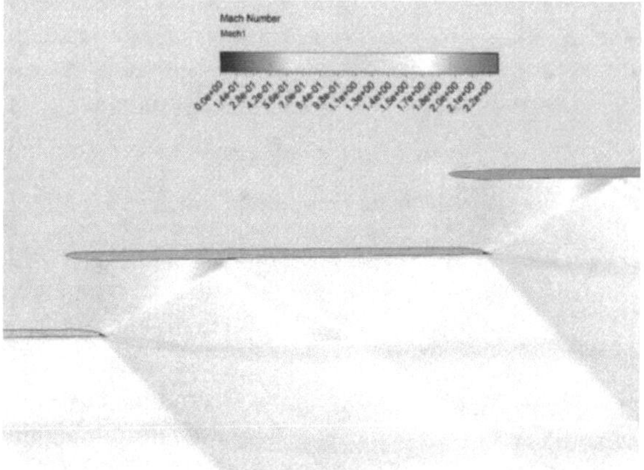

Fig. 10. Field of the Mach number isolines for $M_{2is} = 1.6$.

As can be seen from Fig. 10, the flow field for M_{2is} = 1.6 is very similar to the field for M_{2is} = 1.8, only the differences of Mach numbers are not so noticeable. The decisive influence on the flow structure in the inter-blade channel has again the interaction of the inner branch of the exit shock wave with the boundary layer on the suction side of the blade. The value of the outlet Mach number affects mainly the angle of the output shock wave. As the Mach number increases, this angle decreases. This leads to a shift in the position of the shock-wave/boundary-layer interaction with the increasing Mach number downstream towards the trailing edge.

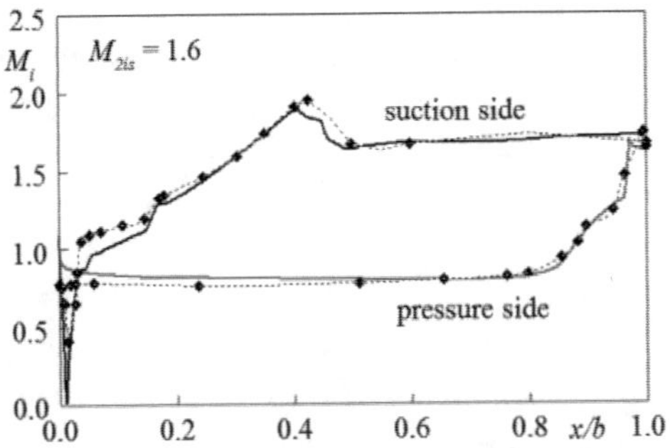

a) Isentropic Mach number

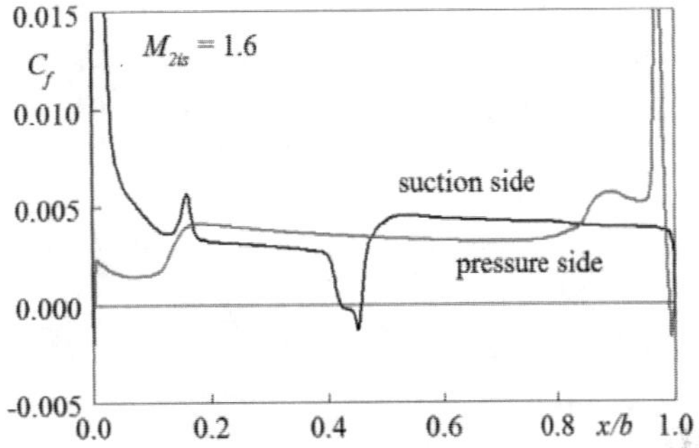

b) Skin friction coefficient

Fig. 11. The distribution of the isentropic Mach number and the skin friction coefficient for M_{2is} = 1.6.

The distribution of the isentropic Mach number and the skin friction coefficient for M_{2is} = 1.6 shown in Fig. 11 practically corresponds to the results for M_{2is} = 1.8. There is only a decrease in the Mach number M_i on the trailing edge from the value of M_i = 1.96 to M_i = 1.66. The transition onset on the pressure side is for the Mach number M_{2is} = 1.6 at the distance x/b = 0.1 as well as for the higher value M_{2is}. The position of the interaction of the shock wave with the boundary layer is shifted from x/b = 0.55 for the lower Mach number M_{2is} = 1.6 to the distance x/b = 0.425. The flow separation and the transition in the detached flow on the suction side of the profile are shifted in the same way.

4 Conclusions

Numerical simulations of 2D transonic flow through the turbine blade cascade consisting of flat profiles were realized. The cascade is one of variants of the tip section of ultra-long blades, which were designed for the last stage of the steam turbine. CFD simulations were realized by means of the ANSYS CFX commercial software using the γ-$Re_{\theta t}$ bypass transition model completed by the two-equation SST turbulence model. Some simulations were made only by the SST turbulence model for comparison. Numerical results were compared with experimental data. Calculations were carried out for two nominal regimes and two computational domains. In addition to the standard computational domain, the calculation was performed for a domain with an extended output part for the suppression of reflected shock waves. Due to the blade profile, the flow near the leading edge on the pressure side can be influenced by the short laminar separation bubble. Similarly, a short flow separation occurs just upstream the asymmetric trailing edge.

The interaction of the inner branch of the exit shock wave with the boundary layer on the blade suction side leads in the both flow regimes to the flow separation followed by the transition to turbulence. This shock-wave/boundary-layer interaction with flow separation and the transition in the detached flow on the suction side of the profile are shifted downstream with increasing outlet isentropic Mach number M_{2is}. For the extended computational domain the agreement of numerical simulations with experimental data is acceptable.

This research was supported by the Czech Science Foundation (GACR) under the grant No. 20-11537S. Institutional support RVO 61388998 is also gratefully acknowledged. Institutional support RVO 61388998 is also gratefully acknowledged.

References

1. Luxa, M. (2017). *Effects of the design of the last-stage rotor of the low-pressure part of the high-power steam turbine with very long blades on the transonic flow field,* Habilitation Thesis, Czech Technical University in Prague, Prague, (in Czech)
2. Musil, J., Příhoda, J., Fürst, J. (2019). Simulation of supersonic flow through the tip-section turbine blade cascade with a flat profile, *Proc. Conference Topical Problems of Fluid Mechanics 2019,* Prague, 169-174, 2019, https://doi.org/10.14311/TPFM.2019.023
3. Luxa, M., Příhoda, J., Šimurda, D., Straka, P., Synáč, J. (2016). Investigation of the compressible flow through the tip-section turbine blade cascade with supersonic inlet, *Journal of Thermal Science,* 25, 2, s. 138-144, https://doi.org/10.1007/s11630-016-0844-0
4. Ligrani, P. M., McNabb, E. S., Collopy H., Anderson, M., Marko, S. M. (2020). Recent investigations of shock wave effects and interactions, *Advances in Aerodynamics*, 2, 4, 23 p., https://doi.org/10.1186/s42774-020-0028-1
5. Ganapathisubramani, B., Clemens, N. T., Dolling, D. S. (2007). Effects of upstream boundary layer on the unsteadiness of shock-induced separation, *Journal Fluid Mechanics*, 585, 369-394, https://doi.org/10.1017/S0022112007006799
6. Giepman, R. H. M., Schrijer, F. F. J., Oudheusden van, B. W. (2015). High-resolution PIV measurements of a transitional shock wave–boundary layer interaction, *Experiments in Fluids,* 56, 1-20, https://doi.org/10.1007/s00348-015-1977-8
7. Luxa, M., Šimurda, D., Hála, J. (2020). *Measurements on TR-U-9 cascade*, Institute of Thermomechanics, Czech Academy of Sciences, Prague, Research Report Z1610/2020
8. Menter, F. (1994). Two-equation eddy-viscosity turbulence model for engineering applications, *AIAA Journal,* 32, 1598-1605, https://doi.org/10.2514/3.12149
9. Langtry, R. B., Menter, F. R. (2009). Correlation-based transition modeling for unstructured parallelized computational fluid dynamics codes, *AIAA Journal,* 47, 2894-2906, https://doi.org/10.2514/1.42362
10. Cutrone, L., De Palma, P., Pascazio, G., Napolitano, M. (2007). An evaluation of bypass transition models for turbomachinery flows, *International Journal of Heat and Fluid Flow*, 28, 161-177, http://doi:10.1016/j.ijheatfluidflow.2006.02.031
11. Louda, P., Straka, P., Příhoda, J. (2018). Simulation of transonic flows through a turbine blade cascade with various prescriptions of outlet boundary condition, *EPJ Web of Conferences*, 180, 02056, 6 p., https://doi.org/10.1051/epjconf/201818002056
12. Louda, P., Příhoda, J. (2021). Study on the numerical model of transonic wind tunnel test section, *Journal of Thermal Science*, 30, 231-241, https://doi.org/10.1007/s11630-020-1277-3

Preparation of a Real Model of Nasal Cavities from Computed Tomography for Numerical Simulation

Filip Trnka[1,*], *Hana* Schmirlerová[1], *Martin* Májovský[2], *David* Netuka[2], and *Michal* Schmirler[1]

[1]CTU in Prague, Faculty of Mechanical Engineering, Department of Fluid dynamics and Thermodynamics, Technická 4, 160 00 Prague, Czech Republic
[2]Charles University and Military University Hospital, First Faculty of Medicine, Department of Neurosurgery and Neurooncology, U Vojenské nemocnice 1200, 169 02 Prague 6, Czech Republic

Abstract. With improves in computing hardware and CFD methods, it is possible to work with more complex geometries. The aim of this study is to describe the preparation and creation of a model from CT scans for a numerical study of air flow into the nasal cavities.

Research background: The influence of pituitary tumor surgery on the change of geometric proportions of the nasal cavity is investigated in cooperation with the Neurosurgical Department.
Purpose of the article: In pituitary tumor surgery, the nasal cavity is used as an access route to the Sella turcica where the pituitary gland is located. Geometric changes occur during surgery. These changes in the geometry of the nasal cavities affect the air flow into the nasal cavities and the sense of the smell of the operated patients.
Methods: Based on CT scans of one patient, a procedure was created for creation of models of nasal cavities before and after surgery of pituitary tumor. The open-source software 3D Slicer was used for processing CT scans. Furthermore, the model before and after surgery was modified in CAD program Autodesk Inventor and program Ansys Space Claim. Meshing and subsequent solution of the finite volume method was solved in program Ansys Fluent 2021 R1.
Findings & Value added: The models were created considering the anatomical structure of the cavities and were aligned into corresponding coordinate systems. Despite of the efforts to align the planes, inaccuracies occur between the planes. However, the results are comparable.

Keywords: *Nasal Cavity; CT scans; CFD; Numerical study; Adenomas*

1. Introduction

The expansion of the possibilities of using real geometries in fluid flow research have been possible as a consequence of the development of computational technology. Thanks to more powerful computational technology, it is possible to process more complex geometries,

* Corresponding author: f.trnka@fs.cvut.cz

which would have been too complicated and laborious to create just a few years ago. This can be used not only in numerical studies, but also in experimental measurements.

In the framework of cooperation with the Neurosurgical Department of the Central Military Hospital, a problem arose with the creation of real models from Computed Tomography (CT) scans. In the case of pituitary adenoma surgery, the geometry of the patient's nasal cavities is changed given that the nasal cavity is used as an access road to the Sella turcica where the pituitary gland is located. The influence of surgery on the change of geometry through the passage of surgical instruments is therefore crucial. The change of the nasal cavity after surgery affects the change in air flow through the nasal cavities and may also affect olfactory functions [1, 2].

In previous years, studies have been developed to reach a general model of nasal cavity. These articles were based on a statistical approach and the processing of images of a large number of patients (16 and more). This approach is interesting, but it is not possible to use it for the purpose of determining the effect of adenoma surgery on air flow through the nasal cavities. This article will therefore describe the process of designing two models for numerical study where the two models will be compared before and after surgery of one patient [3-5].

2. Methods

2.1 CT images processing

The patient's data from which the model was created were obtained from a Neurosurgical Department. Both Magnetic Resonance Imaging (MRI) and CT images were processed by the open-source 3D Slicer program. This program was used for imaging, segmentation of the model and changing the coordinate system of the images.

First, both MRI and CT images were imported to the 3D Slicer program. Comparing the data from different sources, the CT images were better. This is due to the properties of the investigative methods. In general, the MRI data had worse resolution compared to the CT and contained noise.

After importing the CT data into the Slicer, a threshold function was used to create a three-dimensional body. The threshold function works with intensity range, and it is possible to select a part of the skull, brain, soft tissue, or cavity. In this study, cavities were selected to be further worked with. The volume created was adjusted by simple functions of erase, scissors, and paint. The erase and paint functions remove (create) parts of the model in the individual slices of which the visualization is made of. Conversely, the scissors' function can remove or add volume in several slices at once. These functions allow to remove secondary nasal cavities and other parts, that do not affect the change in geometry before and after surgery. Modifications of the model in the Slicer program and individual removal of anatomical parts in both models were consulted with the Neurosurgery Department.

An important part of the process of creating the nasal cavity model is the comparison of the coordinate systems between model before and after the adenoma surgery. The results of the pressure and velocity fields are presented in the coronary planes. In order to compare the pressure and velocity fields of the two models, the coordinate systems need to be unified.

Although the effort to compare coordinate systems is great, the process has not yet been improved enough to achieve an identical comparison. Figure 1 shows the best result of the comparison that has been achieved. The two models differ in some parts even from an anatomical point of view. This is due to the change in geometry after the surgery.

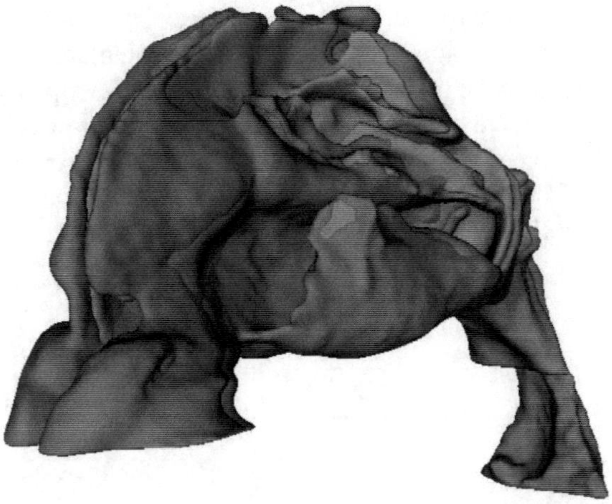

Fig. 1. Transformation of the coordinate system (green – before surgery; red – after surgery).

The last part of creating a model of nasal cavities in Slicer is smoothing the model. The 3D Slicer automatically smooths surfaces when displaying the created volume. Gaussian smoothing is used for this purpose. At this point it is necessary to mention an article [6] in which the authors compare various errors in the creation of the models. The study is concerned with comparing the experimental model and the model that is created for the numerical study. They compare the wall roughness, caused by the inaccuracy of the casting used for the experimental model, and the surface smoothing. Surface smoothing is caused by the resolution of MRI or CT data and the approximation of the model to real geometry. The model in the article is smoothed by two smoothing factors. The article shows that small smoothing can help reduce pressure resistance, however, with high smoothing the pressure loss of the flowing air decreases too much. In the case of this article, the model uses minimal smoothing of the model by Gaussian smoothing of 0.3, so that there is as little difference as possible from the real nasal cavities.

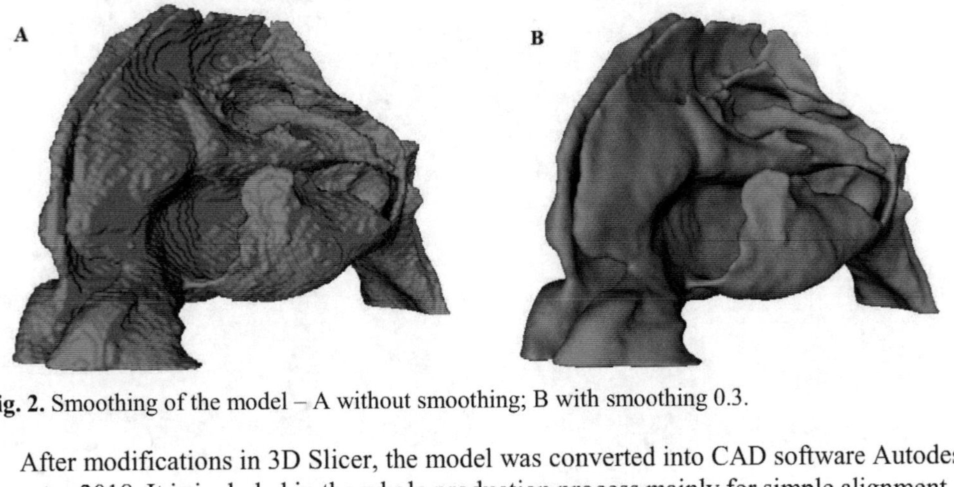

Fig. 2. Smoothing of the model – A without smoothing; B with smoothing 0.3.

After modifications in 3D Slicer, the model was converted into CAD software Autodesk Inventor 2018. It is included in the whole production process mainly for simple alignment of the nostril surfaces and the surface in the nasopharynx area. For the nostril area, only planes parallel to the coordinate system of the model were used. In the area behind the nasopharynx, the alignment of the model was a bit more complicated. CT scans are mainly influenced by the position in which the patient is located during the examination. For unity between the models before and after the surgery, a plane parallel to the Chamberlain line was selected. Chamberlain line is a radiologically defined plane between the posterior part of the hard palate and the posterior point of foramen magnum. More detailed description is in article [7]. The next step after alignment of the nostrils and the plane in the nasopharynx area is to create a starting length to meet the theoretical conditions of fluid mechanics. The calculation thus creates a speed profile, corresponding to the second-order curve. To create the starting length of the model, the plane of the nasopharynx area is used, which is subsequently labelled as outlet. The created areas in the nostril area are further labelled as inlet.

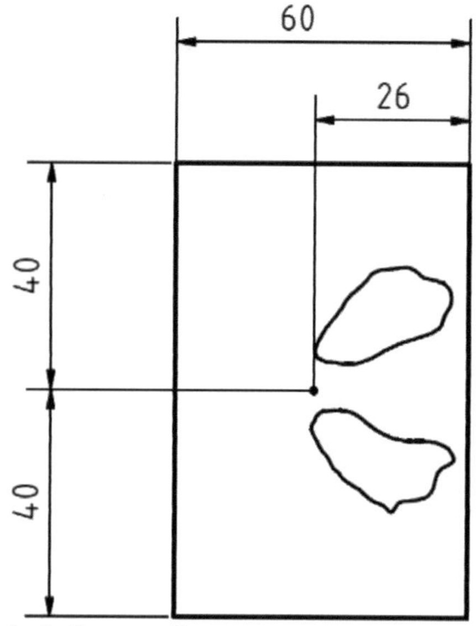

Fig. 3. Sketch for creating cube before nostrils.

A cube is created in front of the nostrils, on which a pressure boundary condition is prescribed in the solution. A sketch of this object is shown in Figure 3. The sketch is extruded in the next step and a cube is created. The top wall of the cube is referred as wall boundary condition and the side walls are referred to as the inlet boundary condition.

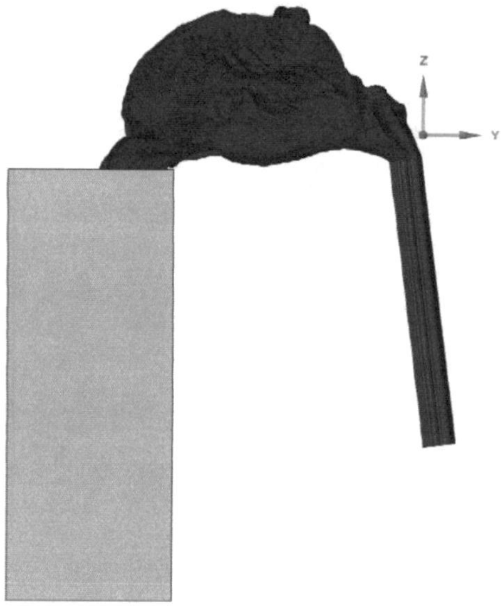

Fig. 4. Model of nasal cavity.

2.2 Meshing

The meshing of the nasal cavity model was done in Ansys Fluent 2021 R1. Prior to the meshing, the model was modified in Space Claim, where minor geometry errors were corrected, and areas were divided into facet zones. Space Claim was mainly used for compatibility with program Ansys Fluent, which is further used for meshing and solution. The program is also suitable for faster managing the model, thus faster division of the model into facet zones and corrections of errors such as intersecting areas or holes in the model.
For meshing the models before and after the pituitary tumour surgery were used the same parameters. The global cell size of the surface mesh was 5-10 mm. The local cell size of 1 mm was used for the surface mesh of nasal cavity zone. The volume mesh was constructed with polyhedral cells, the size of which was set to 5 mm. The model before the surgery contained 0.72 million cells and the model after the surgery was made of 0.745 million cells. The two mesh were selected based on a previous study comparing the results of four meshes with different number of cells. The model after the surgery is expected to have more cells. The anatomy was affected in the area of sphenoidal sinus and some passages were enlarged. Therefore, these parts of the model could not be erased from the model after the surgery in the affected area as they were in the model before the surgery.

2.3 Boundary condition

As described at the end of the 2.1 CT images processing, parts of the model have been created in Autodesk Inventor to prescribe boundary conditions. A cube has been created in front of the nostrils, on which the side and bottom walls are marked as inlet. On the inlet planes, a pressure inlet boundary condition is prescribed with zero difference from total pressure. After creating a starting length behind the nasopharynx, the extruded plane is marked as outlet. On the outlet, a mass flow outlet boundary condition is prescribed. On all other surfaces, wall boundary conditions are prescribed, where the flow rate is zero. On the Figure 5, a model of nasal cavity with a schematic representation of these boundary conditions is illustrated.

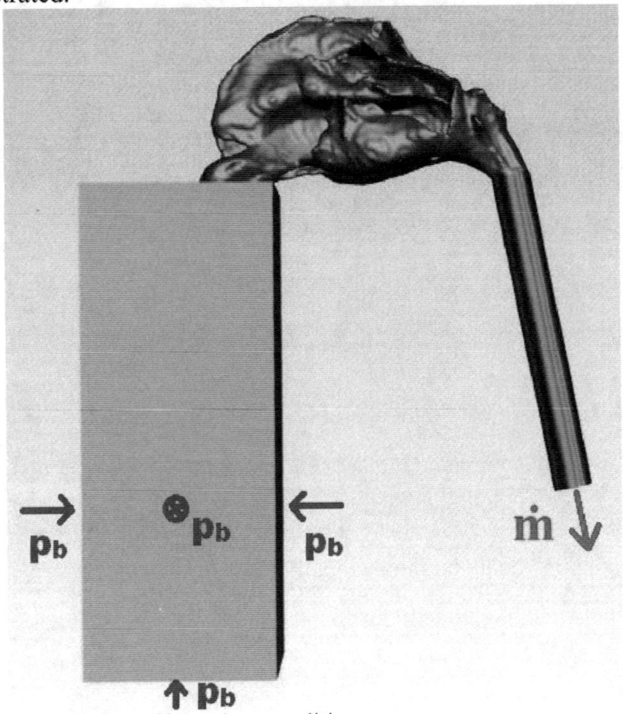

Fig. 5. Schematic representation of boundary conditions.

The numerical calculation was performed for steady flow. The Figure 6 shows the mass flow during the human respiratory cycle. Individual red points indicate the mass flow values that were prescribed for individual calculations on the outlet.

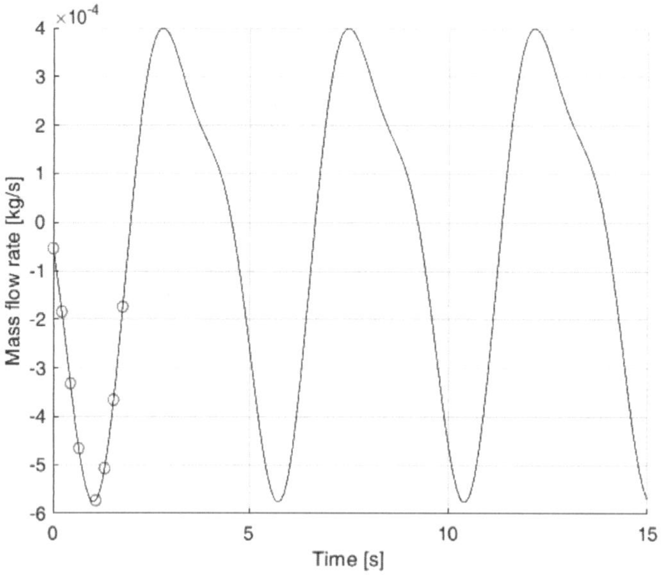

Fig. 6. Mass flow rate during human respiratory cycle.

2.4 Solution

The numerical simulation of air flow in the inspiration part of breathing cycle was performed under steady flow conditions. A viscous laminar computational model was used on both models before and after the adenoma surgery and the used solver was pressure based. The pressure gradient was calculated by second order function and the momentum gradient by a second order upwind scheme [8-10].

3. Results

In model were created planes to plot the results of the numerical study. The Y-planes are created parallel to the coronary planes. The schematic representation of the created planes can be seen on the Figure 7. The blue area is the olfactory area.

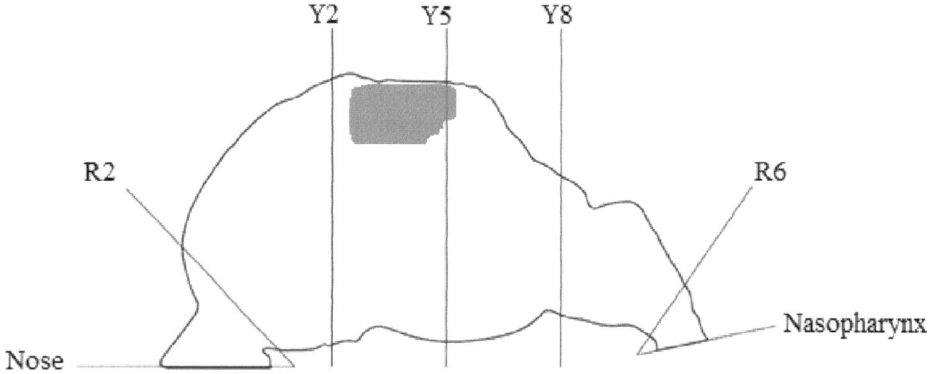

Fig. 7. Schematic representation of planes for displaying result.

This paper presents the results of a numerical study on the plane Y5 for a steady solution at maximum mass flow rate. The planes were created in the model before and after adenoma surgery, in the same distance from the beginning of the coordinate system.

Pressure loss, or pressure resistance, is a very significant factor in fluid dynamics. Therefore, pressure loss was investigated in the nasal cavity model. In both models, static pressure values were calculated on the planes of the nasopharynx and nostrils and subtracted from each other. This determined the pressure loss of the entire model.

Table 1. Pressure loss in nasal cavity model for maximal mass flow rate.

Plane	**Nose**	**Nasopharynx**	
	Static pressure [Pa]	**Static pressure [Pa]**	**Pressure difference [Pa]**
Before surgery	-0.07027	- 1.0210	- 0.950730
After surgery	- 0.04848	- 0.7989	- 0.750420

On the Figure 8 and Figure 9 pressure and velocity fields are displayed. Of the left, there are fields of model before surgery and on the right side are results on plane Y5 after surgery. When comparing pressure fields before and after surgery, it is possible to notice an increase in static pressure throughout the cross-section. For the results of the velocity fields, it is possible to notice much smaller differences in values. The most noticeable differences are in the middle part of the velocity field. Maximum values of both models are located there, although velocity values in the model before surgery are higher. Furthermore, there is a decrease in velocity values in the upper area of the plane Y5 in model after adenoma surgery. This occurs probably due to an increase of volume in the model after the surgery.

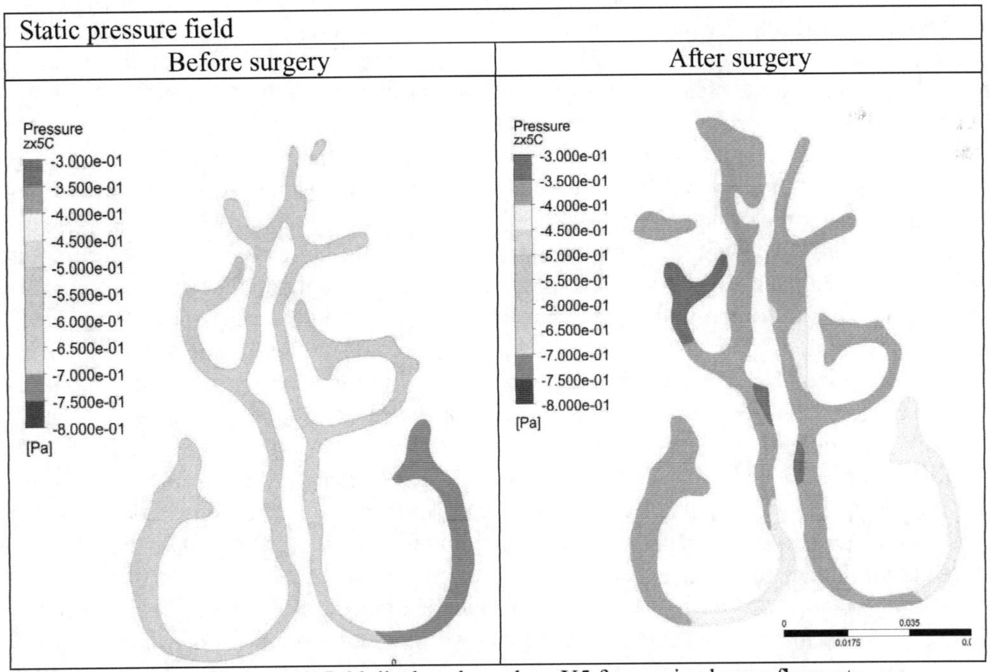

Fig. 8. Results of static pressure field displayed on plane Y5 for maximal mass flow rate.

Velocity field	
Before surgery	After surgery

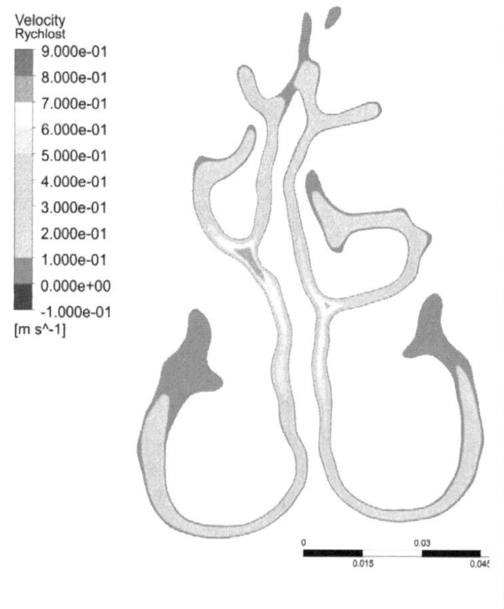

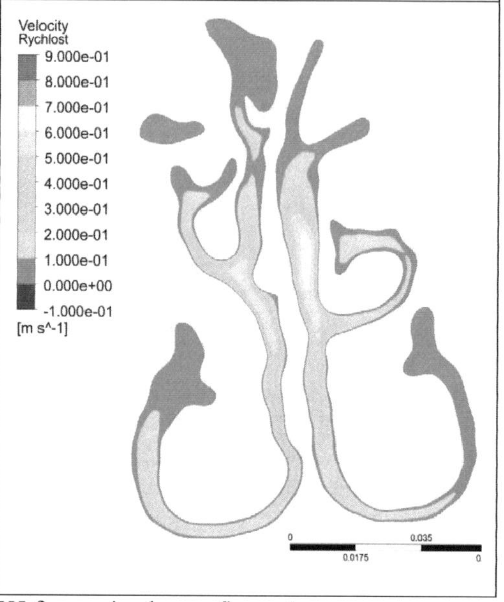

Fig. 9. Result of velocity field displayed on plane Y5 for maximal mass flow rate.

4. Discussion

The main objective of this article is to create a procedure for the creation of realistic models of nasal cavities. Models of nasal cavities were created according to the above procedure regarding the change in the anatomy of nasal cavities after the adenoma surgery. In chapter 3. Results are presented the results of numerical study in the steady flow mode for maximum value of mass flow rate during inhale part of breathing cycle on the plane Y5. In the Results section it is possible to observe not only the velocity and pressure fields, but also the changes in geometry between models.

From the presented results, it may appear that there are large changes in both pressure and velocity fields. It is true that the static pressure in the whole model increased significantly after surgery, but it is important to note the change in cross-section area between the models. In the model after surgery, there is an expansion of the middle meatus due to the passage of surgical instruments. That means the local dimensions of models differ, but when comparing the global dimensions of the model before and after the surgery, there is no greater deviation than 2.5 % in the basic dimensions.

Changes in the nasal cavity appear day by day. These changes are hormonally affected, they are not perceived. This can also affect the differences between the models of patients before and after surgery, because when examined by CT or MRI, patients can be at different stages of this cycle.

An important part of the research on the influence of adenomas surgery on the change in geometry is to compare the results between the two models. Without comparing the models into the same coordinate systems, it would be very difficult to create planes at the same place of the models. Among other things, setting the same coordinate systems provides a faster possibility of creating planes for presenting the results.

This work was supported by the Grant Agency of the Czech Technical University in Prague, grant No. SGS22/099/OHK2/2T/12.

References

1. NETUKA, D., MASOPUST, V., FUNDOVÁ, P., ASTL J., ŠKOLOUDÍK, D., MÁJOVSKÝ, M., BENEŠ, V. (2019). Olfactory Results of Endoscopic Endonasal Surgery for Pituitary Adenoma: A Prospective Study of 143 Patients. *World Neurosurgery*. 129, e907-e914.
2. MAJOVSKY, M., ASTL, J., KOVAR, D., MASOPUST, V., BENES, V., NETUKA, D. (2019). Olfactory function in patients after transsphenoidal surgery for pituitary adenomas—a short review. *Neurosurgical Review*. 42(2), 395-401.
3. LIU, Y., JOHNSON, M. R., MATIDA, E. A., KHERANI, S., MARSAN, J. (2009). Creation of a standardized geometry of the human nasal cavity. *Journal of Applied Physiology*. 106(3), 784-795
4. BRÜNING, J., HILDEBRANDT, T., HEPPT, W., et al. (2020). Characterization of the Airflow within an Average Geometry of the Healthy Human Nasal Cavity. *Scientific Reports*. 10(1)
5. TRETIAKOW, D., TESCH, K., MEYER-SZARY, J., MARKIET, K., SKOREK, A. (2021). Three-dimensional modeling and automatic analysis of the human nasal cavity and paranasal sinuses using the computational fluid dynamics method. *European Archives of Oto-Rhino-Laryngology*. 278(5), 1443-1453
6. SCHROETER, J. D., GARCIA, G. J. M., KIMBELL, J. S. (2011). Effects of surface smoothness on inertial particle deposition in human nasal models. *Journal of Aerosol Science*. 42(1), 52-63
7. CRONIN, C. G., LOHAN, D. G., MHUIRCHEARTIGH, J. N., MEEHAN, C. P., MURPHY, J., ROCHE, C. (2009). CT evaluation of Chamberlain's, McGregor's, and McRae's skull-base lines. *Clinical Radiology*. 64(1), 64-69.
8. ZHAO, K., SCHERER, P. W., HAJILOO, S. A., DALTON, P. (2004). Effect of Anatomy on Human Nasal Air Flow and Odorant Transport Patterns: Implications for Olfaction. *Chemical Senses*. 29(5), 365-379.
9. XIONG, G. X., ZHAN, J. M., JIANG, H. Y., LI, J. F., RONG, L. W., XU, G. (2008). Computational Fluid Dynamics Simulation of Airflow in the Normal Nasal Cavity and Paranasal Sinuses. *American Journal of Rhinology*. 22(5), 477-482.
10. Zubair, M., Riazuddin, V., Abdullah, M., Ismail, R., Shuaib, I., Ahmad, K., (2017). Computational fluid dynamics study of the effect of posture on airflow characteristics inside the nasal cavity. *Asian Biomedicine,* 7(6), 835-840.

Numerical Study of Flow Field Characteristics in the Trachea During Growth of Human Upper Airways

Jakub Elcner[1,*] and *Miroslav* Jícha[1]

[1]Brno University of Technology, Faculty of Mechanical Engineering, Department of Thermodynamics and Environmental Engineering, Technicka 2986/2, 616 69, Brno, Czech Republic

Abstract. The development of organs in the human body does not end at birth. During the first five years of life, changes occur in the respiratory tract, not only in terms of its dimensions but also in the way it is used. Efforts to provide non-invasive treatment in the form of medical aerosols administered to children´s lungs during this period must be supported by knowledge of the flow pattern that significantly influences their transport and deposition. Research related to flow patterns in the adult human respiratory tract is quite widespread and the phenomena that occur during inhalation in different parts of the respiratory tract have been widely documented. In the case of the paediatric respiratory tract, research is relatively scarce due to the age of the patient and the desire to minimise interference with the paediatric organism. At the Brno University of Technology, we have the geometry of the airway of a ten-month-old infant, a scaled model of an adult to match the geometry of a five-year-old child based on scientific knowledge and also an adult model of the human respiratory tract. These geometries, together with knowledge of respiratory physiology were used to compare the changes in airflow behaviour that occur in the trachea during the first five years and compare it to fully developed adult human geometry. Computational Fluid Dynamics was used to investigate the model using a Large Eddy Simulation approach. The periods of life captured by the geometries differ not only in their dimensional difference but also in their approach to airway use. The impact of these differences has been captured in the paper.

Keywords: *Human upper airways, Flow pattern, CFD, LES*

1 Introduction

Development of the human respiratory tract starts in utero, but considerable remodelling appears after birth and continues until the age of approximately 5 years. During this period the respiratory tract is quite different from the adult one. The differences can be seen not only in airway dimensions but also in airway morphology. As stated by [1], younger subjects have smaller nostrils, a shorter turbinate region, a narrower nasopharynx, and a narrower pharynx-larynx. Of particular interest is the turbinate region in the nasal cavity that experiences fast growth from birth to the age of 5 and is much simpler in morphology than in older children and adult models. Children also have a much smaller and narrower nasopharynx lumen.

* Corresponding author: elcner@fme.vutbr.cz

Differences are also in the nostril-tovalve (the end of the nose vestibule) distances, which are much shorter in infants and increase from about 8 mm in several-days-old to 27 mm in adults. Breathing mode is also different. As stated by [2], up to the age of 20 months the nasal breathing exceeds that of oral breathing and becomes equivalent just at the age of 20 months when infants stop sucking. Up to this age, nasal breathing is more intense than mouth breathing due to the different development stages of nasal and oral cavities. Both routes, i.e. Nasal or oral breathing strongly depend on the geometrical features of nasal and oral routes and their hydraulic resistances which are significantly different from adults. Therefore, children can't be taken as "down-sized" adults. The ventilation rates in infants and small children are also different from the adults and create significantly different flow fields and aerosol routes in case of its transport and deposition. The differences in the geometry, breathing patterns, the nonexistence of similarities between infant and adult lungs in the developing phase and considerable remodelling after birth present difficulties in any studies of airflow and particle deposition in infants and children [3]. Currently, there are very few studies that systematically illustrate the airflow patterns in the infant's and children upper airways [1, 2]. For infants who suck, breathing by nose prevails up to the age of 20 months. After this age, the breathing is more influenced by the activities of children and the development phase of the lungs is terminated at the age of approximately 5 years. From this age the geometrical features of "mean" children's lungs can be "down-sized" from the adult lungs. We currently know very little about how airflow patterns change throughout early infants and children's life. We have almost no information/data on how the aerosol penetrates through the mouth/nose and upper airways in infants and children developing lungs to more distal airways and mainly how the deposition is linked with different inhalation profiles [4] or administered by different techniques. Despite the number of studies devoted to flow field and aerosol transport and deposition in human airways, there are almost no information/data on these properties in developing lungs of infants and children including nasal and oral passages, linked with different inhalation profiles. There are important differences in airway anatomy, air-flow dynamics, and aerosol deposition between subjects' ages. Unfortunately, so far published articles are always limited and don't provide a complete image of the flow field and aerosol deposition for different breathing profiles linked with different administration techniques. [1], studied four age categories of infants and children but only nasal inhalation without the tracheobronchial tree. [4] studied only the airflow in one infant and one child geometry but without extra-thoracic airways. This study focuses on the changes that occur in the trachea during the first five years of life and attempts to analyse these differences concerning their possible influence on aerosol transport and deposition.

2 Methods

2.1 Models

Three types of geometry corresponding to an infant at 10 months of age, a child at 5 years of age and an adult at age of 25 were used in this paper (Fig.1). All geometries represent a model of the upper airway and part of the tracheobronchial tree. The nasal cavity, larynx, trachea, and tracheobronchial tree are branched into the second (sixth in the case of the adult human geometry) generation because of the realistic formation of the flow field in the upper airway. Each of the geometries is described below. Since the models capture different stages of human development, when not only the shape of the geometry but also the breathing pattern, lung volume and respiratory characteristics changes, the so-called resting mode of breathing was chosen as a unifying parameter, which represents a different flow through the model in each period. The inspiratory flow values are shown in Table 1.

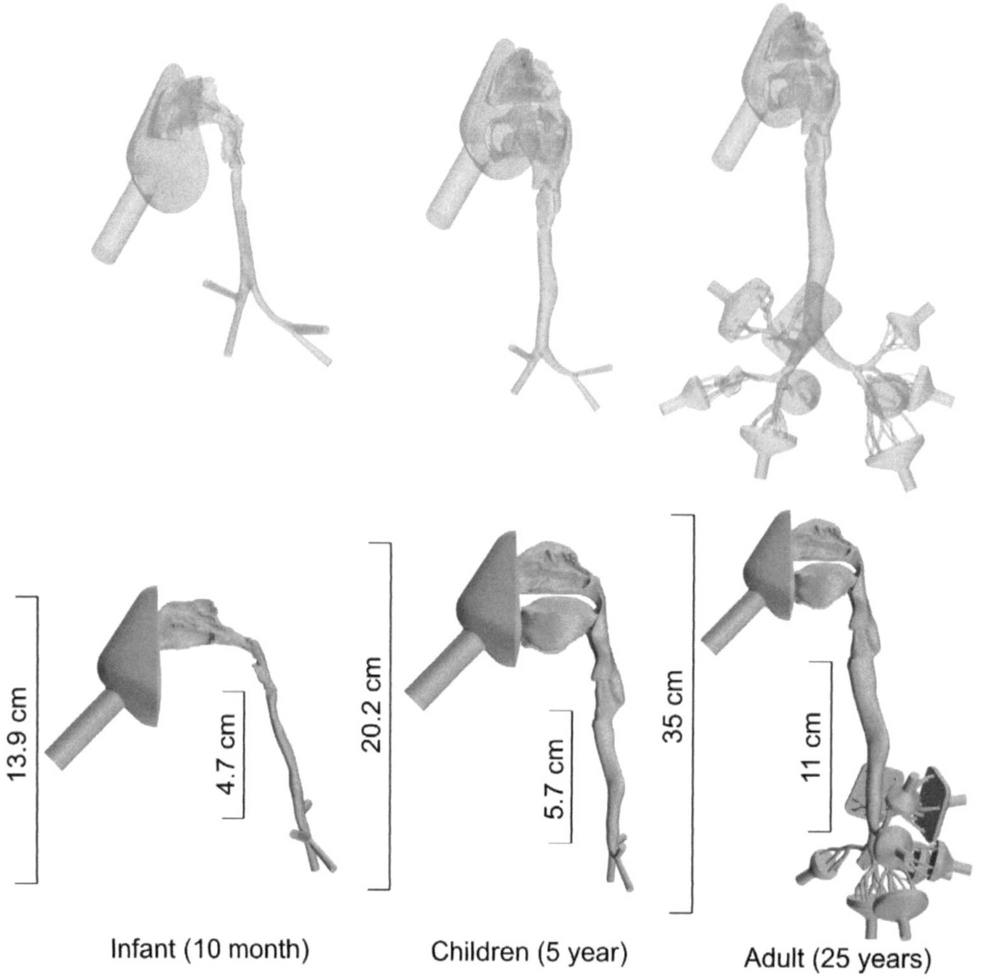

Fig. 1. Geometries used for numerical simulations.

2.1.1 Geometry of 10-months old infant

At this age, breathing is mainly performed through the nasal cavity, so the model does not include the oral cavity. The model was made based on cooperation with Brno university hospital and the geometry of a 10-months old child comes from high-resolution computed tomography, which was further processed in the laboratories of CEITEC (Brno, Czech Republic). The part of the geometry containing the nostrils was modelled in Rhinoceros after consultations with an otorhinolaryngologist. The diameter of the suction attachment mounted on the mask is 13 mm and the height of the area replicating the face is 69 mm and its width is 40 mm. The trachea contained in the model is 47 mm long and its diameter is 7.9 mm. At the end of the tracheobronchial tree, there are artificial extensions 20 mm long and 4.5 mm in diameter.

2.1.2 Geometry of 5-year-old child

This model was developed on the assumption that there are no significant developmental changes in individual parts of the respiratory tract (change in the shape of the larynx, nasal cavity, etc.) after the age of five and that all changes that occur from this time are scalable based on known sizes [5]. The adult respiratory tract model that BUT has and has been developing for a long time was used for scaling. Due to the change of ratio in the head-to-thorax that occurs during human growth, the scaling factor chosen for the tracheobronchial tree differs from the scaling factor of the upper airway region. The method of scaling is described in [5]. The model contains a mask with a height of 77 mm and a width of 46 mm. The diameter of the suction attachment is 16 mm. The length of the trachea is 57 mm, and its diameter is 8.5 mm. At the end of the tracheobronchial tree, there are artificial extensions of a length of 20 mm and 4.5 mm in diameter.

2.1.3 Geometry of adult

This geometry has long been expanded and used in research at BUT [5]. The geometry includes an oral and nasal cavity brought into an idealized face and fitted with a mask of dimensions (96 mm height and 58 mm width). The oral and nasal cavities are further brought by a soft palate into the larynx, which is connected to a trachea 109 mm long and 15.1 mm in diameter. The tracheobronchial tree is then brought into ten leads, allowing connection to the air supply during the experiment. The diameter of these leads is 10 mm. This model exists in three different variants [6], reflecting the different regimes of breathing (nasal, oral, and combined). For this research, the variant allowing combined breathing was chosen because it will allow comparison with the geometry corresponding to a five-year-old child.

2.2 Numerical setting

Numerical simulations were performed for the case of stationary inspiration, under conditions corresponding to the tidal volume at a given age (see Table 1). The boundary conditions for all models were identical. Inspiration was initiated from below using the inlet velocity boundary condition with negative values of velocity. The velocity parameters prescribed for the models of infant and child models were described in Table 1. The parameters of the boundary condition prescribed on the adult model are in [6]. The model wall was treated with the wall with the no-slip condition. A pressure outlet BC with zero boundary resistance was prescribed on the suction attachment fitted to the mask of each model. The numerical simulations were performed by using the Star-CCM+ commercial solver (version 2019.2, by Siemens company). The simulation methodology is the same for all three models. Using the steady RANS (Reynolds averaged Navier-Stokes) model with turbulence model realisable k-epsilon, the initial flow field was calculated, and it was verified that the cell size of the computational grid complies the requirements for Large-Eddy Simulations (LES) by analyzing the average size of Taylor micro scale inside the computational domain. Based on this information, the base size of the element for each model's computational mesh was selected. The base size of the element on the geometry of a 10-month-old infant was set to 0.4 mm and the size of generated mesh was 4.9 million cells. Mesh on the geometry of a 5-year-old child has a base element size of 0.5 mm and consists of 7.3 million cells. Mesh generated on adult upper airways geometry consists of 14.2 million cells and the base size of elements was set to 0.5 mm. Each of the meshes consists of polyhedral cells and the near-wall layer was treated with ten rows of prismatic cells. Turbulence was modelled by the LES approach with the Wall Adapting Local Eddy (WALE) viscosity subgrid-scale model. The constants of the WALE model were Cw = 0.544, Ct = 3.5 and Kappa = 0.41. The nonstationary solver operated with a time step of 0.00005 sec with a second-order upwind temporal discretization scheme. This setup ensured that the average Courant number in each domain was around 0.3 and did not exceed 5 in more than 0.01 per cent of cells. A segregated solver with a SIMPLE algorithm with a bounded central-differencing convection scheme was applied for pressure-velocity coupling. This approach to solving numerical simulations was validated by experiments in the paper [6]. Values of velocities in the whole domain were averaged in each time step for three seconds to achieve its mean values.

Table 1. Characteristics of airflow inside model's geometries.

	Infant (10 mo)	Children (5 yr)	Adult (25 yr)
Inspiratory flow rate (lpm)	8.4	12.5	15
Inlet velocity (lpm)	1.06	1.02	0.79
Tracheal velocity (m/s)	2.86	3.67	1.4
Avg. Reynolds number in trachea	1440	1992	1345
RUL flowrate (lpm)	1.76	2.61	x
RLL flowrate (lpm)	2.79	4.15	x
LUL flowrate (lpm)	2.13	3.16	x
LLL flowrate (lpm)	1.73	2.58	x

3 Results and discussion

Numerical simulations showed clear differences in the velocities within the chosen models. Due to the different characteristics of each model (different dimensions and flow rate conditions), it is necessary to compare the flows based on dimensionless numbers. If we focus only on the trachea and derive its Reynolds number (Re) based on tracheal diameter and flow rate (Table 1) we see that for all ages the values of Reynolds numbers inside the trachea correspond to laminar flow, with the largest Re coming out for the geometry at five years. However, the results are not significantly different global. However, Reynolds number values are dependent on local velocity values and since the laryngeal jet is created at the inlet into the trachea, turbulent behaviour cannot be judged based on averaged dimensions of the trachea.

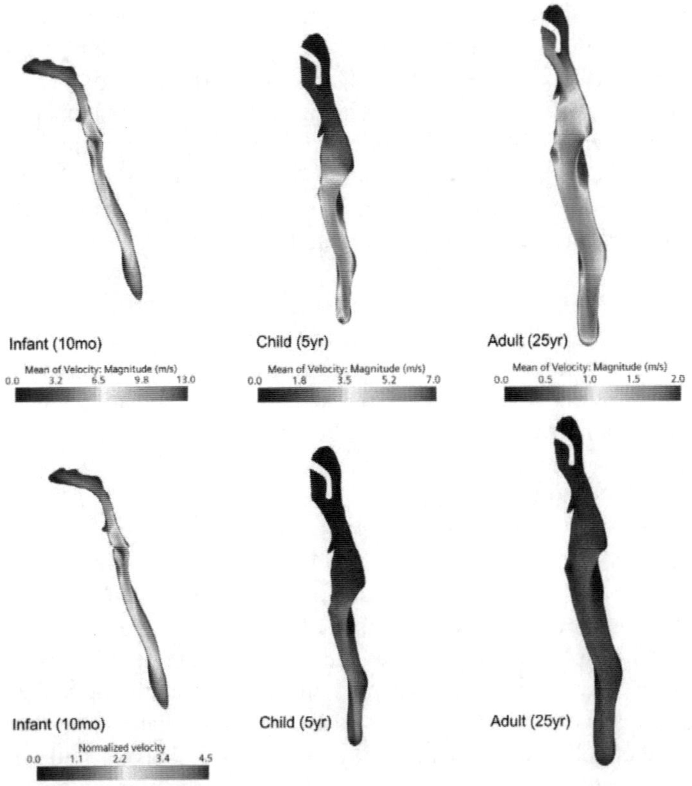

Fig. 2. Airflow patterns inside of trachea of investigated models.

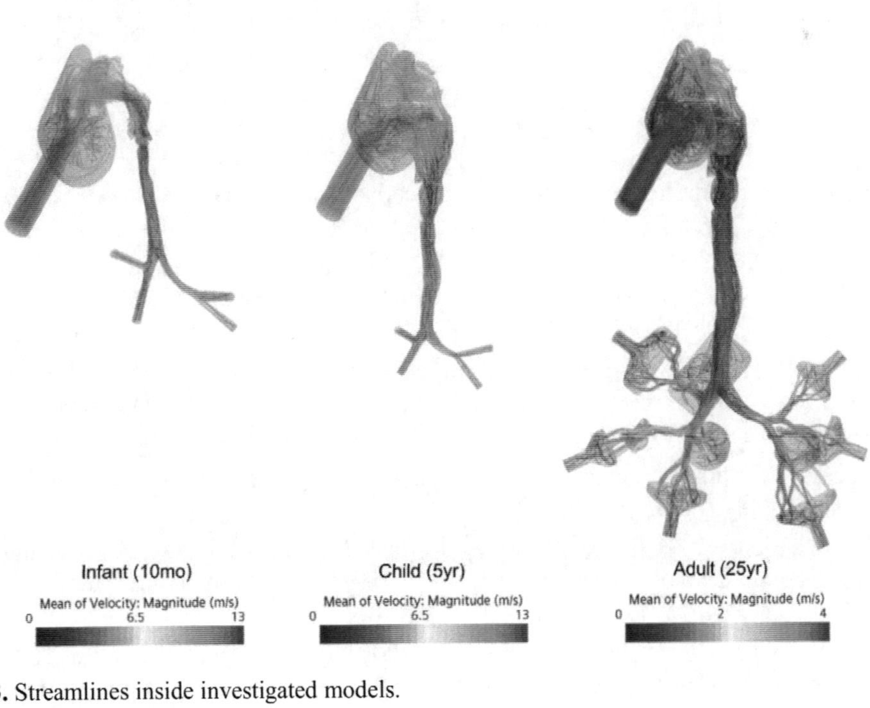

Fig. 3. Streamlines inside investigated models.

The results of the mean velocity across the trachea show that the laryngeal jet is presented in each investigated model. This jet is characterized by a significant increase in velocity values caused by a constriction of the geometry of the trachea in the region of the vocal cord region. Figure 2 shows that while in the case of the 5yr and 25yr models the velocity maxima are biased towards the posterior side of the trachea, in the case of the 10mo model the laryngeal stream forms anteriorly. This may be because inspiration is only initiated through the nasal cavity in the case of the 10mo model or the development of the oesophagus during this period. Although inspiratory flow rates are lower in children, we find significantly higher rates within the 10mo and 5yr models than in the resting adult. This is due to the lower dimensions of the branches of the tracheobronchial tree in the case of children and probably also to the smaller cross-sectional area in the vocal cords, where the velocity increases significantly due to constriction (to 12 m/s in the case of the 10mo and 7 m/s in the case of the 5yr model). If we normalize the velocities in the trachea by dividing the mean velocity by the tracheal velocity, we obtain a similar flow pattern in the case of the 5yr and 25yr models, which is consistent with the statement [1] that airway development ends at age 5 and although the dimensions of the two models differ (especially in the nasopharynx) the normalized velocities approximately correspond. However, the 10mo model exceeds the other two models in normalized velocity.

While in the case of the 10mo model the maximum velocity value is in the vocal cords, the maxima in the 5yr and 25yr models are located in the tracheobronchial tree. If we relate this information to the prediction of aerosol deposition in the upper airways, then particles of the same diameter are more likely to settle in the 10mo and 5yr models due to the higher flow velocity affecting the Stokes number value. This number then determines the ability of the particle to follow the flow. For higher velocities, the Stokes number value then increases, and for values greater than one, the probability of deposition by inertial mechanism increases. This information then tells us that larger particles are more likely to settle in the upper part of the child's lungs, and in the case of particle delivery deeper into the lungs, finer particles must be used.

This work was supported by the Czech Science Foundation project GA 20-27653S.

References

1. Xi, J., Si, X., Zhou, Y., Kim, J. and Berlinski, A. (2014) "Growth of nasal and laryngeal airways in children: implications in breathing and inhaled aerosol dynamics." *Respiratory Care*, 59(2), pp. 263–273.
2. Amirav, I., Borojeni, A. A. T., Halamish, A., Newhouse, M. T. and Golshahi, L. (2015) "Nasal versus oral aerosol delivery to the 'lungs' in infants and toddlers.*," Pediatric Pulmonology*, 50(3), pp. 276–283.
3. Fleming S, Thompson M, Stevens R, Heneghan C, Plueddemann A, Maconochie I, et al. Normal ranges of heart rate and respiratory rate in children from birth to 18 years of age: a systematic review of observational studies. *Lancet* 2011;377(9770):1011-1018.
4. Oakes, J.M., Roth, S.C., and Shadden, S.C. (2018), *Annals of Biomedical Engineering*, 46, No. 3, 498–512.
5. Lizal F, Elcner J, Belka M, Jedelsky J, Jicha M. Experimental and computational modelling of aerosol transport using BUT benchmark human lung model (2021). *ERCOFTAC Bulletin*, 128,36-43.
6. Lizal F, Elcner J, Jedelsky J, Maly M, Jicha M, Farkas A, Belka M et. Al. 2020, The effect of oral and nasal breathing on the deposition of inhaled particles in upper and tracheobronchial airways, *Journal of Aerosol Science*, vol. 150.

CFD Flow Analysis of the Influence of Rotor Seals in a Cantilever Microturbine

Tomáš Syka[1,*], *Andreas* P. Weiß[2], and *Richard* Matas[1]

[1]New Technologies - Research Centre (UWB), Univerzitní 2732/8, Pilsen, Czech Republic
[2]OTH Amberg-Weiden, Kaiser-Wilhelm-Ring 23, Amberg, Germany

Abstract. The article describes the results from the research and development of a cantilever microturbine designed for ORC power plants. The main objective of the paper was to compare different types of rotor seals and their effect on the turbine efficiency and rotor forces. The turbine stage and the numerical models used in the CFD system NUMECA FINE/Turbo are briefly described. The results are compared with the values obtained from measurements and previous simulations. The results show a relatively limited influence of the seal design on the turbine efficiency. The results also show the effect of the seal on the rotor axial force.

Keywords: *CFD simulation; NUMECA; ORC; leakages; rotor seal; cantilever micro-turbine*

1 Introduction

The number of decentralized small-scale stationary electrical energy sources (< 100 kW_{el}) connected to electricity distributing networks is growing and their development requires adjustments to the operation of these systems [1]. An example of such a source is the Organic Rankine Cycle (ORC) power plant, heated by e.g. waste heat from industry or from internal combustion engines, geothermal heat or even solar radiation.

An ORC power plant works like a coal-fired steam power plant but the difference between ORC and the conventional steam cycle is that instead of water, another fluid like alcohol, a refrigerant or siloxanes is used. More information about micro-turbines in general can be found in [2, 3]. This paper deals mainly with CFD simulations of geometric modifications of a small cantilever turbine with a radial inlet direction into the stator channels and a radial outlet from the rotor.

1.1 The turbine and the ORC test plant

The design of the cantilever microturbine is simple compared to a generic steam turbine and to keep costs low it is advisable to design a single stage turbine, despite its lower efficiency [4, 5, 6]. It is also possible to use multi-stage turbines but the efficiency benefits of this solution are not conclusive and it is more complex and therefore more expensive [7]. It is also necessary to develop a very flexible "construction kit" due to the wide range of

* Corresponding author: tsyka@ntc.zcu.cz

temperatures and pressures of the operating conditions depending on the specific heat source [8].

The turbine design, including the generator, is really compact and the dimensions of the whole unit are relatively small – see Fig. 1 left. The maximum diameter of the rotor wheel is 120 mm. The main parts of this unit are the inlet casing, stator wheel with nozzles, rotor wheel, outlet casing and generator. The concept, design and application of this turbine is proven and this work only focuses on its fine-tuning [4]. Another option is the ELEKTRA type re-entry turbine. This research is intensively underway. The ELEKTRA turbine has the advantage of a lower operational speed (leading to a cheaper electrical generator) and the aim is now to achieve an efficiency comparable to the "classic" turbine type presented in [9, 10].

In order to investigate real characteristics of the cantilever turbine, it was tested in the ORC research plant at the Centre for Energy Technology (ZET) at the University of Bayreuth. The research plant was designed to investigate the waste heat recovery from a 250kW biogas internal combustion engine with an exhaust gas temperature of about 500 °C – see Fig. 1 right. The microturbine works with hexamethyldisiloxane as a working fluid [11, 12, 13]. More information about the turbine and the ORC research plant layout and its parameters has been published in paper [14]. There are no seals to lower leakage losses in the basic design of the microturbine.

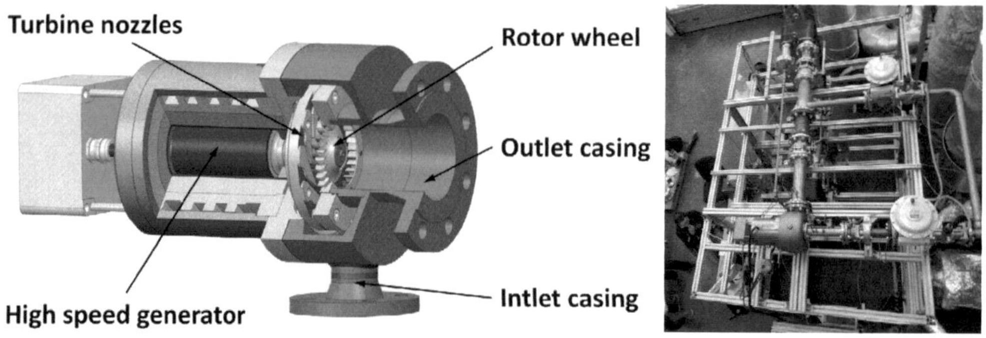

Fig. 1. Layout of the radial inflow cantilever turbine (left) and the ORC test rig (right) [14].

1.2 The CFD simulations and their goal

This work and the presented results are based on CFD simulations published in paper [14]. The work describes the comparison of measured and CFD results and an investigation of the influence of the outlet geometry on the turbine working parameters. As mentioned above, there are no seals in this basic turbine design. Therefore, the main objective of the follow-up work was to build on the simulations and to compare the different types of rotor seals and their possible influence on the turbine efficiency and the magnitude of the axial force acting on the rotor.

2 CFD setup and design of the rotor seal

CFD simulations of microturbines have specific features due to their functional principle. The working fluid enters the turbine with low velocity and high pressure values. The expansion to very low pressure occurs in the stator nozzles. This is associated with a high increase of the flow velocity. The flow velocity at the outlet of the stator nozzles is supersonic and the Mach number is usually higher than 2.5. The high Mach number and the presence of shock waves introduce complications in the modelling of the flow transition between the static and rotating regions of the microturbine. All the simulation issues, options and model settings used are also described in [14].

2.1 Various seal designs

A simplified model of the microturbine stage was created. Five design modifications of the rotor seal were prepared. It is known that the design of the labyrinth seal can affect the parameters of turbomachines [14] such as efficiency and forces.

The model without seals was represented by the basic design (a). The most complicated seal design was the staggered labyrinth seal with six teeth (b). The simplified design was the see-through labyrinth with three teeth (c). The last design prepared in two variants was the staggered labyrinth seal with only one tooth – the first one located at the smaller diameter of the rotor (d) and the second one located at the larger diameter (e). All design variants are shown in Fig. 2.

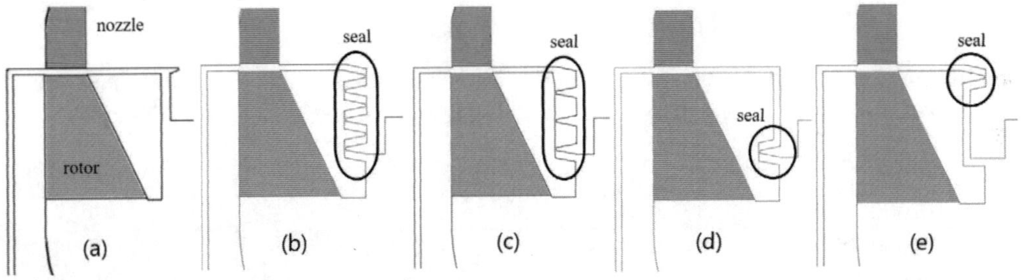

Fig. 2. Design of the seals geometry variants.

2.2 CFD model

NUMECA FINE/Turbo 11 [15] software was used to prepare, compute and evaluate the cases. The CFD model consists of the following parts/domains: microturbine stage (stator, rotor, outlet), labyrinth seals (leakage) and dead space between rotational and static parts. The stator consists of 12 nozzles and non-axisymmetric channels. The rotor consists of 30 blades. The geometry of the model is shown in Fig. 3.

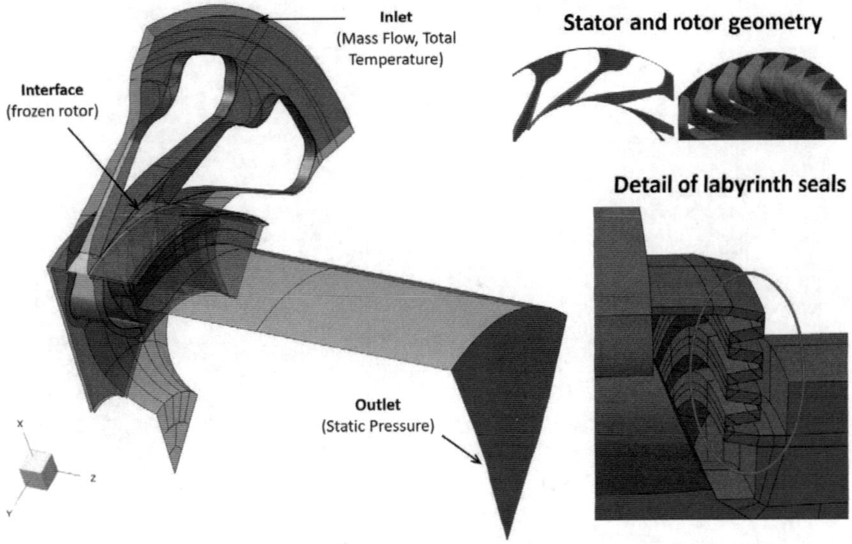

Fig. 3. CFD model of the micro-turbine stage – boundary conditions and geometry details.

The grid was created in NUMECA AutoGrid5 software. The mesh was hexahedral and multi-block structured with about 35 million cells. The mesh resolution had been studied and evaluated in previous simulations and it is sufficiently fine for our CFD model. The first cell at the wall is 0.0003 mm high and the maximum y^+ value in computations was 6.83.

A more complex EARSM turbulence model was used because the simulation results gained by using this model for complex geometries and complex flow structures are better with the NUMECA package than the results gained with the more commonly used Spalart – Almaras or SST k-ω models in "engineering" cases.

The working medium was set as the condensable gas model of hexamethyldisiloxane (MM). Properties of the fluid based on the NIST-REFPROP [16] are defined by tables generated in the NUMECA TabGen.

Boundary conditions were set as follows: at the inlet the variations of the total pressure p_{t0} and the total temperature T_{t0} at the outlet the area weighted average of the static pressure p_{s2} – see Table 1. The shaft revolution speed varied from 22.000 to 29.000 rpm. All walls were considered adiabatic. Steady solution was assumed.

Table 1. Inlet and outlet boundary conditions.

Revolutions [1/min]	Inlet total pressure [Pa]	Inlet total temperature [K]	Outlet pressure [Pa]
22,000	613,104	455.33	25,700
23,000	613,870	455.48	25,449
24,000	614,082	455.73	25,639
25,000	614,538	456.05	25,830
26,000	615,416	456.35	25,400
27,000	614,998	456.57	25,800
28,000	615,445	456.82	26,200
29,000	615,560	457.04	26,500

2.3 Evaluation

The planes and areas used for the evaluation of results are depicted in Fig. 4. The results were averaged during the last 1.500 iterations to obtain mean values for the evaluation.

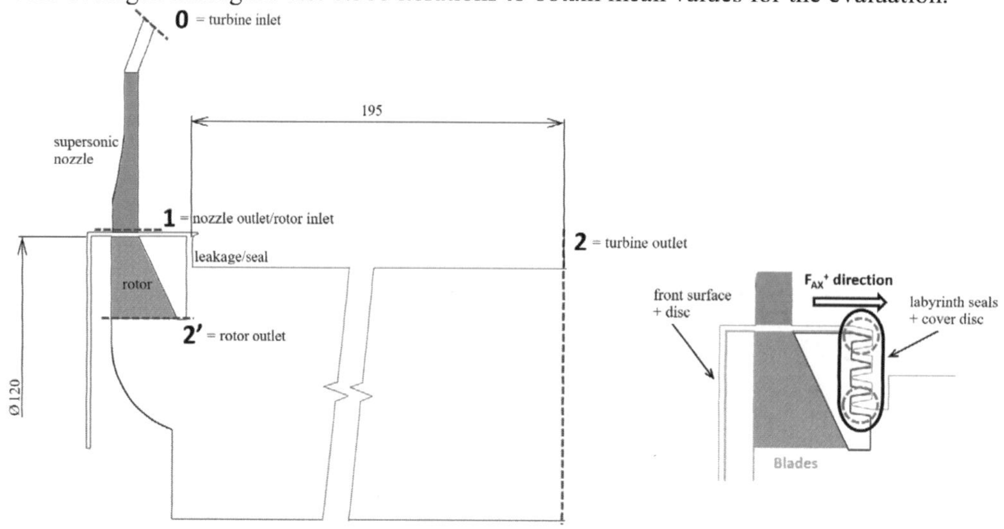

Fig. 4. Definition of evaluation planes and the areas for friction torque evaluation.

A very important value for turbomachines is efficiency. The total-to-static turbine efficiency (1) was evaluated from the computed torque M_T, the rotational speed n, the mass flow rate $\dot{m}$, the inlet total enthalpy h_{0T} and the outlet static enthalpy h_{2Sis} (which is a function of the inlet total entropy s_0 and the outlet static pressure p_{S2}).

$$\eta_{TS} = \frac{M_T \cdot 2 \cdot \pi \cdot n}{\dot{m} \cdot (h_{0T} - h_{2Sis})}, \tag{1}$$

where $$h_{2Sis} = f(s_0, p_{S2}).$$

3 Results

Numerical simulations were performed for all points of the working characteristic of the basic design (a) and for the staggered labyrinth seal with six teeth (b). The simulations were relatively time-consuming so the other design variants (c), (d) and (e) were simulated only for three speed values. This approach proved sufficient to obtain the results of these variants.

3.1 Efficiency

Table 2 shows the total-to-static turbine efficiency η_{TS} and the pressure ratio $\Pi = p_{t0}/p_{s2}$. The difference in the efficiency from 2 % to 4 % between the measurement and the simulations has already been discussed in [13]. It may be caused by the inaccuracy of the outlet pressure measurement and also by local inaccuracies on the rotor-stator interface. But the shapes of the characteristics are similar in all cases and it is possible to use CFD simulations to investigate the influence of various geometry changes on the turbine parameters.

Fig. 5 also shows that the efficiency of the designs with labyrinth seals is very similar to the basic design with only small differences.

Table 2. The total-to-static turbine efficiencies and the pressure ratios, experimental data from [4].

Revolutions [1/min]	ηTS [%]	Π [-]	ηTS [%]	Π [-]	ηTS [%]	Π [-]
	Experiment bas. des.		Basic design (a)		Labyrinth 6 teeth (b)	
22,000	69.37	23.87	71.77	23.87	71.97	23.87
23,000	71.30	24.12	73.59	24.13	73.75	24.13
24,000	73.36	23.96	75.18	23.96	75.55	23.97
25,000	74.29	23.81	76.55	23.81	77.17	23.81
26,000	75.32	24.23	77.72	24.24	77.81	24.24
27,000	75.80	23.85	78.84	23.87	78.87	23.87
28,000	76.09	23.52	79.52	23.54	79.37	23.54
29,000	74.71	23.27	79.80	23.26	79.16	23.27
	Labyrinth 3 teeth (c)		Labyrinth 1 s. t. d. (d)		Labyrinth 1 l. t. d. (e)	
22,000	72.13	23.87	72.09	23.87	72.16	23.87
25,000	76.83	23.81	76.79	23.81	76.77	23.81
28,000	79.40	23.54	79.44	23.54	79.46	23.54

Only a small increase in the efficiency can be observed at lower speeds, which gradually disappears. At the maximum revolution speed the effect is even rather negative – this effect is relatively surprising because better sealing should lead to the higher efficiency of the turbomachine.

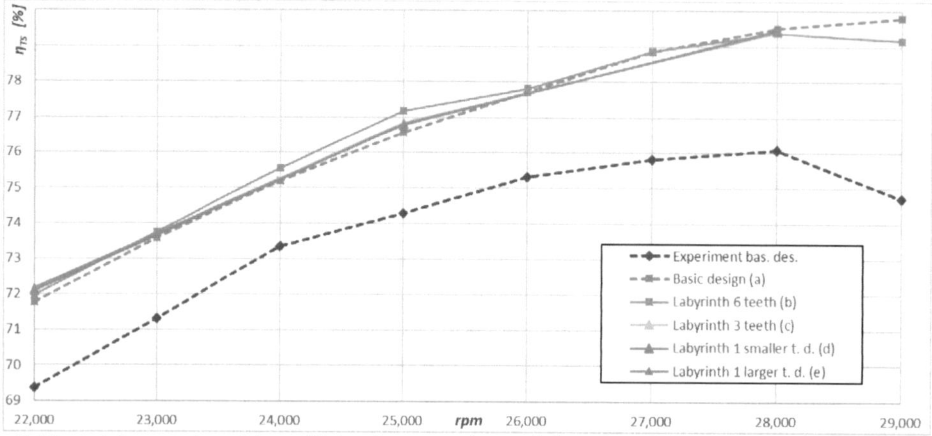

Fig. 5. The total-to-static turbine efficiencies, experimental data from [4].

3.2 Mass-flow rate

The computed total turbine mass-flow rate was about 0.33 kg/s for all cases and the rates were almost constant due to the choked flow in the supersonic stator nozzle. The value was about 2.8 % higher than the measured values. This is probably caused by the ideal geometry used in the CFD model.

Figure 6 shows the leakage mass-flow rate for the computed design geometries. The leakage mass-flow rate rises with the shaft revolutions and the values are significantly higher in the basic design than in the newly designed geometries with labyrinth seals. With the best sealing (b) the leakage mass-flow rate is only half that of the basic design. However, this positive and expected result has only a minimal impact on the predicted efficiency as is shown above. The benefit from reduced leakage flow is countered by the higher friction torque from the increased wetted area of the complex sealing geometries as shown below.

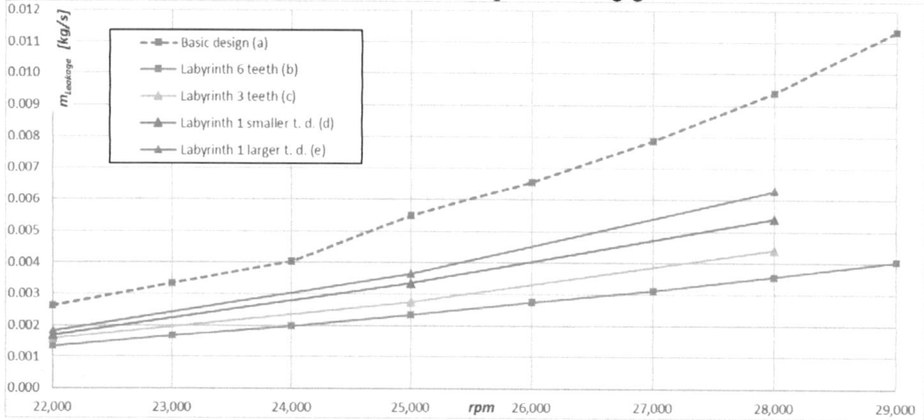

Fig. 6. The leakage mass flow rate.

3.3 Torque

The results mentioned above in 3.1 and 3.2 are followed by the analysis of the rotor torques. The total torque T_t decreases from about 6.8 Nm (22,000 rpm) to 5.6 Nm (29,000 rpm). This torque can be divided into individual components – the blade torque T_{blades} and the loss friction torques T_{front} (front surfaces/disc) and T_{seals} (seals/cover disc) according to relation (2). The calculated torque values can be found in Table 3. The T_{front} component has an almost constant T_{front} value around -0.026 Nm in all cases.

$$T_t = T_{blades} + T_{front} + T_{seals}. \quad (2)$$

Table 3. Evaluation of torque values.

Revolutions [1/min]	T_{blades} [Nm]	T_{seals} [Nm]	T_t [Nm]	T_{blades} [Nm]	T_{seals} [Nm]	T_t [Nm]
	Basic design (a)			Labyrinth 6 teeth (b)		
22,000	6.708	-0.004	6.679	6.744	-0.021	6.697
23,000	6.607	-0.003	6.579	6.641	-0.021	6.594
24,000	6.463	-0.003	6.434	6.513	-0.021	6.466
25,000	6.313	-0.001	6.286	6.374	-0.021	6.337
26,000	6.203	0.001	6.178	6.234	-0.023	6.185
27,000	6.032	0.003	6.009	6.06	-0.024	6.010
28,000	5.850	0.003	5.827	5.869	-0.025	5.817
29,000	5.647	0.004	5.624	5.632	-0.025	5.580
	Labyrinth 1 s. t. d. (d)			Labyrinth 1 l. t. d. (e)		
22,000	6.756	-0.018	6.708	6.76	-0.017	6.715
25,000	6.344	-0.014	6.306	6.343	-0.013	6.304
28,000	5.861	-0.013	5.821	5.86	-0.011	5.823

The value of the T_{seals} is very low in the basic design (a). The values of torque for all the labyrinth seals show that the T_{blades} increases in cases with higher rotor flow rates. T_+ is lower than the frictional loss torque. T_{seals} increases on the labyrinth surfaces, T_- especially at higher speeds, see Figure 7.

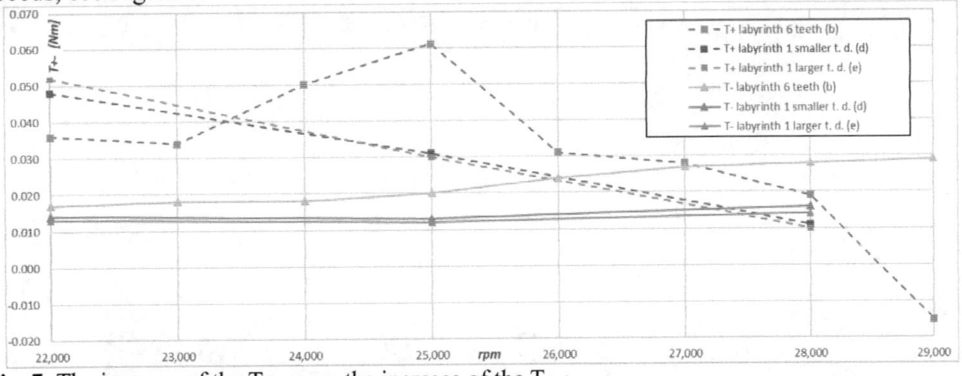

Fig. 7. The increase of the T_{blades} vs. the increase of the T_{seals}.

3.4 Axial force

The axial force value is important for the axial bearing design. The rotor axial force was evaluated and the values are shown in Fig. 8. The changing values for different seal designs can be seen.

While the forces are almost identical in the basic design (a) and the complex sealing (b), the use of one tooth and its position on the rotor (d) or (e) can partially influence the magnitude of the force. The position of the tooth influences the pressure distribution in the sealing channel. The smallest axial force was predicted for design (d) where the location of the seal is near the smaller diameter of the rotor.

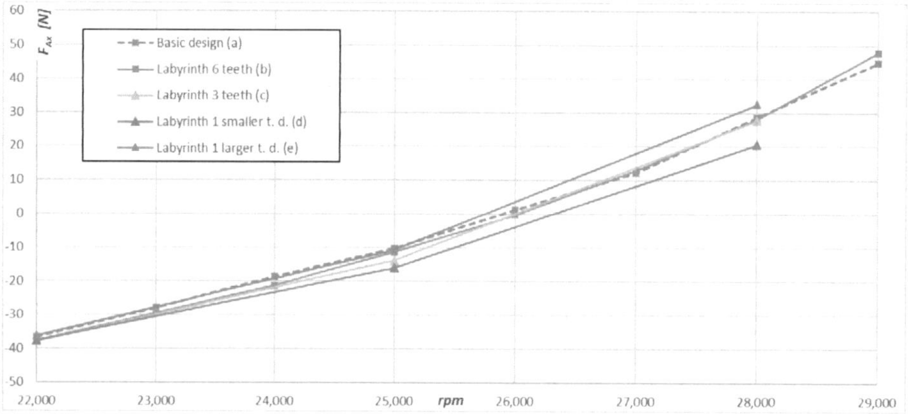

Fig. 8. The axial force acting on the rotor wheel.

3.5 Flow field

Fig. 9 shows the fields of the averaged velocity magnitude in ZR direction for 28,000 rpm for two designs. It shows that the presence of the labyrinth seal can influence the structure of the flow in the outlet part, but not significantly. The velocity fields in the other cases were similar to the labyrinth design with 6 teeth.

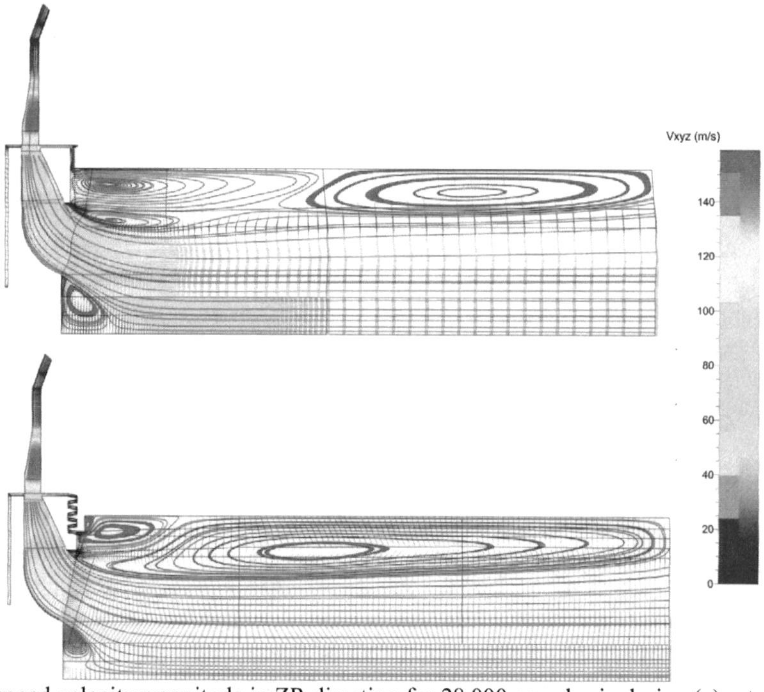

Fig. 9. Averaged velocity magnitude in ZR direction for 28,000 rpm, basic design (a) – top, labyrinth 6 teeth (b) – bottom.

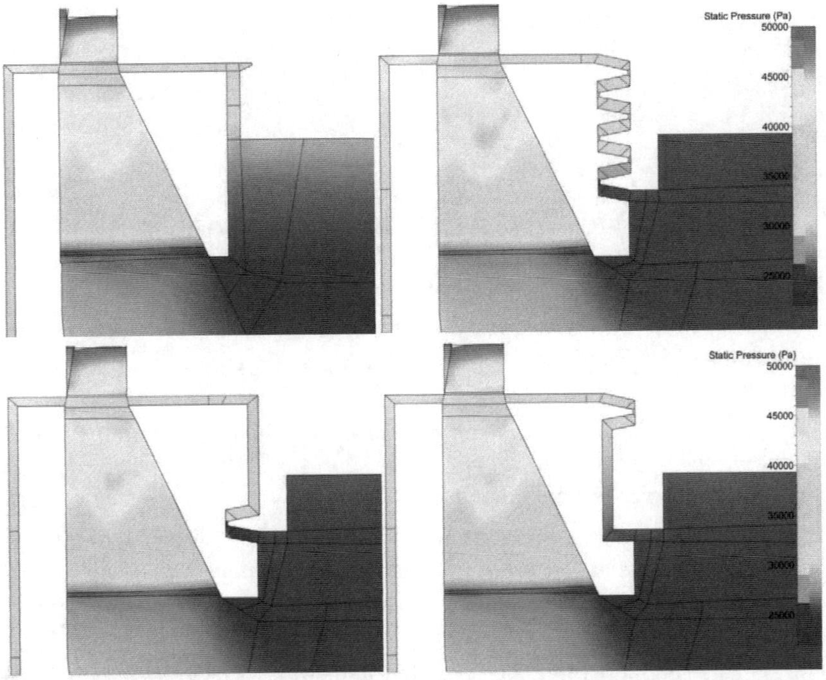

Fig. 10. Meridional averaged static pressure for 28 000 rpm, basic design (a) – top left, labyrinth 6 teeth (b) – top right, labyrinth 1 s. t. d. (d) – bottom left and labyrinth 1 l. t. d. (e) – bottom right.

The fields of the meridional averaged static pressure in Fig. 10 are more interesting. It is possible to see the changes in the pressure distribution in the turbine for various seal designs.

4 Discussion and conclusions

The numerical case study was carried out to determine the effect of the seals on the ORC micro-turbine parameters. Labyrinth seals used in the cantilever turbine design influence the working characteristics in several ways.

The leakage mass flow corresponds to the number of teeth used in the seals. The seal design (b) with six teeth reduces the mass flow to at least a half of the original value.

The most complicated staggered labyrinth seal with six teeth (b) has a positive effect on the turbine efficiency in the area of lower rotation speeds (about 0.3 %) but the effect is smaller than expected. The effect on the efficiency is zero in the area of the higher rotating speed (including the design point). The results of simulations thus showed that the increased friction eats up the gain in the blades. In the area of higher rotation speeds the influence of the friction on the larger surfaces negates the seal effect. All the efficiency values for the different labyrinth designs confirm this result, but it is necessary to consider the uncertainty of the results from the numerical solution for such complex flow fields.

Labyrinth seals with three or six teeth – (b), (c) have no practical effect on the axial force acting on the rotor wheel in comparison to the original geometry. This is due to the similar pressure distribution behind the rotor wheel. The labyrinth seals with one tooth (d) and (e) have an effect on the acting axial force in the order of ±6 N thus about 20 % higher or lower than the value of the basic design (a)) for 28,000 rpm. This effect is positive or negative depending on the tooth diameter.

The work has been supported by the project "Ziel - ETZ INTERREG V Project 53" and by the specific research – grant SGS-2022-023.

References

1. Cui, S., Zeng, P., Wang, Z, Zuo, Y. (2021). *Adaptive Current Protection Technology for Distribution Network with Distributed Power Sources Based on Local Information.* Mobile Information Systems, vol. 2021, Article ID 5137749. https://doi.org/10.1155/2021/5137749
2. Weiß, P. (2016). *Micro turbine generators for waste heat recovery and compressed air energy storage. 15th conference on Power System Engineering, Thermodynamics & Fluid Flows - ES2016.* Pilsen, Czech Republic.
3. Weiß, P. (2016). *Volumetric expander versus turbine – which is the better choice for small rc plants? 3rd International Seminar on ORC Systems* (22):301–310. http://www.asme-orc2015.be/content/publication.
4. Weiß, A.P., Popp, T., Müller, J., Hauer, J., Brüggemann, D. (2018). *Experimental characterization and comparison of an axial and a cantilever micro-turbine for small-scale Organic Rankine Cycle.* Appl Thermal Eng, 140 (2018), pp. 235-244 https://doi.org/10.1016/j.applthermaleng.2018.05.033
5. Fiaschi, D., Innocenti, G., Manfrida, G., Maraschiello, F. (2016). *Design of micro radial turboexpanders for ORC power cycles. From 0D to 3D.* Appl. Therm. Eng. ISSN 1359-4311 99 (2016) 402–410. https://doi.org/10.1016/j.applthermaleng.2015.11.087.
6. Da Silva, E.R., Kyprianidis, K.G., Camacho, R.G.R., Säterskog, M., Angulo, T.M.A. (2021). *Preliminary design, optimization and CFD analysis of an organic rankine cycle radial turbine rotor.* Appl. Therm. Eng. ISSN 1359-4311 195 (2021) 117103. https://doi.org/10.1016/j.applthermaleng.2021.117103.
7. Kaczmarczyk, T.Z., Żywica, G., Ihnatowicz, E. (2019). *Experimental study of a low-temperature micro-scale organic Rankine cycle system with the multi-stage radial-flow turbine for domestic applications.* Energy Convers Manage, 199 (2019), p. 111941, 10.1016/j.enconman.2019.111941
8. Weiß, A.P., Novotný, V., Popp, T., Streit, P., Špale, J., Zinn, G. et al. (2020). *Customized ORC micro turbo-expanders - From 1D design to modular construction kit and prospects of additive manufacturing.* Energy, 209 (2020), p. 118407, 10.1016/j.energy.2020.118407
9. Weiss, A.P., Streit, P., Popp, T., Shoemaker, P. (2020). *Uncommon turbine architectures for distributed power generation -development of a small velocity compounded radial re-entry turbine.* Archives of Thermodynamics, vol. 41, no. 4, pp. 235–253. DOI: 10.24425/ather.2020.135862
10. Weiß, A.P., Stümpfl, D., Streit, P., Shoemaker, P., Hildebrandt, T. (2022). *Numerical and Experimental Investigation of a Velocity Compounded Radial Re-Entry Turbine for Small-Scale Waste Heat Recovery.* Energies 2022, 15, 245. https://doi.org/10.3390/en15010245
11. Preißinger, M., Brüggemann, D. (2016). *Thermal stability of hexamethyldisiloxane (MM) for high-temperature organic rankine cycle (ORC).* Energies 9(183). Doi:10.3390/en9030183.
12. Preißinger, M., Schwöbel. J. (2017). *Multi-criteria evaluation of several million working fluids for waste heat recovery by means of organic Rankine cycle in passenger cars and heavy-duty trucks.* Applied Energy (206):887–899.

Doi:10.1016/j.apenergy.2017.08.212.

13. Turunen-Saaresti, T. (2017). *Design and testing of high temperature micro-orc test stand using siloxane as working fluid.* Journal of Physics: Conference Series (821). Doi:10.1088/1742-6596/821/1/012024.
14. Syka, T., Weiß, A. (2018). CFD flow analysis of the cantilever micro-turbine. *Acta Polytechnica CTU Proceedings 20:108–113.* Doi: 10.14311/APP.2018.20.0108
15. Wu, T., San Andrés, L. (2019). *Gas labyrinth seals: On the effect of clearance and operating conditions on wall friction factors – A CFD investigation.* Tribology International, Volume 131, 2019, 363-376. ISSN 0301-679X. https://doi.org/10.1016/j.triboint.2018.10.046.
16. NUMECA-Online, https://www.numeca.com/home (retrieved on 09.12.2019)
17. NIST-REFPROP, https://www.nist.gov/srd/refprop (retrieved on 09.12.2019)

Hydraulic Design of an Axial Runner Using Classical Methods and CFD Simulation

Peter Hlbočan[1,*], *Marta* Murgašová[1], *Zoltán* Csuka[1], and *Michal* Varchola[2]

[1]ZTS – Výskum a Vývoj, a.s., Ul. Lieskovec 598/75, 018 41, Dubnica nad Váhom, Slovenská republika
[2]Sjf STU v Bratislave, Námestie slobody 17, 812 31, Bratislava 1, Slovenská republika

Abstract. The paper deals with the principles of hydraulic design of runners of axial turbines or pumps. The procedures of classical methods of hydraulic design of primary runner geometry and the possibilities of subsequent optimization of this geometry by CFD simulation are presented. In the paper, the advantages of the procedure of hydraulic design of the primary runner geometry of a tubular Kaplan turbine with the application of the classical methods and the subsequent CFD simulation are shown on a specific example. The purpose of the CFD simulation is to verify the achievement of the required parameters of the designed runner geometry and its possible fine-tuning and optimization.

Keywords: *primary geometry; axial runner; hydraulic turbine; hydraulic design; CFD simulation*

1 Introduction

Hydraulic design of the axial turbine (or pump) runner (axial machine) is a challenging task that can be approached in several ways. The requirement of the hydraulic design is to achieve the required values of flow rate (discharge) and head at the given rotating speed at the best efficiency point (BEP). In addition, e.g.: runner diameter, cavitation properties, noise and vibration level, etc. may be required. Several hydraulic design methods have been developed based on simplified models of two-dimensional flow of an ideal liquid using the theory of flow around aerodynamic (hydrodynamic) profiles, for which experimentally determined courses of lift and drag coefficients are known. The measured lift and drag values refer to an isolated airfoil (they are obtained by measurements in wind tunnels), therefore it is necessary to apply correction coefficients processed in the form of diagrams, through which the lift coefficient of an isolated airfoil is corrected to a profile in an axial flow blade cascade. The mentioned methods strongly depend on the results of previous experimental research and optimization tasks. Based on them, the initial (preliminary) design of the main dimensions of the runner (diameter of the hub and chamber, blade span angle, etc.) and the efficiency of the machine is obtained. Calculation of the geometry of blade cuts (profiles) is carried out in the next step using the methods described above. In this way, the geometry of the runner is obtained, the parameters of which, as a rule, approximately correspond to the design requirements. We call this geometry primary geometry.

[1] Corresponding author: peter.hlbocan@ztsvav.eu

Subsequently, it is necessary to optimize the primary geometry in order to get closer to the design requirements or in order to achieve maximum efficiency values, etc. In the past, the optimization of the primary geometry was mainly carried out using experimental research, whereas the very important role played the experience of the researchers performing the task or knowledge regarding general relationships between hydraulic properties (performance and effective parameters) of the machine and geometry.

Currently, the optimization of the geometry is usually carried out on the basis of CFD simulation. A larger number of optimization algorithms are known, which are commonly applied to solve these tasks. Many of them are today a common part of commercial CFD (CAE) software (gradient methods, response surface DOE optimization, six sigma analysis, genetic algorithms, etc.). Of course, the essential question remains how to design the preliminary (i.e. initial – primary) geometry of the designed machine. In principle, it is possible to choose an "ad-hoc" primary geometry and, on the basis of an optimization algorithm, gradually and systematically modify the geometry to "find" the desired shape. The number of design points – geometric alternatives (and thus the optimization time) can be extremely high in this case. The second option is to start from the known geometry of the machine with similar parameters as specified in the hydraulic design requirements. The third option is to carry out the hydraulic design of the primary geometry using the classical method, which significantly reduces the number of design points (geometric alternatives) within the subsequent optimization algorithm. With the application of the mentioned optimization procedures, it is possible to almost "automatically" design the optimal geometry of the axial runner. A common feature of all three listed approaches to "automatic" optimization of the runner geometry is a relatively high number of geometry design points, which usually ranges from tens to hundreds.

So-called genetic algorithms are a specific group of current optimization algorithms. Multi-objective genetic algorithms (MOGA) are included, for example, in commercial software CAESES. This environment can be directly connected to the ANSYS Workbench software environment.

An example of the optimization of a Kaplan turbine with a three-blade runner using MOGA algorithms in the CAESES and ANSYS environment is described in the paper [14]. The primary geometry was designed and fully parameterized in the CAESES environment. Sensitivity analysis in the first phase of optimization revealed a greater impact of changing some basic parameters of the geometry. The SOBOL algorithm included in the CAESES environment was chosen for the sensitivity analysis. Totally 16 parameters defining the runner blades and 1 parameter describing the position of the distributor vanes were analyzed. As part of the sensitivity analysis, a total of 300 simulations took place, on the basis of which the dependencies of changes in individual parameters and their mutual correlations were determined. The most important parameters were the angle of the distributor vanes, the inlet and outlet angle of the runner blade and the runner blade chord length [14].

The MOGA algorithm was implemented for the optimization. Totally 9 parameters were optimized. Two objective functions were defined, for which the global minimum value was sought in the optimization process. They were F_1=1-eta (hydraulic efficiency) and F_2=$sigma_{,ref}$ (cavitation number at reference flow). The goal was to find the optimal ratio of the values of both objective functions. The initial hydraulic efficiency (of the primary geometry) was 90 %. The cavitation properties reached a value of $sigma_{,ref}$ = 1.25. A similar hydraulic efficiency was achieved with a significant improvement of cavitation properties to $sigma_{,ref}$ = 0.83 already by the sensitivity analysis procedure. The optimization process further increased the hydraulic efficiency to eta = 90.7 % while achieving the cavitation coefficient $sigma_{,ref}$ = 0.77. Approximately 700 simulations were carried out as part of the optimization process. The last phase of the optimization process was the verification of the final geometry in the wider operating range of the heads and rotating speeds of the machine [14].

Paper [15] has a similar character. MOGA algorithms were used in the work [15] in the CADRUN-opt environment (originally designed to optimize Francis turbines). Some problems are mentioned, because of which the automatic optimization of the blade geometry of the Kaplan turbine can be more difficult in the work [15]. One of them is dual turbine regulation, which increases the number of simulations needed. The second problem is the presence of a thin gaps between the runner blade tips and the chamber. The flow in these gaps has a significant effect on the hydraulic efficiency of the machine. That is why these gaps in CFD simulations are important, which is supposed to increase the complexity of creating a computational mesh and at the same time increases the need for CPU time or computing performance.

The first step of the optimization is the parameterization of the geometry (Fig. 1). In a given optimization phase, the geometry of the blades is changed, while the geometry of the runner hub and the turbine chamber remains unchanged. On the one hand, the parameterization of the shape of the runner blade should allow fully flexible variation of the complex hydraulic design. On the other hand, the parameterization should contain the smallest possible set of parameters to reduce the number of design points. Totally 24 geometric parameters were optimized: 16 parameters of blade geometry, 8 parameters determining the overall dimensions of the blade [15].

The objective functions were the requirements for hydraulic efficiency and cavitation properties, the value of which was maximized or minimized. The hydraulic efficiency value was monitored both in BEP (Q_11^((1)), n_11^((1))) and in full load mode (Q_11^((2)), n_11^((2))). The cavitation properties of the runner were estimated by calculating the torque increment in the region where the pressure p is lower than the saturated vapor pressure p_V. The goal was to minimize the relative area of cavitation W_{cav} on the suction side of the blade. The paper presented the achieved values of the hydraulic efficiency and the cavitation parameter described above after optimization for two different Kaplan turbines. The first of them had 5 blades and was designed for a head of 20 m. The optimization was solved using a multi-objective genetic algorithm (MOGA). Totally 25 populations were calculated, each consisting of 120 unique design points. In the next step, several suitable geometric designs (design points) are selected from the Pareto front for further analysis [15].

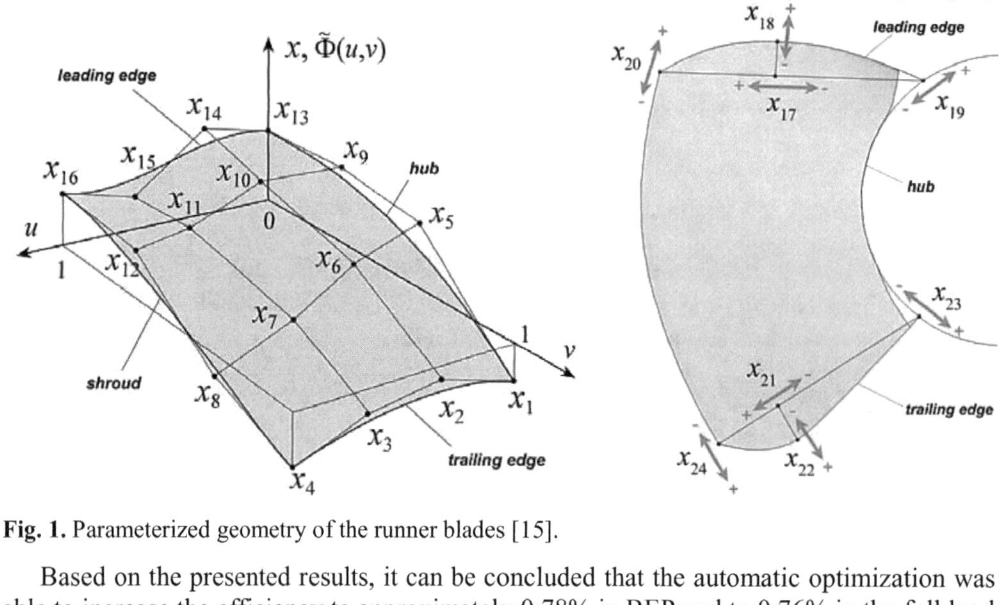

Fig. 1. Parameterized geometry of the runner blades [15].

Based on the presented results, it can be concluded that the automatic optimization was able to increase the efficiency to approximately 0.78% in BEP and to 0.76% in the full load operating point. The cavitation properties of the optimized blade were also improved compared to the original blade [15].

The paper also presents the results of the automatic optimization of the Kaplan runner for a turbine with a head of 40 m and a runner blade number of 6. The efficiency results were again presented for the BEP and for the full-load operating point of the turbine. There were 29 populations during the optimization. From the calculated Pareto front, a specific geometry was chosen for further analysis that met the set of requirements. Based on the presented results, it can be concluded that the efficiency in BEP increased by approximately 0.15 % and the efficiency for the full load operating point increased by approximately 0.9 %. The cavitation properties have also improved significantly [15].

An alternative to automatic optimization algorithms is the so-called "engineering" approach which also utilizes CFD simulation as the tool for predicting parameters or hydraulic properties of the designed machine. Previous knowledge about the relationship between hydraulic properties (power and efficiency properties) of the machine and its geometry is utilized to a significant extent. The design of the initial - primary geometry is basically the same as for the automatic optimization approaches mentioned above. Geometry modifications in specific design points are carried out "manually". In this, the experience and knowledge of the designer play an important role. Analyzes of the internal flow pattern (distribution of velocity and pressure fields) can be an additional design tool. This is implemented as part of the post-processing of the individual geometry alternatives (design points) of the proposed machine, and based on them, it is possible to identify possible critical places (eddies, flow separation areas, etc.). Energy dissipation usually occurs in these places which results to reducing the efficiency of the proposed machine. Next geometry alternatives (design points) might be adjusted to eliminate or at least reduce the extent of identified problematic areas. The advantage of this approach is a significantly smaller number of design points compared to the "automatic" optimization algorithm.

The aim of the contribution is to show (using a specific example) the advantage of the design procedure of a tubular Kaplan turbine runner primary geometry using the classical method with the subsequent CFD simulation aimed at verifying the fulfillment of the required parameters and possible "fine-tuning" of the machine geometry. The main advantage of applying procedure in the design of the runner geometry is a principal reduction of the number of design (shape) alternatives of the proposed machine.

2 Classical methods of hydraulic design of runners of axial turbines or pumps

Let's consider a specific example from practice. The request was to design a turbine for a specific location according to the assignment (Fig. 2). The red frame in Fig. 2 indicates the required operating range of the turbine, and the purple line indicates the head value at which it is necessary to guarantee the performance parameters (power and efficiency). The required rotating speed is n = 360 min^{-1} and the diameter of the runner is D = 1100 mm. Co. ZTS VaV, a.s. has its own portfolio of model turbines. When applying the "most suitable" model turbine from the current portfolio, it is possible to "cover" the expected operating range as shown in Fig. 2 on the right. The problem is low efficiency or failure to meet the required performance parameters with a guaranteed head and small flow rates (efficiency at the minimum flow rate should be at least 87%). From this reason, there was a request to modify the geometry of the original turbine, or to design a new turbine, which would have shifted the operating range towards smaller heads (flow rates) and increased efficiency in the lower left part of the operating chart. One of the ways to achieve this is to ensure a shift of the best efficiency point (and thus the entire operating chart) in the above-mentioned direction.

The best efficiency point of the original turbine is at a head of approx. H = 10.75 m. On a modified turbine, the best efficiency point should be approximately at a head of approx. H = 8.2 m, while the flow rate at the best efficiency point may remain unchanged, or may be a few percent smaller. It is important to increase the efficiency at the guaranteed head so that the efficiency at the minimum operating flow rate is at least 87 %.

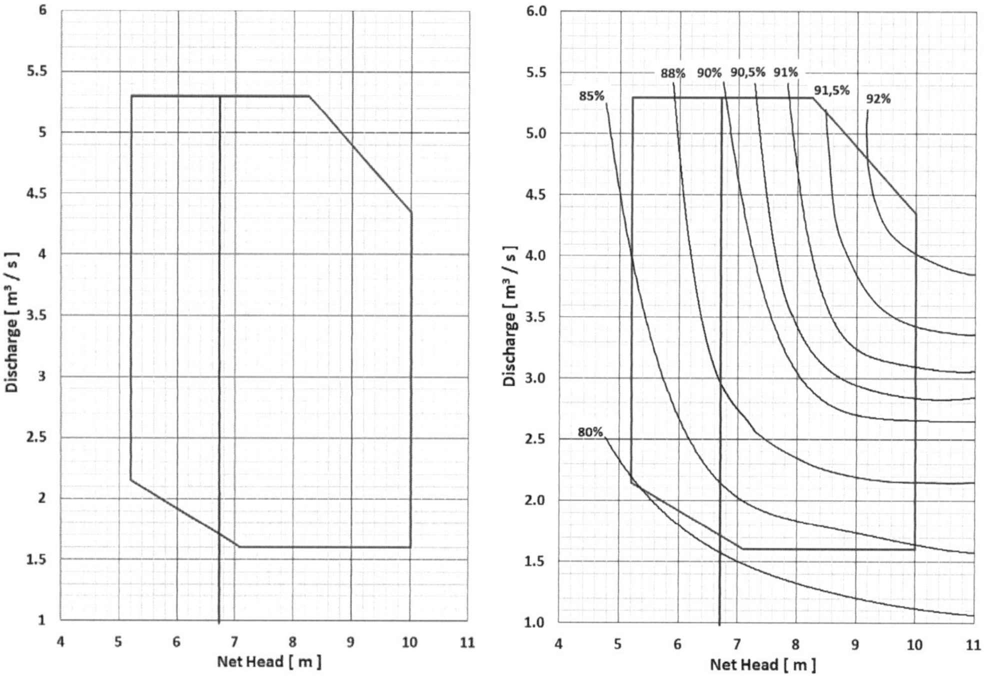

Fig. 2. Required operating range (left), operating chart of the original turbine with marked operating range (right).

In this case, the modification of the geometry of the original turbine was performed by the hydraulic design of the new runner using the "classical" approach with the application of the Voznesensky method. The geometry of all stator parts of the turbine remained original. Subsequently, CFD analysis was applied for the purpose of more detailed verification of the achievement of the required performance parameters.

It was based on the following required parameters at the best efficiency point: Q = 4.36 m^3/s, H = 8.2 m, n = 360 min^{-1}, D = 1100 mm.

The uniqueness of the presented procedure is that the runner of the Kaplan turbine was designed as the impeller of an axial pump. The fact that every hydrodynamic machine is in principle reversible and can work in a pump or in a turbine operating mode was used. The performance parameters at BEP of the machine are different for the pump and for the turbine operation mode. The ratio between the nominal flow rate or the nominal specific energy of the machine in pump and turbine operation mode varies in principle from machine to machine. Nevertheless, it is possible to observe certain tendencies that this ratio acquires depending on the achieved hydraulic efficiencies of the machine as well as on the specific speed [1, 2, 5].

If we can successfully estimate the specific values of the ratio between the nominal parameters of the machine in pump and turbine operation mode, we can perform a recalculation of the design (required) parameters of the machine in turbine mode to pump mode and apply these parameters as input parameters for the calculation of the axial runner geometry [1, 2]. This procedure was also applied in this case.

2.1 Hydraulic design of the axial runner meridional cut (flow path)

The procedure of the hydraulic design of the meridional cut is explained in more detail in the works [1 – 13]. The scheme of the meridional cut of the axial runner with the main dimensions is shown in Fig. 3 (left). The calculation is performed according to calculation relations in the Table 1.

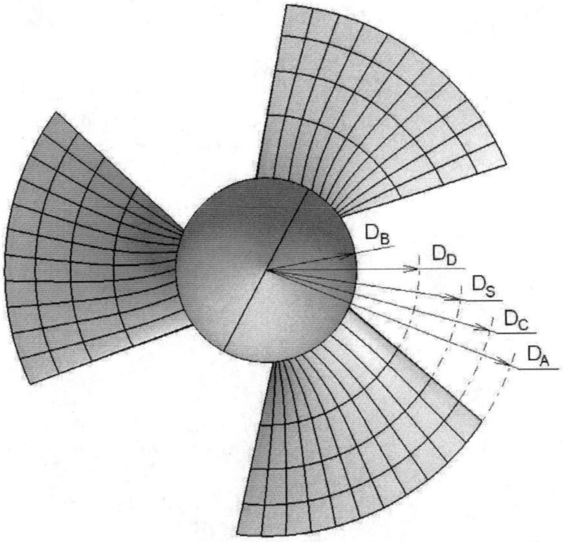

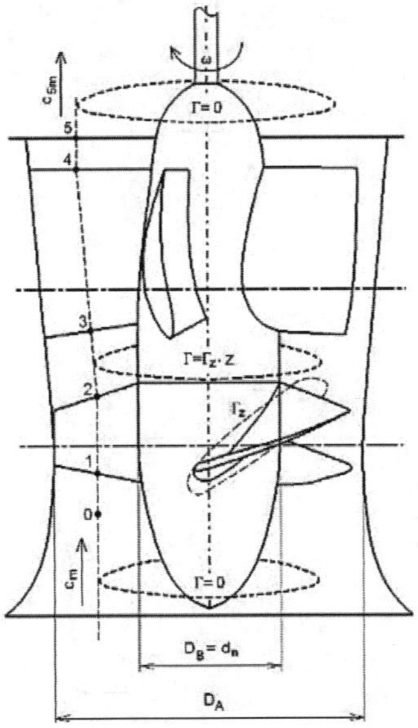

Fig. 3. Meridional cut, runner blades and streamline-cylindrical surfaces, on which the design of blade profiles is performed [3].

Table 1. Equations applied in the design of the meridional cut.

Coefficient of specific speed	$n_b = n \cdot \frac{\sqrt{Q}}{(g \cdot H)^{0.75}}$	(1)
Coefficient	$K_m = 0.0688 + 0.733 \cdot {n_b}^{1.1}$ $valid\ for: 0.33 < n_b < 1$	(2)
Meridional velocity	$c_m = K_m \cdot \sqrt{2 \cdot g \cdot H}$	(3)
Runner chamber diameter	$D = D_A = \sqrt{\frac{4}{\pi} \cdot \frac{1.04 \cdot Q}{c_m} \cdot \frac{1}{1 - \left(\frac{D_B}{D_A}\right)^2}}$	(4)
Ration between the runner hub and the chamber	$\frac{d_n}{D} = \frac{D_B}{D_A} \doteq 0.63 - 0.346 \cdot (n_b - 0.25)$	(5)

2.2 Hydraulic design of axial pump runner blade profiles

The hydraulic design of blade profiles is usually realized on several cylindrical surfaces evenly distributed in the space of the runner (Fig. 3 on the right). The calculation itself utilizes the experimental results obtained in the wind tunnel on isolated profiles. The details of the calculation are explained in more detail in the works [1 – 6].

Blade number (z) is chosen mainly according to the specific speed of the machine (parameter n_b). Regarding the design of the number of blades, it is necessary to calculate a suitable blade span angle φ_r (Fig. 4 and Fig. 5).

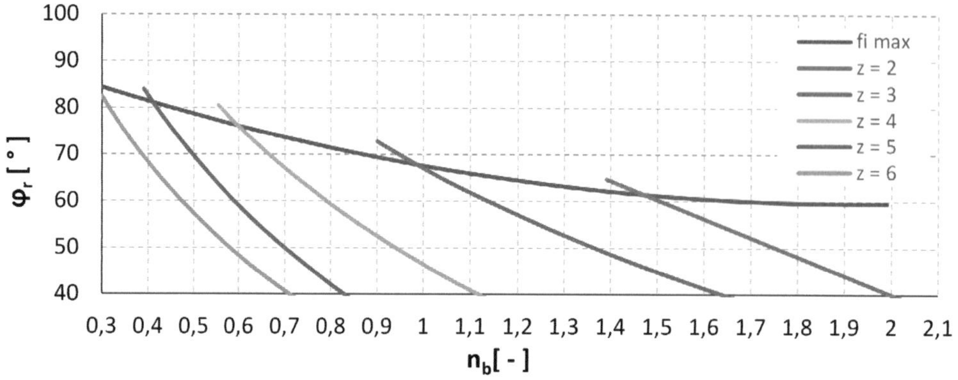

Fig. 4. Nomogram for determining the number of blades and the blade span angle with the marking of the upper limit curve.

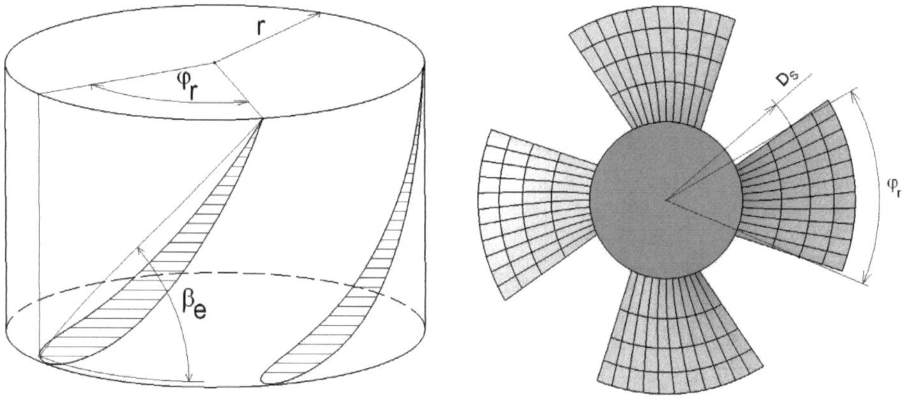

Fig. 5. View of the cylindrical surface of the blade cut [3].

Fig. 6 shows the velocity triangles, kinematic parameters and basic geometric parameters of the axial blade cascade of the runner, essential from the point of view of the achieved machine parameters. The task is to design these parameters as well as some other geometrical parameters of the blade profile so that the required performance or hydraulic parameters of the machine are achieved. The design itself is done iteratively.

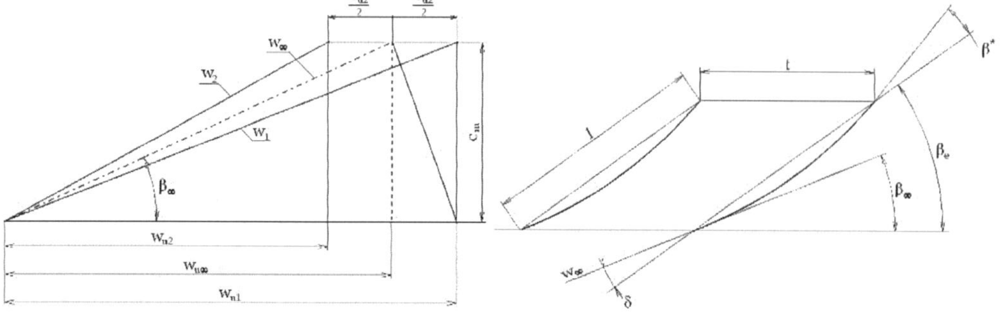

Fig. 6. Velocity triangles and blade cascade of axial pump runner [3].

The procedure is as follows. The peripheral component of the mean relative velocity $w_{u\infty}$ (6), angle β_{∞} (7) and angle of the cascade's inclination β_e (8) are determined, while the angle of attack δ is considered in the zero approach equal to zero. Parameter η_H in equation (6) is the hydraulic efficiency, which is determined on the basis of diagrams compiled according to experimental data. Next, the length of the blade chord (l) is calculated according (9) and relative pitch as well (10).

Coefficient of cascade's impact ξ, can be calculated from (11), but it is possible po determine it by the diagram in the Fig. 7.

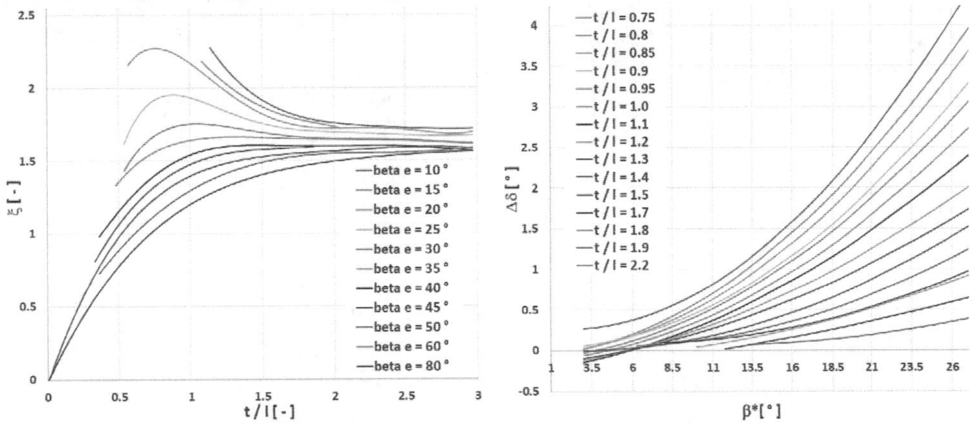

Fig. 7. Dependency of the coefficient of cascade's impact ξ and changing of the angle of attack $\Delta\delta$.

Value of ξ depends on relative pitch t/l and angle of the cascade's inclination β_e (Fig. 7). It is one of the most important coefficients that determines the ratio between the lift coefficients of an isolated airfoil in a wind tunnel $(c_z)_1$ (14) and in a blade cascade c_z. We need to know this ratio because experimentally determined values (c_z) of a NACA profile were obtained on an isolated profile in a wind tunnel. Therefore, we have to recalculate the lift coefficient of the airfoil in the cascade to the value that the same airfoil would have if it were isolated. In addition, the difference in the lift coefficients of the isolated airfoil and the airfoil in the cascade, due to the fact that the airfoil in the cascade can be considered as an airfoil of infinite length, must be taken into account. This difference is compensated by changing the angle of attack of the profile $\Delta\delta$. The value of the changing of the angle of attack $\Delta\delta$ is shown in the diagram in the Fig. 7. This value depends on the relative pitch t/l and the angle of profile's deflection (camber) β^* (13) (Fig. 6). Then the mean geometric velocity (w_∞) is calculated according to equation (12).

After determining the change in the angle of attack $\Delta\delta$ a detailed calculation of $(c_z)_1$ must be performed. This calculation takes into account the value of $\Delta\delta$ in the calculation of the angle of cascade's inclination β_e (Fig. 7). This means that it is necessary to recalculate the angle of cascade's inclination according to (15) and then repeatedly perform the calculations (9) – (14).

Subsequently, it is possible to make a comparison with the experimentally determined lift coefficient (c_z) of the NACA profile, with the certain blade chord length l, relative pitch t/l, angle of attack δ and the maximum blade profile thickness t^*. Subsequently, all parameters (16) – (21) are recalculated.

A control of the lift coefficient (c_z) of the NACA profile is performed according to equation (16). The selected parameters are the relative maximum thickness of the profile t^*/l and the ratio of distance of maximum camber to chord length L^*/l. With the application of these parameters and the calculated values of ratio of maximum camber to chord length m/l and β_e it is possible to determine the geometry (coordinates) of the designed NACA airfoil. The angle of attack must be iterated to achieve an approximate match (c_z) and $(c_z)_1$.

Table 2. Equations used in the design of the blade profile.

Peripheral component of the mean relative velocity	$w_{u\infty} = u - \frac{c_{u2}}{2} = u - \frac{g \cdot H}{2 \cdot \pi \cdot D \cdot n \cdot \eta_H}$	(6)
Angle	$\beta_\infty = \text{arctg}\frac{c_m}{w_{u\infty}}$	(7)
Angle of the blade cascade's inclination	$\beta_e = \beta_\infty + \delta$	(8)
Blade chord length	$l = \frac{\varphi_r \frac{D}{2}}{\cos\beta_e}$	(9)
Relative pitch	$\frac{t}{l} = \frac{\pi \cdot D}{z \cdot l}$	(10)
Coeficient of cascade's impact	$\xi = \frac{(c_z)_1}{(c_z)}$	(11)
Mean geometric velocity	$w_\infty = \sqrt{c_m{}^2 + w_{u\infty}{}^2}$	(12)
Angle of profile's deflection (camber)	$\beta^* = \frac{\frac{g \cdot H}{z \cdot n \cdot \eta_H}}{\xi \cdot w_\infty \cdot l}$	(13)
Lift coefficient of the isolated profile	$(c_z)_1 = \frac{\frac{2 \cdot g \cdot H}{z \cdot n \cdot \eta_H}}{l \cdot w_\infty} \cdot \frac{1.6}{\xi}$	(14)
Angle of the blade cascade's inclination with the changing the angle of attack	$\beta_e = \beta_\infty + \delta + \Delta\delta$	(15)

Table 3. Equations used for determining of the lift coefficient of the NACA profile.

Lift coefficient on the NACA profile	$(c_z) = \frac{\partial c_z}{\partial \delta} \cdot (\delta_\lambda - \delta_0)$	(16)
Ratio	$\frac{\partial c_z}{\partial \delta} = 0.079 - 0.037 \cdot \frac{t^*}{l}$	(17)
Angle of attack by zero lift	$\delta_0{}^\circ = -83.3 \cdot \frac{m}{l} \cdot \left(0.74 + \frac{L^*}{l}\right)$	(18)
Ratio of maximum camber to chord length hrany $(m/l)_{40\%}$ (distance $L^* = 0.4 \cdot l$)	$\frac{m}{l} = \frac{\sin^2\left(\frac{\beta^*}{2}\right)}{\sin\beta^*} - \frac{1}{20} \cdot \text{tg}\frac{\beta^*}{5}$	(19)
The angle of attack for the final span	$\delta_\lambda = \delta_i + \Delta\delta$	(20)
The induced angle of attack	$\delta_i = 57.3 \cdot \frac{(c_z)_1}{\pi \cdot \lambda}$	(21)

Calculated parameter values l, β_e, $\frac{m}{l}$, t^* are used to calculate the coordinates of points of specific blade cuts – blade profiles on streamline surfaces "B – A" marked in Fig. 3. The calculation is carried out on the basis of equations describing the relative values of the coordinates of a specific NACA profile.

The mentioned procedure was implemented in our own "in – house" calculation software (created in the Excel VBA environment). This ensured the automatic design of the runner geometry. The designed geometry of the runner is subsequently exported to the simulation software environment via text files containing the coordinates of the blade cuts (profiles).

3 CFD simulation of a turbine with a new runner

CFD simulation was performed on a full turbine flow domain. Each simulated operating mode took into account the specific position of the runner blades and the distributor vanes.

The computational mesh was purely hexahedral with a total number of cells of approximately 7 million (Fig. 8).

A steady – state, incompressible, isothermal flow model was applied, including the solution of the Navier – Stokes equations, the continuity equation and the k-ε RNG turbulence model equations. "Mixing – plane" interfaces between stationary and rotating subdomains were implemented.

The 2nd order discretization schemes were applied. Boundary conditions: total pressure and intensity of turbulence – inlet, static pressure – outlet, rotating speed.

CFD simulation was performed for 53 operating modes with different positions of the distributor vanes and the runner blades. The simulated operating modes included modes related to "guaranteed" head, operating range boundaries and operating points near BEP.

Integral values of capacity, torque of the entire rotor, power, efficiency were evaluated. The comparison of the required and calculated parameters (of the turbine with the new runner – new design) for the guaranteed head is shown graphically in the Fig. 9. CFD simulations were performed identically on the original turbine (with the original runner). The Fig. 10 shows the calculated parameters of the original turbine and calculated parameters of the turbine with the new runner. The difference in parameters between the original and modified turbine (new design) is obvious.

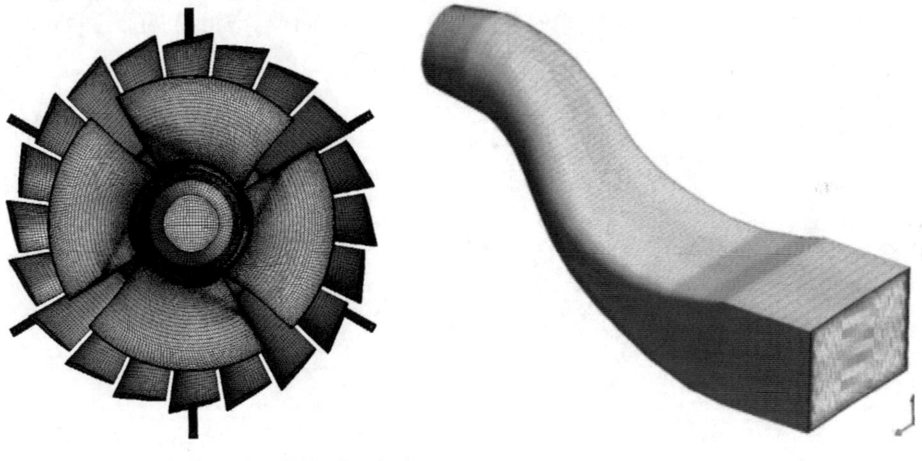

Fig. 8. Computational mesh for CFD simulation.

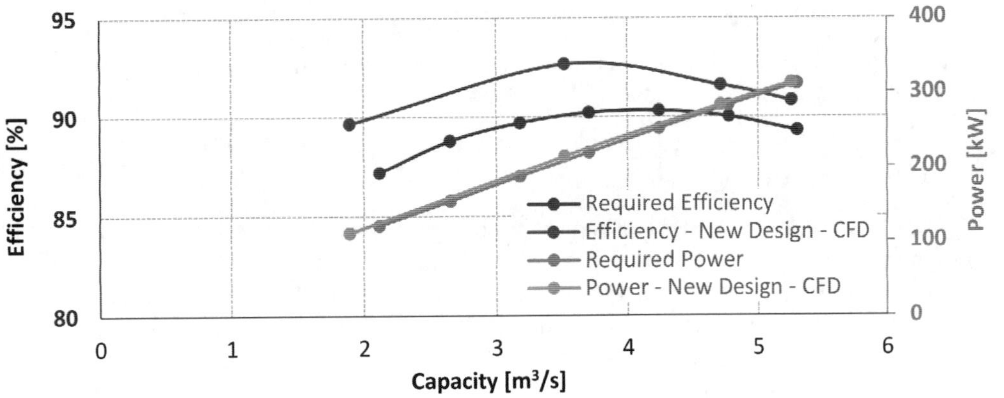

Fig. 9. Comparison of required (contracted) efficiency and power values (blue or red curve) with calculated (CFD) efficiency and power values – new design (purple or green curve), H = 6.7 m.

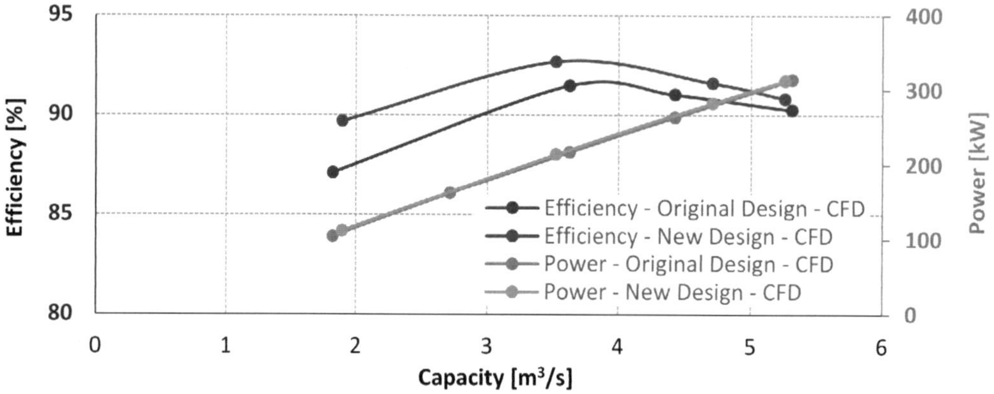

Fig. 10. Comparison of calculated (CFD) efficiency and power values – original design (blue or red curve) with calculated efficiency and power values – new design (purple or green curve), H = 6.7 m.

4 Validation of CFD simulation

Results of model tests performed on the original turbine model were compared to the results of CFD simulation of the turbine with original geometry with model size (Fig. 11).

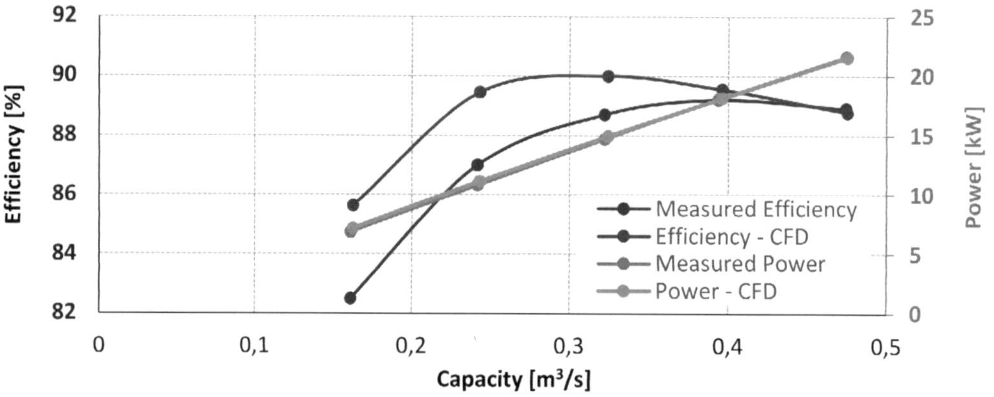

Fig. 11. Comparison of measured efficiency and power values – model test (blue or red curve) with calculated efficiency and power values (purple or green curve) – original geometry, H = 5.23 m.

It can be seen from the Fig. 11 that a relatively good agreement is achieved between the results of the measurements and the CFD simulation.

5 Conclusion

A CFD simulation showed that the runner designed using the classical method meets all the required parameters. It should be remarked that the new runner geometry is still the primary geometry, which could be optimized using one of the approaches stated in the introduction of this paper. It turned out that the hydraulic design carried out by the classical approach with the verification of parameters by CFD simulation is a practical and time-saving way to fulfill the assignment. The advantage of the classical approach to the design of a primary geometry is its simplicity and the explicit design of the geometry of the runner based on the required parameters at the BEP. Subsequent more complex optimization algorithms ensure automatic optimization of the hydraulic shape of the runner, while a greater number of design variants are analyzed. Automatic optimization can provide slightly better parameters compared to the "engineering" approach to a runner optimization, but at the cost of increased computing time while the achieved benefit may not always be sufficient.

Authors would like to thank to project: Regenerácia použitých batérií z elektromobilov / Regeneration of used batteries from electric vehicles ITMS2014+: 313012BUN5

References

1. Varchola, M., Hlbočan, P. (2016). *Hydraulický návrh axiálneho stroja. 1. vyd.* STU v Bratislave.
2. Hlbočan, P., Varchola, M. (2015). CFD modelovanie v procese návrhu axiálneho stroja. *In ESPM 2015. Technické vedy a výrobný manažment 2015 : zborník abstraktov z medzinárodnej vedeckej konferencie*. S. 49.
3. Varchola, M., Bielik, T., Hlbočan, P. (2013). Methodology of 3D hydraulic design of a impeller of axial turbo machine. *In Engineering Mechanics. Vol. 20, No. 2, s.107-118.*
4. Hlbočan, P., Varchola, M. (2016). CFD simulation in process of axial turbo machine design. *In Production Management and Engineering Sciences. CRC Press. Taylor & Francis Group*, S. 383-389.
5. Hlbočan, P., Varchola, M. (2013) Numerical simulation on a mixed-flow pump operating in a turbine mode. *In Engineering Mechanics. Vol. 20, No. 2, s.97-105.*
6. Hailu, G., Varchola, M., Hlbočan, P. (2011). *Design of Hydrodynamic Machines Pumps and Hydro-Turbines*. CRC Press. Taylor & Francis Group, 266 Pages.
7. Bielik, T. (2010). *Hydraulický návrh axiálneho stroja v 3D prostredí*, Diplomová práca SjF STU v Bratislave.
8. Vrábel, M. (2014). *Hydraulický návrh axiálneho stroja s výstupom pre CFD softvér*, Diplomová práca SjF STU v Bratislave.
9. Strýček, O. (1994). *Hydrodynamické čerpadlá*, STU v Bratislave
10. Brůha, O., Felber, V. Smolař, V. (1939). *Letecký Průvodce Díl 2*, Matice česká.
11. Nechleba, M. (1962). *Vodní Turbíny jejich konstrukce a príslušenství*, SNTL Praha.
12. Lewis, R., I. (1996) *Turbomachinery performance analysis*, Elsevier Science and Technology books.
13. Lobanof, V., S., Ross, R., R. (2013). *Centrifugal Pumps – Design and Application.* Gulf Professional Publishing.
14. Kantor, M., Chalupa, M., Souček, J., Bílková, E., Nowak, P. (2010). Application of Genetic Algorithm Methods for Water Turbine Blade Shape Optimization. *Manufacturing Technology, Vol. 20, No. 4.*
15. Semenova, A. et al. (2014). Multi-objective shape optimization of runner blade for Kaplan turbine, *IOP Conf. Ser.: Earth Environ. Sci. 22 012025.*

Optimal Design of Battery Storage Systems for RES Using Simulation Methods

Zoltán Csuka[1,*], *Jozef* Matušov[2], *Peter* Hlbočan[1], *Zuzana* Kolková[1], and *Marta* Murgašová[1]

[1]ZTS - VÝSKUM A VÝVOJ, a.s., Lieskovec 598/75, 018 41 Dubnica nad Váhom, Slovak Republic
[2]ENERGO-AQUA a.s., Trenčianske Biskupice 7051, 911 04 Trenčín 4, Slovak Republic

Abstract: The use of battery storage systems (BSS) is an increasingly common topic in the context of the operation of various types of renewable energy sources (RES). One of the applications may be to solve the problem of energy-free operation of small hydropower plants, where quite specific requirements are placed on the design of the BSS. In the context of optimal BSS design for a given application, this paper discusses the computer analysis capabilities that are essential in the design and performance verification phase of a BSS. The paper presents the possibilities of simulation methods in the field of electrical analysis of cell current load during charging and discharging processes, thermal analysis related to the selection of a suitable cooling system with respect to the operating mode of the battery system.

Keywords: Battery storage system, renewable energy sources, lithium-ion battery, battery thermal management system, optimization, CFD simulation

1 Introduction

During the operation of renewable energy sources (RES) various situations arise when energy storage is required, at least for a short period of time. For example, when using solar energy, its availability during 24 hours a day needs to be dealt, where battery storage system (BSS) is an ideal solution. In the operation of a small hydropower plants the power grid may fail for various reasons. This means a sudden disconnection of the electric generator. The relieved turbine (no load from the generator) is then spun up to runaway speed and has to be shut down in an emergency. This is so-called energy-free operation and it is an undesirable condition where the structure is overstressed. In some locations, such outages happen more frequently and can last from a few minutes to several hours. The energy-free operation problem can be solved by using a battery storage system - the battery modules will be charging at the time of disconnection from the grid. Such a BSS is ideal for the use of "second-life" batteries. These are lithium-ion batteries retired from use in electric vehicles, which are reused in industry after the regeneration process.

BSSs are composed of battery packs, which are made up of an array of lithium-ion cells. Lithium-ion batteries (LIB) store a certain amount of electrical energy and generate a significant amount of heat during the charging and discharging processes. The heat generated by lithium-ion batteries is divided into two main components, the heat generated by reversible reactions q_{re} and the heat generated by irreversible reactions q_{irr}. The heat generated by

* Corresponding author: zoltan.csuka@ztsvav.eu

irreversible reactions is directly related to the discharge rate of the batteries, i.e. the amount of so-called irreversible heat is higher the higher the discharge rate, while the amount of heat generated by reversible reactions is predominant at lower discharge rates. The total heat generated by the battery can be determined by equation (1) [1].

$$q_{tot} = q_{re} + q_{irr} \tag{1}$$

The reversible heat generation is related to the entropy change during charging and discharging due to the intercalation and deintercalation processes at the electrodes. The entropy changes at the electrodes depend on the electrode materials and the state of charge. The state of charge (SOC) is known to be a factor that has a large influence on reversible heat generation. The calculation of the amount of heat generated by reversible reactions is described in equations (2) and (3) [1]:

$$q_{re} = S_{(a,i)} j_{(loc,i)} \frac{T\Delta S}{n \cdot F} \tag{2}$$

$$\Delta S = n \cdot F \frac{\delta U_i}{\delta T} \tag{3}$$

where $S_{a,i}$ is the specific surface area (m^{-1}), $j_{loc,i}$ is the local current density (Am^{-2}), ΔS is the entropic change, U_i is the breakdown voltage, which describes the equilibrium potential of the electrodes as a function of SOC, F is the Faraday constant $(F = 96487 Cmol^{-1})$, n is the number of electron moles [1].

Irreversible heat generation is divided into two categories, Joule heat generation, which occurs due to the internal resistance of the battery, and polarization heat generation due to the electrochemical reaction that occurs during the charging and discharging process. The irreversible heat is calculated according to equation (4) [1].

$$q_{irr} = q_p + q_j \tag{4}$$

The generation of polarization heat is calculated according to the following equation (5).

$$q_p = S_{(a,i)} \cdot j_{(loc,i)} \left(\varphi_{(1,i)} - \varphi_{(2,i)} - U_i\right) \tag{5}$$

U_i is the breakdown voltage, which describes the equilibrium potential of the electrodes as a function of SOC. Since temperature influences the battery discharge process, it is necessary to include the so-called Taylor expansion at the reference temperature in the equation (6) [1].

$$U_i = U_{ref,i} + \left(T - T_{ref}\right) \frac{dU_i}{dT} \tag{6}$$

Joule heat generation in battery cells is based on three different factors, the resistance of the material, the movement of ions through the electrolyte and the heat generated through the collector. Joule heat can be calculated using equation (7) [1].

$$q_j = I^2 R \tag{7}$$

where I is the discharge current and R is the internal resistance of the cell. The above equation shows that the magnitude of the discharge current affects the amount of heat generated quadratically [1].

2 Problems in LIB operation

During the charging and discharging processes of LIB batteries, a certain amount of heat is generated due to electrochemical reactions. If the generated heat is not transferred properly, conditions unfavourable to LIB batteries may occur.

2.1 Impact of high temperatures

Due to the high temperature of the cell, loss of capacitance and reduction of operating voltage may occur. A large amount of heat is generated at elevated operating values, such as a cell operating at a high discharge rate. Capacity loss occurs when the active material inside the battery is converted to inactive material. Due to raised temperatures, the internal impedance of the cell will increase, which will reduce the operating voltage at different discharge rates. The capacity of a lithium-ion cell depends on the cell temperature. LIB have an optimum operating temperature range set by the manufacturers, usually 15-35 °C. Operating battery cells outside this range can have several negative effects on capacity and lifetime (Fig. 1.) [2].

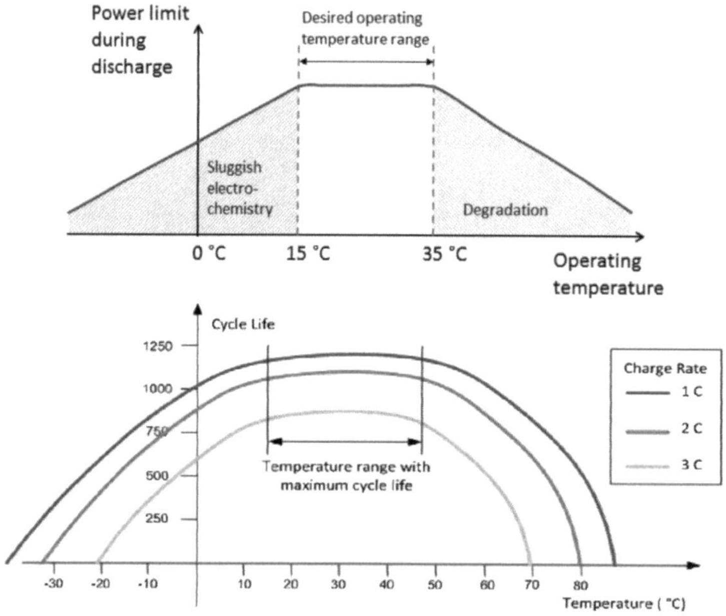

Fig. 1. Optimal operating temperature of the cell in terms of its performance during the discharge process (up) and in terms of lifetime at different discharge currents (down) [1].

2.2 Thermal Runaway (TR)

TR of lithium-ion cells can be initiated for various reasons such as internal short circuit, overheating, overcharging or over-discharging. These reasons cause the temperature of the cells to rise further and trigger chemical reactions that lead to even faster temperature rise of the cells. TR is a sequence of reactions that results in irreversible damage to the battery cell or its surroundings; a violent exothermic reaction can end in an explosion and a fire that is difficult to extinguish (Fig. 2.) [2].

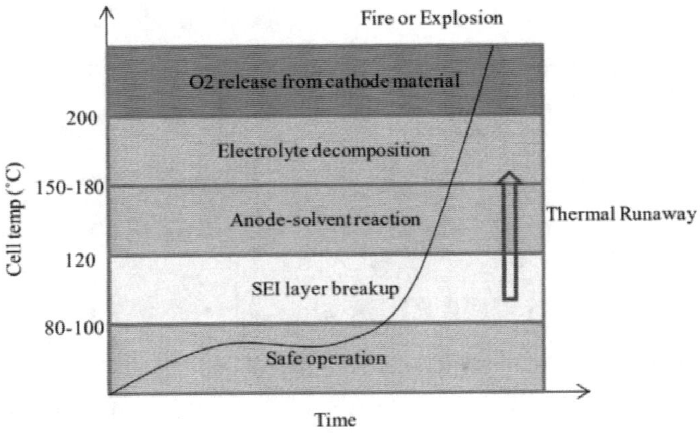

Fig. 2. Thermal Runaway [2].

2.3 Temperature maldistribution

If the charging or discharging rate is increased, more heat will be generated in the cells. If this heat is not properly dissipated, it will concentrate inside the battery cells. In addition, heat transfer by convection takes place only on the outer surfaces of the battery units. This will result in uneven temperature distribution inside the battery packs. As mentioned in the previous section, the performance of a cell is dependent on its temperature. This means that sustained inappropriate temperature distribution will lead to variability in capacity between cells. This will create a chain reaction where cells with a suitable operating temperature are forced to deliver higher power to compensate for cells with lower power, which in itself leads again to an increase in the temperature of the more stressed cells. In addition, lithium cells have a low tolerance to overcharging, so the total capacity of the batteries when charged is limited to the cell with the lowest power output. In principle, it states that the max. allowable temperature variation between cells in a battery module should be less than 5 °C [2]. Based on the above problems, it is necessary to ensure sufficient cooling of the battery system. Due to the large number of cooling system concepts and the resulting cooling system parameters, the cooling design of the BSS needs to be solved at the conceptual and design stage.

3 LIB cooling options

Several types of different cooling system concepts are used to cool the BSS (Fig. 3.).

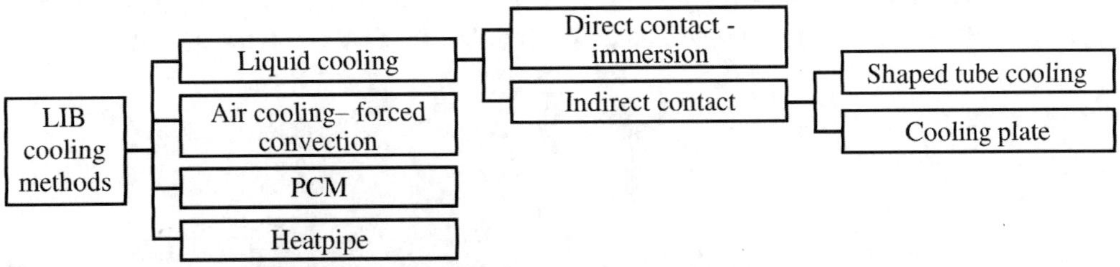

Fig. 3. Cooling systems used for LIB batteries.

3.1 Liquid cooling with direct contact

In this method, the coolant is "forced" to flow through the battery module and dissipate the heat generated, as shown schematically in Fig. 4. If the battery module is directly immersed in the coolant, the coolant must have dielectric properties. The heated fluid is then cooled by a heat exchanger. This method is preferable to forced air cooling due to the higher heat capacity and thermal conductivity of the coolant. There is no need for an additional cooling system to dissipate heat even at high discharge or charge rates (C-rate) [3], [6].

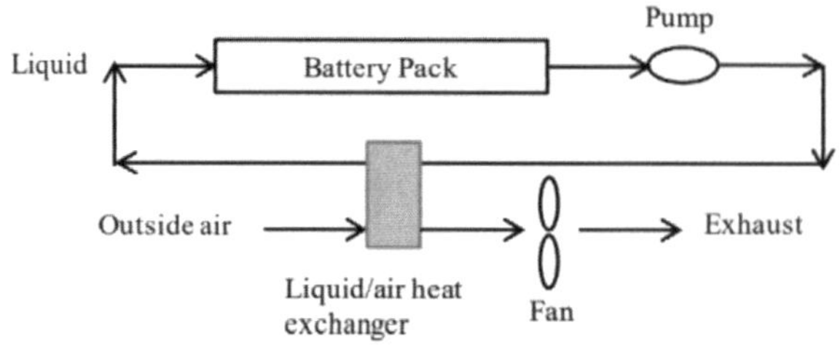

Fig. 4. Liquid cooling system with direct contact [2].

3.2 Liquid cooling with indirect contact with a shaped tube

In this indirect cooling method, ethylene glycol most often acts as the cooling medium. Between the individual battery cells, a shaped tube is placed. The heat from the battery is transferred into the tube material and then by convection into the coolant. Many automotive companies use a method of indirect cooling using so-called shape tubes. The coolant is cooled back either by a compressor cooling circuit or by a heat exchanger [3], [6].

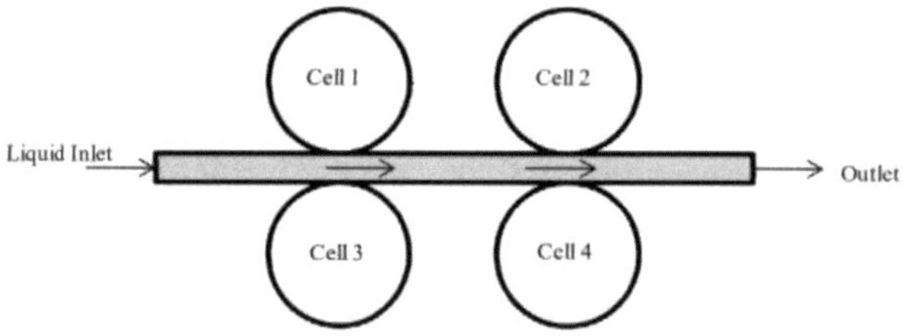

Fig. 5. Liquid cooling system with indirect contact - shaped tube [2].

3.3 Liquid cooling with indirect contact with the cooling plate

In this method of cooling the battery cells, a plate with coolant flowing from one end of the battery pack to the other is placed under the battery in the case of cylindrical cells (Fig. 6.). The heat from the battery is transferred to the plate material, then the coolant in the plate channels removes heat due to convection. The cooling method shows poor performance in terms of temperature distribution within a single cell (applies exclusively to cylindrical cells). However, the temperature distribution within the battery module is relatively uniform. Many automotive companies use this method of cooling batteries by means of so-called cooling plates, but mainly for flat or prismatic cells, where the contact area between the cell and the exchanger plate is many times larger compared to the cylindrical cell type [3], [6].

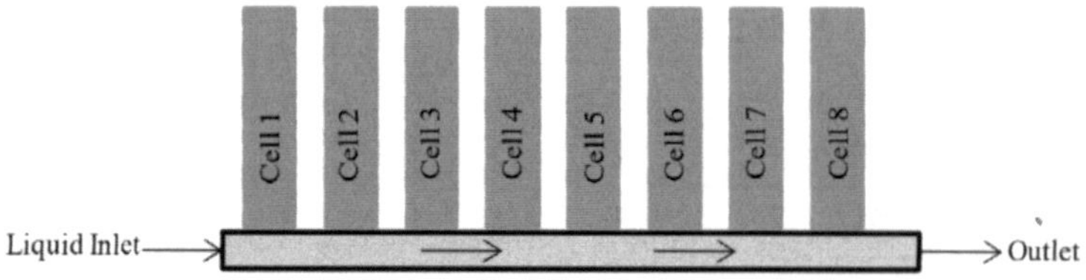

Fig. 6. Liquid cooling system with indirect contact - cooling plate [2].

3.4 Air cooling with forced convection

In this method, air is forced to flow through the battery pack and dissipate the heat generated, as shown in Fig. 7. The air cooling system is mostly used in hybrid cars such as the Toyota Prius and Nissan Leaf. The air-cooled battery cooling system is widely used due to its simple design, high reliability and low acquisition cost. However, the low heat capacity and low thermal conductivity of air may not be suitable for high performance applications.

For forced convection, a fan or blower is required to extract the heated air to the outside and draw in the cold air. Due to the lower thermal capacity and conductivity, higher temperature differences may occur between the individual cells in the battery module. As a disadvantage, it is also necessary to mention a certain noise level of the air-cooled system [3], [4], [5].

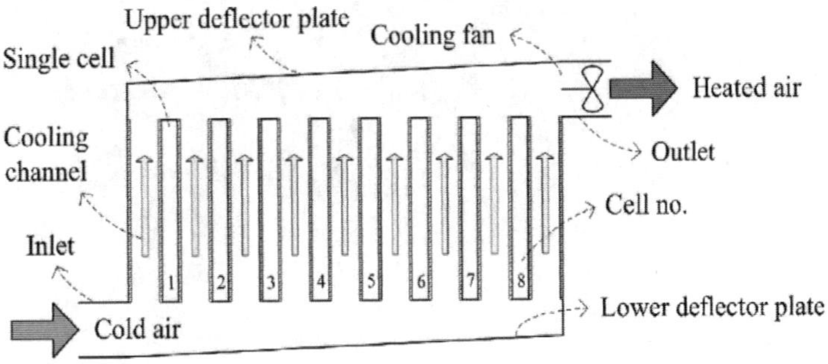

Fig. 7. Air cooling system - forced convection [5].

3.5 Heatpipe cooling system

Heatpipe is a self-acting device designed for heat transfer, combining the principles of thermal conduction and phase transition for efficient heat transfer between two solid surface interfaces. The main parts of the Heatpipe device are: the evaporator side (hot side), the condenser side (cold side), the porous material and the working fluid. The body is usually copper or aluminium. The operating temperature range is approximately 25° to 150 °C. Based on the application, the working fluid is suitably selected, typically water, acetone or liquid butane is used. The heatpipe is usually placed between the battery cells as shown in Fig. 8. The hot side of the Heatpipe is attached to the battery from which it extracts heat. The other (cold) side of the Heatpipe is placed outside the battery so that it can be cooled. The working fluid removes heat on the battery side (hot side) while evaporating. Due to the increase in static pressure, the vapour passes to the cold side where it condenses. The condensate leaks spontaneously back to the hot side through the porous material, due to its own viscosity, pressure differences and possible gravity [5], [7], [8].

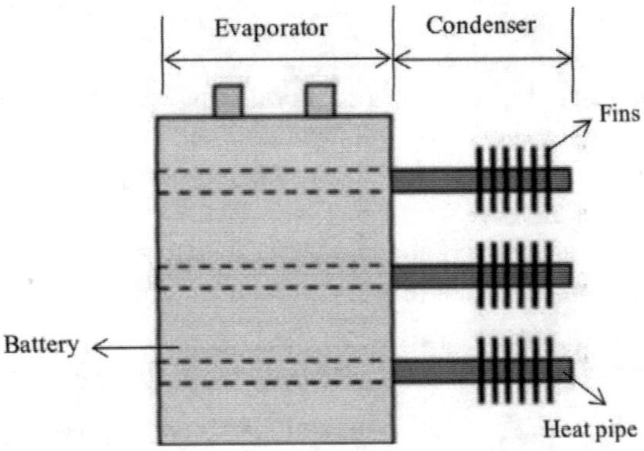

Fig. 8. Heatpipe cooling system [2].

3.6 PCM cooling

PCM is a material that accumulates or transfers heat through a phase change process from one state to another at a specific temperature. Phase change materials used for cooling are divided into organic (paraffin wax), inorganic and eutectic compounds. PCM materials are often mixed with graphite powder, metal foam and carbon fibre to improve the thermal conductivity of the resulting composite. This cooling system often requires an additional system to remove heat from the PCM material [7], [8].

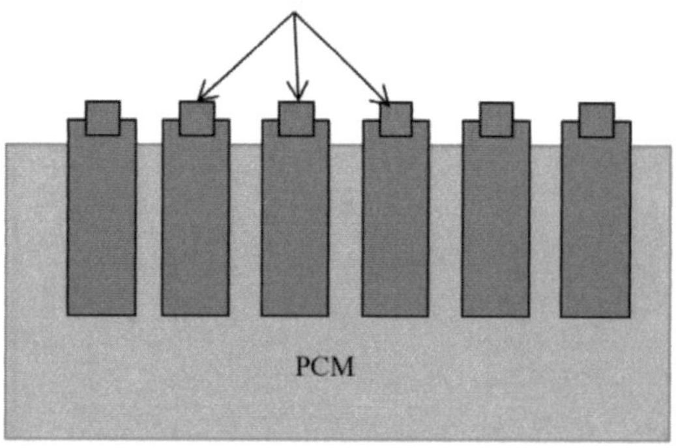

Fig. 9. PCM cooling system [2].

Both the suitability of the individual systems must be assessed in terms of operation, design and other requirements, and the specific cooling system must be optimised (in terms of geometry, shape, dimensions, etc.). The above tasks can be solved with the application of CFD methods. In general, it can be stated that CFD simulation of a BSS can be used to calculate the temperature of cells and other components of the battery pack and to assess the safety of operation with respect to the maximum temperature. The importance of CFD methods in BSS design can be divided into several groups:

1. deciding or assessing the suitability of specific types of cooling systems,
2. selection of suitable materials for construction,
3. sizing, i.e. shape and dimension optimization of the battery system.

A properly constructed and tuned CFD model should be characterised by or take into account the following attributes:

1. Geometry of the whole pack (multiple packs with which the coolant flow channels).
2. Heat generation inside the battery cells (due to electrochemical reactions, Joule heat, etc.) during charging/discharging.
3. Conduction of heat through the solid walls of cells, modules, packs, etc.
4. Heat transfer between solid walls and coolant.
5. Coolant flow and heat transfer in coolant.
6. Heat transfer to the surroundings.
7. Results should include coolant flow rate distributions and temperature distributions throughout the pack space (evolution over time to be calculated).

4 Numerical modelling capabilities of LIB cells in ANSYS Fluent

ANSYS Fluent has implemented several different methods for simulating physical processes in LIB batteries (including cooling modelling). Since cooling simulations of battery modules are a rather challenging, essentially multi-physics task, the CFD setup will be built sequentially from the simplest cases (heat transfer on the surface of a single cell without modelling coolant flow) to complex tasks (cooling of the whole pack with coolant flow). Some approaches for modelling heat generation in cells or for modelling electro-thermal phenomena (Conjugate Heat Transfer, Ciruit Networ Method, Multi Scale Multi-Dimensional approach) as well as some sub-models related to the modelling of electro-chemical phenomena (Equivalent Circuit Model, NTGK model, etc.) will be tested step by step. Several ways of defining boundary conditions (mainly related to the calculation of heat transfer to the surroundings) will also be tested. Experimentally determined discharge characteristics will be applied in the calculations.

The aim will be to firstly debug the different approaches to CFD simulation and then (based on comparison with experimental results) to select the optimal approach for modelling the above phenomena. The final simulation model will be used to calculate the heat transfer and temperature distribution on the surface of the batteries or the cooled battery system during battery discharge. At the same time it will serve as a tool for the development of the cooling device of the developed battery system and for the shape-dimensional optimization of some structural parts of the system. A detailed description and individual electro-thermal models and electro-chemical sub-models and their equations can be found in literature sources [9], [10] and [11].

5 Numerical modelling options – summary

The following Tables 1, 2 and 3 provide examples of the basics of setting up a CFD for a proper battery solution from an electrical, thermal and cooling perspective.

Table 1. CFD simulation on a single battery cell (least complex approach).

Access to simulation	**Calculation of 1 battery cell**		
	Conjugate Heat Transfer Model (CHT)	Circuit Network + Equivalent Circuit Model	MSMD + Equivalent Circuit Model
Simulated physical processes	Heat generation (fixed heat output), Joule heat generation in the passive zone, heat conduction through the solid cell material, heat transfer by convection to the surroundings	Heat generation (ECM method - takes into account electrical - chemical reactions), Joule heat generation in the passive zone, heat conduction through the solid cell material, heat transfer by convection to the surroundings	Heat generation (ECM method - takes into account electrical - chemical reactions), Joule heat generation in the passive zone, heat conduction through the solid cell material, heat transfer by convection to the surroundings
Equations describing given processes	Energy equation, "User scalar 0"	Energy equation, ECM equations	Energy equation, ECM equations
Outputs monitored	Temperature of the cell	Temperature of the cell	Temperature of the cell

Table 2. CFD simulation on multiple connected battery cells.

Access to simulation	**Calculation of multiple connected battery cells**			
	Circuit Network + Equivalent Circuit Model	Circuit Network + NTGK Model	Circuit Network + Equivalent Circuit Model + Coolant Flow	Circuit Network + NTGK Model + Coolant Flow

Simulated physical processes	Heat generation (ECM method - takes into account electrical - chemical reactions), Joule heat generation in the passive zone, heat conduction through a solid cell material, heat transfer by convection to the surroundings	Heat generation (NTGK method - takes into account the electro-chemical reactions), Joule heat generation in the passive zone, heat conduction through the solid cell material, heat transfer by convection to the surroundings, capacity drop as a function of temperature	Heat generation (ECM method - takes into account the electrical - chemical reactions), Joule heat generation in the passive zone, heat conduction through the solid cell material, coolant flow with turbulen-ce, heat transfer by convection into the coolant, heat transfer in the coolant and ambient heat transfer	Heat generation (NTGK method - takes into account the electrical - chemical reactions), Joule heat generation in the passive zone, heat conduction through the solid cell material, coolant flow with turbulence, heat transfer by convection into the coolant, heat transfer in the coolant and ambient heat transfer
Equations describing given processes	Energy equation, ECM equations	Energy equation, NTGK equation	Continuity equation, N-S equations, turbulence model equations, energy equation, ECM equations	Continuity equation, N-S equations, turbulence model equations, energy equation, NTGK equations
Outputs monitored	Temperature of the cell	Temperature of the cell	Temperature of the cell	Temperature of the cell

Table 3. CFD simulation on the whole battery module (pack).

Access to simulation	**Calculation of the whole module (pack)**	
	Circuit Network + Equivalent Circuit Model + Coolant Flow	Circuit Network + NTGK Model + Coolant Flow
Simulated physical processes	Heat generation (ECM method - takes into account the electrical - chemical reactions), Joule heat generation in the passive zone, heat conduction through the solid cell material, coolant flow with turbulence, heat transfer by convection into the coolant, heat transfer in the coolant and ambient heat transfer	Heat generation (NTGK method - takes into account the electrical - chemical reactions), Joule heat generation in the passive zone, heat conduction through the solid cell material, coolant flow with turbulence, heat transfer by convection into the coolant, heat transfer in the coolant and ambient heat transfer
Equations describing given processes	Continuity equation, N-S equations, turbulence model equations, energy equation, ECM equations	Continuity equation, N-S equations, turbulence model equations, energy equation, NTGK equations
Monitored outcomes	Temperature of the cell	Temperature of the cell

6 Conclusion

The paper describes a CFD approach to optimize the proposed BSS cooling system, which will be verified in the near future on a physical model of the battery module from which the BSS is formed, as high agreement between simulation and measurement results is a prerequisite. Initially, the simulations of the electrochemical processes that take place in the cells in the context of heat generation will be debugged, under different discharge and charging conditions. Subsequently, the numerical model will be extended to include the battery module and the cooling system, possibly a part of it, resulting in a rather complex multiphysics numerical model. For the CFD calculations, ANSYS Fluent simulation software will be used. The results of the research will later be applied for the complex design of the cooling system, which is a key part of the BSS.

Authors would like to thank to project: Úložiská energie z batérií druhej životnosti / Second life battery energy systems ITMS2014+: 313012BNG

References

1. Frank, K., Qasemi, R., Samvin, D. (2020). Thermal Management of Lithium-Ion Battery Pack, *Product Development and Materials Engineering.*
2. Mylaudy, S. R., Ananth, S. (2020). Battery Thermal Management in Electrical Vehicle - Review Article, *IJISET – International Journal of Innovative Science, Engineering & Technology, Vol. 7 Issue 2, ISSN 2348 – 7968.*
3. Chunyu, Z., Beile, Z., Yuanming, Z., Shunyuan, H., Tongtong, Y., Xiufang, L. (2020). Hybrid Battery Thermal Management System in Electrical Vehicles: A Review, *Energies 2020, 13, 6257; doi:10.3390/en13236257.*
4. Mambazhasseri, D. A., Hamilton, D., Manjunatha, K. N., Minakshi, M. (2020). Design, Development and Thermal Analysis of Reusable Li-Ion Battery Module for Future Mobile and Stationary Applications, *Energies 2020, 13, 1477; doi:10.3390/en13061477.*
5. Henke, M., Hailu, G. (2020). Thermal Management of Stationary Battery Systems: A Literature Review, Energies 2020, 13, 4194; doi:10.3390/en13164194.
6. Bhattacharjee, A., Mohanty, K. R., Ghosh, A. (2020) Design of an Optimized Thermal Management System for Li-Ion Batteries under Different Discharging Conditions, *Energies 2020, 13, 5695; doi:10.3390/en13215695.*
7. Buidin, T. I. C., Mariasiu, F. (2021) Battery Thermal Management Systems: Current Status and Design Approach of Cooling Technologies, *Energies 2021, 14, 4879. https://doi.org/10.3390/en14164879.*
8. Xingxing, W., Shengren, L., Yujie, Z., Shuaishuai, L., Hongjun, N., Yelin, D., Yinnan, Y. (2022). A Review of the Power Battery Thermal Management System with Different Cooling, Heating and Coupling System, *Energies 2022, 15, 1963. https://doi.org/10.3390/en15061963.*
9. ANSYS Fluent Theory Guide, Release 2021 R1, *ANSYS, Inc., Southpointe, 2600 ANSYS Drive Canonsburg, PA 15317, ansysinfo@ansys.com, http://www.ansys.com*
10. Madani, S. S., Swierczynski, M. J., Kaer, S. K. (2017). The discharge behaviour of lithium-ion batteries using the Dual-Potential Multi-Scale Multi-Dimensional (MSMD) Battery Model, *Twelfth International Conference on Ecological Vehicles and Renewable Energies, ISBN: 978-1-5386-1692-5/17/$31. 00.*
11. Paccha-Herrera, E., Calderón-Muñoz, W. R., Orchard, M., Jaramillo, F., Medjaher, K. (2020). Thermal Modeling Approaches for a LiCoO2 Lithium-ion Battery-A Comparative Study with Experimental Validation, *Batteries 2020, 6, 40; doi:10.3390/batteries6030040.*

https://doi.org/10.1051/matecconf/202236901010

Design of the Discharging Electrode of the Electrostatic Precipitator Depending on the Applied Voltage

Juraj Drga[1,*], *Michal* Holubčík[1], *Jozef* Jandačka[1], and *Andrej* Klačko[1]

[1]Department of Power Engineering, Faculty of Mechanical Engineering, University of Žilina, Univerzitna 1, 010 26 Zilina, Slovakia.

Abstract. The article deals with the design of the diameter and the simulation of the distribution of the intensity of the electric field of the charging electrode of the electrostatic precipitator, given the supplied DC voltage. The purpose was to calculate the value of the critical intensity of the electric field, which must be exceeded when trying to achieve the state of corona discharge on the electrode, which is a condition for electrostatic separation. Subsequently, CFD simulations of two 3D models with electrode diameters of 1 and 4 mm were created, on which the distributions of the electric field intensity were observed at a DC voltage of 20 kV. The simulations confirmed the results of the calculations that the corona discharge at a voltage of 20 kV will occur only on an electrode with a diameter of 1 mm.

Keywords: *electric field, electrostatic precipitation, discharging electrode, corona discharge, CFD simulation*

1 Introduction

The article deals with the study of the influence of the diameter of charging electrodes on the operating parameters of a tubular electrostatic precipitator. An electrostatic precipitator is a device used for the secondary capture of solid particles (PM) using a corona discharge [1-5]. Such a device consists of a positive and a negative electrode, connected to a source of high DC voltage. The electrode connected to the positive pole of the voltage source is formed by a metal chimney pipe, and the electrode connected to the negative pole of the voltage source is formed by a conductive wire of circular diameter. PM flowing through the separator is charged by a negative electric charge, thanks to which they are attracted to the surface of the collecting (positive) electrode.

The function of electrostatic separation is conditioned by the formation of a coronary discharge on the charging electrode area. The formation of this type of discharge must be ensured by the supply of voltage with a value higher than the initial critical voltage. The size of this value affects the diameter of the charging electrode and the distance between the positive and the negative electrode. Likewise, the formation of a coronary discharge affects

* Corresponding author: juraj.drga@fstroj.uniza.sk

the shape of the charging electrode. Sharp edges on the surface of the electrode create favorable conditions for the formation of coronary discharge [6-8].

This paper investigates the influence of the diameter of the charging electrode on the operating parameters of a tubular electrostatic precipitator with a voltage of 20 kV. The essence was to calculate the value of the initial critical voltage using generally known relations and then to verify the occurrence of a corona discharge using numerical CFD simulation.

Numerical simulations using CFD software are commonly used tools in the design of various devices [9-12]. In the past, several works were devoted to numerical simulations of electrostatic precipitators, respectively charging electrodes, which were also the inspiration for this article.

2 Methods

2.1 Numerical calculations

Two 3D models are investigated in the work, Fig.1 consisting of an identical positive electrode with a diameter of 120 mm and a length of 500 mm, Fig. 2a.). The negative electrodes were different with a length of 300 mm and diameters of 1 mm and 4 mm, Fig. 2b.). The simulated voltage applied to the electrodes was set at 20 kV. All input parameters of the electrostatic precipitator are given in Tab. 1.

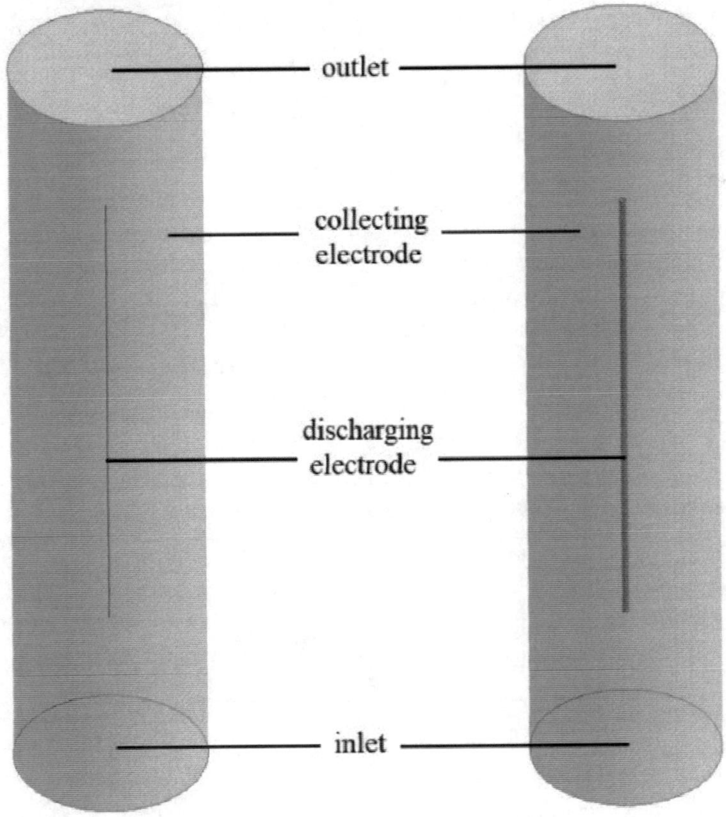

Fig. 1. 3D geometry of models.

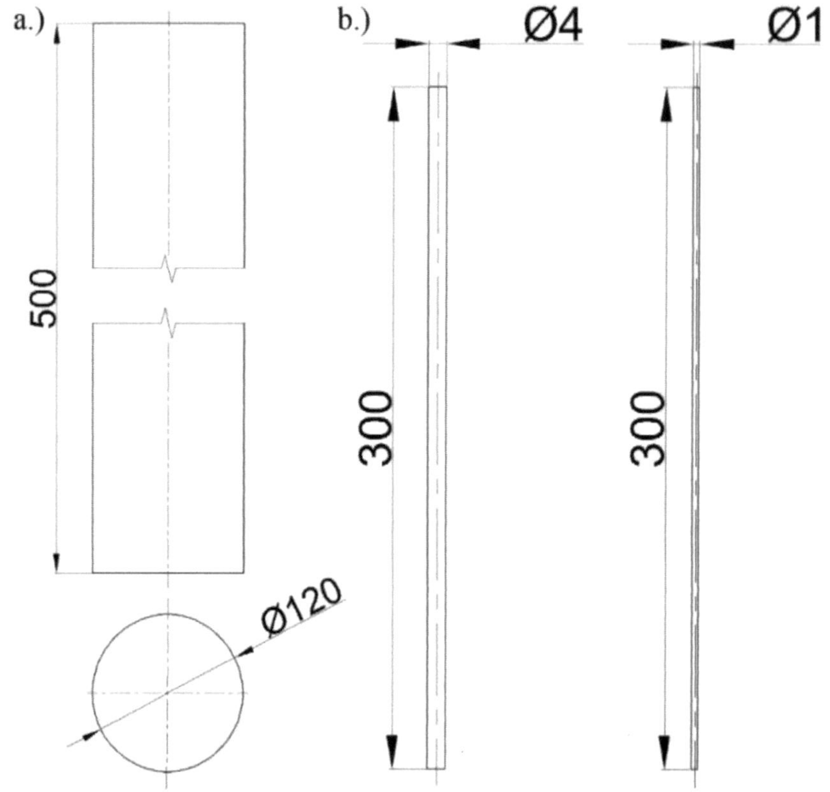

Fig. 2. Sheet of: a.) collecting electrode; b.) discharging electrodes.

First, numerical calculations of basic electrical quantities determining the parameters of the precipitator were performed. An important parameter is the value of the critical initial voltage U_{0c}, which is calculated from the value of the initial critical intensity of the electric field E_{0c}. The initial critical voltage is the limit value of the applied voltage, which, if exceeded, corona discharge occurs at the charging electrode. This effect lasts until the value of the applied voltage does not exceed the value of the jump voltage U_p. The course of voltage and current is displayed using V/A characteristics.

Table 1. Input parameters of the electrostatic precipitator.

Quantity	Value	Units	Name
R	0.06	m	collecting electrode radius
r_1	$0.5 \cdot 10^{-3}$	m	discharging electrode radius
r_2	$2 \cdot 10^{-3}$	mm	discharging electrode radius
U	$20 \cdot 10^{3}$	V	applied voltage
ε_0	$8.85419 \cdot 10^{-12}$	$F \cdot m^{-1}$	vacuum permittivity

ε_1	1.00054	-	The relative permittivity of gas (air)
l	0.5	m	length of precipitator
$l_{r,R}$	0.3	m	length of discharging electrode
S	0.1885	m^2	collecting electrode area
k_1	$3.102 \cdot 10^6$	$V \cdot m^{-1}$	quantity of W-B equation (for T = 298K and p_{atm})
k_2	0.0308	$m^{-0,5}$	quantity of W-B equation
ϑ	1	-	rel. gas density (air)

The operating parameters were calculated based on generally available relationships, which were divided into 2 parts, a 1 mm diameter electrode and a 4 mm diameter electrode [13].

The critical initial field strength was calculated according to equation (1)

$$E_{0c} = k_1 \vartheta \left[1 + \frac{k_2}{\sqrt{(\vartheta r)}}\right] \; [\mathrm{V} \cdot \mathrm{m}^{-1}] \tag{1}$$

where *k1, k2* are quantities of the Whitehead-Brown equation, ϑ is the relative gas density, *R* is the radius of the collection electrode and *r* is the radius of the discharge electrode.

The critical initial voltage was calculated according to equation (2)

$$U_{0c} = r\, E_{0c} \ln\left(\frac{R}{r}\right) [\mathrm{V}] \tag{2}$$

The radius of the ionization region at pure gas was calculated according to equation (3)

$$z = \frac{1}{2\vartheta}\left[\frac{2U}{k_1 ln\left(\frac{R}{r}\right)} + k_2^2 - k_2\sqrt{\left(\frac{4U}{k_1 ln\left(\frac{R}{r}\right)} + k_2^2\right)}\right] [\mathrm{m}] \tag{3}$$

where *U* is the applied voltage.

2.2 Numerical simulations

In this part, attention is paid to the verification of the value of the electric field intensity on the proposed electrodes using CFD simulations in the ANSYS Fluent software. Using simulations, it is possible to verify whether a given electrode will be suitable for the occurrence of a corona discharge at a given applied DC voltage.

The mesh of the model with a 1 mm electrode consisted of 1 802 684 elements, with tetrahedrons cells on the collecting electrode and the multizone method on the discharging electrode. The quality of the mesh was considered adequate based on an average value of the orthogonal quality at 0.78 and the skewness at 0.22. The simulation was based on the k-omega SST turbulence model. In terms of boundary conditions, the inlet velocity of 0.5 m/s was applied at the inlet as air. The outlet was defined as a pressure outlet with a gauge pressure of 0 Pa. Turbulence was defined as 5% for turbulent backflow intensity and 10 for turbulent backflow turbulent viscosity ratio for both inlet and outlet. The boundaries of precipitators were defined as stationary walls with no-slip shear conditions.

The mesh of the model with a 4 mm electrode consisted of 1 431 263 elements, with tetrahedrons cells on the collecting electrode and the multizone method on the discharging electrode. The quality of the mesh was considered adequate based on an average value of the orthogonal quality at 0.78 and the skewness at 0.22. The simulation was based on the k-omega SST turbulence model. In terms of boundary conditions, the inlet velocity of 0.5 m/s was applied at the inlet as air. The outlet was defined as a pressure outlet with a gauge pressure of 0 Pa. Turbulence was defined as 5 % for turbulent backflow intensity and 10 for turbulent backflow turbulent viscosity ratio for both inlet and outlet. The boundaries of precipitators were defined as stationary walls with no-slip shear conditions.

The electric potential was applied by MHD module as conducting walls for collecting and discharging electrodes. 20 kV was considered applied voltage on the discharging electrode.

3 Results

The electric field intensity values of both solved electrodes were calculated and verified according to the mentioned procedure. The results of numerical calculations are recorded in Tab. 2.

Table 2. The calculated values of the individual quantities.

Quantity	$r_1 = 0.5$ mm		$r_2 = 2$ mm	
	Calculated value	**Units**	**Calculated value**	**Units**
E_{0c}	$7.37 \cdot 10^6$	$V.m^{-1}$	$5.24 \cdot 10^6$	$V.m^{-1}$
U_{0c}	$17.65 \cdot 10^3$	V	$35.63 \cdot 10^3$	V
z	$6 \cdot 10^{-4}$	m	$9.5 \cdot 10^{-4}$	m

CFD simulations confirmed that at the considered supply DC voltage, the corona effect will occur only on an electrode with a diameter of 1 mm. In Fig. 3. the 2D plane on which the field intensity was monitored is shown.

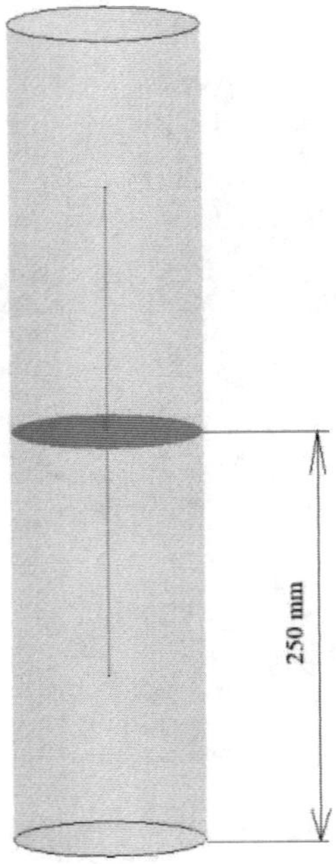

Fig. 3. Plane of measuring electric field intensity.

Fig. 4 and Fig. 5 show the distribution of the intensity of the electric field of individual electrodes. As expected, the distribution of the electric field strength is located close to the electrode and exceeds the required initial critical value of the electric field strength near the electrode surface, in the ionization region z. The intensity of the electric field in the case of an electrode with a diameter of 1 mm reached a maximum of $9.47 \cdot 10^6$ V·m^{-1}, which exceeds the value of the initial critical intensity, and a corona discharge will occur on the electrode at the given voltage. On the contrary, with an electrode with a diameter of 4 mm, there will be no corona discharge, because the value of the field intensity reached a maximum of $3.09 \cdot 10^6$ V·m^{-1}. To achieve the desired condition, a voltage higher than the calculated value of the initial critical voltage would have to be applied.

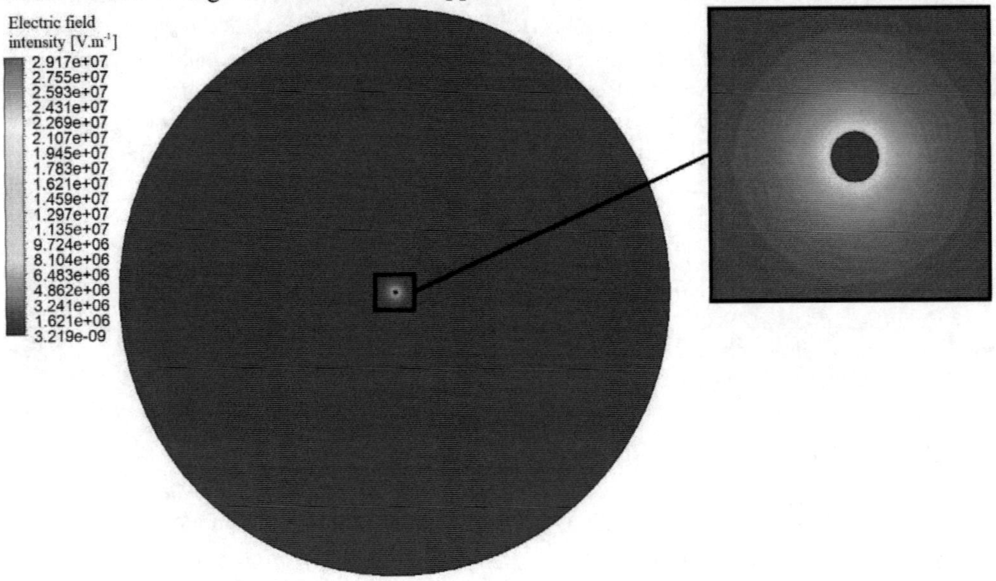

Fig. 4. Electric field intensity of 1 mm electrode.

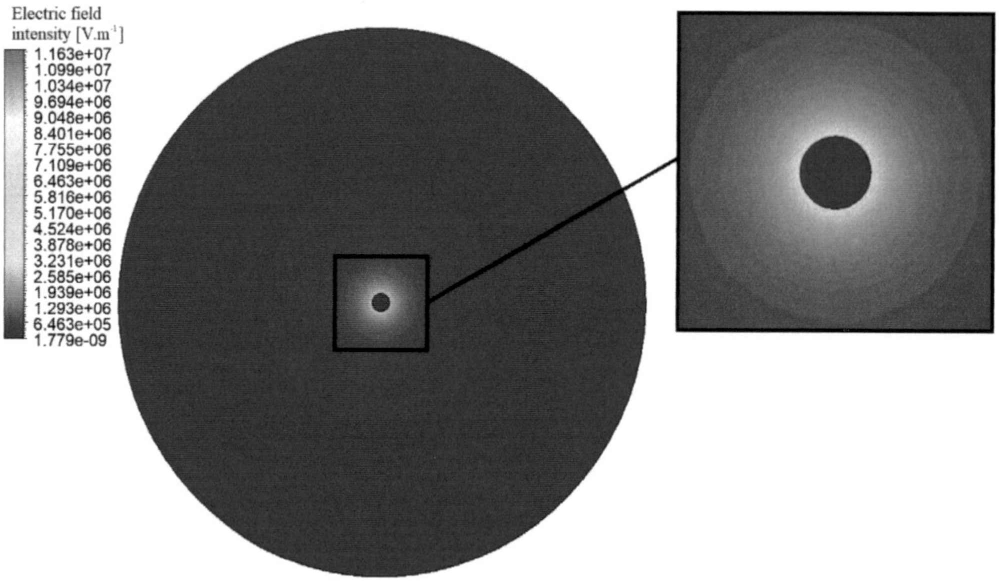

Fig. 5. Electric field intensity of 1 mm electrode.

4 Conclusion

The work was focused on determining the diameter of the charging electrode of the electrostatic precipitator using numerical calculation and numerical simulation. Two 3D models with electrode diameters of 1 and 4 mm were designed. In both cases, the supplied voltage of 20 kV was simulated, based on which the value of the critical intensity of the electric field was calculated. With the help of CFD simulations, it was verified whether in the given cases corona discharge will occur on the surfaces of the electrodes. The simulations confirmed the results of the calculations that the corona discharge will occur at a voltage of 20 kV only on an electrode with a diameter of 1 mm. The value of the critical intensity of the electric field was calculated at $7.37 \cdot 10^6$ V.m^{-1}. The calculated value of the electric field intensity using CFD simulation was 9.47·106 V·m-1. With a 4 mm diameter electrode, the applied voltage would have to exceed approximately 36 kV for a corona discharge to occur. Such numerical verification is an effective tool that can be used in the design of electrostatic precipitator electrodes for capturing solid pollutants in the stack. In this way, it is also possible to optimize the electrodes into different shapes, in which basic numerical calculations would be difficult to implement.

This work has been supported by the projects: KEGA 032ŽU-4/2022 Implementation of knowledge about modern ways of reducing environmental burden in the energy use of solid fuels and waste into the pedagogical process, VEGA 1/0233/19 "Construction modification of the burner for combustion of solid fuels in small heat sources" and APVV-21-0452 Impact of using small electrostatic precipitators to reduce particulate pollutants in domestic fuel combustion.

References

1. Schmatloch, V., Rauch, S. (2005). Design and characterisation of an electrostatic precipitator for small heating appliances. *Journal of Electrostatics*, 63, 85-100.
2. Dastoori, K., Holhe, M., Mallard, C., Makin, B. (2011). Electrostatic precipitation in a small scale wood combustion furnace. *Journal of Electrostatics*, 69, 466-472.

3. Brunner, T., Wuercher, G., Obernberger, I. (2018). 2-Year field operation monitoring of electrostatic precipitators for residential wood heating systems. *Biomass and Bioenergy*, 111, 278-287.
4. Carroll, J., Finnan, J. (2017). Use of electrostatic precipitators in small-scale biomass furnaces to reduce particulate emissions from a range of feedstocks. *Biosystems energineering*, 163, 94-102.
5. Yan, K., Li, S., Huang, Y., Zheng, Q., Deng, G. (2019). A numerical model for predicting particle collection efficiency of electrostatic precipitators. *Powder technology,* 347, 170-178.
6. Everson, R., Arif, S., Branken, D., Neomagus, H., Noras, M., le Grange, L., Arif, A. (2021). An experimentally validated computational model to predict the performance of a single-channel laboratory-scale electrostatic precipitator equipped with spiked and wire discharge electrodes. *Journal of electrostatics,* 112, 103595.
7. Gou, X., Zhang, Q., Zhang, J. (2017). Improvement of Corona Discharge Model and Its Application on Simulating Corona Discharge in the Presence of Wind. *Mathematical Problems in Engineering.* 2017, 10, 9856439.
8. Kasdi, A., Zebboudj, Y., Yala, H. (2015). Simulation of Corona Discharge in DC Electrostatic Precipitator. *Journal of Electrical Systems.* 11-4, 384-396.
9. Dong, M., Zhou, F., Zhang, Y., Shang, Y., Li, S. (2018). Numerical study on fine-particle charging and transport behaviour in electrostatic precipitators. *Powder technology,* 330, 210-218.
10. Dong, M., Zhou, F., Shang, Y., Li, S. (2019). Numerical study on electrohydrodynamic flow and fine-particle collection efficiency in a spike electrode-plate electrostatic precipitator. *Powder technology,* 351, 71-83.
11. Zhao, L., Knight, R., Zhu, H. (2021). Modeling and optimisation of a wire-plate ESP for mitigation of poultry PM emission using COMSOL. *Biosystems energineering*, 211, 35-49.
12. Branken, D., Everson, R., Arif, S., Neomagus, H., Noras, M., le Grange, L., Arif, A. (2016). CFD modeling of particle charging and collection in electrostatic precipitators. *Journal of Electrostatics.* 84, 10-22.
13. Böhm, J. (1977). *Elektrické odlučovače.* SNTL. Nakladatelství technické literatúry Praha.

Measurement of Fluid Flow with Different Velocity Profiles by New Built Ultrasonic Flowmeter

Veronika Barraclough[1,*], *Miroslav* Strob[1], *Jaroslav* Pulec[2], *Jan* Novosád[2], *Petra* Dančová[2], and *Jan* Čížek[1]

[1]4Jtech Ltd., Ringhofferova 115/1, 15521 Prague, Czech Republic
[2]Technical University in Liberec, Faculty of Mechanical Engineering, Department of Energy Equipment, Studentská 1402/2, 46117 Liberec, Czech Republic

Abstract. The paper describes the development of a non-invasive flowmeter for lower flow rates and its first tests. This gauge is physically based on the interaction of fluid flow with an ultrasonic signal that passes through the fluid from the transmitter to the receiver. Ultrasonic flowmeters are currently relatively commonly used gauges, whose advantages such as non-invasiveness (i.e. zero pressure losses) and the ability to seamlessly measure the flow rates of any (for example opaque) liquids, without contact with the liquid, are widely known. However, there are still parts of the ultrasonic flowmeter measurement chain that are undergoing research and development. It can be signal processing itself (mainly), its design solution, measurement for different flow cases (measurement in a flow field with a uniform velocity profile, in a flow field with an axisymmetric velocity profile, in a flow field with a general velocity profile), validation of the applied signal processing approaches, evaluation of uncertainties. The flowmeter itself, which development is described in the paper, will be used for trouble-free measurement in air engineering, but also serves as a training device for building a more complex ultrasonic gauge. Therefore, this flowmeter contains more signal transmitters and receivers than it is usual and all transmitter-receiver combinations are captured during the measurement. This gauge is called ultrasonic tomograph and its principle is also outlined in the paper. Here, so far, without a reconstructed vector field.

Keywords: *experimental methods, mechanics of fluids, ultrasonic flowmeter, ultrasonic tomograph*

1 Introduction

This paper deals with the development of two gauges for use in fluid mechanics, an ultrasonic flowmeter and an ultrasonic tomograph. These gauges are based on non-contact measuring principle in which ultrasonic transmitters and receivers are used as sensors, and the sound signal travelling through the flowing medium and the interaction of both of these moving

* Corresponding author: veronika.barraclough@4Jtech.cz

entities is considered for measurement. The measured quantity is the time delay of the travelling signal. From this quantity, velocity/ flow rate is evaluated in case of flowmeter. In case of the tomograph, the signal processing is considerably more complex and includes, among other quantities, the determination of the angle of incidence of the signal, while only the same transmitters and receivers are available as for flowmeter.

The very nature of these flowmeters has been known for decades and used with success; there is a whole range of solutions for these gauges, one can purchase meters from indicative to highly accurate, with the possibility of completely non-contact measurement, based on various principles. Their fifty years of history overview is clearly described in paper by Lynnworth, 2006 [1]; together with the article by Furness, 1991 [2], who summarizes all available flowmeters and their areas of application, these two papers form an introduction to the issue of flow measurement. Today, the development of this gauge is directed towards special areas of fluid mechanics, such as cryogenic applications or the area of compressible flow (Otero, 2019 [3]).

A more advanced application of the principle of interaction of the ultrasonic signal and the flowing medium is called the ultrasonic tomograph.

The principle of measuring the 3D velocity field using an ultrasound tomograph has also been known for quite a long time (see Teerawatanachai, 1991 [4], Komiya, 1993 [5]), but its implementation is not very frequent. From the beginning, the limiting factor for complex development was apparently the fact that it was believed that ultrasonic transmitters and receivers needed to be supplemented with yet another gauge (Schlieren technique; Braun and Hauck, 1991 [6], Johnson et al, 1977 [7]) and all the theory remained rather on paper. Nowadays, when, for example, Jovanovic showed in her dissertation in 2008 [8] that she could reconstruct a 3D velocity field from only known velocities and ultrasound signals, the development of new - and thus spatially dimensional - experimental methods in areas of fluid mechanics are probably outside the area of interest of the respective institutions, or beyond their capacity.

1.1 Ultrasonic flowmeter and tomograph development

The ultrasonic gauge development, as it was conceived for the project described in this paper, involved following tasks: A) selecting and testing ultrasonic transmitters and receivers, while several aspects had to be taken into account in the selection, like their design (small dimensions are more suitable for intended applications, but one cannot forget a reach considering), physics (frequency characteristics, directional characteristics and wave range), microelectronics (control and power supply). B) The next step was the construction of gauges (to choose suitable production technology, especially with regard to the exact geometry demanded) and C) building the whole electronics using the simplest and most affordable components.

One of the main tested attributes of the flowmeter being developed is the invariance to the velocity profile of the measured fluid. At the same time, the whole system is designed as variable as it can be used on pipes of any cross-section (circular, rectangular, general shape) in an effort to build in as little space as possible, i.e. to construct a gauge with the shortest possible length.

At the moment, the development of the flowmeter has completed the structural part and the assembled electronics and also few validation measurements have been provided. (Simultaneously with the construction of the gauge, an experimental line was also built, on which its parameters were and are being verified. The test line is fitted with a standard flowmeter and, as necessary, additional verification measurements are made using other experimental methods.)

The tomograph development is a logical step that other research teams who are building their own ultrasonic flowmeter are also trying to achieve as far as possible. It means to take the data from flow measurements and to reconstruct a 2D vector velocity field using these data. It does not appear to be a complicated problem from a design point of view. If a pipe of circular cross-section and a configuration where two rings of transmitters and receivers are placed around the pipe are taken into account, then the non-axial components of the flow velocity vector in the channel can be obtained by such arranged measurements. For example, this is roughly how the team Liu 2015 [9] proceeded.

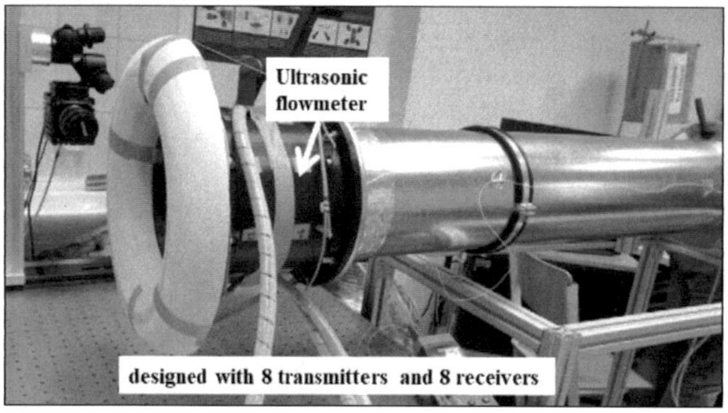

Fig. 1. Ultrasonic flowmeter designed for testing a 2D tomograph principle.

2 Measurement setup

The task of the project, as already mentioned in the introduction, was to build an ultrasonic flowmeter used (and validated) especially for air flows of the order of 1 m^3/s or lower units of m^3/s. At the same time, it was necessary to determine the overall accuracy of the gauge for a uniform, axisymmetrically deformed and generally deformed velocity profile.

2.1 Ultrasonic flowmeter Prototype

The ultrasonic flowmeter will be used mainly for measuring air substances, from the point of view of fluid dynamics in the area of incompressible flow. The temperature of the flowing medium is assumed to be within the range usual for our climate zone. Due to the expected use of the gauge (for example, in air engineering), the main requirement for production was the variability of the geometry; the pipe can have a circular, rectangular or general cross-section. The velocity profile at the point of measurement can be laminar or turbulent, but it can also have a completely general, deformed shape, which usually affects the uncertainty of the measurement. The proposed device is designed as a contact device, i.e. transmitters and receivers will be placed directly in the pipeline as a part of it and come into contact with the flowing medium.

2.1.1 Construction, electronics and data processing proposal

Taking into account the variability of production while maintaining the high accuracy of the geometry, 3D printing was chosen. A circular cross-section test piece was designed with a diameter of 200 mm. The transmitters and receivers are distributed around the perimeter evenly in two concentric circles so that the distance between the centers of the sensors' cavities is 200 mm. The axes of the cavities in the case of this product make an angle of 45 degrees with the axis of the pipe. A fundamental part of the design of microelectronics is the selection of suitable transmitters/receivers. The requirements for sensors for the intended use (and especially for the proposed test prototype) are contradictory: on the one hand, they should be subtle products with a low supply voltage, on the other hand, ideal transmitters/receivers should work at frequencies higher than 100 kHz, so that the frequency of the sensors moves above the frequency of possible non-stationarities in the flow. Finally, sensors designated MCUST and MCUSR working on frequency 40 kHz, suitably small and having a good noise property (and, also available and expected to be available in the future, which is an aspect that also needs to be taken into account at present) were selected.

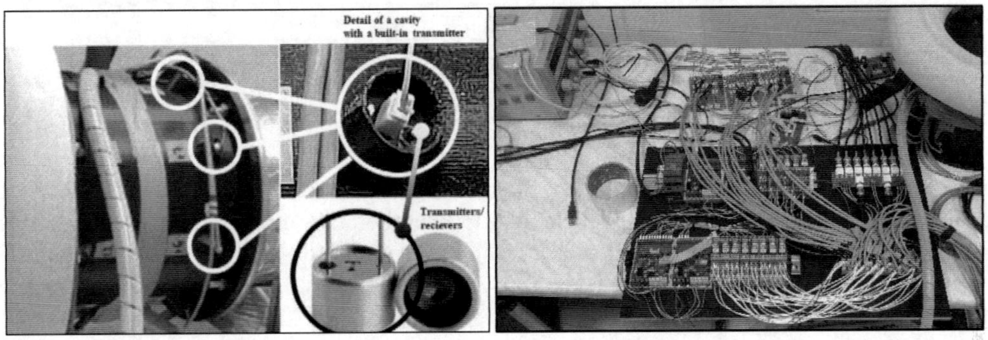

Fig. 2. A) Detail of the installation of transmitters/receivers of the gauge. B) Control and switching unit of the gauge.

For the construction of the control unit, emphasis was once again placed on the simplest possible instrumentation, composed of easily available, common and possibly replaceable components. The control unit provides the operation of excitation and reception of ultrasonic signals, high-speed sampling of these signals, their temporary buffering into the buffer and transmission for further processing.

Conceptually, there are three different approaches to ultrasonic flowmeter measurement and signal processing: Doppler shift, direct time-of-flight measurement, and cross-correlation. For applications under consideration for Prototype flowmeter, only time-of-flight measurements and cross-correlation are considered (and appropriate). Determining the exact time-of-flight directly is quite a difficult task. Fig. 3 shows an example of a raw outgoing/ incoming signal from the testing measurement of the gauge Prototype. A crucial problem of this task can be seen there: to accurately find out the origin of either the transmitting or the received signal is necessary (in general, this is a separate topic for research on its own; the exact determination of the origin of the reception of an ultrasound signal is solved, for example, by Fang et al., 2017 [10]). In addition to determine the beginning of the signal on the receiver side, it is also necessary to find the exact moment of excitation on the transmitter side, which exact outset is apparently emerged in noise.

For the reasons mentioned above, the method of mutual comparison of signals was finally chosen, therefore the cross-correlation method. Again, the application of this method may vary: the cross-corelation can be performed either in a frequency domain (Sullivan and Wright, 2002 [11] have shown the flow velocity-dependent frequency shift for the receivers working on the frequency reaching 1Mhz) or a time domain ([12]).

For the Prototype, the time domain was selected.

Basic schema for measurement is depicted in Fig. 3.

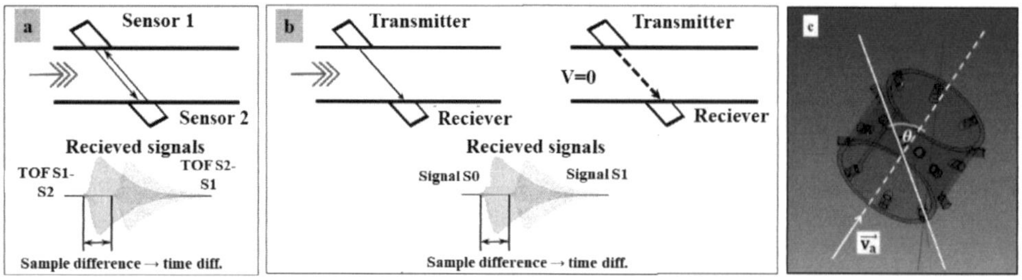

Fig. 3. Measurement setup schema and geometry: a) usual type, b) cross-corelation proceeds in a manner that only one signal is taken under flow condition and is corelated with no-flow measurement. C) Geometry of the gauge included in the velocity and flow rate calculation.

During the testing of the Prototype, one change was made in the usual cross-correlation procedure against the scheme in Fig. 3a., where the signals measured up the stream and down are correlated; this arrangement was replaced (apparently temporarily) by the correlation of a single signal measured with flow and a signal without flow in the pipe. This is a conceptually less convenient setup, since it is not always possible to immediately measure the regime with zero flow velocity in the pipe. However, the first tests of the Prototype showed that the signals measured downstream and upstream are every now and then difficult to correlate. Fig. 3b shows the diagram that is used in the measurement of the Prototype.

The time-of-flight of the signal between the transmitter and the receiver is expressed by a curve integral:

$$\tau = \int_{\Gamma} \frac{1}{\mathbf{u_g}} d\mathbf{s} = \int_{\Gamma} \frac{1}{(\mathbf{cn} + \mathbf{v}).\mathbf{s}} d\mathbf{s} \tag{1}$$

where "." denotes the scalar product, τ is time-of flight, Γ is sonic ray trajectory, $\mathbf{u_g}$ is the sonic wave group velocity, $\mathbf{s}$ is vector tangent to the Γ trajectory, $\mathbf{c}$ is speed of sound, $\mathbf{n}$ is vector normal to the wave front, $\mathbf{v}$ is the flow velocity.

The simplified equation (1) applied to the flowmeter Prototype leads to a determining the axial velocity as:

$$\mathbf{v_a} = \frac{\mathbf{c} - \frac{L}{\Delta\tau + \frac{L}{c}}}{\cos\theta} \tag{2}$$

where $\mathbf{v_a}$ is the axial velocity, L is the distance between transmitter and receiver, θ is the tilt angle of transmitters/receivers (the angle between their axis and the pipe axis) and $\Delta\tau = \tau_1 - \tau_0$ the time difference between Prototype downstream and no-flow downstream soundwave "time-of-flight".

As it was mentioned, the cross-correlation of the signals is proceeded in the time domain. Time delay of two signals is determined as the maximum of the correlation function (eq. 3) of these signals, computed with help of Fourier transform (eq. 4):

$$w(t) = u(t) \otimes v(t) \tag{3}$$

$$w(t) = F^{-1}(F(u(t))F^{*}(v(t))) \tag{4}$$

The result of the cross-correlation is number of samples converted to the time, and together with the geometry of the gauge and the speed of sound (evaluated from the temperature) this quantity fully determines the flow rate.

2.2 Ultrasonic tomograph proposal

The ultrasonic flowmeter was designed with emphasis on the invariance to the measured velocity profile. One of the design aspects that helped this feature is the fact that the flowmeter can work as 8 independent flowmeters. At the same time, this design allows the flowmeter to be used as a tomograph, where the non-axial components of the velocity vector are evaluated. The principle of this tomograph is based – again – on the time delay measurement. It requires the assembly of tripoles - see diagram in Fig. 4 - and determining the time delay between transmitter and receivers R0, R1 and R2, which corresponds to the angle between the velocity vector and the normal of the incident sound wave vector. From the known distance between the sensors, from the detected time-of-flight in the tripole configuration, their differences, and from the detected TOF between T0-R0, the velocity field can be reconstructed via inverse problem.

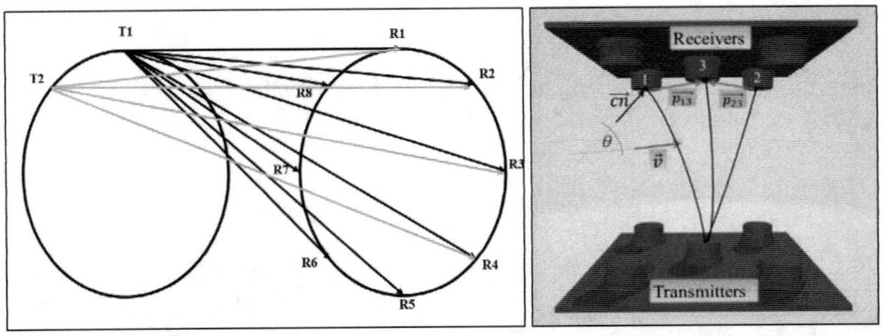

Fig. 4. Scheme of the system of transmitting and receiving the ultrasonic signal in the tomograph.

3 Measurement results

Measurements have been performed and evaluated so far for a uniform, an axisymmetric and an uneven velocity profile, plus for different humidity values in the laboratory. These configurations were achieved by inserting elements to deform the velocity profile at the beginning of the gauge, just after the lead. Their sketch for an axisymmetric and non-uniform velocity profile is marked in the corresponding graph. The following graphs in Figs. 5, 6 & 7 show and compare the most important data: calculated velocity for different velocity profiles and calculated flow rates from anemometer and from ultrasonic flowmeter measurements in the comparison (ideally, the corresponding values should meet on the y=x line). For the ultrasonic flowmeter, one sample value is taken from one transmitter and receiver for two cases (in the configuration in which the gauge is designed, it actually works as 8 independent flowmeters, since theoretically only 1 transmitter and 1 receiver are sufficient). Other values are calculated as averages from all 8 transmitter-receiver pairs.

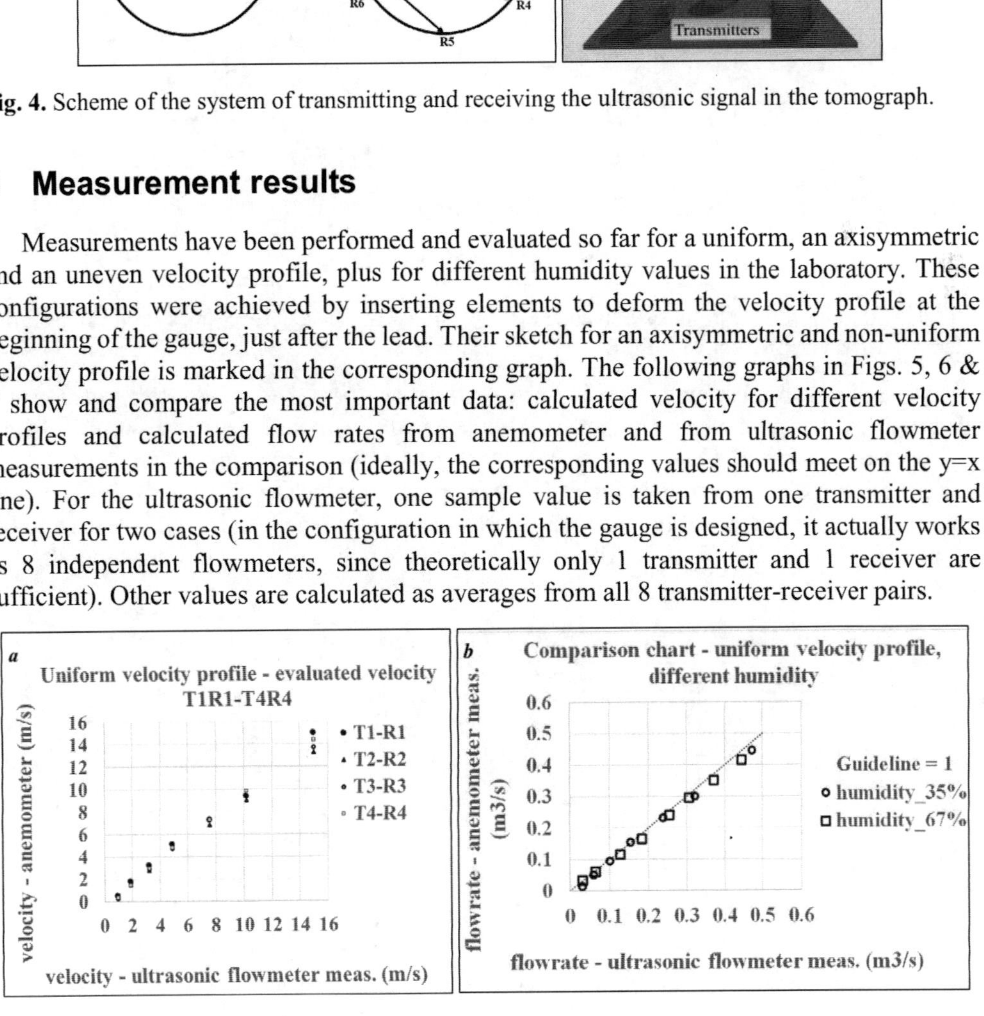

Fig. 5. (a) Ultrasonic velocity measurement in comparison with the reference measurement from the anemometer, uniform velocity profile. **(b)** Evaluated flow rate in comparison with the reference measurement, ultrasonic gauge used in different humidity.

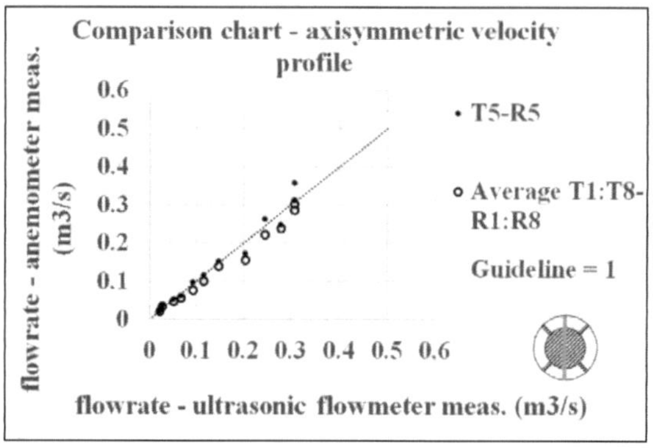

Fig. 6. Ultrasonic flow rate measurement in comparison with the reference measurement from the anemometer, axisymmetric velocity profile.

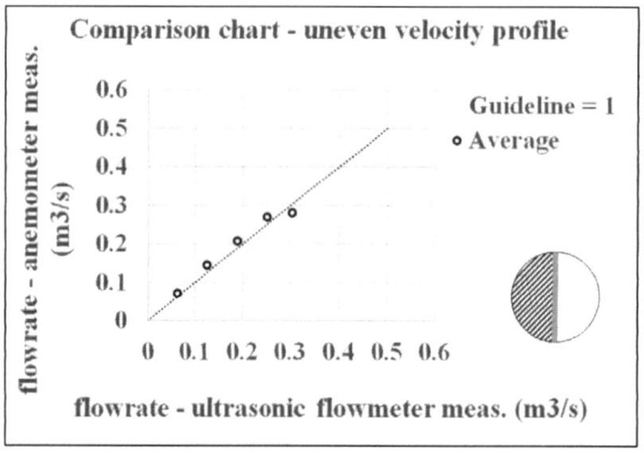

Fig. 7. Ultrasonic flow rate measurement in comparison with the reference measurement from the anemometer, uneven velocity profile.

As an almost perfect match for the measured values from the anemometer and from the ultrasonic gauge was expected (and was not achieved), more detailed statistics was evaluated as well. For future experiments, more than 20 signals from one transmitter-receiver pair are to be taken to get more accurate results. The course of the calculated values of the velocity (and flow rate) for an increasing number of measurements can be seen in Fig. 8. For a number of measurements higher than 20, the average value changes only in the order of tenths of a percent.

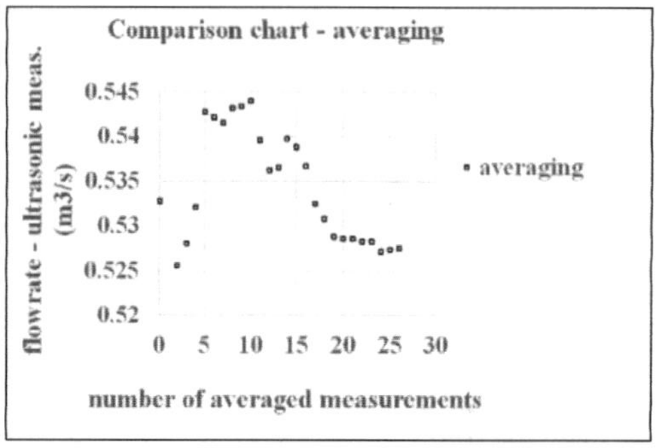

Fig. 8. Ultrasonic flow rate measurement – averaging of T1-R1 signal, uniform velocity profile.

For 2D ultrasonic tomograph, the first results were provided so far; their results were supposed to show, whether the positions of the sensors are appropriate or not; especially those where the transmitter-receiver pair is located next to each other. Whether the receiver could hear the corresponding transmitter. The results are depicted in the Fig. 9.

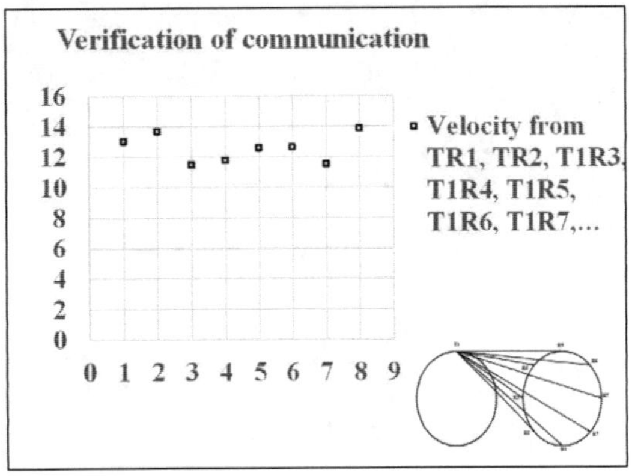

Fig. 9. Tomograph first tests.

4 The uncertainty analysis

The uncertainty analysis of the axial velocity in the pipe is based on default equation (2).

The speed of sound uncertainty is calculated as a standard deviation of a uniform distribution of a temperature gauge. The determination of the temperature by the reference thermometer is with an error of ±0.5 °C, which makes the width of the interval, i.e. the variance will appear within the limit Δ = 1°.

The uncertainties of measuring the geometry of the flowmeter were estimated resulting from the meters and their capabilities; values are quantified in Table 1.

The uncertainty of Δτ was determined as the uncertainty of type A of repeated measurements and combined with autocorrelator uncertainty of B-type. In the autocorrelator, an error uniform over an interval of ± 10 samples was assumed.

For the overall determination of the velocity uncertainty, the partial derivation of Eq. 2 according to all quantities and the uncertainties u_B and u_B is used.

Table 1. Overview of uncertainties.

Quantity		Uncertainty calculated as	Value
Speed of sound		$u_B(\mathbf{c}) = \sqrt{\left(\frac{\partial f_c}{\partial T}\right)^2 \cdot \left(u_B(T)\right)^2} = \sqrt{\left(\frac{8.3}{\sqrt{T}}\right)^2 \cdot \frac{\Delta}{\sqrt{12}}}$	$0.04\frac{m}{s}$
Geometry	angle	$u_B(L) = \frac{\Delta}{\sqrt{12}}$	0.14 mm
	dimensions	$u_B(\Theta) = \frac{\Delta}{\sqrt{12}}$	0.29 °
Time	autocorrelator	$u_B(\Delta\tau) = \frac{\Delta}{\sqrt{12}} = \frac{20}{100.10^6.\sqrt{12}}$	57 ns

	from repeated measurement, uncert. of A type evaluated	$u_A(\Delta\tau) = \sqrt{\frac{\Sigma(xi-\bar{x})^2}{N.(N-1)}}$	$45\ ns$
	combined – C uncertainty	$u_c(\Delta\tau) = \sqrt{u_A(\Delta\tau)^2 + u_B{}^2(\Delta\tau)}$	$72.6\ ns$
Velocity (calculated for velocity 7,6 $\frac{m}{s}$)	Combined – C uncertainty evaluated for uniform velocity profile, for velocity 7,6 m/s	$u_c(v_a) =$ $= \sqrt{\left(\frac{\partial f_{va}}{\partial c}\right)^2 \cdot [u_B(c)]^2 + \ldots + \left(\frac{\partial f_{va}}{\partial \Delta\tau}\right)^2 \cdot [u_B(\Delta\tau)]^2} =$ $= \sqrt{\left\{\frac{1}{\cos\Theta}\cdot\left(1-\frac{L\cdot(\Delta\tau\cdot c+L)-L+c}{(\Delta\tau\cdot c+L)^2}\right)\right\}^2 \cdot [u_B(c)]^2}$ $+\cdots+\frac{-L\cdot c}{\cos\Theta\cdot\left(\Delta\tau+\frac{L}{c}\right)^2}$	$0.076\frac{m}{s}$

5 Conclusions

This paper introduced the development and testing of the ultrasonic flowmeter and the further aiming of the project to the ultrasonic tomograph. The flowmeter was conceived as the invariant to the flow field measured, geometrically variable, available in terms of electronic components, designed for lower flow rates and non-compressible flow. Together with the gauge Prototype the experimental line was built. The components for changing the flow velocity profile, turbulence rate, humidity were designed and the measurements with changing all these parameters were provided. The paper contains the first test results for the ultrasonic flowmeter for different velocity profiles. For the uniform velocity profile, the reference anemometer and the Prototype differed about maximum 6 %. For the axisymmetric velocity profile, above a speed of 4 m/s, the reference and Prototype showed an acceptable agreement with a difference of max. 6 %, for lower flows the difference was close to 10 %. The measurement with the uneven velocity profile ended up with the similar result. The dependence of the measurement on humidity was not proven. Unfortunately, the results from measurements with a high degree of turbulence are still confusing. On the contrary, measurements to determine the suitability of the construction geometry proved that the tomograph might be operated in this configuration. The next logical step will be to model what resolution and what uncertainty of vector field determination can be achieved with a given number of transmitters and receivers.

Acknowledgement to Technology Agency of the Czech Republic for supporting FW02020070 project.

References

1. Lynnworth, L.C., and Liu, Y. (2006). Ultrasonic flowmeters: Half-century progress report, 1955-2005. *Ultrasonic.* Vol. 44, pp. 1371-1378.
2. Furness, R.A. (1991). BS 7405: The principles of flowmeter selection. *Flow measurement and Instrumentation.* Vol. 2, pp. 233-241.

3. Otero R. Jr et al. (2019). Coupled velocity and temperature acoustic tomography in heated high subsonic Mach number flows. *Measurement Science and Technology.* Vol. 30 105901.
4. Teerawatanachai, S., Komiya, K., and Sasamoto, H. (1991). Estimating the velocity profile of air flow by means of the ultrasonic translation time computed tomography method. *Proceedings IECON '91: 1991 International Conference on Industrial Electronics, Control and Instrumentation*, pp. 2379-2384, vol.3, doi: 10.1109/IECON.1991.238972.
5. Komiya, K., Teerawatanachai, S. (1993). Ultrasonic tomography for visualizing the velocity profile of air flow. *Flow Measurement and Instrumentation.* Vol. 4, Issue 2, pp. 61-65.
6. Braun, H., Hauck, A. (1991). Tomographic reconstruction of vector fields. *IEEE Transactions on Signal Proceeding*, vol. **39**, no. 2.
7. Johnson, S. A., Greenleaf, J. F., Hansen, C. R., Samayoa, W. F., Tanaka, M., Christensen, D. A., and Wooley, R. L. (1977). Reconstructing threedimensional fluid velocity vector filds from acoustic transmission measurements. *Acoustical Holography.* L. W. Kessler, ed. New York: Plenum, vol. 7, pp. 307–326.
8. Jovanović, I. (2008). *Inverse Problems in Acoustic Tomography: Theory and Applications.* Doctoral thesis, Ecole Polytechnique de Lausanne, Switzerland.
9. Liu, J.-N., Wang, B.-X., Cui, Y.-Y., Wang, H.-Y. (2015). Ultrasonic tomographic velocimeter for visualization of axial flow fields in pipes. *Flow Measurement and Instrumentation*, Volume 41, pp. 57-66.
10. Fang, Z., Hu, L., Qin, L., Mao, K., Chen, W., and Fu, X. (2017). Estimation of ultrasonic signal onset for flow measurement. *Flow Measurement and Instrumentation.* Vol. 55, pp.1-12.
11. Sullivan, I.J., Wright, W.M.D. (2002). Ultrasonic measurement of gas flow using electrostatic transducers. *Ultrasonics.* Vol. 40, pp. 407-411.
12. Nguyen, T. H. L. and Park, S. (2020). Multi-angle liquid flow measurement using ultrasonic linear array transducer. *Sensors.* vol. 20, no. 2, pp. 388.

Flow and Vibration in the Small Steam Turbine Last Stage

Michal Hoznedl [1,*], *Tereza* Dadáková[2], and *Jindřich* Bém[1]

[1]Doosan Škoda Power s.r.o., Experimental Research Dpt., 30100, Tylova 1/57, Plzeň, Czech Republic
[2]Doosan Škoda Power s.r.o., Measurement & Diagnostics Dpt., 30100, Tylova 1/57, Plzeň, Czech Republic

Abstract. The paper deals with evaluation of experimental data obtained from the real-size 30 MW output Waste to Energy steam turbine. For many turbine regimes pressures in the last stage area were obtained. At the same time amplitudes of blade tips vibrations were measured of last rotor blades using the blade tip timing system and axial velocity of the steam at the last stage outlet. The obtained data are mutually correlated and relations are analysed between increased vibration values and pressure ratios at the last stage. Increased tip vibrations occur mainly in the area when the pressure ratio over the last blade tip and root is higher than one. In this case there is no expansion, but compression in the last stage of the turbine. The presented results contribute to understanding of processes in the last stage of the steam turbine, mainly for those regimes of turbine operation when its output is lower than 20 % of the nominal output, e.g. during the island regime.

Keywords: *steam turbine, last stage, low load regime, vibration, experiment*

1 Introduction

Due to changes in the energy mix steam turbines often serve as sources of electric energy in industrial applications or in biomass or municipal waste incineration plants: the so called „Waste to Energy – WtE“. It is also valid for the turbine described in this article. Operation of WtE turbines is specific by variable input and output steam parameters. At the turbine inlet mainly steam temperature changes by tens of °C depending on the burnt waste. As a result, at the turbine outlet the steam humidity and, in combination with frequently used air-cooled condensers, also the steam pressure changes. During the day even significant changes of the output occur. The turbine output can, in the long run, be at the level of 15 % or less of the nominal output value.

For the above described reasons the flow of water steam in the last stages and exhaust hoods of small steam turbines is still a current and evolving topic. When designing the last blades, great emphasis is placed on the strength and dynamic design of the last rotor blades. [1-3]. For turbine designers it is important to define well the output, and for low volume conditions even the input of last stages [4-6] using numerical or sometimes even experimental methods. In recent articles [7, 8] aerodynamic instability of the flow has been discussed in the turbine last stages. In this case the flow is solved using demanding DES numerical

* Corresponding author: michal.hoznedl@doosan.com

methods. The paper follows a series of works carried out on first stages of turbo compressors with the rotating stall phenomenon. In [9] the outlet part of the steam turbine is examined for low outputs using experiments and non-stationary numerical calculations.

In the ventilation regime compression heat is generated in the last stage area. In this case under certain circumstances it is necessary to cool the last blade using waterjets located on the outer wall of the axial exhaust hood. The influence of cooling water on the flow at the turbine outlet is examined in detail in [10]. Excessive erosion of leading edges often occurs for ventilation regimes. Due to cooling of steam flow with water jets there is also a risk of erosion of trailing edges. The erosion of the rotor blades is dealt with by authors of article [11].

The turbine blades of last stage which are currently being developed are subject to fundamentally different requirements than the blades developed in the past. Current blades must be mechanically robust and quality anti-erosion protection must be addressed while maintaining a low price and good efficiency. A summary of goals of the current blades development is given in article [12].

However, none of the mentioned articles presents the dynamic behaviour of the blade and aerodynamics of the steam flow on the real (i.e. not experimental) turbine for nominal and low regimes. The submitted paper aims to appropriately supplement the information published so far and discuss the conclusions.

2 Experiment description

2.1 Steam turbine description

Experiments were carried out on a 30 MW nominal output Škoda steam turbine. During the experiments the turbine output was changed in the range of 1 to 30 MW, where the power of 1 MW means island operation of the turbine. In this case the turbine produced energy only for its own waste incinerator. During the measurement, temperature at the turbine inlet and outlet pressure continued to change. The last stage has root diameter 0.928 m, the length of last blade is 0.578 m. Circumferential velocity at the mean blading diameter is 433.7 m/s. The nominal outlet pressure is 6.1 kPa and steam dryness then 0.92.

2.2 Static pressure measurement

Static pressures (index „p“) were measured using wall samples at the root (index „r“) and tip (index „t“) limiting wall. All samples were doubled. It means that pressures were measured at two circumferentially different places on the left (index „l“) and right (index „r“) side of the machine, always15° above the dividing plane. Altogether 12 static pressures were measured before (plane 0) and behind (plane 1) the stator blades and behind rotor blades (plane 2) of the last stage. Average static pressure sample was 2 mm. The reason for the relatively large sample diameter was the fear of clogging the sample with impurities. Impulse pipes were automatically blown every 10 minutes for 5 seconds with atmospheric air. The aim of blowing was to remove possible condensed water steam from pressure pipes. Pressure measurements were done using Netscanner 9022 with pressure sensors Netscanner 9401 in the scope 0-15 psi(a). Combined expanded uncertainty of the measurement for 2σ reached values ±110 Pa at the pressure level of 5 000 Pa. Data were scanned every minute.

2.3 Temperature measurement

In the last stage area altogether ten temperatures were measured at the same positions as for the static pressure samples. For construction reasons only temperatures t_{2rl} and t_{2rr} were not measured, i.e. both temperatures at the root behind the last rotor blade. For temperature measurement flexible K type thermocouples were used with the highest accuracy class. Even for temperatures data were read every minute using cDAQ system from NI. The combined expanded measurement uncertainty was ± 0.4 °C. The last stage scheme including the measurement positions of all pressures and temperatures is shown in Fig. 1.

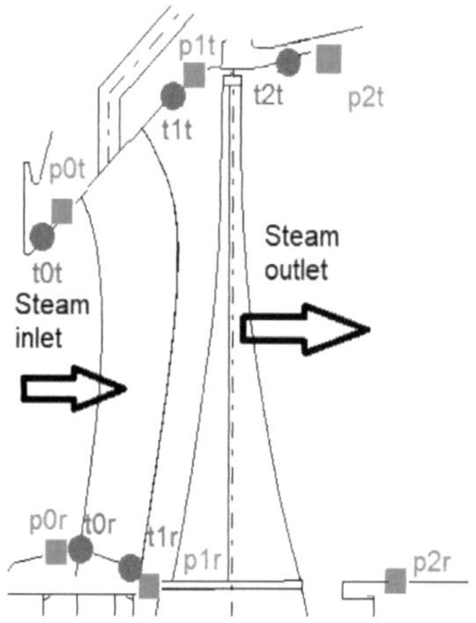

Fig. 1. Last stage.

2.4 Measurement of blade tip timing (BTT)

Measurement of amplitudes of blade tips vibrations was done using in-house developed eddy-current sensors. Data collection was organized using HW from HoodTech Corporation based on NI solutions. Data were processed using special in-house SW that can work with under-sampled data and multiple non-equidistant space sensors. For this measurement altogether 8 sensors were used located in two groups of 3 and 5 sensors. The larger group was placed above the trailing edges and the group with 3 sensors above the leading edges. Two axial positions were chosen for better accuracy of the measurements of vibration amplitudes.

Evaluating in-house SW provides amplitudes of all nodal diameters for each monitored blade. BTT measurement was carried out at the same time as pressure and temperature measurements, that is in a variety of operating regimes of the turbine.

3 Results and discussion

3.1 Nominal and ventilation regime

Experimentally obtained data of vibration amplitudes are evaluated dependent on pressures or temperatures ratio over the root and tip of the rotor blade or the whole stage. At the same time data from BTT are evaluated dependent on the ventilation coefficient φ, when:

$$\varphi = c_{2x} / u \tag{1}$$

In relation (1) c_{2x} is the axial velocity at the last stage outlet and is calculated from the volume flow of the stage. Quantity u is the circumferential velocity at the mean blading diameter (u = 433.7 m/s). The last turbine stage gets to the ventilation regime when $\varphi_{lim} \cong 0.17$. In this case the output of the last stage is zero. For $\varphi < \varphi_{lim}$ the output of the last stage is negative. The stage, in certain cases even several last stages, consume energy generated by the previous stages. If $\varphi \rightarrow 0$, the machine works with minimal output, it is the island operation. In the ventilation regime of the turbine ventilation coefficient φ must be defined not only for the last stage, but also for the previous stages. However, the value of φ_{lim} is different for each stage. An example of two expansion lines of the last five stages in the nominal and ventilation regime is shown in Fig. 2. For the nominal regime (Fig. 2a) it is valid that $\varphi \cong 0.45$, while for a certain ventilation regime (Fig. 2b) $\varphi \cong 0.15$. In the ventilation regime there is a significant compression at the last rotor blade row accompanied by increased entropy and thus also temperature in the area behind the last stage.

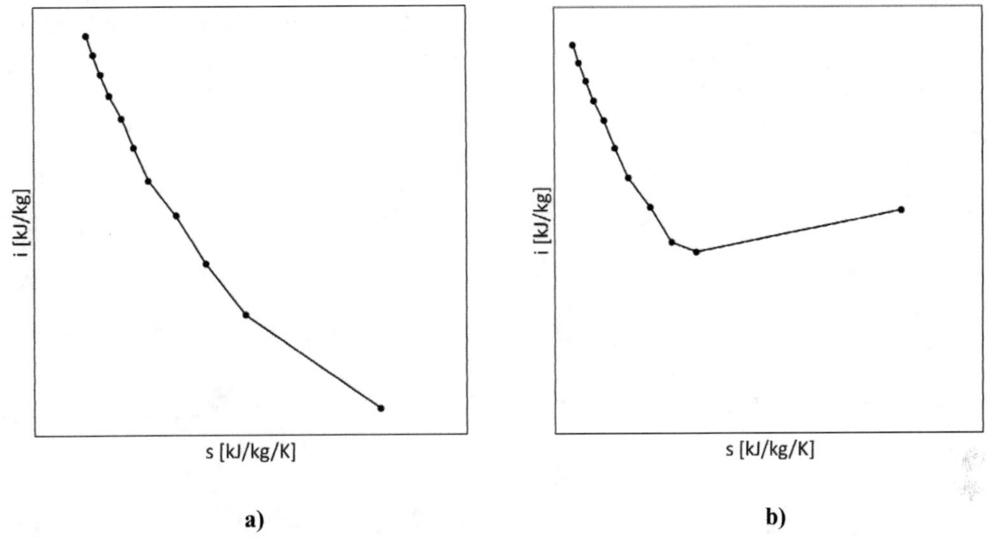

Fig. 2. Expansion line of last five stages in a) nominal and in b) ventilation regime.

For both nominal and ventilation regime even the flow field in the exhaust hood differs, see Fig. 3. For the nominal regime the exhaust hood is evenly filled with steam flow and velocities at the last stage outlet reach the values about 180 to 250 m/s. In contrast, for the ventilation regime at the tip limiting wall the velocity is high including its circumferential component, while at the root cross-section the steam flow returns to the blading and by the root limiting wall there is a significant separation of the steam flow from the wall. The axial velocity is obviously low.

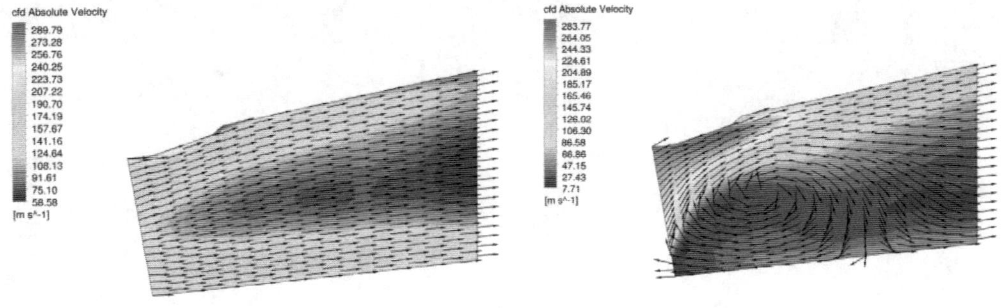

a) b)

Fig. 3. Flow field in the exhaust hood for a) nominal and for b) ventilation regime.

3.2 Flow ratios and vibrations

Pressure ratios at the root of the last stage rotor blade are shown in Fig. 4. Both curves have a similar character. For $\varphi \geq 0.35$ there is no further decrease in the pressure ratio. It stays at value $p_{2r} / p_{1r} = 0.5$, the blade is aerodynamically clogged even in the root area. Rotor blades vibrations are for $\varphi \geq 0.35$ at the minimum as well as in the interval $0.1 \leq \varphi \leq 0.35$. In this interval the blade is gradually loaded, the pressure ratio decreases with increasing φ. Roughly at level $\varphi = 0.15$ the negative pressure ratio changes to positive. The steam flow starts to suck through the root cross-section towards the blading. At the same time amplitudes of blade vibrations start to increase. For $\varphi < 0.1$ pressure ratio further increases up to the value of 1.1. The side differences in the pressure ratio are beginning to show, given mainly by an uneven steam inflow to pressure samples and by measuring the partial dynamic component of the pressure. For $\varphi = 0.03$ significantly higher amplitudes of blade vibrations suddenly occur.

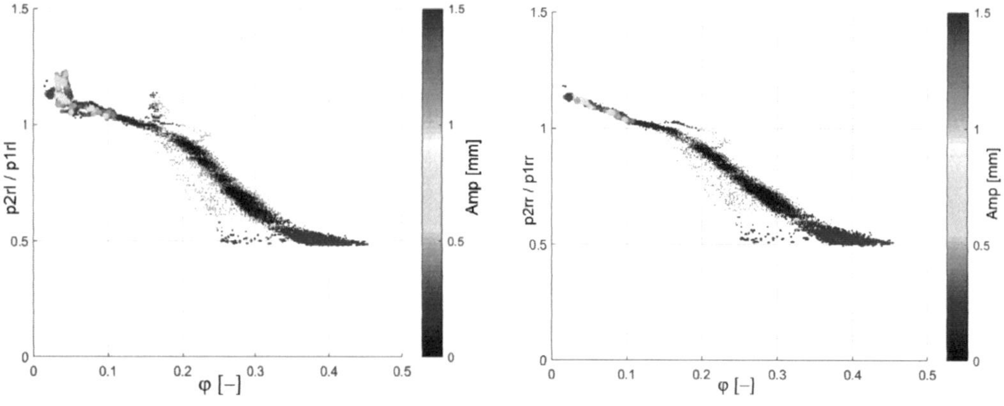

a) b)

Fig. 4. Pressure ratio at the last stage root on the a) left and b) right side of the machine.

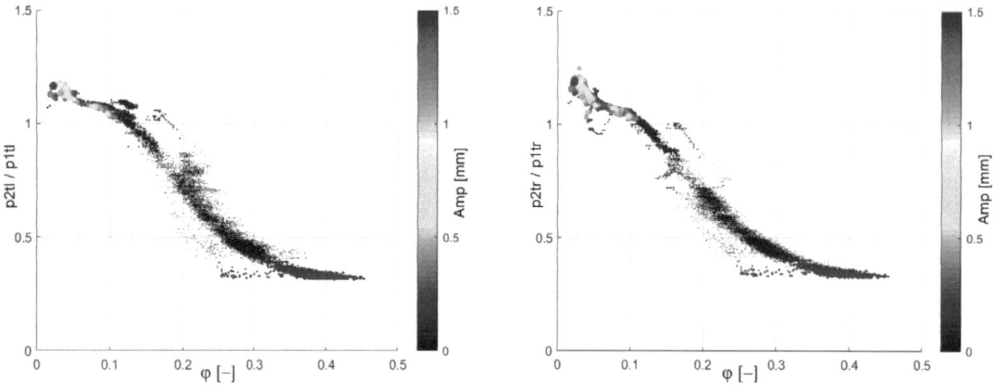

a) **b)**

Fig. 5. Pressure ratio at the last stage tip on the a) left and b) right side of the machine.

Similar behaviour of flow and vibrations as at the root can be observed even for pressure ratios at the tip of the last stage, see Fig. 5. The pressure ratio at the rotor blade tip increased for $\varphi \cong 0.03$ up to the value 1.2. At the maximal value of pressure ratio even a significant increase of vibrations occurs. Value $\varphi = 0$ was not reached, as it would practically mean a state when the turbine is stopped. In the case $\varphi = 0$ it would be valid that $p_2 / p_1 = 1$, thus the pressure in the last stage area (and the whole turbine) is the same.

It should be noted here that in Fig. 4 and Fig. 5 the value of circumferential velocity is related to the mean blading diameter while the pressure ratios are defined at the root and tip limiting section. This is because c_{2x} is a calculated velocity that cannot be defined in individual parts of the outlet cross-section, but only as a mean value over the whole cross-section.

The measurement campaign took place in four periods (four days) numbered below 1 to 4. In Fig. 6 there are graphs that show pressure ratios over the root and the tip of the last stage and the amplitude of vibrations dependent on φ and the measurement period. The course of pressure ratios p_{2r} / p_{0r} in period 1 has a large hysteresis given by the behaviour of the front part of the turbine, namely by gradually increasing and subsequently decreasing of the flow to the regeneration samples of the turbine. Slightly increased vibrations appeared during this period. The pressure in the axial exhaust hood did not change significantly. The most significant increase in vibrations occurred in Measurement #2, when there was a short-term shutdown of the turbine and a subsequent island operation with turbine output of about 0.5 MW for $\varphi = 0.03$. In the next two days no significant anomalies in the character of the measured curves were recorded. The last rotor blade can be excited both by pressure changes at the turbine input and by pressure changes given by the behaviour of the condenser. However, the designed blading can be safely operated throughout the operating range of the machine.

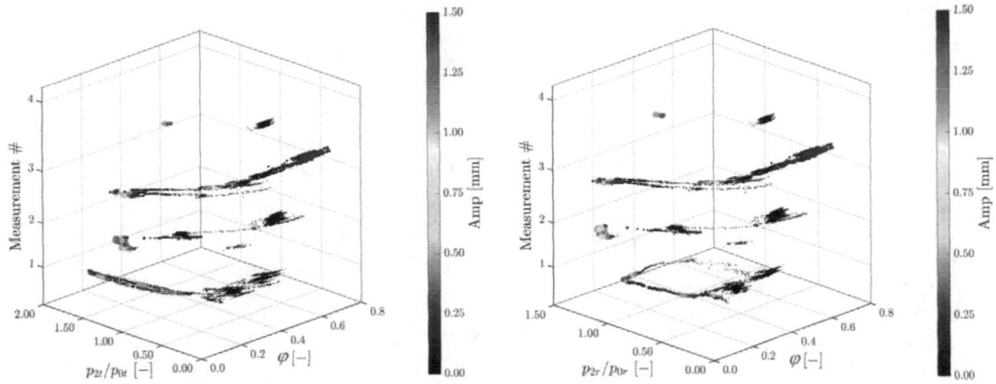

a) **b)**

Fig. 6. Pressure ratio at the tip a) and the root b) of the last stage.

The behaviour of the temperature ratios over the rotor blade tip t_{2t} / t_{1t}, or over the tip of the whole last stage t_{2t} / t_{0t}, is more complicated and is shown for all four measurement periods in Fig. 7. Mainly for low outputs or during turbine shutdown the steam temperature behind the stage may increase up to 1.8 times more than the steam temperature before the rotor blade or the whole stage. It is confirmed that the most significant heating occurs in the area of the last rotor blade tip. At the same time, however, the steam heating does not occur in conditions with maximum vibration amplitudes, but especially in conditions that precede maximum vibration amplitudes. In other words, the highest heating occurs when the turbine is in island operation for φ = 0.02 and also at about the level φ = 0.07. Temperature ratios in the tip area are again influenced by both the pressure value in the condenser and the input temperature to the turbine.

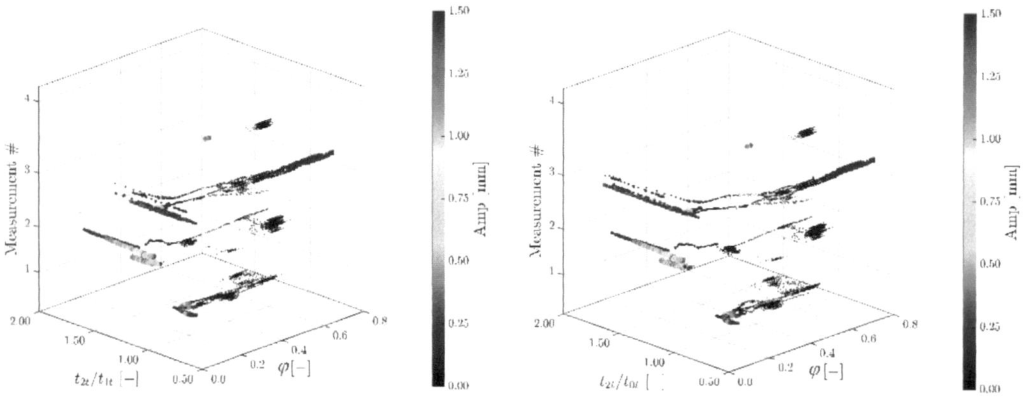

a) b)

Fig. 7. Temperature ratio over the rotor blade tip a) and over the whole stage b).

The courses of selected values of pressure and temperature ratios together with vibration amplitudes as well as calculated last stage outputs for Measurement #2 are in Fig. 8. The picture also shows the ventilation limit φ_{lim} of the last stage and the ratio of pressures and temperatures that equals one. The pressure in the condenser was changing from 0.04 up to 0.29 bar during the measurement.

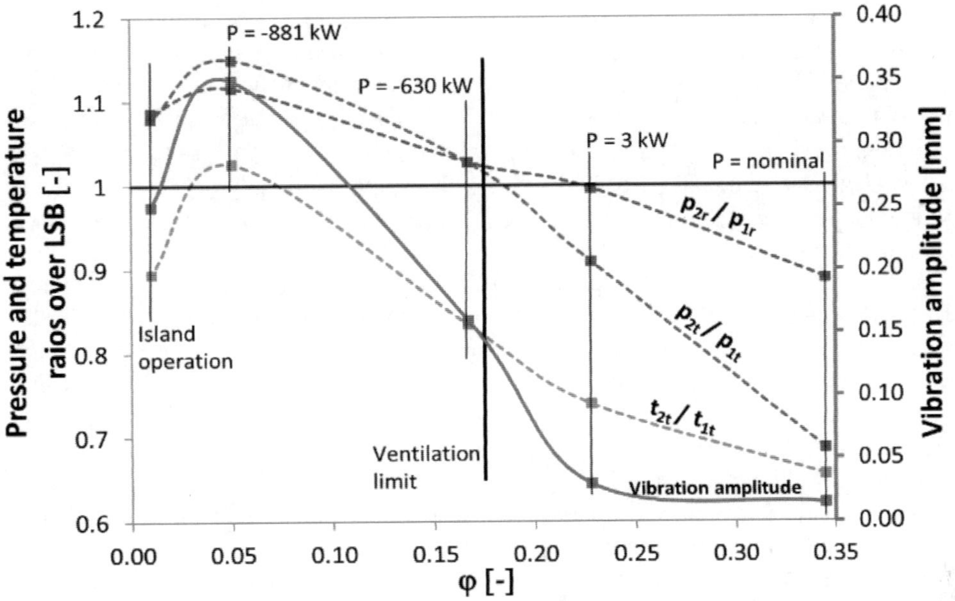

Fig. 8. Courses of pressure and temperature ratios and vibrations in dependence on the ventilation coefficient

When reducing the coefficient φ the pressure ratio over the rotor blade is exceeded first at the root, for the value φ = 0.225. At this point the flow is detached from the rotor blade root. The blade output is approximately zero. At the same time, however, there is an increase in vibrations. For ventilation limit φ_{lim} = 0.18 the steam flow over the blade tip towards the condenser will also stop, ratio p_{2t} / p_{1t} ≅ 1. At this time the rotor blade already functions as a centrifugal compressor and it has a significant negative output. The value of vibration amplitudes further increases. The maximal value of pressure and temperature ratio and at the same time the maximal value of vibrations and last stage input is for φ = 0.04. At this value of the ventilation coefficient the steam flows significantly radially in the inter-blade rotor channel, centrifuges and dynamically stresses the rotor blade tip. As the flow through the last stage further decreases due to the transition to island operation, the stage is relieved and vibrations are reduced. Pressure ratios are higher than one, the steam is compressed. However, the steam flow is so low that it does not fill the whole inlet cross-section and the steam cools down. It is also possible that the thermometers are not washed at all by the steam flow.

For turbine ventilation regimes it is difficult to measure mean values of pressures and temperatures in the area of turbine outlet at low outputs. There are different values of pressures and temperatures in the root, middle or tip area of the exhaust hood flow profile. When fitting the turbine with operational or research measurement, it is necessary to keep in mind the phenomena associated with the steam flow at low outputs.

4 Conclusions

Four series of experiments were performed on a steam turbine with the output of 30 MW operated in the municipal waste incinerator. At the same time pressures, temperatures and vibration amplitudes of last stage rotor blades were measured. Experiments were carried out in a broad range of the turbine and last stage operation - from the island operation to the overload regime.

For each of four series of measurements the dependences of pressure ratios over the tip and the root of the rotor blade and the whole stage on the turbine operation were determined. Furthermore, the temperature ratios over the tip of the rotor blade and the whole stage for various turbine operations were also determined. Turbine operation means the so-called ventilation coefficient $\varphi = c_{2x} / u$. The limit value is $\varphi_{lim} \cong 0.17$. This is the value for which the last stage output equals zero. For $\varphi < \varphi_{lim}$ the output is already negative and the stage operates in the compressor regime, for $\varphi > \varphi_{lim}$ the stage output is positive. The stage operates in the turbine regime.

A significant increase in the values of vibration amplitudes occurred for the monitored turbine when the steam flow was detached from the rotor blade root, which was indicated by the reversal of the steam flow at the root back to the turbine. When the steam flow was reversed even from the rotor blades tip ($p_{2t} / p_{1t} > 1$) value φ_{lim} was subceeded and the last stage operated in the compressor regime. Maximum values for vibration amplitudes of pressure and temperature ratios were measured for $\varphi = (0.02, 0.07)$. Steam temperature behind the rotor blade tip reached temperatures 1.8 times higher than the temperature before the tip. At the place t_{2t} temperatures up to 210 °C were recorded. For values $\varphi \cong 0.01$ (island operation of the turbine) the steam flow through the turbine was so low that the steam did not fill the whole volume of the exhaust hood and pressure and temperature ratios began to decrease significantly again.

The experiments further verified that the last stage behaviour from the point of view of dynamic loading is influenced both from the front (opening and closing of regeneration samples) and the back (pressure change in the condenser) part of the turbine. When evaluating the measured data it is more appropriate to monitor pressure ratios over the rotor blade than the absolute values of pressures before and behind the blade. The steam flow in the outlet part of the turbine is significantly uneven at low regimes. Mean values of pressures and temperatures can be only determined by suitably selected positions of thermowells and pressure samples at several places of the exhaust hood.

The work was financially supported by the Technology Agency of the Czech Republic; project FW03010515 Ultra flexible operation regimes of steam turbines and by European Regional Development Fund-Project "LoStr" (No. CZ.02.1.01/0.0/0.0/16_026/0008389).

References

1. Cao, L., Xu, M., Luo, H., Si, H. (2021). Strain-life estimation of the last stage blade in steam turbine during low volume flow conditions. In *Engineering Failure Analysis*, Volume 125.
2. Xu, M., Cao, L., Song, L. (2022). Structure and modal analysis of the last stage blade in steam turbine under low volume flow conditions. In *Engineering failure Analysis,* Volume 131.
3. Zhu, X., Hu, P., Lin, T. (2021). Numerical investigations on non-synchronous vibration and frequency lock-in of low-pressure steam turbine last stage. In *Proceedings of the Institution of Mechanical Engineers, Part A: Journal of Power and Energy.*
4. Mambro, A. et al. (2019). Experimental study and modelling of the ventilation power and maximum temperature of low-pressure steam turbine last stages at low load. In *Applied Energy*, Volume 241, 59-72.
5. Mambro, A., Galloni, E., Congiu, F. (2020). Assessment of the heat transfer coefficient for the prediction of ventilation power in steam turbine last stages operating at low load. In *Thermal Science and Engineering Progress*, Volume 18.

6. Mambro, A., Congiu, F., Galloni, E. (2022). Influence of stage design parameters on ventilation power produced by steam turbine last stage blades during low load operation. In *Thermal Science and Engineering Progress*, Volume 28.
7. Ercan I., Vogt, D. (2021). Detached Eddy Simulation of Rotating Instabilities in a Low-Pressure Model Steam Turbine Operating Under Low Volume Flow Conditions. *Turbo Expo: Power for Land, Sea, and Air,* GT2021-58704, V008T22A006, 11 pages.
8. Hu, P, Lin, T., Yang, R. (2021). Numerical investigation on flow instabilities in low-pressure steam turbine last stage under different low-load conditions. In *Proceedings of the Institution of Mechanical Engineers, Part A: Journal of Power and Energy,* Volume 235 issue 6, 1544-1562.
9. Tanuma, T. et al. (2015). Aerodynamic and Structural Numerical Investigation of Unsteady Flow Effects on Last Stage Blades. In *ASME TurboExpo*, GT2015-43848, V008T26A035; 13 pages.
10. Cao, L., Du, H., Si, H. (2021). Influence of water spray parameters on aerodynamic characteristics and temperature drop characteristics of exhaust passage in steam turbine. In *International Journal of Heat and Mass Transfer*, Volume 180.
11. Synáč, J., Rudas, B., Luxa, M. (2021). Erosion of steam turbine last stages. In *AIP Conference Proceedings 2323, 060008.*
12. Lubbe, B., Aschenbruck, J., Putz, O., Theidel, M. (2021). Design and Validation of a Large Steam Turbine End-Stage Blade to Meet Current and Future Market Demands. *Turbo Expo: Power for Land, Sea, and Air,* GT2021-59315, V008T22A014, 12 pages.

Application of Hydrostatic Knowledge in the Construction of Forming Tools

Ján Moravec[1,*]

[1]University of Zilina, Faculty Mechanical Engineering, Univerzitna 8215/1, 010 26 Zilina

Abstract. Not only traditional (conventional) but also non-traditional (unconventional) methods are used in metal forming. Unconventional methods will, as it turns out, have more scope than in the past, especially in relation to environmental constraints. One of the methods of unconventional forming is forming in a liquid environment, more precisely – sheet metal forming by applying hydrostatic pressure. This is the subject of the present.

Keywords: *forming; fluid; tool; deep drawing; application*

1 Introduction

The principle of hydrostatic forming consists in the action of hydrostatic all-round pressure of liquids in the forming tool and the punch. Among the methods used for sheet metal drawing, methods using a flexible - plastic environment for this purpose occupy an important place. In this case, it is the so-called unstable tool. It has a number of advantages over fixed tools - savings in production costs for the tool, especially in the production of smaller series of moldings of complex shapes. The forming process is also simplified by eliminating the need to fit the active parts of the tool and by eliminating complex holders. In the following text, attention is paid to the issue of deep drawing of sheet metal using a special tool for laboratory conditions.

2 Theoretical findings

At a constant temperature, the volume of the liquid changes minimally under the influence of pressure, and this is essentially a phenomenon and the consequent knowledge that liquids are minimally compressible. If the pressure acting on the liquid increases, its volume decreases by a certain, very small value. When the pressure is reduced to the original value, the liquid immediately occupies the original volume. The liquids are therefore perfectly flexible. The compressibility of liquids is evaluated by a known compressibility coefficient, which is defined by the following relation :

* Corresponding author: jan.moravec@fstroj.uniza.sk

$$\gamma = -\frac{1}{V}\frac{dV}{dp} \qquad (1)$$

where: V is the initial volume of the liquid, - dV is the elementary decrease in volume caused by the elementary increase in pressure by the value dp.

The compressibility coefficient is different for different liquids. At high pressures it decreases, with increasing temperature it usually increases. Its value for water at 20 ° C is: γ = 4,998. 10-10 (Pa^{-1}). Forming in a liquid medium uses, resp. physical knowledge, known as Pascal's law, is applied to it. According to FIG. 1: let there be an incompressible liquid in the vessel with a density ρ which is in equilibrium in the gravitational field of the earth.

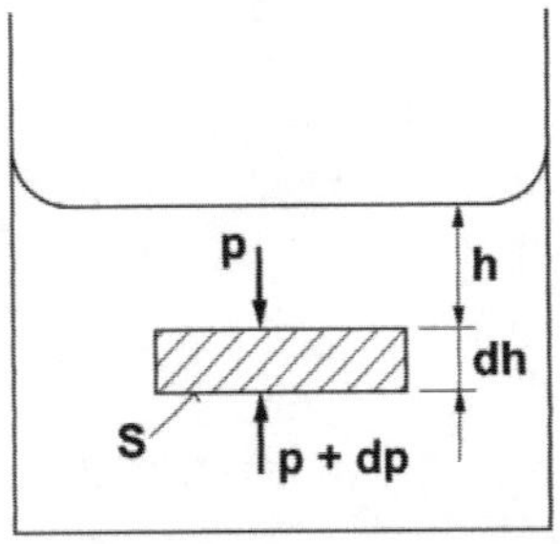

Fig. 1 To derive relationships

At a depth h below the surface of the liquid, an element with the volume of an elementary cylinder with a base S and a height dh is extruded. The selected elementary amount of the liquid is affected by its own weight (volume force), as well as the compressive forces (area force) of the remaining liquid. The compressive forces acting on the upper and lower bases of the elementary cylinder have a vertical direction. The compressive forces acting on the cylinder shell have a horizontal direction. Equilibrium conditions require that the vector sum of all forces be zero. Since the weight of the selected amount of liquid has the value $dG = \rho\ Sg\ dh$, resulting compressive force of the vertical direction oriented upwards (buoyancy):

$$df = (p+dp)\,S - pS = Sdp. \qquad (2)$$

The equation condition follows from the equation:

$$\rho\ Sg\ dh - Sdp = 0. \qquad (3)$$

After modifying and integrating equation (3) will be:

$$p - \rho\ gh\ = const. \qquad (4)$$

The potential of the Earth's gravitational field at depth h below the surface of the liquid can be expressed by the relation $V = - gh$, and equation (4) goes to the form:

$$p + \rho V = const. \qquad (5).$$

Equation (5) has general validity. It can be written: if the incompressible liquid is at rest in any force field, the sum of the potential energy of its volume unit and the pressure is the same everywhere in the liquid. If the liquid is under higher pressure and does not occupy too large a volume, so that the potential changes in the liquid are small, the term ρV in equation (5) can be considered practically constant and from equation (5)
the relation $p = const.$, which is actually an expression of *Pascal's law* of uniform pressure distribution in fluids. According to *Pascal's law*, it is: the pressure in a fluid that is in

equilibrium spreads integrally, that is, in all directions and with the same magnitude. This knowledge is applied in the following text in the field of sheet metal forming. [5,6,7,8]

3 Application

In the field of surface forming, materials based on sheet metal semi-finished products are processed. For this reason, forming in a hydraulic environment has a great perspective. With this type of forming, forming tools working precisely with the use of a liquid, which in most cases is used as a replacement for one active element of the tool. It is therefore also an economic reduction in the cost of their production. Liquid forming itself is not new at present, but somehow still remains in the shadow of all-metal tools, despite its advantages. This fact is surprising because the method is able to largely eliminate the negative effects of production processes, because it uses water as an active element, i.e. an element that does not burden the environment. Figures 2 and 3 show several tool solutions that certainly meet the required parameter imposed on them.

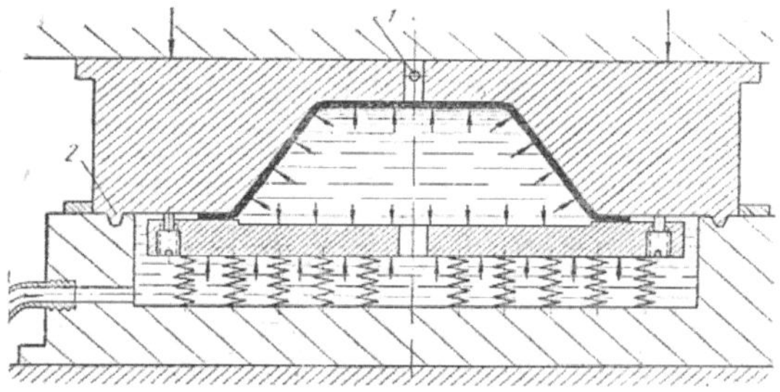

Fig. 2 Hydraulic drawbar 1- venting, 2- labyrinth sealing groove

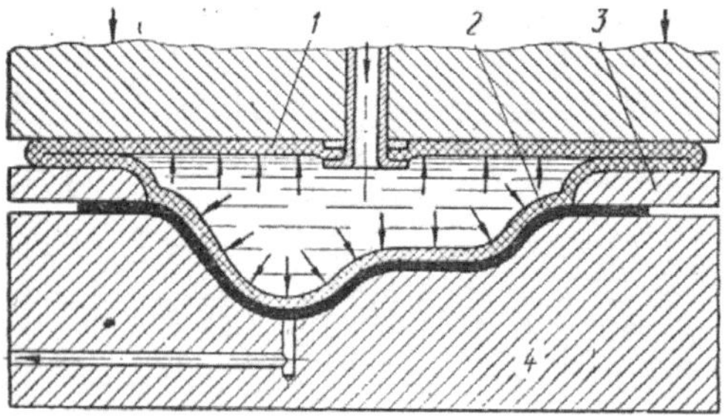

Fig. 3 Hydraulic towing toll 1 – sealing rubber cushion, 2 – molded part, 3 – holder, 4 – rod

4 Experimental part

A forming tool was constructed for experimental work, in which relevant knowledge of fluid mechanics is applied. In FIG. 4 and 5 are photographs of the tool. The suitability of the hydraulic solution and the comfort of the tool operator, which is at the appropriate level, was verified. The findings are sufficient and it is only a matter of time before the solution will be optimized, respectively. supplemented by the necessary peripheral elements to make the operator more comfortable and to reduce the time required for the production of the sheet metal part.

5 Description of the toll

The forming tool for hydro-drawing is constructed of two basic parts. The lower part is a block of plexiglass. M6 screws are pressed into the block to secure the upper part of the tool. An opening for the supply and discharge of pressurized fluid is led out at the side. The entire lower part is assembled with screws and stands on four legs. A rubber seal is pushed onto the M6 screws. The upper part is a duralumin plate into which the necessary shape for the extract is milled. The cavity must not have sharp edges and the transitions must be rounded. The part is drilled with the lower part and slides on the M6 screws screwed into the tank. A pressure fluid is admitted into the tool and by its action the sheet metal cut is evenly pressed into the recess in the upper part of the tool. In FIG. 6 is a diagram of the connection of the tool to the hydraulic system.

Fig. 4 Experimental hydraulic towing tool (assembly)

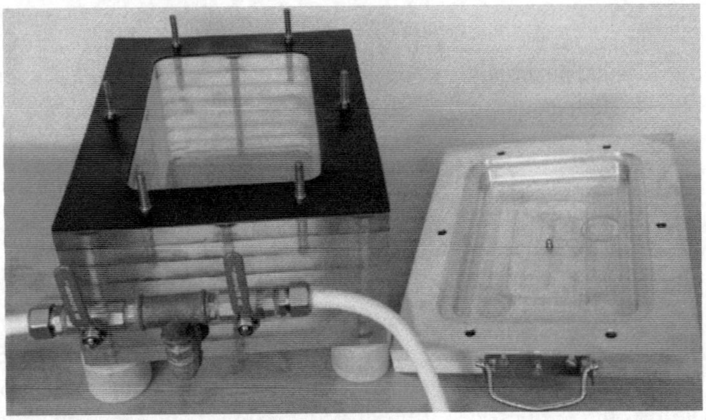

Fig. 5 Experimental hydraulic towing tool (disassembled)

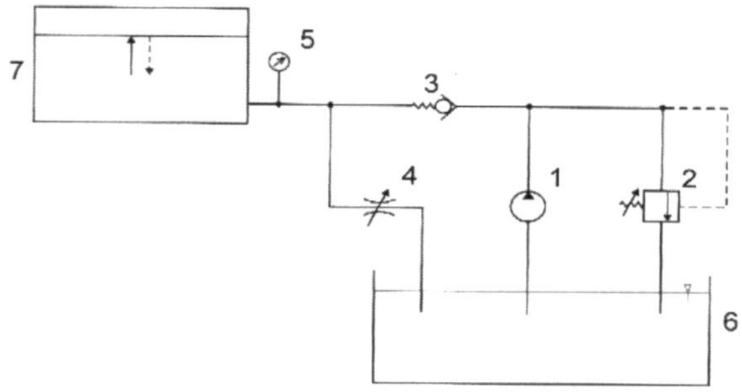

Fig. 6 Wiring diagram for the hydraulic towing tool in the hydraulic system 1 – gear pump, 22 – PV safety valve, 3 – JV check valve, 4 – SV throttle valve, 5 – Manometer, 6 – tank, 7 – toll-forming chamber

5.1 Technological procedure for working in the tool

- *Insert the sheet metal cut on the top of the tool,*
- *slide the upper part onto the M6 screws,*
- *screw in M6 nuts,*
- *switch on the pump, check the value of the liquid pressure, the formation of the extract,*
- *switch off the pump, drain the liquid from the tank,*
- *dismantle the upper part, put it aside,*
- *select the art and visually check the surface of the extract,*
- *place the part on a pallet.*

When water is used as a medium, although it is relatively most available, there is a risk of surface corrosion of the extracts. Its advantage is that it does not represent an ecological burden on the environment. The oil is optimal for this work in terms of tribological characteristics. The appropriate oil viscosity should be tested. The oil clearly takes precedence over another medium. Disadvantage - in the event of an accident, oil leaks into the environment.

6 Discussion

Physical knowledge in the field of fluid mechanics is a great source from which it is possible to draw almost indefinitely. When applying (generally) any finding from this sphere - reflected in the technical solution - there is almost always one limitation. It is the impact of using the solution on the environment. Nowadays, this condition will probably become limiting in the application of any new solution. It will be necessary to respect the fact that the earth is one huge ecosystem, regardless of political boundaries. For the sphere of production in the field of engineering, it will be necessary to focus on ecological solutions. It is for this reason that the submitted contribution was also created. Solutions that can respect environmental constraints will find their application. The use of water as a basic element of living nature in the production process is one of the ways leading to this goal.

7 Conclusion

The present paper addressed a topic which is unfortunately not in the forefront of the interest of sheet metal parts manufacturers to the extent it deserves. The biggest advantage of hydromechanical drawing is the achievement of the coefficient $m = 0.3$, in one operation, which has not yet been achieved by another procedure. This will be reflected in a reduction in the number of operations and consequently in a reduction in the price of tools. The price is up to ten times lower for some moldings. Related to this is the possibility of using hydromechanical drawing even with a smaller number of yields. The fact that only the die has a precise shape of the molding and the die is actually formed by a pressurized fluid greatly helps to reduce the cost of the tools. The forming process takes place in a liquid medium, therefore the surface of the material, which can be surface-treated, is not damaged. The method allows the drawing of intricate moldings even with a hole in the bottom. The drawing depth is practically not limited from the point of view of procedure. The undeniable advantages of hydromechanical drawing are: - savings in pulling conical and parabolic parts, - the sheet is thinned by only 2-3% during the process, ie the same sheet thickness is maintained in the whole molding, - the corrugation during drawing causes a greater deformation, which improves the mechanical properties of the material. The disadvantage of the described procedure is the increase in pulling force, which results in the need to use more powerful presses. The above also applies to the holder. Within the workplace at KTI FME of the University of Žilina in Žilina, the following issues concerning the area of the use of hydromechanical traction are also solved. Experimental work has been performed using this method, although not yet in the field of volume forming. [4] The intention is to expand the scope and focus intensively on surface forming with the application of knowledge from hydromechanics. Experimental work has confirmed the suitability of this type of forming, especially for piece production, and the application of the tool expands the sphere of physical knowledge usable in sheet metal forming.

References

1. Moravec, J.(2011). *Unconventional metal forming technologies*, EDIS, Zilina
2. Smetana, J.(1987). *Hydraulics I.* , CSAV Praha
3. Krempasky, J.(1982). *Physics,* ALFA Bratislava 1982
4. Moravec, J.(2006). Contribution to the issue of forming in the hydropower environment, *Hydraulics and tire*, VIII./ 1-2, pp.59-60.
5. Mielnik, E.M.(1991). *Metalworking Science and Engineering*, McGraw- Hill Book Company, New York,
6. Hosford, W. F. (2011). *Metal Forming, Mechanics and Metallurgy*, Cambridge University, pp. 277-283
7. Boljanovic, V. (2004). *Sheet Metal Forming Processes and Die Design*, Industrial Press, New York, pp. 85-100
8. Kundar, S. (2008). *Technology of Metal Forming Processe*, Asoke U. Ghosh, Prentice-Hall of India, New Delhi, pp. 105-107.
9. Marciniak, Z. , Duncan,J. L. , Hu, S. J. (2005). *Mechanics of Sheet Metal Forming,* Butterworth Heinemann, Oxford, pp. 152-163.
10. Pearce, R. (1991). *Sheet Metal Forming*. OP Publishing Ltd. Bristol.
11. *Metal Forming Handbook*, (2009). Schuler GmBH Gőppingen.
12. Moravec, J. (2003). *Usinf Liquid enviroment in process forming*, 18th Symposium on Anemometry, Úvaly u Prahy, pp. 19-24.
13. Moravec, J. (2012). Construction of hydrostatic forming tool, *Technolog IV/ 4,* pp. 227-232.

https://doi.org/10.1051/matecconf/202236902003

Pool boiling heat transfer performance of the surface modified with laser

Łukasz J. Orman[1,*] , *Norbert* Radek[1], *Andrej* Kapjor[2], and *Luiza* Dębska[1]

[1]Kielce University of Technology, al. Tysiaclecia P.P.7, 25314 Kielce, Poland
[2]University of Zilina, Univerzitna 2, 01026 Žilina, Slovakia

Abstract. The paper deals the problem of pool boiling heat transfer on the surface produced by the modification with the laser beam. The laser enabled to produce grooves and roughness which have a positive impact on heat transfer. The distilled water and ethanol boiling performance of a horizontally located specimen is discussed and the test results are compared with the model of boiling heat transfer selected from the literature. The laser technique proved to be a valuable tool for producing surfaces that improve thermal performance during pool boiling.

Keywords: boiling, heat transfer, laser melting

1 Introduction

Both boiling and condensation provide the opportunity to exchange large heat fluxes, which is of interest due to the fact that modern technological devices require the dissipation of considerable heat loads. The further increase in heat transfer can be obtained with the modification of the surfaces.

Kaniowski and Pastuszko [1] focused their paper on the modification of the heater surface with specially prepared microchannels (of the depth 0.2 – 0.5 mm) and tested boiling of water. These authors found out that the maximum heat flux for a modified surface was about 2.5 times bigger in comparison to the heater without any grooved on its surface. During the testing, the mean diameters of the departing bubbles were determined with high speed camera and it was reported that the diameter rose and the frequency of departure fell with the increasing heat flux. A later paper by the same authors [2] contains data of FC – 72 boiling. Here the improvement of the modification of the heater was also evident. It was stated that heat transfer coefficient was more then five times the value of the surface with no grooves. The researchers also pointed out that many elements influence boiling performance such as surface roughness, wettability and etc. Maciejewska et al. [3] focused on the boiling enhancement in the flow mode in minichannels, however the concept of the phenomenon is consistent with the pool boiling mode. The modification of the heater surface with the lasers has been the subject of some papers already. Piasecka et al. [4] considered laser texturing assisted with vibrations as well as the electromachining technique as the means to intensify boiling. In [5] six samples were tested in water and ethanol pool boiling conditions. The specimens were produced in such a way that

* Corresponding author: orman@tu.kielce.pl

longitudal fins were developed. The authors claimed that the modification with the help of the laser beam was very efficient and the heat flux dissipated from such surfaces was a few times higher than from the reference surface. The grooved generated with the laser were up to 0.55 mm deep and up to 1.15 mm wide.

It needs to be noted that the enhancement of heat transfer can be very important for many areas of engineering such as the ventilation systems [6] and others [7 - 11].

Due to the significant possibilities offered by the laser modification, the present paper will be focused on the analysis of nucleate pool boiling on the sample produced with the laser beam.

2 Testing method and sample characteristics

Laser beam enables to produce various shapes on the surface. The present study deals with a specimen whose diameter is 3 cm. It was made by SPI G3.1 SP20P pulsed fiber laser at the set impulse frequency of 60 Hz and scanning velocity of 200 mm/s. Figure 1 shows examples of specimens generated with this technology.

Fig. 1. Examples of specimens made with the laser.

The advantage of using the laser beam is the creation of the grooves (longitudal fins), which extend the surface available for heat exchange. However, the additional and very important aspect is the creation of the rough surface at the bottom (due to the thermal interaction with the metal). Roughness can also significantly increase the heat transfer during boiling due to the increased density of nucleation sites (locations where bubbles grow) on such a modified surface. Figure 2 presents the morphology of the surface at the bottom of the groove. It can be seen that it is not uniform and dips as well as hills are present (of the height up to 85 μm).

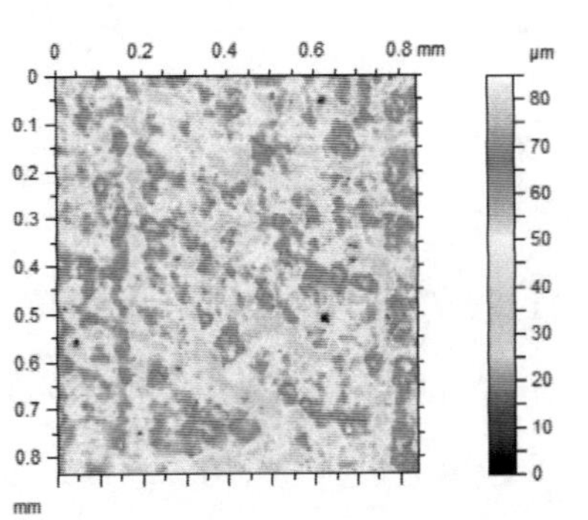

Fig. 2. Morphology of the surface subjected to laser (the bottom of the grooves of samples in Fig. 1).

The measurements of heat flux have been made on the stand, where the sample is located on top of the heater and attached horizontally through soldering. Above it the vessal with the liquid (water and ethanol) is located. The details of the testing method have been provided by the authors in [5].

3 Results and discussion

Apart from measurements of heat flux dissipated by the samples, the boiling performance investigations are also focused on the actual processes occurring on the specimens. This task is arranged with high speed digital cameras that can take pictures at very high frequencies. Figure 3 presents the photos of the fully developed nucleate boiling phenomenon of distilled water.

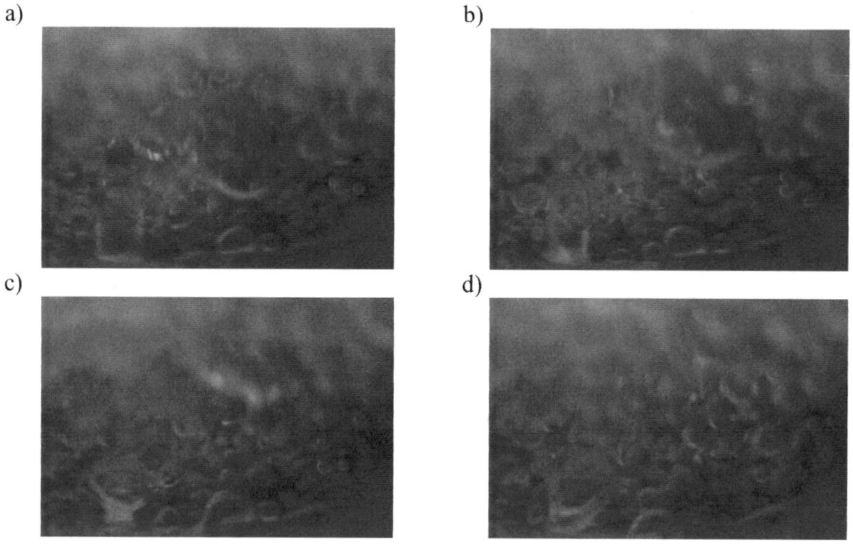

Fig. 3. Boiling of water on the horizontal sample: a) t = 0, b) t = 0.01 s, c) t = 0.02 s, d) t = 0.03 s.

Two laser - made samples were selected for comparison, they both had the same groove depth of 0.55 mm, but sample A had the groove width 0.60 mm, while sample B the groove width of 1.15 mm. Figure 3 shows the ratio of the heat flux dissipated from sample B to the heat flux from sample A (data adopted from [5]).

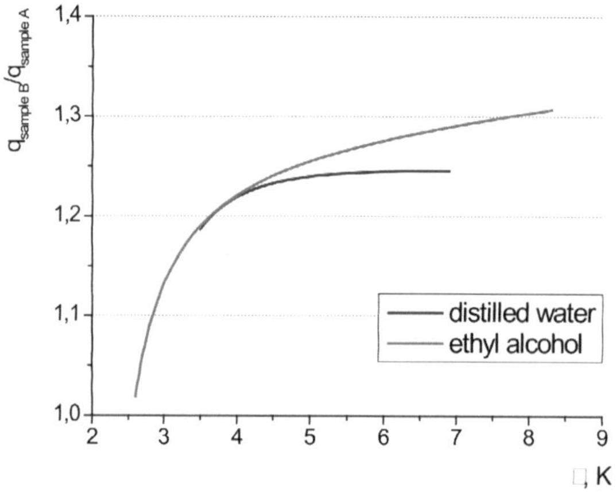

Fig. 4. Ratio of heat flux from selected laser treated samples.

As can be seen the performance of sample B is slightly better than sample A and very similar for both the boiling agents. The heat flux values of sample B become larger with increasing superheat (difference between the surface and saturation temperatures). This might be linked with the larger groove width of sample B and, thus, larger surface area of increased roughness.

The proper operation of phase – change heat exchangers and their design requires that the heat flux and general performance can be correctly calculated with the adequate correlations as well as models. For this aim, various methods are used in terms of modeling (as given, for instance in [12 – 14]). In the present paper, a simple model based on the conduction heat transfer, will be used to compare the experimental data with the calculated ones. This model was developed by Nishikava et al. [15]. The test results of samples A and B (adopted from the authors' work [5]) were compared with the results obtained with the application of the model [15] and it has been graphically shown in Figures 5 and 6.

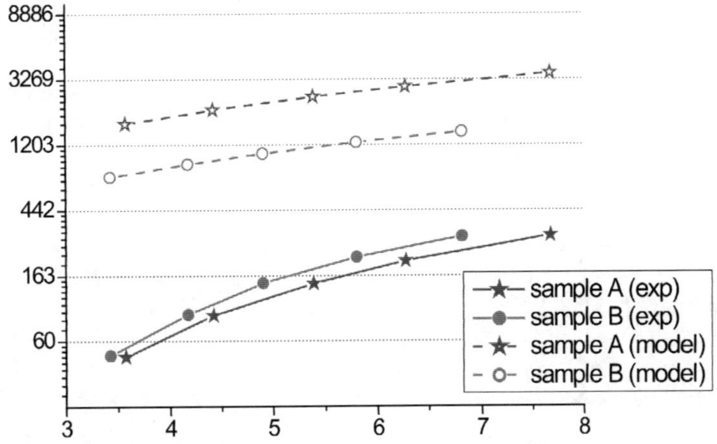

Fig. 5. Comparison of the experimental and model calculated data (water).

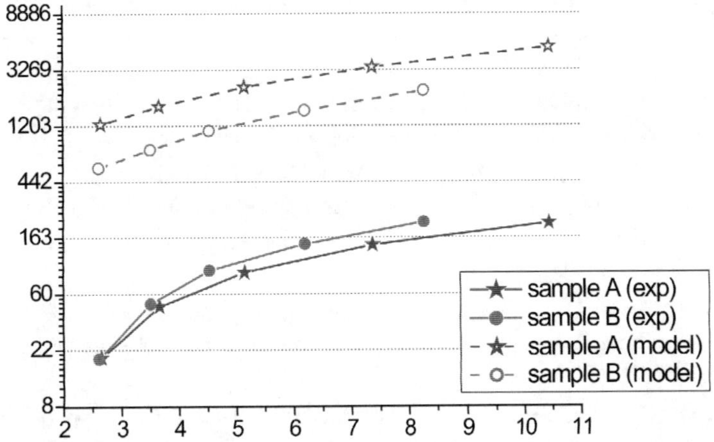

Fig. 6. Comparison of the experimental and model calculated data (ethanol).

Nishikava et al. [15] model is based on the concept of heat transfer conduction and does not take into account a complex process of bubbles' formation and departure. Thus, the results obtained with this model proved to be quite distant from the experimental data for both the samples and both the fluids (though it was especially visible for water). However, the general trend of heat flux increase with superheat has been maintained. It seems that this model should be used with caution for the analyses of surfaces modified with the laser beam.

An interesting feature is the fact that the experimental data are relatively close to each other for both the specimens, while the respective calculation results according to [15] are quite apart from each other. It indicates that the assuptions taken in the model about the large impact of conduction might not be true. However, more data is needed to be certain regarding the usability of the Nishikava et al. model in the assessment of heat transfer mechanisms on the laser treated heaters.

4 Summary and conclusions

The laser treatment technique can generate surfaces of modified morphological features. This has a large influence on thermal performance of specimens during nucleate boiling. The extension of the surface area, on which heat exchange occurs as well as the creation of roughness by the laser beam can lead to increased heat fluxes. The model by Nishikava et al. failed to provide correct calculation results that would match the experimental data. This can be caused by the mechanisms of boiling related to convection such as bubble and liquid movements and not solely conduction.

"The project is supported by the program of the Minister of Science and Higher Education under the name: "Regional Initiative of Excellence" in 2019 - 2022 project number 025 / RID / 2018/19 financing amount PLN 12,000,000."

References

1. Kaniowski, R., Pastuszko, R. (2021) Pool boiling of water on surfaces with open microchannels. *Energies,* 14(11) 3062. https://doi.org/10.3390/en14113062
2. Kaniowski, R., Pastuszko, R. (2021) Boiling of FC-72 on surfaces with open copper microchannel. *Energies*, 14(21) 7283. https://doi.org/10.3390/en14217283
3. Maciejewska, B., Piasecka, M., Piasecki, A. (2021) The study of the onset of flow boiling in minichannels – Time-dependent heat transfer results. *Heat Transfer Engineering*, 43(3-5), 223 – 237. https://doi.org/10.1080/01457632.2021.1874181
4. Piasecka, M., Strąk, K., Grabas, B. (2017) Vibration-assisted laser surface texturing and electromachining for the intensification of boiling heat transfer in a minichannel. *Archives of Metallurgy and Materials*, 62(4) 1983-1990. https://doi.org/10.1515/amm-2017-0296
5. Orman, Ł.J., Radek, N., Pietraszek, J., Szczepaniak, M. (2020) Analysis of enhanced pool boiling heat transfer on laser – textured surfaces. *Energies,* 13(11) 2700. https://doi.org/10.3390/en13112700
6. Telejko, M., Zender-Świercz, E. (2017) An attempt to improve air quality in primary schools, *Proceedings of 10th Int. Conf. Environmental Engineering, Book Series: Environmental Engineering - Vilnius Spausdinta.* https://doi.org/10.3846/enviro.2017.051
7. Nemec, P.; Malcho, M.; Polacikova, M.; Jandacka, J.: Heat Dissipation from Power Electronic Components and Electrical Enclosures Using the Gravity Loop Heat Pipe, *In: ELEKTRO 202, 13th international conference. - 1. Edition. - Danvers: Institute of Electrical and Electronics Engineers,* 2020, ISBN 978-1-7281-7541-6. - p. [1-4].
8. Polacikova, M.; Nemec, P.; Malcho, M.; Jandacka, J.: Experimental Investigations of a Passive Cooling System Based on the Gravity Loop Heat Pipe Principle for an

Electrical Cabinet, In *APPLIED SCIENCES-BASEL*, Volume12, Issue3, Article Number1634, 2022, DOI10.3390/app12031634.

9. Kaduchová, K., Lenhard, R. (2021) Numerical simulation of passive cooling beam and its optimization to increase the cooling power. *Processes*, 9(8), 1478. https://doi.org/10.3390/pr9081478
10. Klimko, M., Lenhard, R., Žitek, P., Kaduchová, K. (2021) Experimental evaluation of axial reaction turbine stage bucket losses. *Processes*, 9(10), 1816. https://doi.org/10.3390/pr9101816
11. Lenhard, R., Malcho, M., Jandačka, J. (2019) Modelling of heat transfer in the evaporator and condenser of the working fluid in the heat pipe. *Heat transfer engineering*, 40 (3-4), 215-226. https://doi.org/10.1080/01457632.2018.1426225
12. Krechowicz, M., Krechowicz, A. (2021) Risk assessment in energy infrastructure installations by horizontal directional drilling using machine learning. *Energies*, 14(2), 289. https://doi.org/10.3390/en14020289
13. Krechowicz, M. (2017) Risk management in complex construction projects that apply renewable energy sources: A case study of the realization phase of the Energis educational and research intelligent building. *IOP Conference Series: Material Science and Engineering*, 245(6), 1- 10, https://doi.org/10.1088/1757-899X/245/6/062007.
14. Szeląg, B., Łagód, G., Musz-Pomorska, A., Widomski, M.K., Stránský, D., Sokáč, M., Pokrývková, J., Babko, R. (2022) Development of rainfall-runoff models for sustainable stormwater management in urbanized catchments. *Water*, 14, 1997. https://doi.org/10.3390/w14131997
15. Nishikawa, K., Ito, T., Tanaka K. (1979) Enhanced heat transfer by nucleate boiling on a sintered metal layer. *Heat transfer – Japanese Research,* 8, 65 – 81.

Comparison of Thermal Sensations Under the Same Conditions in the Climatic Chamber and the Lecture Hall

Luiza Dębska[1,*], and *Natalia* Krawczyk[1]

[1]Kielce Uuniversity of Technology, al. Tysiaclecia P.P.7, 25-314 Kielce, Poland

Abstract. For many people, the feeling of thermal comfort becomes an important aspect of helping to increase concentration at work or study. Creating an appropriate indoor microclimate becomes a huge challenge for many designers of modern construction. The most important assumption of the work is to compare the thermal sensations of a group of four, aged 23 to 30, staying in the lecture class of an intelligent building and in a climatic chamber with the same internal parameters. For this purpose, the Testo 400 environmental meter was used, collecting the necessary parameters from the environment, and specially created surveys describing people's thermal feelings during the study. The air temperature and relative humidity were as follows: 21.7 °C and 30.80 %, while for the climatic chamber the same data was applied, but at the end of the study both parameters increased to 21.9 °C and 40.50 %. Moreover, it turned out that people in the lecture hall felt worse than in the climatic chamber with similar internal values. In addition, the average Thermal Sensation Vote (TSV) response showed that the perceived environment is comfortable for the study group staying in the climatic chamber, as opposed to the feeling of influenza in the room. Unfortunately, the Predicted Mean Vote (PMV) for the room and chamber exceeded the set value described in the ISO 7730 standard.

Keywords: thermal comfort, Fanger model, microclimate

1 Introduction

The task of modern buildings is to provide the best possible conditions for the internal environment, as people work more and more and expand their knowledge. For that to be the case, they must feel comfortable indoors. The indicator that determines whether people feel comfortable in a given place is called Thermal Sensation Vote (TSV) and it is obtained by the answers of respondents completing the questionnaire specifying their thermal preferences. On the other hand, another Predicted Mean Vote (PMV) indicator, calculated from the data obtained from the environmental measurement, informs whether the prevailing conditions in the room are compliant with the ISO 7730 standard based on the Fanger model. The next Percentage of People Dissatified (PPD) indicator enables to know the percentage share of people who are dissatisfied with the prevailing conditions, similarly to the previous indicators, by means of questionnaires and the ISO 7730 standard [1, 2].

* Corresponding author: ldebska@tu.kielce.pl

One of the interesting research works on thermal comfort in a public building was proposed by Nakano & Tanabe [3], who examined the thermal sensations of passengers at four railway stations in Tokyo. The authors conducted surveys from 7:00 am to 8:00 pm. The results showed 80 % satisfaction in temperatures from 19 to 29 °C. A similar comparison was made by Krawczyk [4], but on two educational buildings - one intelligent and the other traditional. Moreover, Dębska & Krakowiak [5] analysed the thermal preferences of 83 people in three different educational buildings. The study was carried out using questionnaires and a microclimate device. It turned out that students preferred a temperature of 22.5 °C in an intelligent building than in traditional buildings. The authors of the works [6, 7] primarily focused on thermal comfort by analysing the PMV and PPD indicators in the Energis intelligent building at Kielce University of Technology. In the first study, the Fanger model was modified, while the second study focused on the average thermal assessments of respondents and their dissatisfaction with internal conditions.

However, studies in the climatic chamber were conducted by Kazanci et al. [8]. The chamber has been divided into two rooms in office conditions. 24 people, participated in the study. The temperature was kept constant at 26 °C. The results showed general satisfaction of the subjects. Another study by the authors of Solberg et. al. [9] focused on 21 people in a climate chamber. Efforts were made to recreate similar conditions to those in Norwegian offices, for 24 °C, 26 °C and 28 °C, respectively. Moreover, the authors of Białek & Dębska [10] conducted an analysis of the thermal sensations of 7 people in the classroom and one person in the climatic chamber. The same internal parameters in the chamber that had previously been saved in the classroom were recreated. The study showed that the values of PMV and TSV do not exceed the established range of thermal comfort, while a very high overshoot of the said range was obtained for PMV in the climatic chamber.

On the other hand, not only educational buildings are subject to study, but also nursing homes, hospitals and office buildings as in the authors' works [11-13]. The conclusions drawn by the researchers are the influence of external humidity in the preparation of the data, in the next paper on the inaccuracy of the PMV index, and in the third on the thermal differences of people in air-conditioned rooms with thermal sensations in people who do not have air conditioning in the office, becoming more susceptible to any temperature fluctuations. It needs to be remembered that thermal comfort is closely related to the issues of heat transfer [14-21]

The research conducted so far does not provide any knowledge about people's thermal sensations by comparing ordinary classrooms with a climatic chamber. Therefore, the main goal of the work is to conduct research in the lecture hall and the climatic chamber with the same internal conditions, and then to compare TSV, PMV and PPD.

2 Methods

The research was carried out at the Kielce University of Technology, in one of the educational buildings called Energis, and in a climatic chamber. The Energis building was designed with sustainable construction in mind a decade ago. It has been equipped with Building Manager System (BMS) control systems, mechanical ventilation and underfloor heating. The climatic chamber, on the other hand, has two small rooms with dimensions of 2.70x1.80x2.30 [m^3] with the possibility of setting the temperature from -25 °C to +80 °C and 2.25x1.8x2.30 [m^3] for setting the temperature from +5 °C to +50 °C [10]. Both tested objects are shown in Figure 1.

a) b)

Fig. 1. Research objects: a) Energis, b) climatic chamber.

A microclimate meter and questionnaires were used to carry out the research. The first method allows you to get information from the environment about changes in internal parameters second by second. The Testo meter was used for the test, it has probes that collect data on air temperature, air velocity, carbon dioxide content and relative humidity. The meter was located in the central part of the lecture hall and worked for 15 minutes so that the parameters were as stable as possible. The second method is for students to complete the survey. It includes questions about current thermal sensations, willingness to change a given temperature and humidity, acceptability to internal parameters, well-being, productivity and clothing (to calculate the insulation performance of clothing) and data needed to calculate the Biomass Body Index (BMI), such as height and weight. The Testo 400 measuring device and the tested lecture hall are shown in Figure 2.

a) b)

Fig. 2. Tested room and the device performing the measurements a) lecture room, b) Testo meter.

3 Results and discussion

The study, which was carried out in the climatic chamber and in the classroom, involved 4 women in each, ranging in age from 23 to 30 years. First, a room test was performed to determine the microclimate parameters, which were later recreated in the climatic chamber. The parameters obtained are shown in Table 1 below.

Table 1. Internal microclimate parameters in both tested rooms.

	Lecture hall	Climatic chamber	Units
Air temperature	21.7	21.9	°C
Relative humidity	30.80	40.50	%
Air velocity	0.06	0.07	m/s
Carbon dioxide	875	718	ppm

The air temperature in the climatic chamber slightly increased by 0.2 °C, compared to the lecture hall. Despite the fact that the relative humidity was applied as obtained in the lecture hall, it still significantly increased to 40.50 %. This may be due to the size of the chamber and the fact that the chamber was closed during the test, and people who sat in it increased the humidity by breathing up to about 40 %. Additionally, what you can pay attention to is the fact that the content of carbon dioxide was higher in the lecture hall than in the climatic chamber. This is an interesting phenomenon, all the more that it is in a smaller room that the increase in this value should occur, and not in a room with a much larger area.

In the further part of the work, indicators describing students' feelings about temperature and humidity as well as their dissatisfaction with the prevailing conditions will be discussed. First, attention will be focused on people's thermal sensation at a given moment, known as Thermal Sensation Vote (TSV). It is a question that determines whether the examined person is "too cold (-3)", "too cool (-2)", "pleasantly cool (-1)", "comfortable (0)", "pleasantly warm (+ 1) "," too warm (+2) ", and" too hot (+3) ". Figure 3 shows the obtained results of the conducted research.

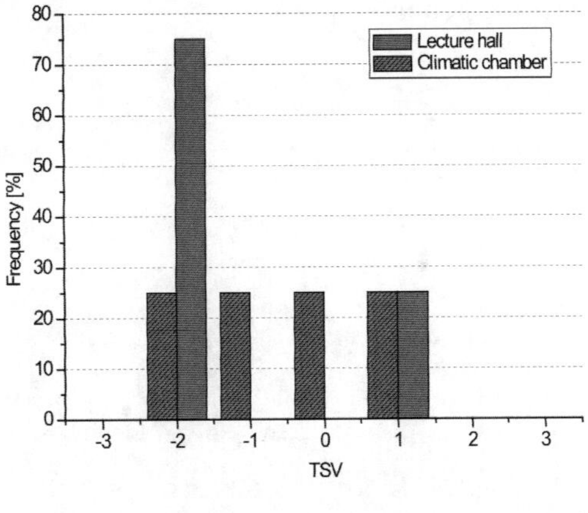

Fig. 3. Thermal Sensation Vote - questionnaire responses of students in the climatic chamber and lecture hall.

The set temperature for three people in the chamber was within the generally recognized comfort range from -1 to +1. Only one person felt that the room was too cool. 75 % of people in the lecture hall had similar feelings. Only one person said it was "pleasantly warm". The author's [10] research, which was also carried out in the room and in the climatic chamber, showed that despite 25 °C, almost 71 % of people described the temperature as "comfortable", and about 29 % of the wasps were too warm. Interpreting the obtained research analysis and comparing it to other studies by the author [10], it should be stated that 21.7 °C is definitely too low air temperature, which may contribute to deterioration of concentration or feeling shivering afterwards. By knowing people's thermal sensations, one can also know their thermal evaluation. The Thermal Assessment Vote (TAV) is responsible for this. It provides information on whether the air temperature according to the participants is "comfortable (2)", "acceptable (1)", "unpleasant (-1)" and "definitely unpleasant (-2)". Therefore, the obtained data is shown in Figure 4.

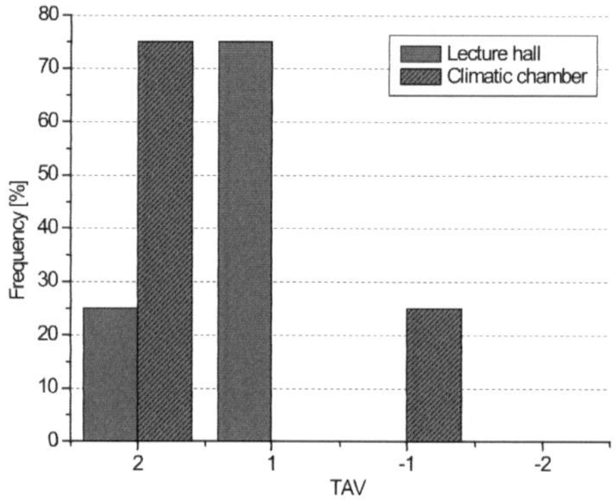

Fig. 4. Thermal assessment of respondents for the tested rooms.

People who were tested in the climatic chamber rated the temperature as comfortable with a value of 75 %, but one person (who probably marked "too cool" in Figure 3) rated it as unpleasant. On the other hand, people studying in the room assessed the temperature as comfortable – 25 % and acceptable for 75 %. The research [10] shows that people in the room positively assessed the temperature, similar to Figure 4. Another indicator that helps to understand people's thermals is the determination of their thermal preferences. Thermal Preferences Vote (TPV) ranges from -2 to +2. Students answer the question whether they would prefer it to be "definitely cooler", "cooler", "no change", "warmer" and "definitely warmer". The results are presented in Figure 5 below.

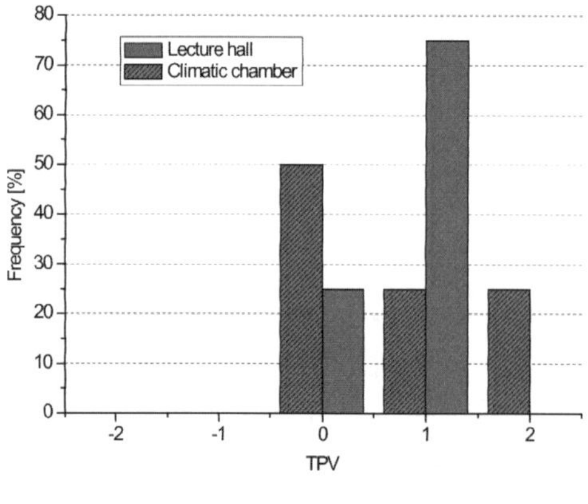

Fig. 5. Thermal preferences according to the respondents.

Based on Figures 3 and 5, it can be seen that people who were in the room and indicated that they were "too cool" in response to TSV, similarly their preferences were directed with the same percentage (75 %) towards warmer temperatures. One person showed no desire to change the temperature to a cooler or higher temperature. On the other hand, in the climatic chamber, a person who is too cold in the chamber would like it to be much warmer. Another person would like it to be warmer, and 50 % would like nothing to change.

Relative humidity is also an important aspect influencing the comfort of people. The room should not be too dry or too humid, because you may feel dry throat or excessive sweating. Therefore, students rated the humidity as "too dry (-2)", "quite dry (-1)", "pleasantly (0)", "quite humid (+1)" and "too humid (+2)" as shown in Figure 6.

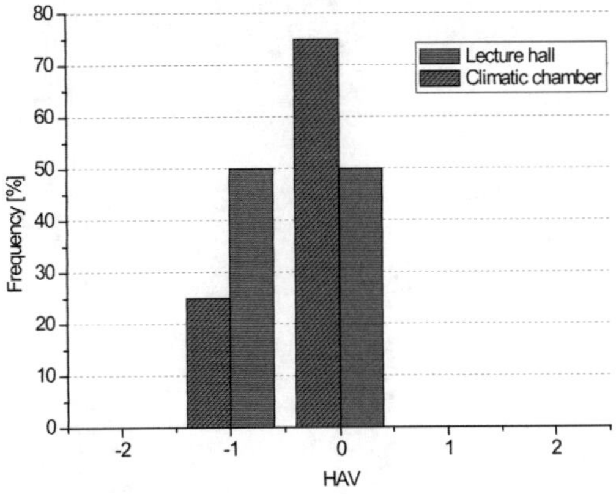

Fig. 6. Humidity Assessment Vote by students.

It turns out that in both tested rooms, people rate the relative humidity as "quite dry" and "pleasantly". Only the difference appears in the percentage share. 75 % of people in the climatic chamber and 50 % of people in the room said "pleasantly", and 25 % of people in the chamber and 50 % of people in the room "quite dry". A difference can be noticed in the assessment of humidity at work [10], where approximately 43 % of people additionally thought that it was quite humid. Figure 7 will discuss Humidity Preferences Vote (HPV). Mainly, the respondents answer whether they would like "more dry (-1)", "no change (0)", "more humid (+1)".

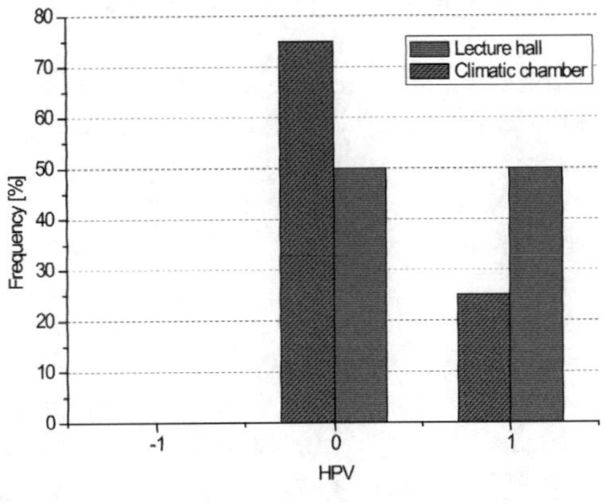

Fig. 7. Humidity Preferences Vote based on student responses.

Taking into account Figure 6 and based on the data in Figure 7, a full analogy can be seen in the respondents' answers and their feelings. There is no percentage change with the "no change" answer (same as in Figure 6). However, you can see that people who indicated in Figure 6 that it is "quite dry" in the rooms under study, in Figure 7, their preference is that the room is "more humid". In general, humidity was felt better in the set conditions than in the lecture hall.

Thermal Sensation Vote and Predicted Mean Vote are no less important indicators. In addition to providing information on the current feelings of people, TSV also enables the average response of the respondents to be found on the basis of surveys, which is then compared with the PMV. It primarily determines whether the prevailing temperature has met the thermal expectations of the respondents. The PMV is calculated on the basis of ISO 7730 [1]. To do this, use the Fanger model [2], which needs parameters such as air temperature, relative humidity, air velocity, clothing insulation, etc. On this basis, it is assessed whether these attractions meet the standard or exceed the established comfort ranges . The comfort range for both is -0.5 to +0.5. Each time it is exceeded, it informs about the feeling of thermal discomfort. Figure 8 shows the calculated TSV from the questionnaires and the PMV from the standard for both test sites.

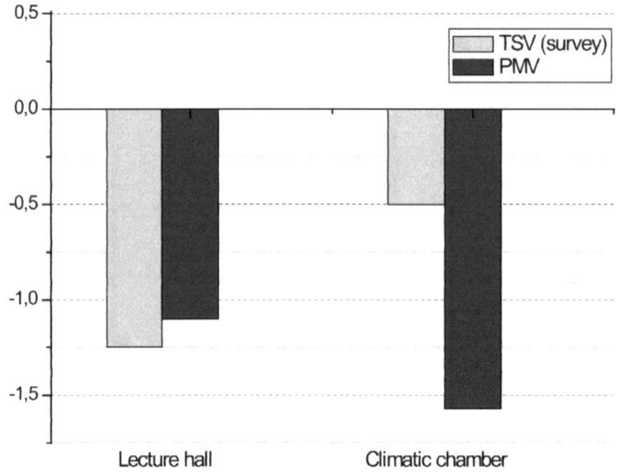

Fig. 8. Comparison of TSV and PMV for the climatic chamber and the lecture hall.

For the lecture hall, TSV from the polls was higher than the PMV. The difference was 0.15. Both indicators were exceeded, which means that the prevailing conditions did not meet the thermal expectations of people and we can say about the feeling of discomfort. Taking into account the climatic chamber, the TSV from the surveys was within the comfort range and was -0.50, but the PMV was already exceeded three tons compared to the TSV. The conclusion that comes to mind is that the temperature in the form set in the climatic chamber was felt much better than in the lecture hall itself. In the study [10] PMV for the climatic chamber was 0.15 and for the room 0.08. Compared to the latest research, you can see the discrepancy in the results. PMV was very low, almost zero. On the other hand, with TSV in the chamber, the mean of respondents' responses was 2, and in room 0.43. Additionally, with the help of questionnaires, we can find out the percentage of people who feel discomfort and are dissatisfied with the prevailing markets (Predicted Percentage of Dissatisfied - PPD). The ISO 7730 standard also makes it possible to find out the value of PPD in line with modern guidelines. It is recognized that 10 % of people may be dissatisfied with the microclimate. Figure 9 shows a comparison of the received values for PPD from the questionnaires and from the standard for the tested rooms.

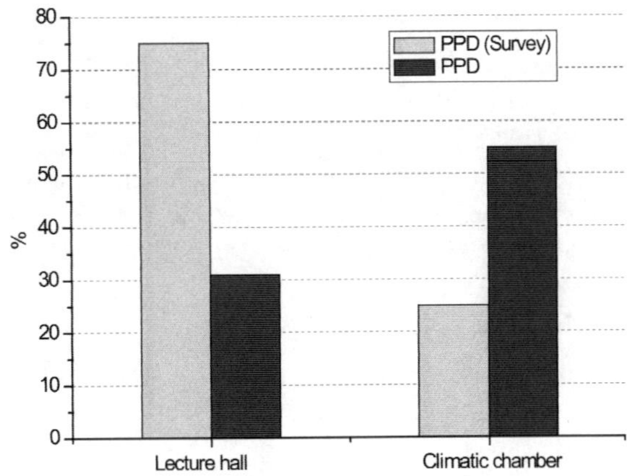

Fig. 9. Comparison of PPD from questionnaires and PPD based on the norm for the climatic chamber and the lecture hall.

For people in the room during the study, the survey showed 75 % dissatisfaction with the indoor climate, while for the chamber it was 25 %. On the other hand, the standardized PPD showed 31 % discomfort for the room, and almost twice as much for the chamber, amounting to about 55 %. This means that none of the examined places can be said to feel thermal comfort.

4 Conclusions

The first of the main conclusions is the fact that, despite similar parameters, the chamber made people feel comfortable in it for 75 % of respondents, while in the lecture room 75% of people marked the answer "acceptable". Moreover, the same percentage of people are satisfied with the relative humidity. The second important finding is that none of the obtained results of PMV and TSV fell within the thermal comfort sensation, except for the TSV for the chamber of 0.50. In the case of PPD, the percentage of dissatisfied people was high especially in the lecture room. This is probably due to the size of the room and the small number of people during the examination. In this case, for the set parameters in the climatic chamber, they turned out to be crucial, because 3 out of 4 people were satisfied with the climatic conditions. This information and data from the tests may have a significant impact in the future on the design of rooms and their size, as well as the possibility of setting specific temperatures for given needs, because it may increase the efficiency at work or the concentration of people in class. Nevertheless, in order to reach an unambiguous conclusion, the research should be continued.

References

1. ISO 7730 International Organisation for Standardization, Ergonomics of the thermal environment – Analytical determination and interpretation of thermal comfort using calculation of the PMV and PPD indices and local thermal comfort criteria, International Standard ISO 7730 (2005).
2. Fanger, P.O. (1974). *Thermal Comfort*, Pub. Arkady, Warsaw ISBN 87-5710-341-0.
3. Nakano, J., Tanabe, S., (2019). Thermal Comfort Condition of Passengers in Naturally Ventilated Train Stations. *Proceedings of the CLIMA 2019 Congress,* 111, pp. 1-6. https://doi.org/10.1051/e3sconf/201911102069

4. Krawczyk, N. (2022). Comparison of thermal comfort in a traditional and intelligent building. *Proceedings of the International Conference on Energy and Green Computing (ICEGC'2021),* 336, pp. 1-6. https://doi.org/10.1051/e3sconf/202233600019
5. Dębska, L., Krakowiak, J. (2021). Thermal environment assessment in selected Polish educational buildings. *Proceedings of the Cold Climate HVAC & Energy* 2021, 246, pp. 1-7. https://doi.org/10.1051/e3sconf/202124615004
6. Krawczyk, N. (2021). Thermal comfort in the low energy building - validation and modification of the Fanger model. *Proceedings of the Cold Climate HVAC & Energy 2021,* 246, pp. 1-6. https://doi.org/10.1051/e3sconf/202124615003
7. Majewski, G., Telejko, M., Orman, Ł., J. (2017). Preliminary results of thermal comfort analysis in selected buildings. *Proceedings of the 9th Conference on Interdisciplinary Problems in Environmental Protection and Engineering EKO-DOK 2017,* 17, pp. 1-7. https://doi.org/10.1051/e3sconf/20171700056
8. Kozanci, O., B., Khovalyg, D., Iida, T., Uno, Y., Ukiana, T., Olecen, B., W. (2019). Human response to the thermal indoor environment created by a radiant, and a combined radiant and convective cooling system. *Proceedings of the CLIMA 2019 Congress,* 111, pp. 1-5. https://doi.org/10.1051/e3sconf/201911102060
9. Solberg, H., Thunshelle, K., Schild, P. (2020). Thermal comfort, thermal sensation and skin temperature measurements using demand-controlled ventilation for individual cooling. *Proceedings of the 12th Nordic Symposium on Building Physics (NSB 2020),* 172, pp. 1-7. https://doi.org/10.1051/e3sconf/202017206001
10. Białek A., Dębska, L. (2022). Verification of the selected thermal comfort model in real and simulated conditions. *Proceedings of the 2nd International Conference on Civil and Environmental Engineering (ICCEE 2022*), 347, 1-9. https://doi.org/10.1051/e3sconf/202234704013
11. Larriva, M., T., B., Mendes, A., S., Forcada, N. (2022). The effect of climatic conditions on occupants' thermal comfort in naturally ventilated nursing homes. *BaE,* vol. 214, 108930. https://doi.org/10.1016/j.buildenv.2022.108930
12. Yuan, F., Yao, R., Sadrizadeh, S., Li, B., Cao, G., Zhang, S., Zhou, S., Liu, H., Bogdan, A., Croitoru, C., Melikov, A., Short, C., A., Li, B. (2022). Thermal comfort in hospital buildings – A literature review. *JoBE,* 45, 103463. https://doi.org/10.1016/j.jobe.2021.103463
13. Rupp, R., F., Parkinson, T., Kim, J., Toftum, J., de Dear, R. (2022). The impact of occupant's thermal sensitivity on adaptive thermal comfort model. BaE, vol. 207, 108517. https://doi.org/10.1016/j.buildenv.2021.108517
14. Hrabovsky, P., Nemec, P., Malcho, M. (2015) Compare cooling effect of different working fluid in thermosyphon, *in Proc of Int. Conf. on Experimental Fluid Mechanics 2014, Český Krumlov, Czech Republic, EPJ Web of Conferences,* 02024. https://doi.org/10.1051/epjconf/20159202024
15. Smitka, M., Malcho, M., Nemec, P., Kolkova, Z. (2013) Use of LHP for cooling power electronic components, *EPJ Web of Conferences*, 45, 01046. https://doi.org/10.1051/epjconf/20146702080
16. Orman, Ł.J. (2014) Boiling heat transfer on single phosphor bronze and copper mesh microstructures, *Proceedings of Int. Conf. "EFM13 – Experimental Fluid Mechanics 2013" (Czech Republic) EPJ Web of Conferences,* 67. https://doi.org/10.1051/epjconf/20146702087

17. Orman, Ł.J. (2014) Boiling heat transfer on meshed surfaces of different aperture, *Proceedings of Int. Conf. on Application of Experimental and Numerical Methods in Fluid Mechanics and Energetics (Slovakia) AIP Conference Proceedings*, 1608, 169-172. https://doi.org/10.1063/1.4892728
18. Chatys, R., Orman, Ł.J. (2017) Technology and properties of layered composites as coatings for heat transfer enhancement. *Mechanics of Composite Materials*, 53 (3), 351-360. https://doi.org/10.1007/s11029-017-9666-8
19. Kaduchová, K., Lenhard, R. (2021) Numerical simulation of passive cooling beam and its optimization to increase the cooling power. *Processes*, 9(8), 1478. https://doi.org/10.3390/pr9081478
20. Klimko, M., Lenhard, R., Žitek, P., Kaduchová, K. (2021) Experimental evaluation of axial reaction turbine stage bucket losses. *Processes*, 9(10), 1816. https://doi.org/10.3390/pr9101816
21. Lenhard, R., Malcho, M., Jandačka, J. (2019) Modelling of heat transfer in the evaporator and condenser of the working fluid in the heat pipe. *Heat transfer engineering*, 40 (3-4), 215-226. https://doi.org/10.1080/01457632.2018.1426225

Aerosol Measurements Before and During Combustion of Various Pellets from Spruce Sawdust and Face Masks FFP2

Nikola Čajová Kantová[1,*], *Alexander* Čaja[2], and *Peter* Hrabovský[1]

[1]Research centre, University of Žilina, Univerzitna 1, 010 26 Žilina, Slovakia
[2]Department of Power Engineering, Faculty of Mechanical Engineering, University of Žilina, Univerzitna 1, 010 26 Žilina, Slovakia

Abstract. Aerosols in the atmosphere affect our health and quality of life. The most commonly measured property is their mass concentration. In this work was measured the mass concentration of aerosols by optical particle sizer in the range from 0.3 μm to 10 μm. It was realized in the laboratory with an automatic heat source for pellet combustion. Aerosols were firstly measured before the combustion process, then during combustion of three types of samples: pellets from pure sawdust, pellets from 5 % of disintegrated face masks FFP2 blended with spruce sawdust and from10 % of disintegrated face masks FFP2 blended with spruce sawdust. The lowest total mass concentration was measured during the combustion of pellets from pure spruce sawdust. However, results indicate that the combustion process does not have an impact on aerosol mass concentration in the laboratory space for the particle range 0.3-10 μm when an automatic heat source for pellets is used.

Keywords: *aerosols, mass concentration, combustion, face masks, pellets*

1 Introduction

Aerosols are microscopic particles that float in the air, which can affect not only visibility and climate but also our health and quality of life. They represent two-phase systems, consisting of the particles and the gas in which they are suspended. Among the most important properties of aerosols is particle size, which is expressed mostly in μm. Another parameter is a shape and it is nearly always spherical for liquid aerosol particles. The most commonly measured property is just mass concentration. It is also most important for health and environmental effects [1, 2].

The primary sources of aerosols in urban regions are vehicle exhausts, airports, thermal plants, industries, and combustion processes [3, 4]. During winter, pollutants are closer to the surface of Earth caused of the lower height of the planetary boundary layer [5].

In the work of Pal et al. were real-time measurements realized by using Scanning Mobility Particle Sizer (SMPS) and Optical Particle Sizer (OPS). The SMPS measures aerosols in the

* Corresponding author: nikola.cajovakantova@uniza.sk

range from 10 nm to 400 nm and the OPS measures in the range from 0.3 μm to 10 μm [6]. Individual ranges and mean diameter of particles measured by OPS are shown in Table 1.

Table 1. Individual ranges and mean diameter measured by OPS.

Range (μm)	Mean diameter (μm)
0.3-0.374	0.337
0.374-0.465	0.419
0.465-0.579	0.522
0.579-0.721	0.65
0.721-0.897	0.809
0.897-1.117	1.007
1.117-1.391	1.254
1.391-1.732	1.562
1.732-2.156	1.944
2.156-2.685	2.421
2.685-3.343	3.014
3.343-4.162	3.752
4.162-5.182	4.672
5.182-6.451	5.816
6.451-8.031	7.241
8.031-10.0	9.015

In this work was measured the mass concentration of aerosols by OPS in mentioned ranges. The measurements were realized in the laboratory with an automatic heat source for pellet combustion. Firstly, aerosols were measured before the combustion process, then they were recorded during combustion. As fuel was chosen pellets consisted of pure sawdust and the other two samples from spruce sawdust blended with disintegrated waste from face masks FFP2. Face masks FFP2 were blended with their weight proportion of 5 % and then 10 %.

2 Material and method

The experimental setup is shown in Figure 1. OPS device was located before the heat source. As the heat source was used automatic boiler for pellet combustion with a heat power of 18 kW and a retort burner. OPS measured the mass concentration of aerosols before combustion and then during the combustion process. During combustion were combusted firstly pellets from pure spruce sawdust, and then pellets from 5 % of disintegrated face masks

FFP2 blended with spruce sawdust and 10 % of disintegrated face masks FFP2 blended with spruce sawdust. Each measurement was performed as a series of ten one-minute measurements. The resulting values of mass concentration represent average values from these series.

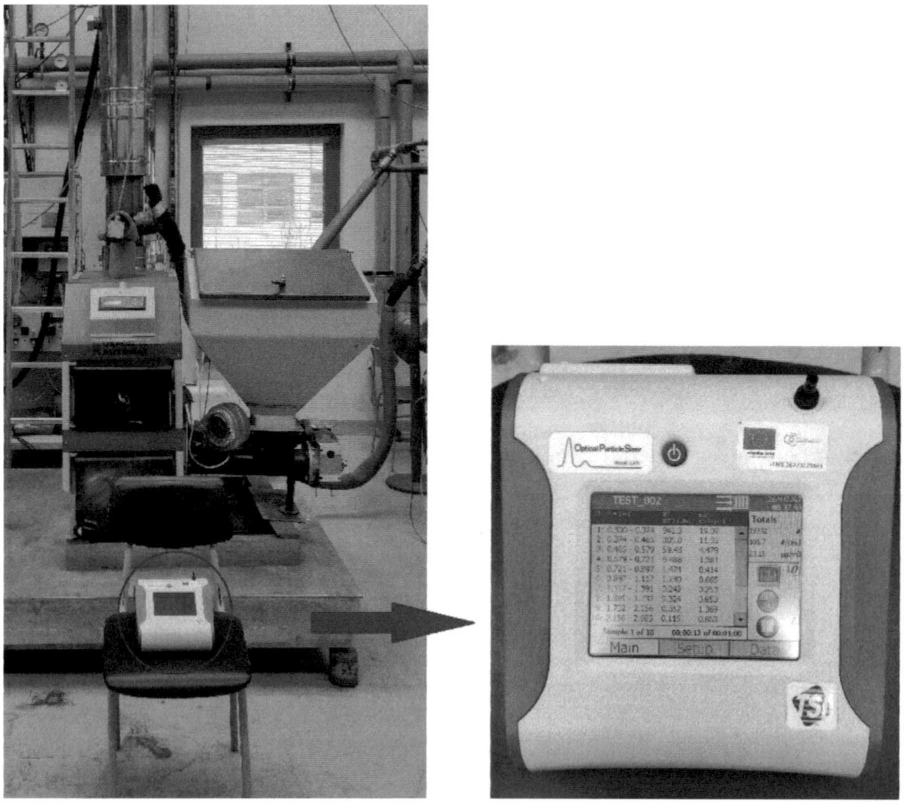

Fig. 1. Experimental setup with OPS device.

3 Results

Results are processed into graphs with the mean diameter of individual ranges and values of average values of measured mass concentrations. Results from the measurement of aerosol mass concentration before combustion are shown in Figure 2. The highest mass concentration was measured with the average value of 78.97 μg/m^3 for the range 8.031-10.0 μm. The total measured mass concentration was 198.67 μg/m^3.

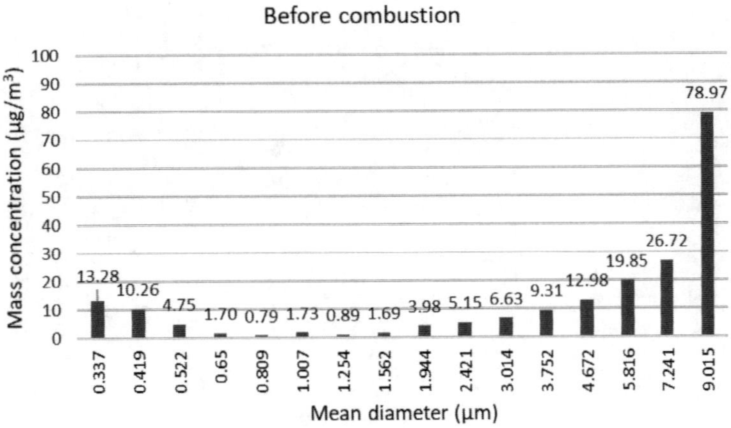

Fig. 2. Aerosol mass concentration before combustion.

Results from the measurement of aerosol mass concentration during combustion of pellets from pure spruce sawdust are shown in Figure 3. The highest mass concentration was measured with the average value of 31.57 µg/m^3 for the range 8.031-10.0 µm. The total measured mass concentration was 94.58 µg/m^3.

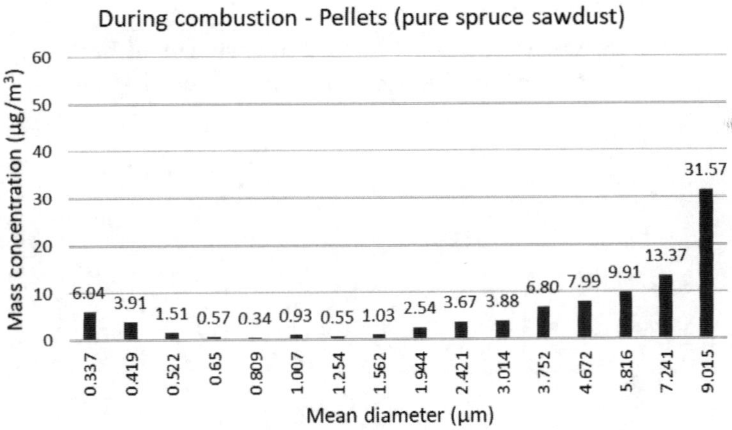

Fig. 3. Aerosol mass concentration during combustion of pellets from pure spruce sawdust.

Results from the measurement of aerosol mass concentration during combustion of pellets from 5 % of disintegrated face masks FFP2 blended with spruce sawdust are shown in Figure 4. The highest mass concentration was measured with the average value of 56.03 µg/m^3 for the range 8.031-10.0 µm. The total measured mass concentration was 164.82 µg/m^3.

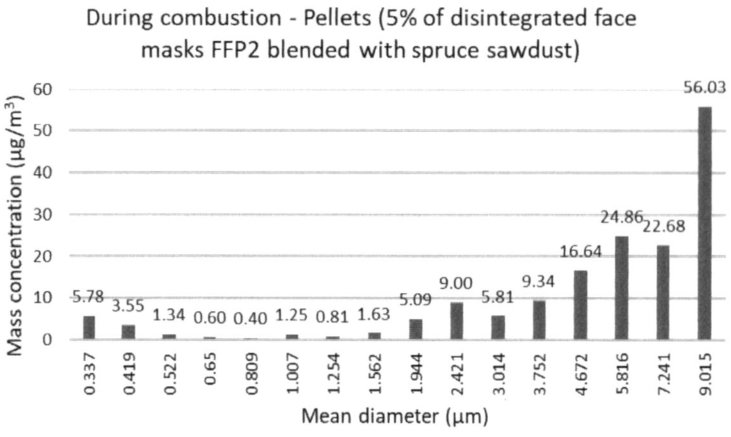

Fig. 4. Aerosol mass concentration during combustion of pellets from 5 % of disintegrated face masks FFP2 blended with spruce sawdust.

Results from the measurement of aerosol mass concentration during combustion of pellets from 10 % of disintegrated face masks FFP2 blended with spruce sawdust are shown in Figure 5. The highest mass concentration was measured with the average value of 52.80 µg/m^3 for the range 8.031-10.0 µm. The total measured mass concentration was 133.86 µg/m^3.

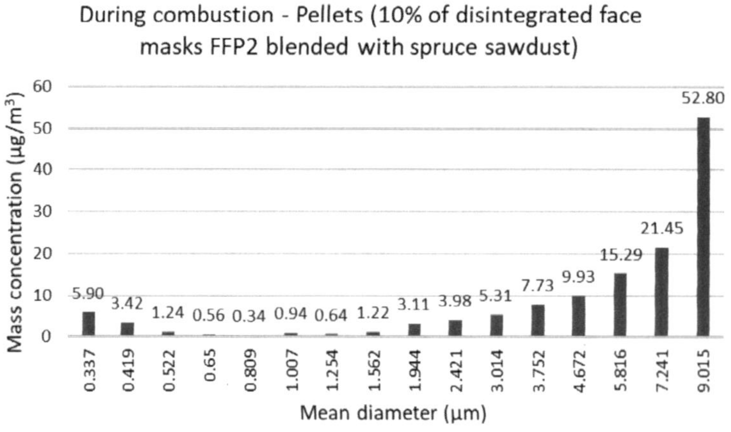

Fig. 5. Aerosol mass concentration during combustion of pellets from 10 % of disintegrated face masks FFP2 blended with spruce sawdust.

4 Conclusion

The most measured property of aerosol and important for health and environmental effects is mass concentration. In this work was mass concentration measured by OPS in laboratory space before and during the combustion process. The lowest total mass concentration was measured during the combustion of pellets from pure spruce sawdust. However, results indicate that the combustion process does not have an impact on aerosol mass concentration in the laboratory space for the particle range 0.3-10 µm when an automatic heat source for pellets is used. It would be interesting to deal with aerosol mass concentration during the filling of pellet hopper or putting wood into fireplaces.

This publication has been produced with the support of the Integrated Infrastructure Operational Program for the project: Creation of a Digital Biobank to support the systemic public research infrastructure, ITMS: 313011AFG4, co-financed by the European Regional Development Fund and VEGA No. 1/0150/22: Energy utilization of produced waste in connection with the COVID-19 pandemic through pellets as an alternative fuel and KEGA No. 021ŽU-4/2021: Primary energy conversion into heat/cold using thermodynamic cycles and compressor cycle with working substance (refrigerant) CO2.

References

1. Hinds, W. C. (1999). *Aerosol Technology. Properties, Behavior, and Measurement of Airborne Particles. Second Edition*. A Wiley – interscience publication : John Wiley & Sons, inc.
2. Pospisil, J., Huzlik, J., Licbinsky, R., Spilacek, M. (2020). Dispersion characteristics of PM10 particles identified by numerical simulation in the vicinity of roads passing through various types of urban areas. *Atmosphere*, 11 (5), 454.
3. Ahlm, L., Liu, S., Day, D.A., Russell, L.M., Weber, R., Gentner, D.R., Goldstein, A.H., DiGangi, J.P., Henry, S.B., Keutsch, F.N. (2012). Formation and growth of ultrafine particles from secondary sources in Bakersfield, California. *Journal of geophysical research,* Vol. 117, 13 p.
4. Taušová, M., Mihalik, E., Čulková, K., Stehlíková, B., Tauš, P., Kudelas, D., Štrba, Ľ., Domaracká, L. (2020). Analysis of municipalwaste development and management in self-governing regions of Slovakia. *Sustainability*, 12 (14), 5818.
5. Järvi, L., Grimmond, C.S.B., Taka, M., Nordbo, A., Setälä, H., Strachan, I.B. (2014). Development of the surface urban energy and water balance scheme (SUEWS) for cold climate cities. *Geoscientific Model Development*, 7, 1691–1711.
6. Pal, D., Dastoor, A., Ariya, P. A. (2020). Aerosols in an urban cold climate: Physical and chemical characteristics of nanoparticles. *Urban Climate,* 34 (2020) 100713.

Effect of Impeller Size on the Performance of a Single Blade Pump

Branislav Knížat[1,*], *Dušan* Šlachtič[1], and *Róbert* Olšiak[1]

[1]STU in Bratislava, Faculty of mechanical engineering, Nám. slobody 17, 81231 Bratislava, Slovakia

Abstract. Changing the impeller diameter is a frequently used method for adjusting pump performance parameters. In the case of conventional multi-blade impellers, this is done by reducing the diameter on the machine tool to the prescribed shape. The other dimensions of the pump (diffuser, inlet) remain unchanged. This method is called trimming. The article deals with the diameter reduction and subsequent modification of the performance parameters of a single blade pump. These pumps are characterized by certain specific features. First of all, it is an unsymmetrical impeller that must be both statically and dynamically balanced. This plays an essential role in the whole modification process. Research results obtained on a pump with an impeller diameter of 138 mm are presented. The change in diameter was monitored on a total of 3 impellers. Experiments were carried out in the laboratory of hydraulic machinery. The results were verified by CFD calculations.

Research background: The article concerns the modification of impellers of single blade pumps. The effects of diameter modifications on multi-blade pumps are currently known. However, these have symmetrical impellers and can therefore be changed without restriction. For asymmetric (single blade) impellers, the problem is more complex as additional mass must be added to provide static and dynamic balance.
Purpose of the article: The aim is to determine the nature of the change in performance parameters when the output diameter is changed. The results of the research can be applied in the prediction of the change in the operating point and the creation of tombstone charts.
Methods: Two kinds of methods were used in the research: experiment and CFD calculation. A total of 4 impeller sizes were investigated.
Findings & Value added: The results of the paper can be divided into two areas. In the experimental area, a device was designed to measure the characteristics of single blade pumps. Four impellers were manufactured and tested. In the area of CFD calculations, simulations of the hydraulic parameters around the best efficiency point (BEP) were performed. The calculation results were verified by experiment. The nature of the change of the BEP when the diameter of the impeller changes up to 87.9 % was found.

Keywords: *single blade pump; trimming; balancing; impeller.*

* Corresponding author: branislav.knizat@stuba.sk

1 Introduction

Parameter adjustment by changing the impeller diameter is a common procedure in the field of turbomachinery, i.e. compressors and pumps and also pumps operating as turbines (PAT). After the modification, a new turbomachine with altered hydraulic parameters is created. By reducing the diameter, the flow rate, head and efficiency are shifted to lower values. In the case of symmetrical impellers (with multiple blades), this is an after-treatment realised after production, carried out on machine tools and is therefore called trimming. In the case of multi-bladed symmetrical impellers, this modification can be carried out without restriction due to the fact that the trimming does not cause mechanical unbalancing. Trimming is an economical method of parameter adjustment and a diameter change of up to a maximum of 15 % is generally recommended. The characteristics of the untrimmed impeller and the maximum trimmed one are defined by the so-called tombstone chart of the particular machine type. The shape of the tombstone chart is determined by the type of turbomachine and the variation of the operating curves depends mainly on the magnitude of the blade angles in the outlet area.

Unsymmetrical impellers of single blade pumps are another case. Here the situation is more complicated, because trimming the impeller results in unbalancing. Performing this operation after manufacture is therefore out of the question. As a rule, manufacturers offer a number of impellers needed to form a tombstone chart of a given type, but these cannot be further modified. They are derived from the original impeller by trimming the outer diameter during design and adding or removing mass as appropriate to ensure overall static and dynamic balance. This paper discusses the effect of reducing the outer diameter of the impeller of a single blade pump on its performance characteristics. The physical nature of the flow as well as the principle of flow and head reduction are the same as for other turbomachines. The difference is that single blade impellers have a large wrap angle and small blade angles, so the sensitivity to diameter change is higher. When making diameter adjustments, it is often necessary to change the blade length to simplify subsequent balancing. However, the basic knowledge is applicable from general turbomachinery theory. Many methods of changing the geometrical parameters by trimming are known. In [10] a so-called pump impeller shroud trimming (PIST) is presented, where the front disk is turned in the range of up to 25 %. The hydraulic parameters as well as the effect of the impeller change on the radial force loading the shaft are monitored. Often the effect of different modifications of the trailing edge is investigated. In fact, the cut does not have to be linear but can be V or U shaped. Such cases are presented in the study [3]. An important observation is the determination and prediction of the change of the position of the best efficiency point (BEP) depending on the change of the output diameter of the impeller. This involves the development of preferably simple relations called trimming laws, derived on the basis of affinity laws. They are presented in papers [1] and [15]. Both experimental methods and CFD simulation methods are used to investigate the effects of trimming. When simulation methods are used, the authors usually verify the results obtained experimentally. In reference [2], the authors investigated the effect of trimming on a hydrodynamic pump with low specific speed. They used CFD methods to investigate the performance parameters (dimensionless head and dimensionless flow) and compared them with the test data. Pressure pulsations and hydraulic noise were monitored depending on the trimming. The results were explained by the shape of the velocity field, which was formed by the interaction of the impeller and the spiral. One of the results is that trimming reduces the pressure pulsations. The authors of the paper [5] investigated the possibility of modifying the impeller of a pump operating as a turbine (PAT). Both experimental research and CFD simulation are presented, and the results of both are verified against each other. The effect of impeller diameter, inlet blade angle, outlet blade angle and wrap angle is analysed. The phenomenon investigated in literature [6] is the axial trimming of the blade. This method is also sometimes used for pump impellers without a front disc. CFD methods have shown that it is possible to reduce the flow rate while maintaining the output pressure and the overall pump efficiency. In [9], the authors analyse the trimming method in terms of its effect on energy losses and the magnitude of radial force. Three impeller trimming schemes are investigated. The pump on which the research was carried out was a double suction pump. The performance parameters calculated by CFD methods were verified by experiment. In this paper, energy dissipation analysis has been carried out to find out the cause of losses and their evolution during the impeller trimming. Another aim of the research was to observe the radial force fluctuations. A significant decrease in forces was found as the impeller diameter decreased. The authors of the paper [13] present an overview of methods to predict the BEP position when the pump is used as a turbine (PAT). One of the methods to adjust the position of the working point is to trim the impeller.

So far, there is not much knowledge in the literature about trimming of single blade pumps. However, this procedure has to be carried out in a different way from the conventional turbomachines described above. Since single blade pumps are being marketed in specific areas such as pumping of solids-containing substances, it is important to deal with this issue. Manufacturers therefore offer tombstone charts for each type. However, the number of impellers in the chart is usually limited to 3-4, as the trimmed impeller has to be designed separately, as it has to be statically and dynamically balanced by adding or removing mass on the rotor body. Balancing is thus an essential part of the design.

There are many articles that deal with the design of single-blade impellers, the analysis of the flow and the resulting consequences such as the radial load on the impeller. These are, for example, the papers [7], [11], [12] and [14]. The effect of the number of blades on the pump operation is discussed in [4].

2 Single blade pump geometry

The machine under investigation is a hydrodynamic single blade pump with a volute diffuser. The impeller has an inclined trailing edge and a maximum output diameter of $D_2 = 138$ mm.

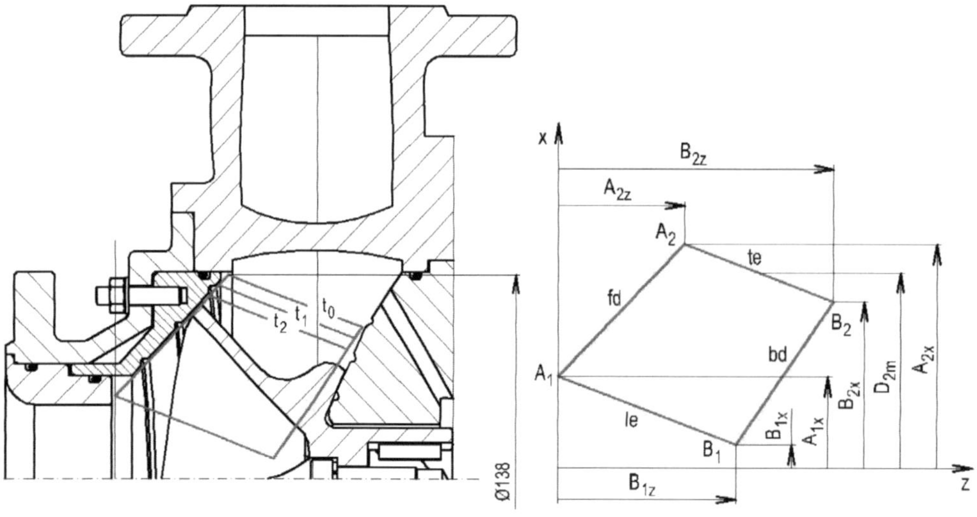

Fig. 1. Dimensions of the tested pump.

Figure 1 shows the basic dimensions of the meridional cut of the pump. The suction and discharge throat of the pump have a diameter of 50 mm. The line markings are as follows: trailing edge (te), leading edge (le), front disc (fd) and back disc (bd). The basic non-trimmed impeller has a blade with a trailing edge marked t_0. From it, the trimmed impellers are then derived with trailing edges marked t_1 and t_2, respectively. Figure 1 shows the shape of the meridional cut as a polygon between the determining points $A_1(A_{1x}, A_{1z})$, $A_2(A_{2x}, A_{2z})$, $B_1(B_{1x}, B_{1z})$ and $B_2(B_{2x}, B_{2z})$ in the coordinate system *x*, *z*. The coordinates of these points are given in Table 1.

Table 1. Parameters of meridional cut.

	t_0	**t_1**	**t_2**
A_{1x}	28	28	28
A_{1z}	0	0	0
A_{2x}	69	65.5	62
A_{2z}	40	36.6	33.2
B_{1x}	7	7	7
B_{1z}	56.6	56.6	56.6
B_{2x}	51.4	47.6	43.8
B_{2z}	87.6	85	82.3

A view of the impeller from the axial direction is shown in figure 2. It shows the streamlines on the front disc (fb) and back disc (bd) and the trailing edge (te).

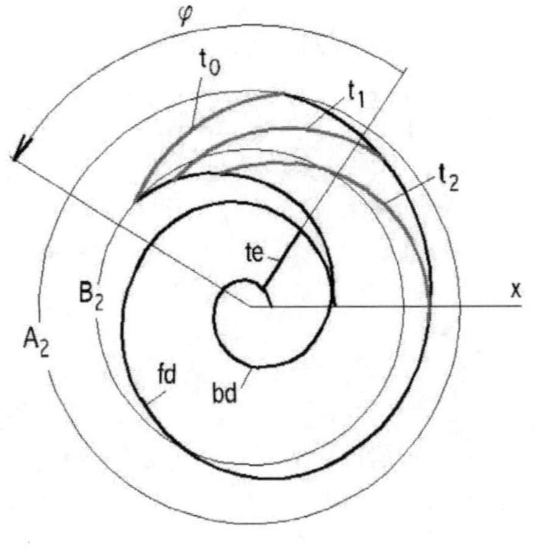

Fig. 2. Axial view on the impeller.

The polar coordinate (angle φ) defining the position of a point on the (fb) and (bd) lines is explained. The shapes of the trailing edges t_0, t_1 and t_2 of each impeller are also indicated in figure 2. Circles A_2 and B_2 correspond to the particular points from the meridional section in figure 1. The same is true for the labels (fd), (bd), (le) and (te). Figure 3 shows the course of the blade angles on the streamlines of the front (fd) and back (bd) discs as a function of the polar coordinate (angle φ from figure 2). The end points of the basic non-trimmed impeller t_0 and the particular trims t_1 and t_2 are marked on the streamlines.

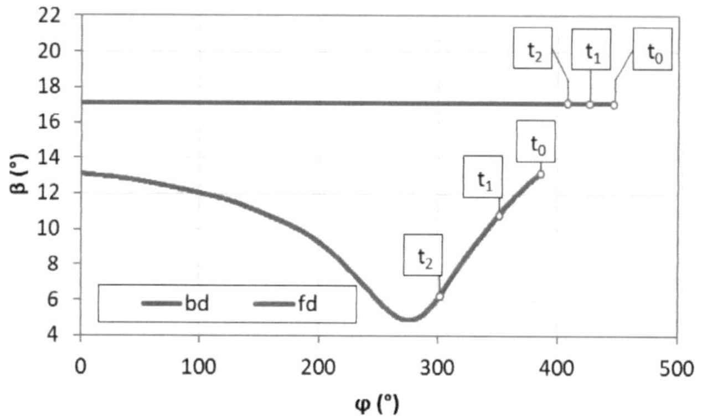

Fig. 3. Blade angle β along streamlines (bd) and (fd).

The surface of the blade is stretched between the front and back disc streamlines and geometrically it is a developable surface that can be unfolded in parts. The impeller had to be rebalanced after each trimming. For an overview of rotor balancing methods, see [8].

3 CFD analysis

The core of the theoretical research was CFD flow analysis. The main objective was the determination of the power curves *Y*(*Q*) and *P*(*Q*). Pump simulations were performed using the Fluid flow CFX module included in the ANSYS 2019R2 software. Before running the actual simulation, it was necessary to prepare a 3D model of the internal part of the pump which was subsequently inserted into ANSYS. For easier resolution, the model was divided into 3 domains: inlet pipe, impeller (OK) and volute. These domains are connected by so-called interface surfaces. The model thus prepared was meshed using the Meshing module and setting the boundary layer thickness to 10^{-5} m. The type of flow analysis was chosen as Steady State together with the turbulent BSL EARSM (Baseline Explicit Algebraic Reynolds Stress Model) applied to the domains and Stage (Mixing-Plane) applied to the interface surfaces. The OK domain was set as rotating with a rotation speed of 2900 min^{-1}. The pressure at both the pump inlet and outlet was set as atmospheric and the simulation was controlled by specifying the flow rate at the pump outlet. By repeating the simulation with the changed flow rate, the *Q*-*Y* dependence of the pump was obtained.

4 Experiments

Figure 4 shows a scheme of the measurement system. The designation of the particular parts of the diagram is as follows: MM – driving motor; AP – auxiliary pump; AM – driving motor for AP; MM – motor; MP – measured pump; TV – throttle valve; SV – shut off valve; FT – feeder tank; CT – collector tank.

The measured parameters speed (n), flowrate (Q), suction pressure (p_1), discharge pressure (p_2), pressure differential ($\Delta p=p_2-p_1$) and torque (M_k) are collected by the DAQ system and evaluated in a PC computer unit. From the measured data, the specific energy Y, the power P and the overall efficiency η are evaluated. The system is characterised by the fact that the liquid (water) is pumped by a metering pump (MP) from the upper tank (FT) to the lower tank (CT). Subsequently, the liquid is pumped back to the upper tank (FT) by an auxiliary pump (AP).

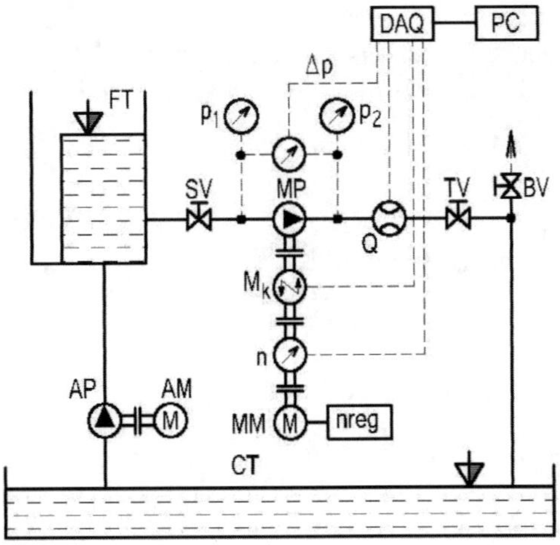

Fig. 4. Scheme of the measurement system.

The suction tank is therefore at a higher geodetic elevation than the discharge tank. This configuration was chosen because the pump has a low height and to compensate for losses in the piping when measuring higher flow rates. The impellers were manufactured from ABS plastic using Rapid Prototyping methods.

5 Results of investigation

All 3 impellers (the original t_0 and the trims t_1 and t_2) shown in figure 2 were tested on the experimental setup and simulated by CFD methods. Thus, by measurements and calculations, the characteristics of the pumps as a function of the flow rate Q were determined: $Y(Q)$, $P(Q)$ and $\eta(Q)$, where Y – specific energy, P – power input and η – efficiency. The simulations were performed as steady state, therefore their results are only relevant in the vicinity of the BEP. The best agreement between the measurement and simulation results was obtained for the non-trimmed impeller t_0. As the impeller was trimmed (i.e., as its outlet diameter decreased), the differences increased.

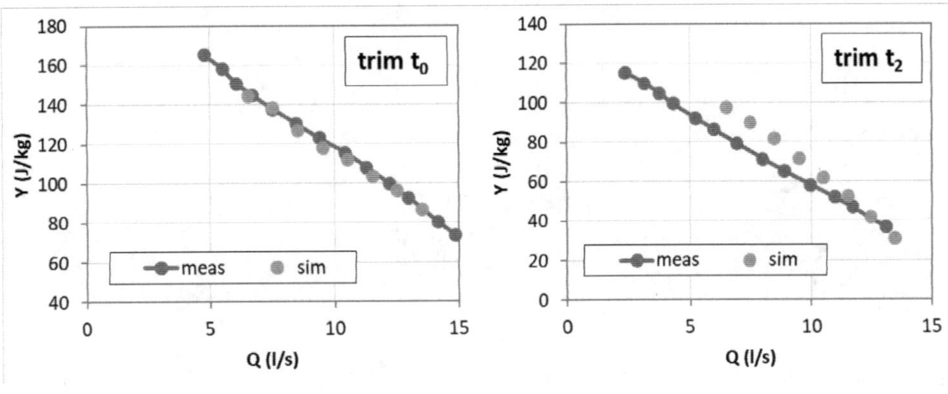

Fig. 5. Measured and simulated specific energy Y. Speed $n = 2900\ \text{min}^{-1}$.

In order to compare the degree of agreement between experiment and CFD, we present the results of only the non-trimmed impeller t_0 and the maximum trim t_2, for which the differences were the largest.
All presented curves were measured or calculated at pump speed n = 2900 min^{-1}. Figure 5 shows the specific energy Y curves. For the variant t_0 (original impeller) we see a very good agreement of the results, indicating a good quality of the computational mesh.

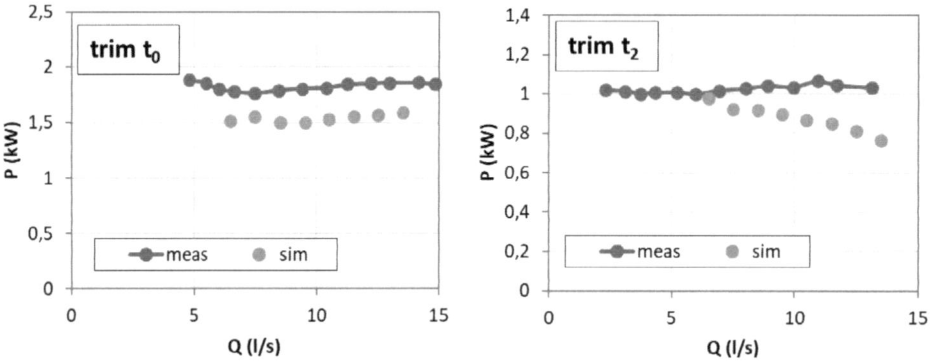

Fig. 6. Measured and simulated input power P. Speed n = 2900 min^{-1}.

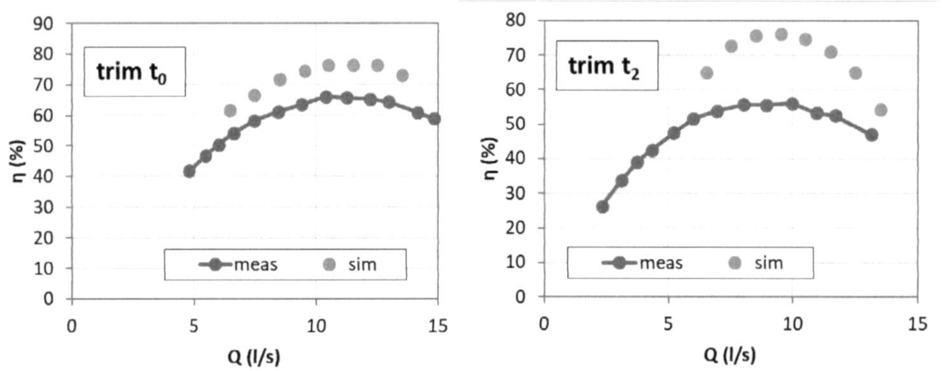

Fig. 7. Measured and simulated efficiency η. Speed n = 2900 min^{-1}.

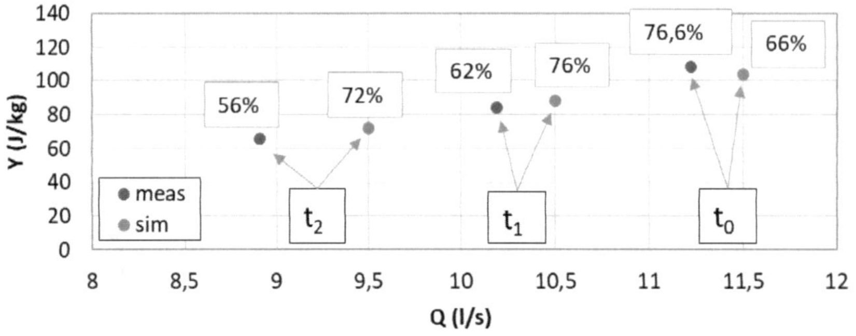

Fig. 8. Best efficiency points with maximum efficiencies. Speed n = 2900 min^{-1}.

In the case of the maximum trim t_2 the difference is larger and especially for smaller flows far from the BEP. The slope of the curves in the case of t_0 is in agreement, for the trim t_2 the calculated curve is steeper than the measured one.

Figure 6 shows the course of the input power P for the variants t_0 and t_2. In the case of the original impeller t_0, the measured power is higher than the calculated one, but the overall character of the course is identical (the values are only shifted by almost the same value with respect to each other). This difference can be explained by the fact that the experimentally determined input power includes the friction in the seal and bearings as well as the disc losses, so it must be higher than the calculated one. In the case of a trimmed impeller t_2 the calculated power input decreases with flow, although the measured power input is kept constant. The above difference is also due to the mechanical losses of the experimental device.

Figure 7 shows the efficiency η for the trims t_0 and t_2. The optimal flow rate determined by experiment and calculation is almost identical. The simulated efficiencies are hydraulic efficiencies and are therefore higher than the measured total efficiencies.

Figure 8 presents the measured and calculated positions of the BEPs of all three investigated impellers (t_0, t_1 and t_2). Numbers indicating the value of the maximum efficiency are attached to the points. The positions of the BEPs lie on a straight line. The largest contribution to the difference between calculation and measurement is the flow rate (for the t_2 trim it is about 7 %). It makes no sense to compare the efficiencies as we would be comparing hydraulic efficiencies with overall efficiencies.

Some studies provide rules that can be used to calculate the performance parameters for impeller trimming. For example, the authors in [15] give the following relations. They introduce the parameter λ as the ratio of the trimmed diameter D_{2i} to the original diameter D_{20}:

$$\lambda = \frac{D_{2i}}{D_{20}} \tag{1}$$

The calculation of the position of the BEP is done by applying the rules, which are modified affinity laws:

$$\frac{Q_i}{Q_0} = \lambda^{1.445} \tag{2}$$

$$\frac{Y_i}{Y_0} = \lambda^{2.090} \tag{3}$$

where Q a Y represent flow rate and specific energy in the BEP.

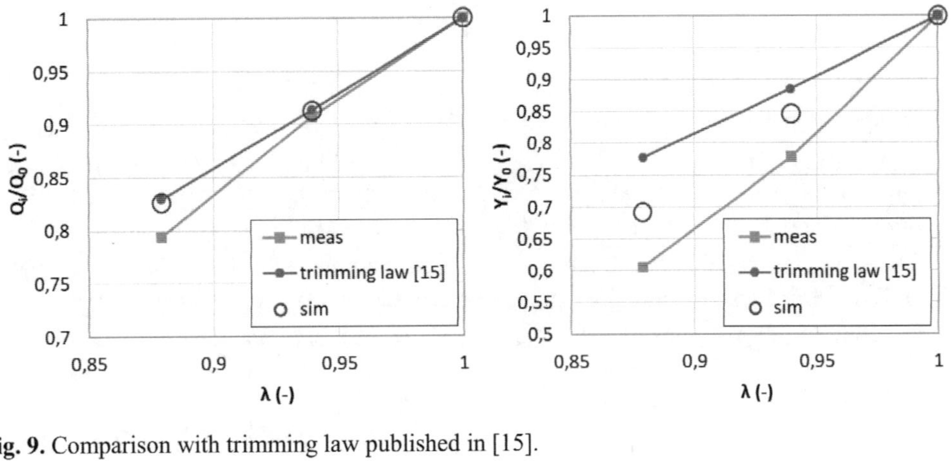

Fig. 9. Comparison with trimming law published in [15].

The index i takes the values 0, 1, 2 according to the impellers t_0, t_1 and t_2. The exponents in relations (2) and (3) are determined experimentally and are valid for multi-blade pumps.

Figure 9 shows a comparison of the measured and calculated ratios Q_i/Q_0 and Y_i/Y_0 versus the parameter λ with the results of relations (2) and (3). Decreasing the diameter of the single-blade impeller has almost the same effect on the change of the optimum flow rate as that of the multi-blade pump. However, the change in the optimum specific energy is more pronounced. Therefore, care must be taken when trimming the impeller of a single blade pump to ensure that the drop in height is not too great after the diameter has been reduced.

6 Conclusions

Analyses have been carried out regarding the influence of the single blade impeller trimming on the pump performance parameters. The research was carried out on a single blade pump with a maximum output of 2 kW. Experimental research methods were used as well as CFD methods. A total of three impellers with different trim sizes made of ABS plastic by method of Rapid Prototyping were tested.
The experimental and CFD simulation results are compared at the level of power curves. A good agreement between the specific energy curves for all impellers is achieved. The best agreement was in the case of the base size t_0 (non-trimmed impeller). The deviation increased with decreasing impeller diameter and was largest for the trim t_2. The measured power reached higher values (about 15 %) than the power simulated by CFD methods. This difference is due to the fact that neither mechanical losses nor hydraulic losses due to friction between the discs and the stator are included in the simulated power input. For the same reason, the efficiency found by simulation is then higher than the efficiency found by measurement.
The nature of the decrease of the parameters flow rate Q and specific energy Y as a function of the relative trim λ is presented in the paper. These results are compared with previously known results for multi-blade pumps. As far as the flow rate is concerned, the decrease in its proportional value is in agreement with the multi-blade pumps. However, the drop in the relative specific energy is much steeper than for conventional pumps. This result is due to the physical conditions of the flow in a single blade impeller, in particular the large slip and small blade angles. It is therefore not possible to automatically adopt the results and experience known from multi-blade pumps.

This work was supported by the Slovak grant agency KEGA, project No 016STU-4/2022.

References

1. Li, CX; Wang, SL and Jia, YK (2011). The performance of a centrifugal fan with enlarged impeller. *Energy Conversion and Management* 52 (2011) 2902–2910.
2. Gangipamula R, Ranjan P, Patil RS (2022). Study on fluid dynamic characteristics of a low specific speed centrifugal pump with emphasis on trimming operations. *International Journal of Heat and Fluid Flow* 95 (2022) 108952.
3. Khoeini D, Tavakoli MR (2018). Flow Characteristics of a Centrifugal Pump with Different Impeller Trimming Methods. *FME Transactions* (2018) 46, 463-468
4. Elyamin G.R.H.A., Bassily M.A., Khalil K.Y., Gomaa M.Sh (2019). Effect of impeller blades number on the performance of a centrifugal pump. *Alexandria Engineering Journal* (2019) 58, 39–48.
5. Sheng Y.S., Yu K.F., Ming J.W., Yun Q.X. (2013). Effects of impeller trimming influencing pump as turbine. *Computers & Fluids* 67 (2012) 72–78

6. Swain D, Engeda A. (2014). Performance impact of impeller blade trimming on centrifugal compressors. *Proc IMechE Part A: J Power and Energy* 2014, Vol. 228(8) 878–888.
7. Tan L, Yang Y., Shi W., Chen Ch., Xie Z. (2021). Influence of Blade Wrap Angle on the Hydrodynamic Radial Force of Single Blade Centrifugal Pump. *Appl. Sci.* 2021, 11, 9052.
8. Li L., Cao S., Li J., Nie R., Hou L. (2021). Review of Rotor Balancing Methods. *Machines* 2021, 9, 89.
9. Deng Q, Pei J, Wang W, Lin B, Zhang C, Zhao J. (2021). Energy Loss and Radial Force Variation Caused by Impeller Trimming in a Double-Suction Centrifugal Pump. *Entropy* 2021, 23, 1228.
10. Shadab M, Karimipour M, Najafi A F, Paydar R, Nourbakhsh S A (2022). Effect of impeller shroud trimming on the hydraulic performance of centrifugal pumps with low and medium specific speeds. *Engineering Applications of Computational Fluid Mechanics* 2022, VOL. 16, NO. 1, 514–535
11. Chen J, Shi W, Zhang D. (2021). Influence of blade inlet angle on the performance of a single blade centrifugal pump. *Engineering Applications of Computational Fluid Mechanics* 2021, VOL. 15, NO. 1, 462–475
12. Benra, F. K. (2006). Numerical and Experimental Investigation on the Flow Induced Oscillations of a Single-Blade Pump Impeller. Journal of Fluids Engineering 2006, 128, 783–793
13. Liu M, Tan L, Ca S. (2022). Performance Prediction and Geometry Optimization for Application of Pump as Turbine: A Review. Optimization for Application of Pump as Turbine: A Review. *Front. Energy Res.* 9:818118.
14. Quan H, Li R, Su Q, Han W, and Wang P (2013). Analysis on Energy Conversion of Screw Centrifugal Pump in Impeller Domain Based on Profile Lines. Hindawi Publishing Corporation *Advances in Mechanical Engineering Volume* 2013, Article ID 512523, 11 pages
15. Li W. G. (2011). Impeller Trimming of an Industrial Centrifugal Viscous Oil Pump. *Int J Advanced Design and Manufacturing Technology*, Vol. 5/ No. 1/ December – 2011

Experimental and Numerical Analysis of Leakage Characteristics of Proportional Directional Valve

Marian Ledvoň[1,*], *Lumír* Hružík[1], *Adam* Bureček[1], and *Filip* Dýrr[1]

[1]VSB-Technical University of Ostrava, Faculty of Mechanical Engineering, Department of Hydromechanics and Hydraulic Equipment, 17. listopadu 2172/15, 708 00 Ostrava-Poruba, Czech Republic

Abstract. This article deals with the experimental and numerical analysis of the leakage characteristics of the proportional directional valve. These characteristics describe the flow through the directional valve in the region of the center position of the spool valve. The element under investigation is a three-position four-way directional valve with zero overlap and feedback from the valve spool position. The actual valve spool position is sensed by an inductive position sensor and processed by integrated electronics. By means of an offset on the integrated electronics, it is also possible to set the center position of the valve spool. By moving the valve spool, the size and direction of the working fluid flow in both directions can be controlled. Internal leakage occurs due to axial and radial clearances between the spool and the sleeve. The magnitude of the axial clearances that occur at the control edges is investigated. The blocked-line pressure sensitivity curve, leakage flow curve and center flow curve are determined by experimental measurements. Depending on the experimental measurement, the correction of the center position of the valve spool is then made using offset. The flow through the control edges at the center position of the valve spool was simulated using Ansys Fluent software. Subsequently, the geometry of the flow simulation model is adjusted. The adjustment is made to take into account manufacturing tolerances. Finally, the simulated dependencies including the effect of manufacturing tolerances are compared with the measured and analytically determined characteristics.

Research background: Spool valve description, mathematical equations, flow analysis through valve, radial and axial clearance.
Purpose of the article: Effect of axial clearance on leakage characteristic.
Methods: Experimental measurements, numerical simulation.
Findings & Value added: Determining the range of the center flow curve as a function of manufacturing tolerances

Keywords: *proportional directional valve; spool valve; axial clearance; leakage characteristic; numerical simulation; radial clearance*

* Corresponding author: marian.ledvon@vsb.cz

1 Introduction

In industrial applications, proportional directional valves are often used to control hydraulic motors. Proportional directional valve allows continuous control of the magnitude and direction of flow in hydraulic systems [1, 2]. From the design point of view of the control element we distinguish between spool and poppet valve. Poppet valve ensures tightness in a certain direction of fluid flow. In spool valve, leak occurs between the spool and the sleeve and internal fluid leakage occurs [3]. Fluid leakage is undesirable but unavoidable. In many applications with directional spool valve, internal leakage is an important factor affecting the design of the hydraulic system. Due to leak, fluid leakage occurs, which can affect energy losses in the system or also undesired movement of the unloaded hydraulic motor [4]. The amount of internal fluid leakage between the valve spool and the sleeve is influenced by the radial clearance and the type of spool valve overlap. The overlap size of the individual control edges of the directional valve may vary depending on the requirements of the hydraulic system [5, 6]. The actual dimensions of the radial clearance and all functional spool surfaces depend on the accuracy of the manufacturing process. The same applies to the overlap of the channels in the axial direction. Internal leakage can also be affected by surface topography errors, which include roughness, corrugation and shape error. These errors lead to an uneven gap field between the spool and the sleeve, resulting in leakage [7]. In spool valve, adhesive and abrasive wear of the functional surfaces and edges of the spool valve occurs during use. This wear causes leakage to increase [8, 9, 10]. Internal leakage can also be affected by thermal deformation of the hydraulic flow orifices caused by viscous heating [11]. Although manufacturing inaccuracies are on the order of a few microns, this can significantly affect the behaviour of the valve in the center position region of the directional valve.

2 Theoretical background

2.1 General valve analysis

The investigated element is a proportional directional valve PRL2-06-32-0-24 from Argo Hytos. This directional valve is mainly used for continuous flow control to hydraulic motors. It is a directly controlled, three-position, four-way directional valve with a sharp edge spool. The overlap of the spool in the center position is stated as zero. The control part of the directional valve consists of a linear motor. The armature of the linear motor is rigidly connected to the valve spool. An inductive spool position sensor is also part of the directional valve. The internal electrical feedback from the position of the valve spool ensures higher accuracy of the directional valve [12].

In general, for the flow through the valve orifices is described by the orifice equation:

$$Q_{Oi} = C_D w_i x \sqrt{\frac{2}{\rho} \Delta p_{Oi}}, \tag{1}$$

where C_D is the flow coefficient, w is the area gradient of the valve, x is the spool stroke, ρ is the fluid density, Δp_O is the pressure drop across the control edges, $i = 1, 2, 3, 4$ is the index corresponding to the given pair of control edges, see Fig. 1.

If a positive overlap of the control edges is achieved, the equation for flow in a narrow cylindrical gap applies [13]:

$$Q_{Gi} = \frac{\pi D_{SP} s^3}{12 \eta x} \Delta p_{Gi}, \tag{2}$$

where D_{SP} is the spool outer diameter, s is the radial clearance, η is the fluid dynamic viscosity, x is the spool stroke corresponding to the length of the narrow cylindrical gap and Δp_G is the pressure drop across the narrow cylindrical gap.

If the valve spool is in the center position, the following equation can be used to calculated the center flow rate [14]:

$$Q_{Ci} = \frac{\pi w_i s^2}{32 \eta} \Delta p_{Ci}, \tag{3}$$

where s is the radial clearance, η the fluid dynamic viscosity, w is the area gradient of the valve and Δp_G is the pressure drop across the control edges.

After adjusting equation (3), the following equations apply to the calculation of the pressure gradient at the control edges:

$$\Delta p_{Ci} = \frac{32 \eta Q_{Ci}}{\pi w_i s^2}, \tag{4}$$

This equation is then used to analytically calculate the center flow rate for comparison with experimental measurement and mathematical simulation.

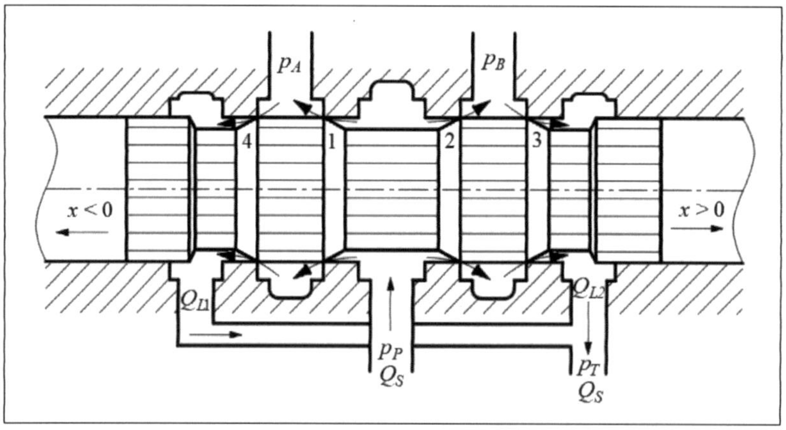

Fig. 1. Analysis of the flow through the 4/3 spool valve in center position $x = 0$.

There are several critical dimensions that must be maintained in manufacture of spool valve. These are the axial and radial clearances between the spool and the sleeve, see Fig. 2. For high performance valves, these dimensions are strictly tolerated in the range $U_{AT} = U_{PA} = (- 2.54 \div 2.54)$ μm. In some cases, a range $U_{AT} = U_{PA} = (- 7.62 \div 7.62)$ μm may be acceptable [14]. These tolerances significantly affect the internal leakage and pressure sensitivity around the center position. The spool outer diameter D_{SP} and the sleeve diameter D_{SL} were measured using a Metrosoft Quartis. The actual radial clearance between the spool and the sleeve is $s = 3$ μm. The amount of axial clearances that occur at the control edges is the subject of research. In the case of valves with zero overlap, the clearance is ideally zero, in reality a positive or negative overlap occurs. The amount of the overlap or underlap is determined by the manufacturing variation.

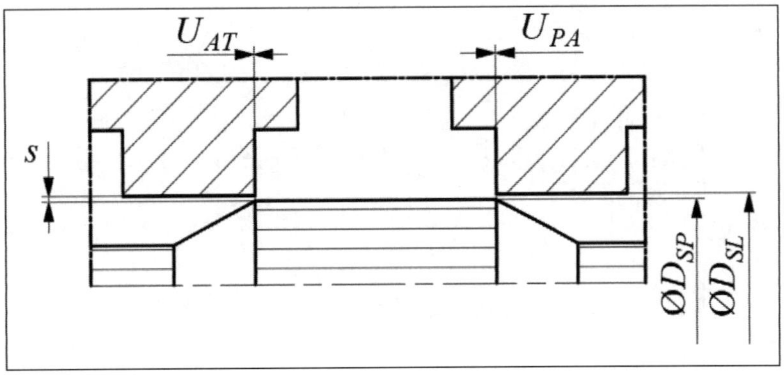

Fig. 2. Control edge geometry with zero overlap.

3 Experimental analysis

Three leakage characteristics are measured and subsequently evaluated in the experimental measurements. These characteristics describe the behaviour of the investigated directional valve around of the center position, when leakage occurs due to radial and axial clearance.

3.1 Experimental test stand

Fig. 3 shows the hydraulic scheme of the experimental equipment that was used to determine the leakage characteristics of the proportional directional valve. The source of pressure energy is a hydraulic unit with a constant pressure control axial piston hydrogenerator. Experimental measurements are performed on the proportional directional valve PRL2 from Argo Hytos. The proportional directional valve is connected to the system by a pressure pipe P. The load ports of valve A and B are blocked by pressure sensors S2 and S3. The load flow between ports A and B is zero. The fluid is drained back to the tank by a return pipe T. Measuring points for connecting pressure sensors S1 and S4 in front and behind of the valve are placed in the hydraulic circuit. The spool position is sensed by an integrated inductive position sensor. The flow meter S5 and temperature sensor S6 are connected in front of the proportional directional valve. The measuring instrument MS 5070 from Hydrotechnik is used to record the measurements. The working liquid is HV46 oil with kinematic viscosity $\nu = 45.3\ mm^2 \cdot s^{-1}$ at oil temperature $t_O = 40$ °C. The measuring range and accuracy of all the sensors used is given in Table 1.

Table 1. Technical parameters of individual sensors.

Sensors	**Measuring range**	**Measuring accuracy**
Pressure sensor S1	(0 ÷ 400) bar	± 0.25% of full scale
Pressure sensor S2, S3	(0 ÷ 250) bar	
Pressure sensor S4	(0 ÷ 60) bar	
Flow meter S5	(0.05 ÷ 5) $dm^3 \cdot min^{-1}$	up to ± 0.4% of reading
Temperature sensor S6	(-50 ÷ 200) °C	0.3 + 0.005 t_O °C

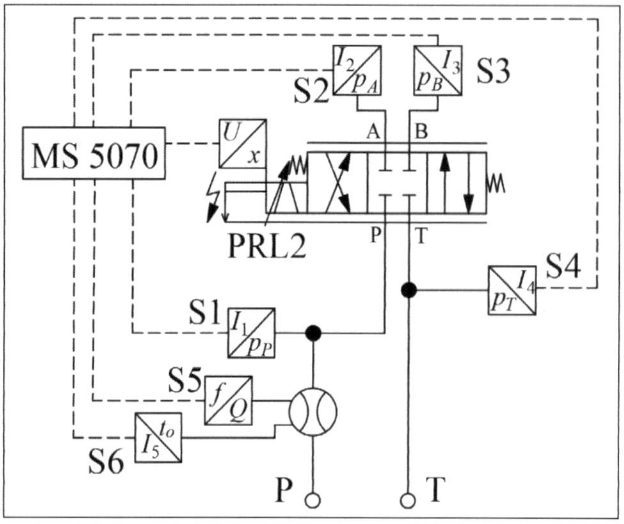

Fig. 3. Scheme of experimental equipment.

3.2 Experimental measurements

Experimental measurements to determine the leakage characteristics are performed with load ports A and B blocked, so the load flow is zero. Supply pressure p_P, pressures p_A a p_B in the load ports, return pressure p_T, flow rate Q, spool position x a working liquid temperature t_O are measured using by a MS 5070. Working liquid temperature is constant t_O = 40 °C. The supply pressure is set to a constant value p_P = 250 bar for determine the blocked-line pressure sensitivity curve a leakage curve. The spool is in the center position and the supply pressure p_P is gradually increased in the range p_P = (50 ÷ 250) bar for determine center flow curve. The measurements were first made with the manufacturer's original setting for the center position of the valve spool. The center position of the directional valve is defined as the state when the pressure difference p_A and p_B at the blocked load ports A and B is zero resp. the pressures are equal. Depending on the determined blocked-line pressure sensitivity curves, the correction of the center position of the valve spool is then performed. Subsequently, the measurements were performed again with a modified center position of the valve spool.

3.2.1 Blocked-line pressure sensitivity curve

Fig. 4 shows the determined blocked-line pressure sensitivity curves in the range of spool stroke x = (-20 ÷ 20) %. For a greater spool stroke in both directions, the pressures p_A and p_B remain constant. From the waveforms of the determined characteristics with the original settings from the manufacturer, it is found that the center position of the valve spool is reached at the input voltage U_x = 0.11 V. Depending on the determined input voltage for the center position, the correction of the center position of the valve spool was then using offset. From the evaluation of the determined characteristics after correction it can be seen that the pressures p_A and p_B are equal at zero spool stroke. The difference in spool stroke before and after correction corresponds to 1.4% of the maximum spool stroke. In the center position of the valve spool, the load ports have the same pressure, which corresponds to 58% supply pressure. It can be said that the pressure drop at the control edges PA and AT is not the same. The difference of pressure drops can be caused by axial clearances on individual control edges that are affected by manufacturing tolerances.

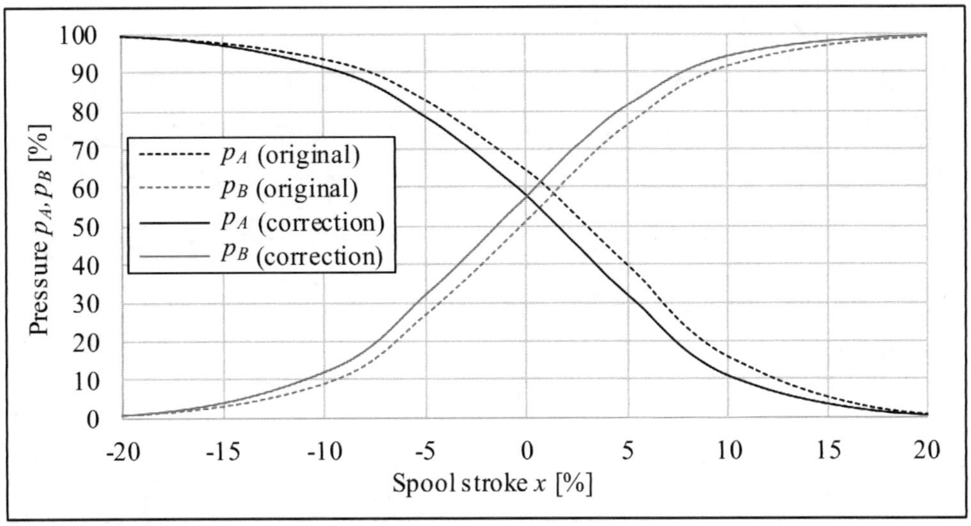

Fig. 4. Blocked-line pressure sensitivity curves.

3.2.2 Leakage flow curve

The leakage curves for the original and adjusted settings of the center position of the valve spool using the correction are shown in Fig. 5. It is clear from the graph that in the center position of the valve spool, the leakage flow Q_L is maximum and decreases rapidly with spool stroke x in both directions. The decrease is due to the increasing length of edge overlap corresponding to the spool stroke.

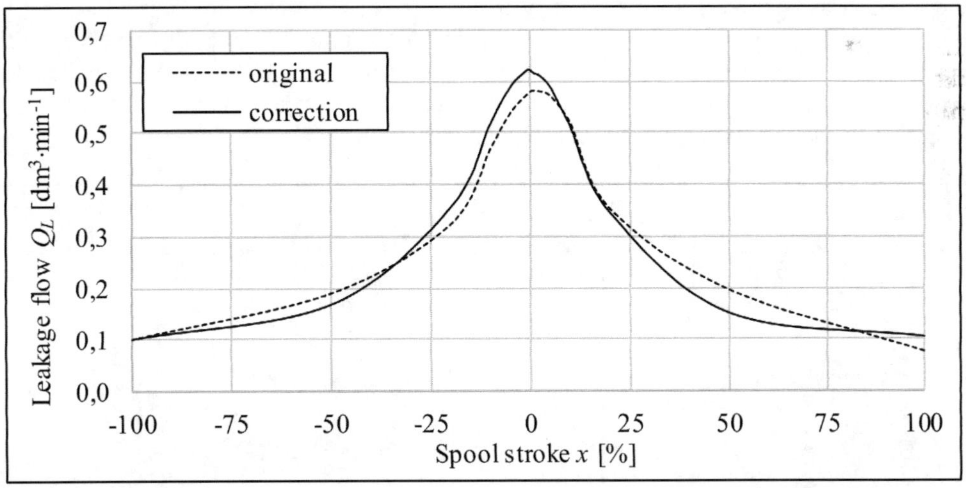

Fig. 5. Leakage curves.

3.2.3 Center flow curve

The center flow curves for the original and adjusted settings of the center position of the valve spool using the correction are shown in Fig. 6. It is clear from the graph that the center flow Q_C as a function of supply pressure p_P increases almost linearly. The center flow Q_C for the maximum supply pressure p_P corresponds to the maximum leakage flow Q_L in Fig. 5.

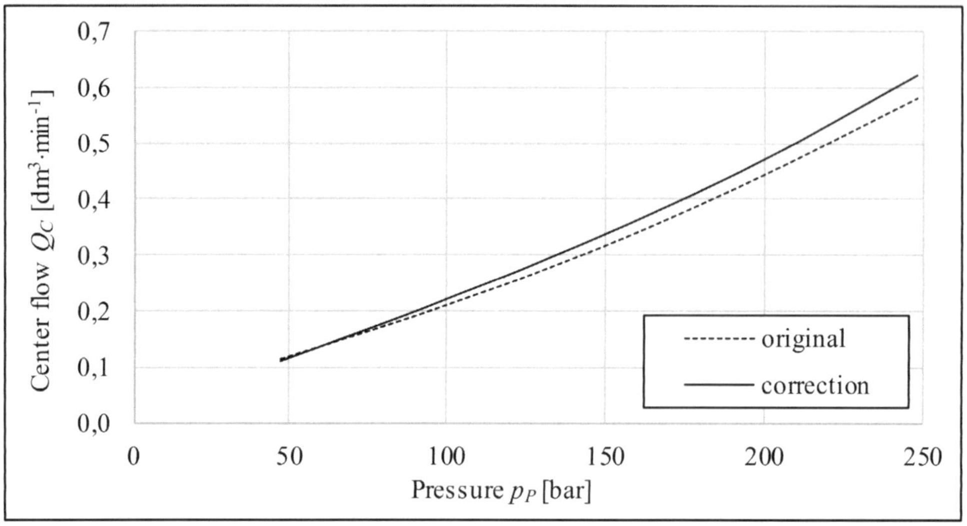

Fig. 6. Center flow curves.

4 Numerical simulation

The simulations of the flow through the directional valve with the valve spool in the center position and with the load ports blocked were carried out. The inverse geometry of the solved region is created in ANSYS Design Modeler. To simplify the calculation, half of the directional valve with spool for channel A is modelled, see Figure 7a. The mesh is created in ANSYS Mesher, see Figure 8. The mesh is refined in the area of the narrowest cross-section. The mesh was subsequently adapted according to the velocity gradient to refine the calculation. The monitored parameter in the mesh adaptations was the pressure drop at the control edges. Further refinement of the mesh at the control edges is achieved by adapting the mesh. The number of elements after the adaptations is approximately 15 million. The definition of each boundary condition is shown in Fig. 7b.

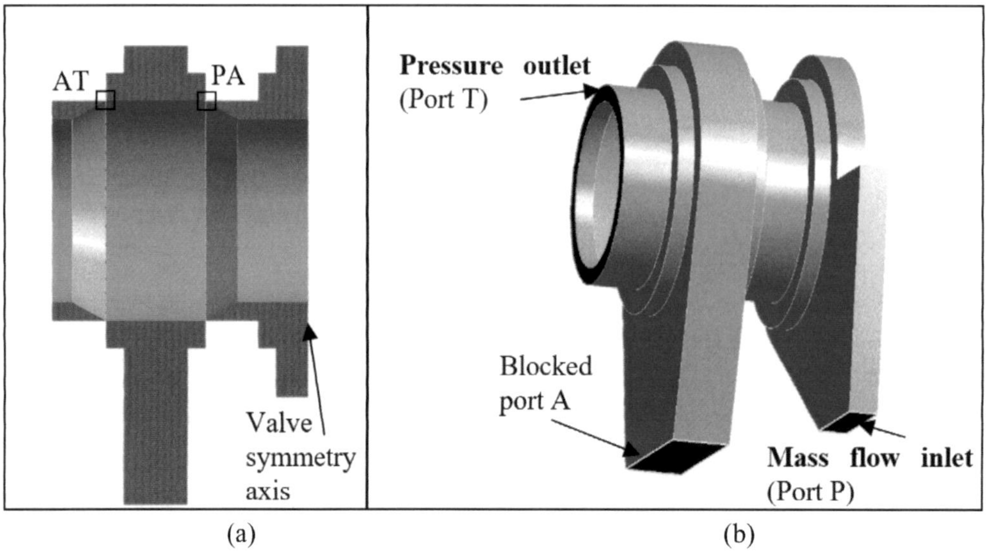

Fig. 7. (a) Cross-section of modelled region for channel A, (b) boundary conditions.

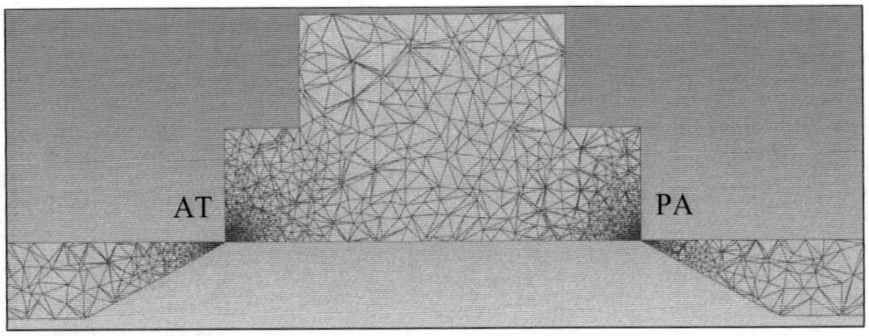

Fig. 8. Mesh of solved region.

Mathematical simulations are performed for geometries directional valve where $U_{AT} = U_{PA} = -2.54$ µm (underlap tolerance), $U_{AT} = U_{PA} = 0$ µm (null tolerance) and $U_{AT} = U_{PA} = 2.54$ µm (overlap tolerance). Symmetric tolerance is considered in the simulations where $U_{AT} = U_{PA}$. The first null tolerance simulation is performed for an ideal directional valve geometry with zero overlap. Underlap and overlap simulations are performed for directional valve geometries that take into account manufacturing tolerances. The RANS model *k-ω* (SST) is used for the numerical simulations. Constant boundary conditions were considered for all simulations performed. The inlet boundary condition is defined by the mass flow rate Q_m and the outlet boundary condition is defined by the outlet pressure p_{outlet}. The mass flow rate is set to half the mass flow rate to simulate flow through only channel A. The settings of the input and output boundary conditions including the fluid parameters are shown in Table 2.

Tab. 2. Boundary conditions and fluids parameters

	Q_m	p_{outlet}
	[kg·s^{-1}]	**[Pa]**
1.	0.000199	0
2.	0.000398	0
3.	0.000667	0
4.	0.001008	0
5.	0.001363	0
6.	0.001732	0
7.	0.002144	0
8.	0.002599	0
9.	0.003081	0
10.	0.003593	0
11.	0.004402	0
ρ_o	852	[kg·m^{-3}]
η_o	0.0385	[Pa·s]

5 Results

Center flow curves determined by mathematical simulations for individual directional valve geometries are shown in Fig. 9. The overlap and underlap tolerances define the range of centre flows that are affected by the size of the axial clearances. The range of centre flow resp. axial clearances is defined for the maximum permitted manufacturing tolerances. In the graph are compared the mathematically simulated dependencies and the experimentally and analytically determined center flow curves. Equation (4) was used for the analytically determined curve. The parameters used for the analytical calculation correspond to the parameters defined for the mathematical simulation. The experimentally determined curve corresponds to the analytically determined curve and is within the defined range.

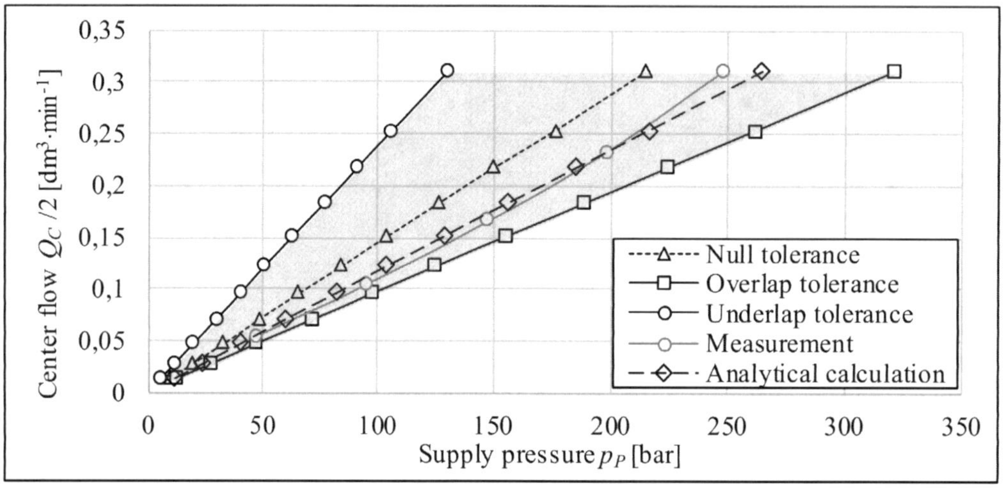

Fig. 9. Mathematically simulated, experimentally and analytically determined center flow curve.

6 Conclusion

This article deals with the experimental and mathematical analysis of leakage characteristics of proportional directional valve. Blocked-line pressure sensitivity curve, leakage flow curve and center flow curve were successively measured and evaluated. As part of the experimental analysis, the correction of the central position of the valve spool was performed. Subsequently, numerical simulations of the flow with the valve spool in the center position are performed for different directional valve geometries taking into account manufacturing tolerances. The determined overlap tolerance and underlap tolerance curves define the range of center flow that is affected by the maximum permitted manufacturing tolerance. Subsequently, the dependencies determined by simulation are compared with experimentally and analytically determined center flow curves.

This work was supported by the European Regional Development Fund in the Research Centre of Advanced Mechatronic Systems project, project number CZ.02.1.01/0.0/0.0/16_019/0000867 within the Operational Programme Research, Development and Education.

This work has been supported by The Ministry of Education, Youth and Sports of the Czech Republic from the Specific Research Project SP2022/32 (Numerical modeling of transient fluid flow problems with the support of experimental research)

References

1. Hružík, L., Bureček, A., Vašina, M. (2017). Effect of Oil Viscosity on Pulsating Flow in Pipe. *Mechanisms and Machine Science*, 44, 137-143.
2. Soon, C. C., Ghazali, R., Jaafar, H. I., Hussien, S. Y. S., Sam, Y. M., Rahmat, M. F. (2016). The Effects of Parameter Variation in Open-Loop and Closed-Loop Control Scheme for an Electro-Hydraulic Actuator System. *International Journal of Control and Automation*, 9(11), 283-294.

3. Gordić, D., Babić, M., Milovanović, D., Savić, S. (2011). Spool valve leakage behaviour. *Archives of civil and mechanical engineering,* 11(4), 859-866.
4. Mao, X., Wu, C., Ding, H., Li, B., & Liu, Y. (2020). Effect Analysis of Leakage in the Middle Position and Improvement of an O-type 3-position-4-way Directional Valve with the Spool Structure. *IOP Conference Series-Earth and Environmental Science*, 508, 012160.
5. Lu, Z., Zhang, J., Xu, B., Wang, D., Su, Q., Qian, J., Yang, G., Pan, M. (2019). Deadzone compensation control based on detection of micro flow rate in pilot stage of proportional directional valve. *ISA transactions*, 94, 234-245.
6. Zhang, L., Fu, W., Yuan, X., Meng, Z. (2020). Research on Optimal Control of Excavator Negative Control Swing System. *Processes*, 8(9), 1096.
7. Tang, W., Xu, G., Zhang, S., Jin, S., & Wang, R. (2021). Digital twin-driven mating performance analysis for precision spool valve. *Machines*, 9(8), 157.
8. Chen, Y., Gong, W., & Kang, R. (2016). Coupling behaviour between adhesive and abrasive wear mechanism of aero-hydraulic spool valves. *Chinese journal of aeronautics*, 29(4), 1119-1131
9. Liu, X., Ji, H., Min, W., Zheng, Z., Wang, J. (2020). Erosion behaviour and influence of solid particles in hydraulic spool valve without notches. *Engineering Failure Analysis*, 108, 104262.
10. Fang, X., Yao, J., Yin, X., Chen, X., & Zhang, C. (2013). Physics-of-failure models of erosion wear in electrohydraulic servovalve, and erosion wear life prediction method. *Mechatronics*, 23(8), 1202-1214.
11. Chen, Q., Ji, H., Zhao, H., Zhao, J. (2020). Optimization Algorithm and Joint Simulation to Micro Thermal Deformation Using Temperature Measurement in the Orifice of Hydraulic Valve. *Processes*, 8(9), 1136.
12. Ledvoň, M., Polášek, T., Bureček, A., Hružík, L. (2019). Modeling and Dynamic Analysis Directional Valve. *AIP Conference Proceedings*, 2118, 030026
13. Manring, N. D. (2005). *Hydraulic control systems*. John Wiley & Sons, Inc.
14. Merritt, H. E. (1967). *Hydraulic control systems*. John Wiley & Sons, Inc.

Experimental Analysis of Flow Through Throttle Valve During Gaseous Cavitation

Tomáš Polášek[1,*], *Lumír* Hružík[1], *Adam* Bureček[1] and *Marian* Ledvoň[1]

[1]VSB-Technical University of Ostrava, Faculty of Mechanical Engineering, Department of Hydromechanics and Hydraulic Equipment, 17. listopadu 2172/15, 708 00 Ostrava-Poruba, Czech Republic

Abstract. The multiphase flow in oil hydraulic systems has a very significant effect on the correct operation of the hydraulic system. Air can be found in various states in hydraulic systems, while free entrained air in the form of bubbles has the potential to be the most problematic. It especially affects the compressibility of the hydraulic liquid resulting in reduced stiffness of the hydraulic system. The actuators of the hydraulic mechanisms then do not achieve the fast response and the precision of movements depending on the input control signals. One possibility for the contamination of hydraulic fluid by air bubbles is through a phenomenon known as gaseous cavitation. This is a phenomenon in which gas is released when the pressure drops below the saturation pressure of the dissolved gas in the liquid. This article focuses on the experimental analysis of the flow through the throttle valve which is affected by the formation of air bubbles at the throttle edge of the valve. The regions of gaseous cavitation were observed at the different flow cross-section of the throttle valve. The throttle valve was placed into the block of transparent material to provide visualization of the individual measurements. The article is supplemented with photographs of the individual measurements showing the gaseous cavitation inception.

Research background: flow cross-section, cavitation phenomenon, discharge coefficient.
Purpose of the article: Effect of flow cross-section size and flow velocity on cavitation development.
Methods: Experimental measurements.
Findings & Value added: The investigation of the gaseous cavitation inception, Visualization of the individual measurements.

Keywords: *throttle valve, cavitation number, discharge coefficient, saturation pressure, visualization*

* Corresponding author: tomas.polasek@vsb.cz

1 Introduction

The throttle valves are variable hydraulic resistances that can be used to control the amount of flow rate. When the liquid passes through these throttle elements, there is significant pressure loss depending on the size of the flow cross-section. These pressure losses have a significant effect on the energy balance of hydraulic systems [1]. Due to the significant change in the flow cross-section, a large increase of the flow velocity occurs on the flow control elements which causes a local decrease of the static pressure. At these location, the gaseous and vapour cavitation can be expected.

All hydraulic liquids, including oil, contain a specific amount of air. The air can be considered as a contaminant in hydraulic equipment which changes the properties of the working fluid and the hydraulic mechanism [2-4]. Therefore, air can exist in different states. While the most discussed is the free entrained air in the form of small bubbles. One of the possibilities of hydraulic equipment contamination by free entrained air is leakage of individual connections in the suction line of the pumps. Due to the insufficient tightness of the sealing surfaces, the air is drawn in by the vacuum in the suction line and entrained into the working areas of the pumps. Another possibility of the hydraulic equipment contamination is due to a phenomenon known as gaseous cavitation. The cause of gas cavitation in hydraulic systems is very similar to the vapour cavitation. The difference between these two phenomena is in the contained gases in the formed cavities. The vapour cavitation occurs when the pressure drops below the saturation vapour pressure of the liquid. The process of evaporation and formation of vapour bubbles occurs including a small amount of non-condensable gas (air) at the given temperature [5]. It is similar in the gaseous cavitation. When the pressure drops below the level of the saturation pressure of the dissolved gas in the liquid, the gas is released at the given temperature. These are different physical processes that occur in liquids. While the vapour cavitation is accompanied by the phase change between liquid and gas phases, gas cavitation is associated with diffusion processes in the liquid [2, 11]. Liquids are affected by both processes [6-9], and both significantly affect the hydraulic mechanisms.

The hydraulic oils also exhibit both gaseous and vapour cavitation [8-11], but the conditions for the release of oil vapour bubbles are more difficult to achieve in comparison to the release of water vapour bubbles. This is due to the different properties of the liquids, which are manifested, among other things, by the very low saturation vapour pressure of the oil. This can vary depending on the operating temperature range, additives added, type of hydraulic oil, etc. However, gaseous cavitation, which is influenced by the amount of dissolved gas in the liquid, is far more common in the oil hydraulic systems. The hydraulic oils can contain much more dissolved air (up to 10 %) in comparison to the water [2]. The amount of dissolved gas in the liquid is directly proportional to the partial pressure of the gas in the liquid which is related by Henry's law. Thus, the solubility of gas in liquids depends on temperature and pressure. Due to the constant changes in pressure and temperature in hydraulic systems, gas is absorbed or released to maintain this equilibrium state.

2 Theoretical background

Contamination of the hydraulic fluid with air occurs already in the tank where the gas dissolves at atmospheric pressure at the interface between the hydraulic liquid and air. The amount of dissolved gas depends on temperature, pressure and it is described by Henry's law. This law states that when a gaseous mixture is in contact with a liquid, the amount of any dissolved gas in the gaseous mixture is directly proportional to the partial pressure of the given gas above the surface. The mathematical description of this law has many forms, and therefore the resulting values and dimensions of Henry's constant vary depending on the formula used. The individual formulas are summarized in the literature [12]. The example of mathematical description of the Henry's law is as follows [12]:

$$x_{G,i} = H^{xp} \cdot p_{G,i}, \tag{1}$$

where $x_{G,I}$ is the molar fraction of the i-th dissolved gas in the liquid, H^{xp} is the Henry's constant which depends on the temperature and the given gas/liquid combination, $p_{G,i}$ is the partial pressure of the i-th gas above the surface. The partial pressure of the given gas is a fraction of the total mixture pressure with respect to the given gas content in the mixture. This is described by Dalton's law:

$$p_{G,i} = p_{total} \cdot x_{G,i}, \tag{2}$$

where p_{total} is the total pressure of the mixture. The gases with a very small fraction in the mixture are neglected in the calculations.

When the hydraulic oil passes through the throttle edges of the valve or orifice, the flow velocity increases and the static pressure drops. This correlation between flow velocity and static pressure can be described by Bernoulli's equation which states that the total energy at any point of the flow is constant. At the location of the narrowest flow area, there is the greatest flow velocity increase and the greatest static pressure decrease at the same time. Therefore, the vapour and gaseous cavitation formation is assumed to occur at this location. The cavitation number has been introduced to quantitatively evaluate the cavitation [13, 14]. The cavitation number is based on the knowledge of the pressure coefficient C_p, where at some point of the flow field, the static pressure is reduced so that the condition is reached [13]:

$$p_{\min} \leq p_v, \tag{3}$$

where p_{min} is the minimum static pressure in the flow field, p_v is the saturation vapour pressure of the liquid. In order to evaluate gaseous cavitation formation when the liquid passes through the throttle valve, the relations were considered where instead of the saturation vapour pressure p_v, the saturation pressure p_{sat} of the dissolved gases was considered. Thus, the dimensionless cavitation number has the resulting relation [13, 14]:

$$\sigma = \frac{2 \cdot (p_1 - p_{sat})}{\rho \cdot v_{CS}^2}, \tag{4}$$

where p_1 is the measured absolute upstream pressure of the throttle valve, p_{sat} is the saturation pressure of the dissolved gases, ρ is the oil density at the given temperature and v_{CS} is the flow velocity at the narrowest flow cross-section. In general, this parameter can be characterized as the ratio of the force trying to suppress cavitation to the force trying to cause it [14]. In the case of flow through the valve, the flow velocity in the dynamic pressure expression can be related directly to the pressure drop Δp [15]. Pressure drop can be measured directly on the valve being investigated. Thus, the cavitation number can be modified into the following form of the cavitation index [10, 14, 15]:

$$\sigma_v = \frac{p_1 - p_{sat}}{\Delta p}, \tag{5}$$

where Δp is the pressure drop of the valve. The assumption was considered in the definition of p_{sat}. When the pressure drops about 40 000 Pa, relative to the atmospheric pressure, there is a violent release of air bubbles [17]. Therefore, the value of the saturation pressure of the dissolved gases was determined to be p_{sat} = 60 000 Pa. Because of the cavitation occurrence, it is necessary to investigate the flow conditions at different valve openings and to determine the impact on the valve function or the device. The 3D model of the investigated throttle valve is shown in Fig. 1.

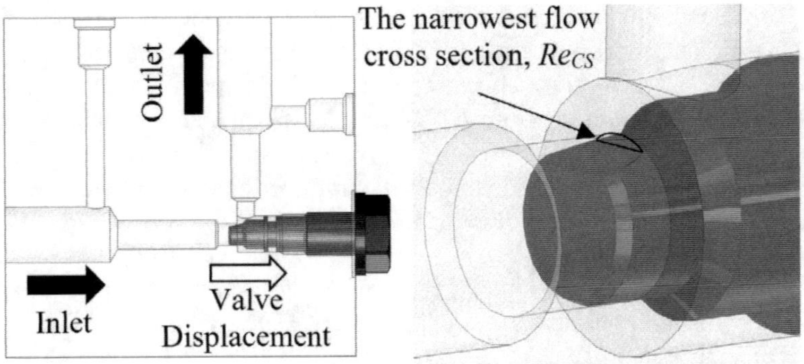

Fig. 1. The 3D model of the investigated valve.

The Reynolds number was determined to evaluate the flow regime through the valve. The Reynolds number was determined for a non-circular flow cross-section in the narrowest flow cross-section of the valve:

$$Re_{CS} = \frac{v_{CS} \cdot d_h}{\upsilon}, \tag{6}$$

where v_{CS} is the mean flow velocity in the narrowest flow cross-section, d_h is the hydraulic diameter and υ is the oil kinematic viscosity at $t_O = 40$ °C. In the case of a non-circular flow cross-section, the hydraulic diameter d_h was determined according to equation:

$$d_h = 4 \cdot \frac{S_{CS}}{o_{CS}}, \tag{7}$$

where S_{CS} is the non-circular flow cross-section, o_{CS} is the wetted perimeter of the flow cross-section. To characterize the loss through the valve, the loss coefficient was determined to the location of the narrowest flow cross-section [16]:

$$\zeta = \frac{2 \cdot \Delta p \cdot S_{CS}^2}{\rho \cdot Q_v^2}, \tag{8}$$

where Δp is the pressure drop of the valve, Q_v is the volume flow rate through the throttle valve, ρ is the oil density at the temperature of $t_O = 40$ °C. The flow through the throttle elements can be characterised by the equation:

$$Q_v = \mu \cdot S_{CS} \cdot \sqrt{\frac{2 \cdot \Delta p}{\rho}}, \tag{9}$$

where μ is the discharge coefficient. The discharge coefficient is in correlation with the loss coefficient and can be expressed by the following equation [16]:

$$\mu = \frac{1}{\sqrt{\zeta}} \tag{10}$$

3 Experimental measurement

The device for experimental analysis of flow through the throttle valve affected by gaseous cavitation was built. Hydraulic oil HV46 was used in the realization of the experimental measurements. For this mineral oil, the dependencies of the oil dynamic viscosity η_O and oil density ρ_O on the temperature t_O were determined. The measurement was performed in the range of oil temperature $t_O = (39.5 \div 43.5)$ °C. The throttle valve was placed into the block of transparent material. Photographic images were recorded during the individual measurements using a camera with a macro lens. The images were used to evaluate the inception of gaseous cavitation and also to evaluate the displacement of the throttle valve relative to the seat. The measurements were performed for 3 variants of the throttle valve opening VSV2-QC2/1 from Argo Hytos company. The experimental measurement was terminated when the upstream pressure p_1 = 40 bar was reached due to the strength of the transparent block. The downstream pressure p_2 was influenced by the pressure loss in the return line. Fig. 2 shows the hydraulic circuit diagram of the experimental device for flow analysis through the throttle valve including the used sensors.

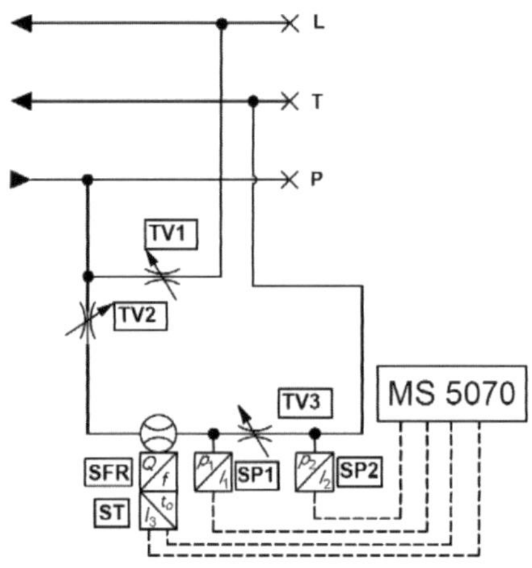

SFR	Hydrotechnik QG100. Measuring range (0.2 ÷ 30) $dm^3 \cdot min^{-1}$. Accuracy ±0.5 % of measured value.
ST	Hydrotechnik TE300. Measuring range (-50 ÷ 200) °C. Accuracy max. 0.3+0.005·t °C
SP1	Hydrotechnik PR400. Measuring range (0 ÷ 60) bar. Accuracy ±0.25 % of full scale.
SP2	Hydrotechnik PR300. Measuring range (-1 ÷ 6) bar. Accuracy 0.5 % of full scale.

Fig. 2. The hydraulic circuit diagram with the used sensors.

The hydraulic unit with an axial piston pump was the source of the flow which distributed the fluid through the high-pressure pipeline P. The amount of the volume flow rate Q_v was controlled by changing the pump speed or by changing the angle of the swash plate. Two throttle valves TV1 and TV2 were placed in the hydraulic circuit to provide flow branching. This layout was chosen to measure the characteristics in the range of the small volume flow rates, but also to achieve greater variability in adjusting the required volume flow rate Q_v. The amount of volume flow rate Q_v was measured using the gear flowmeter SFR. The temperature sensor ST was also placed on the flowmeter to record the oil temperature t_O. To evaluate the pressure drop Δp at the throttle valve TV3, pressures p_1 and p_2 were recorded using pressure sensors SP1 and SP2. The pipelines L and T were used to return the hydraulic oil into the hydraulic unit tank.

4 Experimental results

The evaluation of the displacement was performed for 3 variants of valve opening. In the camera software, the images from the measurements were analysed and the valve displacements were determined to be s = 1.31; 2.31 and 3.31 mm. Based on the displacements, the flow cross-section of the throttle valve were determined, see Fig. 3.

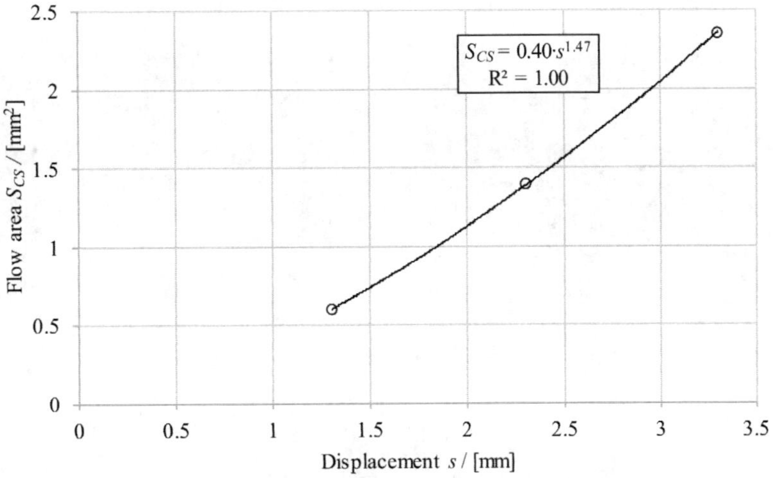

Fig. 3. The dependence of the flow cross-section S_{CS} on the throttle valve displacement s.

Because of the relatively small volume flow rates and the very low saturation vapour pressure of the oil, it can be assumed that the inception of cavitation was mainly influenced by the released air bubbles. The figure below shows the evaluation of the static Δp-Q characteristic of the flow through the throttle valve. In Fig. 4, the equations of the trend curves are evaluated and the location of the gaseous cavitation inception are marked in the form of cavitation index $\sigma_{v,i}$ for each valve displacement s ($\sigma_{v1,i}$, $\sigma_{v2,i}$, $\sigma_{v3,i}$).

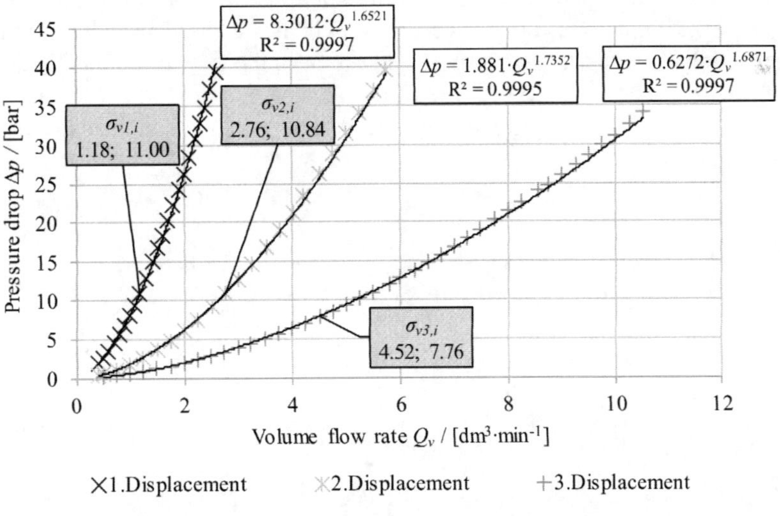

Fig. 4. The static Δp-Q characteristic of the flow through the throttle valve for different flow cross-sections S_{CS}.

In Fig. 4, it is not observed that gaseous or vapour cavitation significantly affects the static Δp-Q characteristic of the throttle valve. However, it can be observed that in the case of the 1st and the 2nd displacement, the gaseous cavitation inception occurred at approximately the same pressure drop. While for the 3rd displacement at the lower pressure drop. The images of the cavitation inception A1 to A3 and the images of the last measurement point B1 to B3 are shown in Fig. 5, also see Fig. 4.

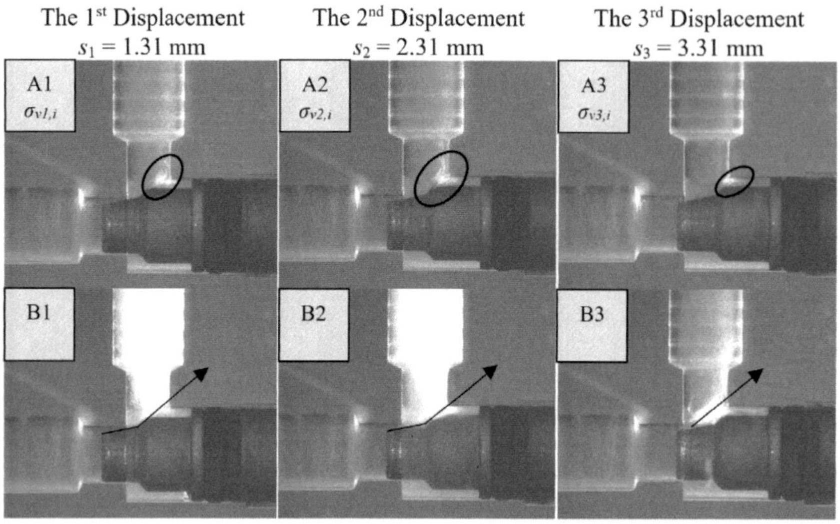

Fig. 5. The graphical evaluation of the gaseous cavitation inception at throttle valve A1, A2, A3 and cavitation at the highest achieved volume flow rates B1, B2, B3 for the individual displacements *s*.

It can be observed in Fig. 5 that for the 1st and the 2nd displacement, there was a significant increase of the gas phase downstream of the valve during the experimental measurements (see Fig. 5 - B1 and B2). While for the 3rd displacement the increase was significantly smaller (see Fig. 5 - B3). At the same time, it can be observed that a significant increase of the gas bubbles occurs in the entrainment where the flow was detached from the valve body (see Fig. 5 - A1, A2, A3). The volume fraction of the gas phase could be affected by the higher downstream pressure p_2, see Fig. 6, which increased with increasing volume flow rate Q_v, but also due to the change in the direction of the flow immediately downstream of the throttle edge of the valve. In the case of the 1st displacement, the effect of the change in flow direction was the most significant and decreased with the further displacements (see Fig. 5 - B1, B2, B3).

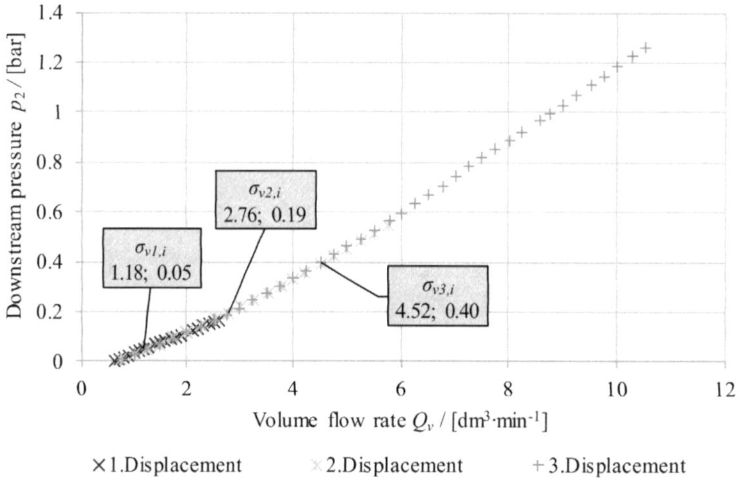

Fig. 6. The dependence of the downstream pressure p_2 on the volume flow rate Q_v.

Next, the mean flow velocities were v_{CS} evaluated and determined to the narrowest flow cross-section of the valve S_{CS}. In Fig. 7, the course of the evaluated dependencies can be observed, where the locations of the gaseous cavitation inception are marked again.

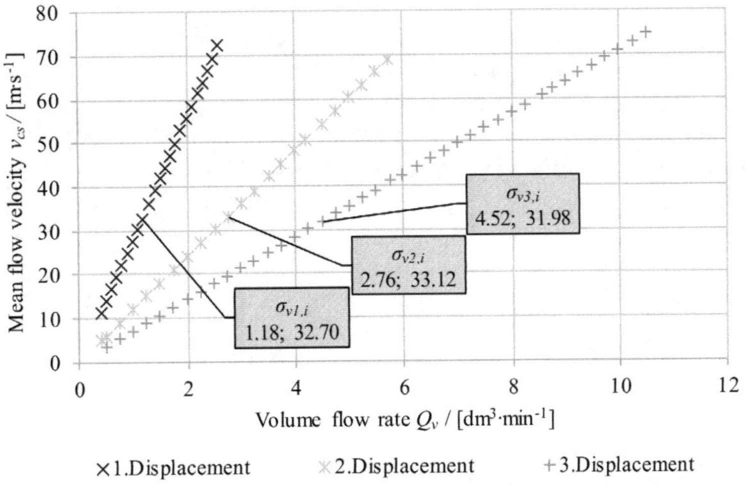

Fig. 7. The dependence of the mean flow velocity v_{CS} on the volume flow rate Q_v.

Fig. 7 shows that the gaseous cavitation inception occurred at approximately the same mean flow velocity v_{CS}. Thus, it can be assumed that the dynamic pressure in the throttle region reaches similar values as the flow cross-section S_{CS} of the throttle valve increases. As the flow cross-section S_{CS} increases, the flow velocity is dispersed more into the space. For this reason, the largest amount of gas phase can be observed at the 1st and the 2nd displacement. After the liquid passed through the throttle edge, a jet was produced which bypassed the valve until the point of break-off. In contrast, the smallest amount of gas phase can be observed at the 3rd displacement when the flow cross-section S_{CS} of the valve is the largest. For these reasons, the different intensities of air bubble release are achieved at the gaseous cavitation inception. The dependence of the discharge coefficient μ on the Reynolds number Re_{CS} was further evaluated to investigate the flow, (see Fig. 8).

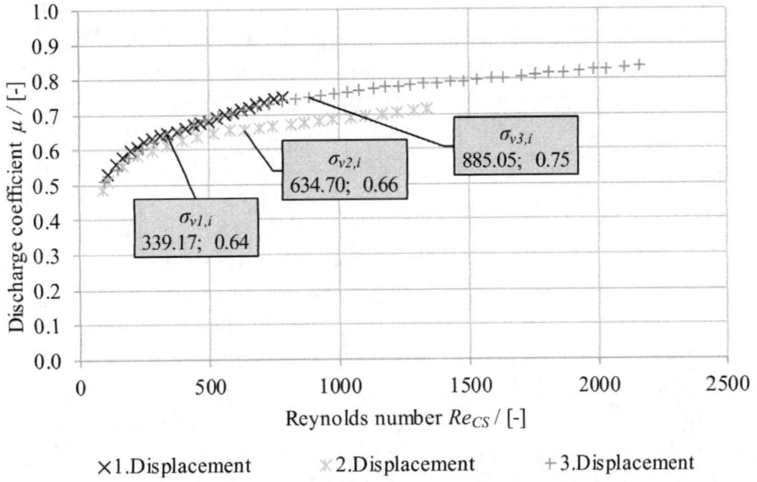

Fig. 8. The dependence of the discharge coefficient μ on the Reynolds number Re_{CS}.

It can be observed in Fig. 8 that at the gaseous cavitation inception, the value of the discharge coefficients in the case of the 1st and the 2nd displacement reached approximately the same value $\mu = (0.64 \div 0.66)$. While in the case of the 3rd displacement, it reached a higher value $\mu = 0.75$. It can be seen from the graphical dependence that in the case of the

1st and the 2nd displacement, the fluid flow was more restricted in comparison to the 3rd displacement (see Fig. 5). This condition may have occurred due to the intensity of the change in flow direction for the individual valve displacement s.

Finally, the cavitation indexes are evaluated. The dependence of the cavitation index σ_v on the pressure drop Δp is shown in Fig. 9. It can be observed that the gaseous cavitation inception occurred at values of the cavitation indexes $\sigma_{v,i}$ = (1.04 ÷ 1.10) depending on the valve displacement s.

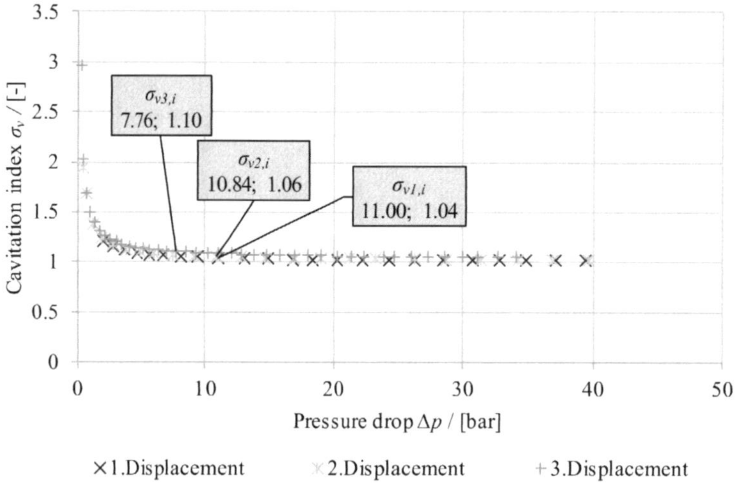

Fig. 9. The dependence of the cavitation index σ_v on the pressure drop Δp.

5 Conclusion

The article focuses on the experimental analysis of the flow through the throttle valve which was affected by cavitation. Due to the small volume flow rates and very low saturation vapour pressure of the oil, it was assumed that the cavitation inception at the valve is mainly influenced by the release of air bubbles. The gaseous cavitation inception was evaluated visually by means of a camera with a macro lens. From the evaluated static Δp-Q characteristic, it can be concluded that the release and formation of the gas cloud did not affect the change in the trend of the resulting characteristic. The inception and intensity of gaseous cavitation depends on the geometrical parameters of the valve. Depending on the position of the throttle valve, the flow was restricted not only in the narrowest cross-section of the valve, but also in the valve entrainment where the fluid flow was detached from the valve body. The largest increase of the gas phase was also observed at these locations where the relatively high flow velocity and the valve geometry resulted in swirling. This location was therefore accompanied by a local decrease of static pressure providing conditions for the air bubble expansion and further bubble release. However, the present experimental analysis does not give an idea of the amount of gas phase that was released. For future analysis, measurements with the Acoustic bubble spectrometer are planned to determine the size and number of bubbles or the volume fraction of gas in the liquid. A pressure control valve is also planned to be placed downstream of the throttle valve to achieve identical conditions at the outlet. The results of the experimental analyses can then be used as input parameters for CFD simulations in order to solve the multiphase flow of hydraulic fluid and air.

This work was supported by the European Regional Development Fund in the Research Centre of Advanced Mechatronic Systems project, project number CZ.02.1.01/0.0/0.0/16_019/0000867 within the Operational Programme Research, Development and Education.

The work presented in this paper was supported by a grant SGS „Numerical modeling of transient fluid flow problems with the support of experimental research" SP2022/32.

References

1. Vasina, M., Hruzik, L., Burecek, A. (2018). Energy and dynamic properties of hydraulic systems. *Tehnicki vjesnik-technical gazette*, 25(2), 382-390.
2. Osterland, S., Muller, L., Weber, J. (2021). Influence of Air Dissolved in Hydraulic Oil on Cavitation Erosion. *International journal of fluid power*, 22(3), 373-392.
3. Holl, E. (2022). Improvement of the modeling of a hydraulic drive for exoskeleton application based on measurements. *Materials Today: Proceedings*, 62(5), 2409-2415.
4. Burecek, A., Hruzik, L., Vasina, M. (2015). Determination of Undissolved Air Content in Oil by Means of a Compression Method. *Strojniski vestnik-journal of mechanical engineering*, 61(7-8), 477-485.
5. Himr, D., Haban, V., Zavorka, D. (2019). Axial check valve behaviour during flow reversal. In *IOP Conference Series: Earth and Environmental Science* (Vol. 405, No. 1, p. 012012). IOP Publishing.
6. Jablonska, J., Kozubkova, M., Mahdal, M., Marcalik, P., Tuma, J., Bojko, M., Hruzik, L. (2021). Spectral analysis of gaseous cavitation in water through multiphase mathematical and acoustic methods. *Physics of fluids*, 33(8), 085128.
7. Suo, XY., Jiang, Y., Wang, WJ. (2021). Hydraulic axial plunger pump: gaseous and vaporous cavitation characteristics and optimization method. *Engineering applications of comupational fluid mechanics*, 15(1), 712-726.
8. Jablonska, J., Kozubkova, M., Bojko, M. (2021). Flow of oil and water through the nozzle and cavitation. *Processes*, 9(11), 1936.
9. Yuan, C., Zhu, LS., Liu, SQ., Li, H. (2021). Examination of viscosity effect on cavitating flow inside poppet valves based on a numerical study. *Applied sciences-basel*, 11(23), 11205.
10. Iben, U., Wolf, F., Freudigmann, HA., Frohlich, J., Heller, W. (2015). Optical measurements of gas bubbles in oil behind a cavitating micro-orifice flow. *Experiments in fluids*, 56(6), 114.
11. Iben, U., Makhnov, A., Schmidt, A. (2018). Numerical study of the effects of dissolved gas release in cavitating flow. In *AIP Conference Proceedings* (Vol. 2027, No. 1, p. 030128). AIP Publishing LLC.
12. Sander, R., Acree, WE., De Visscher, A., Schwartz, SE., Wallington, TJ. (2022). Henry's law constants (IUPAC Recommendations 2021). *Pure and applied chemistry*, 94(1), 71-85.
13. Brennen, C. E. (2014). *Cavitation and bubble dynamics*. Cambridge university press.
14. Tullis, J. P. (1993). *Cavitation guide for control valves* (No. NUREG/CR-6031). Nuclear Regulatory Commission, Washington, DC (United States). Div. of Engineering; Tullis Engineering Consultants, Logan, UT (United States).
15. Martin, CS., Medlarz, H., Wiggert, DC., Brennen, C. (1981). Cavitation inception in spool valves. *Journal of fluids engineering-transactions of the ASME*, 103(4), 564-576.
16. Kudzma, Z., Stosiak, M. (2015). Studies of flow and cavitation in hydraulic lift valve. *Archives of civil and mechanical engineering*, 15(4), 951-961.
17. Sikora, R., Burecek, A., Hruzik, L., Vasina, M. (2015). Experimental investigation of cavitation in pump inlet. In *EPJ Web of Conferences* (Vol. 92, p. 02081). EDP Sciences.

Seasonal Variability of Resuspension

Jakub Linda[1,*], *Klaudia* Köbölová[1], *Ondřej* Uhlík[2], *Jiří* Pospíšil[1], and *Tomáš* Apeltauer Tomáš[2]

[1]Energetický ústav, Fakulta strojního inženýrství, VUT v Brně, Technická 2896, 616 69 Brno, ČR
[2]Ústav automatizace inženýrských úloh a informatiky, Fakulta stavební, VUT v Brně, Veveří 331/95, 602 00 Brno, ČR

Abstract. Particulate air pollution in cities is caused by a variety of sources. One of the less-studied contributors is wind-induced particle resuspension. As the wind speed increases, particles are removed from surfaces. These particles cause an increase in the total concentration in the air. It is known that particles of 10-2.5 μm in size can be resuspended ($PM_{10-2,5}$). Modern emission monitoring in cities also allows the monitoring of fine particles of 10, 2.5 and 1 μm in size. The size fractions can then be sorted into $PM_{10-2,5}$, $PM_{2,5-1}$ and PM_1. When breathed in, particles of different sizes cause various serious health risks. This paper focuses on the identification of the resuspension process of different particle size fractions by a data processing method. Data measured by automatic emission monitoring are used. It is confirmed that the concentration increase can be dominated by the fraction $PM_{10-2,5}$. However, a concentration increase of fractions $PM_{2,5-1}$ and PM_1 is also evident with increasing wind speed. Although the increase in the PM_1 fraction is smaller than $PM_{10-2,5}$, it is more severe due to the respiratory deposition dose. The resuspension of particles of different fractions has different behaviours during the year. $PM_{10-2,5}$ particles are dominantly resuspended in the summer months. In winter, on the other hand, the proportion of $PM_{2.5-1}$ and PM_1 particles increases, which may be related to the heating season

Keywords: *health risk assessment, RDD, PM, particulate matter, road dust*

1 Introduction

PM_{10} air pollution is still a problem in most of the world's metropolises. PM_{10} is the term for the size of particles with an aerodynamic diameter of less than 10 micrometres. There are many sources of particles in the urban environment. Transport is one of the most emissions-intensive activities. Specifically, it accounts for 25 % of particle pollution in cities on a global scale [1]. As the rise of electric vehicles progresses, the focus is shifting from exhaust emissions to non-exhaust emissions [2]. Their amount may already exceed the exhaust particulate emissions [3]. This trend will only increase in the future. Other sources of urban particulate matter include industry, domestic fire burning, natural resources [1]. Study [1]

* Corresponding author: Jakub.Linda@vutbr.cz

points to a high degree of uncertainty in the determination of the source of the particles. The number of particles that cannot be classified as the aforementioned sources are then referred to as "unspecified sources of human origin". Wind-induced particle resuspension is indirectly related to human activity. In this process, already deposited particles are released from surfaces into the air by the wind, as already pointed out by [4]. There are several studies showing the characteristics of the wind-induced resuspension process [4–8]. However, the quantification is different each time, as the studies are based on different assumptions.

Wind-induced resuspension depends on many parameters. Among the main ones are time aspects. The initial number of resuspended particles is significantly greater than the number of particles after a period of higher wind speeds. This is caused by the weakly attached particles being lifted at the beginning of the process [9]. Resuspension is a process that occurs in a matter of seconds. Other factor affecting the resuspension is the humidity of the air. The humidity of the air can substantially affect the detachment of particles from the surface if it exceeds a value of about 60 % [10]. Availability of resuspendable particles varies throughout the year [11]. The total concentration of particles in the city is influenced by background concentrations. However, with respect to health risks, we are interested in the overall increase in particle concentration due to wind. It is known that background concentration is also strongly correlated with wind speed [1]. These factors should be taken into account when identifying wind-induced resuspension in the data.

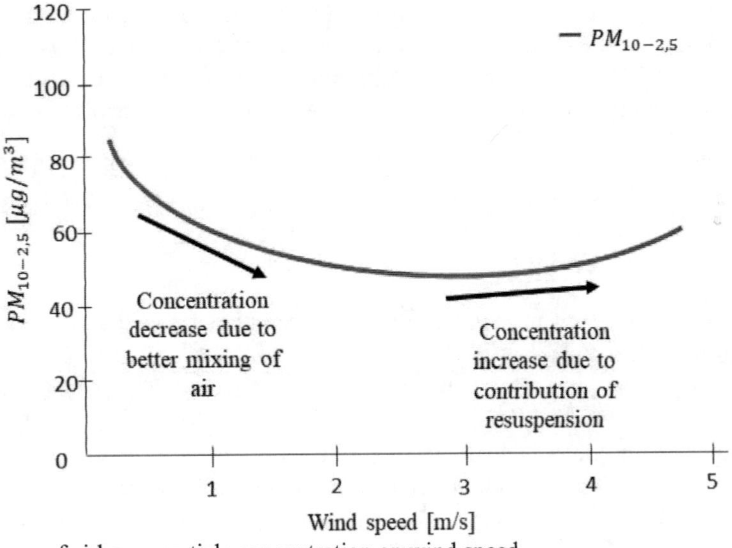

Fig. 1. Dependence of airborne particle concentration on wind speed.

The $PM_{10\text{-}2,5}$ concentration can be divided into two components depending on the wind speed. The first component dilutes with increasing wind speed. The second component increases with increasing wind speed [4]. The authors of the study assume that the $PM_{10\text{-}2,5}$ fraction can be resuspended by the wind and the $PM_{2.5}$ fraction is diluted by the wind. The $PM_{10\text{-}2,5}$ component is diluted by increasing wind speed at the same rate as the $PM_{2.5}$ fraction. Plotting the dependence of $PM_{10\text{-}2,5}$ on wind speed then results in a characteristic U-shape function.

Other studies [6–8], that have dealt with wind-induced resuspension have focused on the $PM_{10\text{-}2,5}$ and $PM_{2.5}$ fractions. Their assumption is that only $PM_{10\text{-}2,5}$particles are capable of resuspension and the concentration of $PM_{2.5}$ particles decreases with increasing wind speed. Many cities, however, are now offering monitoring of PM_1. In Brno, these particles have been monitored since 2016. The question is what results we get if we sort PM_{10} particles into size fractions $PM_{10\text{-}2,5}$, $PM_{2.5\text{-}1}$ and PM_1. Evaluating the dependence of the concentration of each fraction on the wind speed will show how they are able to resuspend. There is a fundamental difference if the wind in a city is lifting larger or smaller particles, especially in terms of health risks.

Due to their size, PM_1 particles enter the human respiratory system very easily and deeply, which has a non-negligible impact on health. Compared to larger particles at the same mass, these fractions have a much higher concentration, active surface area, high deposition efficiency and a greater potential for harmful biological interactions with respiratory tissues [12–14].

Millions of people walk or run daily on streets and in parks as part of their daily routine, either on their way to work or recreationally. During these activities, large amounts of pollutants are inhaled into the respiratory system. Epidemiological studies [15–19] show a link between respiratory diseases and atmospheric PM. Most of those studies look at total deposition, with deposition of different parts of the respiratory system being important for assessing the potential hazard of PM [20]. Inhaled particles may be deposited in different regions of the respiratory system by the complex action of different deposition mechanisms or may be exhaled. The most important mechanisms are impaction, deposition and diffusion [21]. Particles that encounter the walls of the respiratory tract settle there and are not reabsorbed. The extent and location of particle deposition depends on the size, density and shape of the particles, the geometry of the airways and the mode of respiration. Many mathematical models exist to predict total and regional particle deposition. One possibility is to calculate the respiratory deposition dose (RDD), which expresses the entrainment between regional deposition and mass concentration. The aim of this work is to find a link between resuspension and RDD. The possibility of monitoring PM1 particles raises the question of whether these particles are also capable of resuspension and what health risks arise from this.

2 Materials and methods

2.1 Data measurement

The data were obtained from automatic emission monitoring (AIM) in Brno. The AIM in Brno includes five measuring stations:

- Svatoplukova – located in a populated part of the city. The measuring equipment is located at the four-lane road.
- Výstaviště – a location characterised by the presence of a major traffic junction. The measuring equipment is located on a less frequented four-lane road.
- Zvonařka – a site characterised by the presence of significant construction activity in 2019. The measuring device is located along a four-lane road.
- Arboretum – the measuring device is located directly in the center of the city park.
- Lány – a site characterised by the presence of residential development. The measuring equipment is located between houses.

Data were measured throughout 2019. The frequency of recording was 1 second. However, the data is reprocessed into 10-minute averages by default. Measured variables included: temperature, humidity, wind direction, wind speed, pressure and PM_{10}, $PM_{2.5}$ and PM_1 concentrations.

2.2 Data processing

Resuspension is a process that does not occur under all conditions. There are times of the year when, despite increased wind speeds, there is no increase in concentration. In order for the resuspension process to be recognized from the data, the data must be prepared according to predetermined conditions.

Due to the availability of particles at different times of the year, the resuspension process has seasonal variability. Thus, the data were sorted into four seasons. Hourly averages of the measured data are mostly used to indicate the resuspension history [4,5]. Resuspension occurs within seconds [9], the shortest available measured time intervals, 10 minutes, were thus used. Air humidity above 60 % can significantly affect the removal of particles from the surface [10]. Therefore, only data with humidity below 60 % were used. Subsequently, the particle concentrations were sorted into three size fractions:

1) $PM_{10-2,5}=PM_{10}-PM_{2,5}$,
2) $PM_{2,5-1}=PM_{2,5}-PM_1$,
3) PM_1.

The function of concentration on wind speed for particle size fractions, the sorted data were fit with a polynomial. The degree of the polynomial was determined by the k-Fold cross-validation method [22], where the dataset is split into training and test subsets. Subsequently, the root mean square error (RMSE) determining the deviation of the test subset of the real data from the polynomial curve is determined. This process was carried out k times and after its completion, the average of k RMSE values was calculated for a given polynomial degree. Subsequently, the polynomial with the lowest resulting RMSE was selected. The degree of the polynomial can therefore vary for each data filtering condition (season, location and fraction) just based on the RMSE results. Within this paper, the value of k was set to 10, which is the value standardly used in this method across disciplines. Local extremes were then determined on the curves for subsequent analysis of resuspension rates.

2.3 Evaluation of the data

The original methodology for evaluating resuspension characteristics is based on the procedure [4]. The component of particles that can resuspend from the $PM_{10-2,5}$ fraction is obtained by subtracting two functions. The first function is the rate at which particles of the $PM_{10-2,5}$ fraction are diluted with increasing wind speed. This rate is the same as the rate at which $PM_{2.5}$ particles are diluted. The second function is the dependence of the $PM_{10-2,5}$ concentration on wind speed itself. Thus, the wind speed is identifiable when the resuspension event is visible. At the same time, the rate of increase in particle concentration due to resuspension is determined. However, based on the known concentration of PM_1, the behaviour of other size fraction of $PM_{2.5-1}$ and PM_1 can be determined. These classes may have different behaviour depending on the wind speed.

Since the aim of this work is to evaluate how the concentrations of different particle size fractions behave as a function of wind speed, this approach is not applicable. Thus, the presented method of data processing and evaluation provides an indication of which particle size fraction concentration varies the most due to wind in the specified seasons. The increment of the concentration of the size fractions to the total increase in PM_{10} concentration is known. The method itself consists of the following steps:

1. Identifying the best-fit of the relationship between particle concentration and wind speed.
2. Locating the extremes of given functions.
3. Determining the wind speed interval where the PM_{10} increase occurs.
4. Deriving the concentration increase or decrease from the extremes of the functions for each particle size fraction in the specified wind speed interval.
5. Determination of the concentration increment of the individual particle size fractions to the total particle concentration increment.

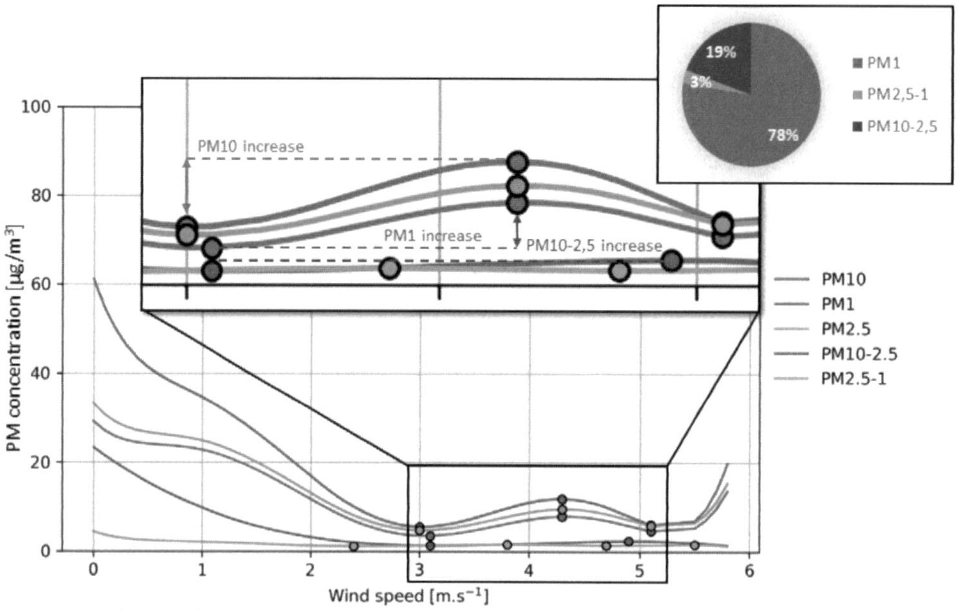

Fig. 2. Dependence of PM concentration on wind speed.

Furthermore, health risks from wind-induced particle resuspension may be identified. At different seasons, particles of different size fractions are resuspended. Regarding health risks, it then makes a difference which size fraction caused the increase in PM_{10} concentration due to increasing wind speed.

2.4 Estimation of the respiratory deposition doses

Atmospheric particles inhaled during breathing are deposited on the lining of the airways and are not exhaled, which can cause serious breathing disorders. A mathematical model based on calculations and data from the International Commission on Radiological Protection (ICRP) has been used to determine the extent of particle deposition in different regions of the respiratory tract. Respiratory deposition dose [23–25] is calculated according to the equation (1):

$$RDD = V_T \cdot f \cdot DF_i \cdot PM_i \quad (1)$$

where V_T is the breathing rate (m^3/inhale), f is the frequency of breathing, DF_i is the deposition fraction of the size fraction i, PM_i is the mass concentration of size fraction i. Breathing volume and respiratory rate depend on physical activity, but also on gender and age. The different values are given in Table 1. People usually perform little activity, mostly walking, when exposed to particles from resuspension, and because of this, the values for light physical activity were chosen and the calculation was calculated for men as well as women and children.

Table 1. Respiratory volume and respiratory rate during light physical activity [20].

	V_t [m^3.inhale^{-1}]	F [min^{-1}]
Man	12.5 . 10^{-4}	20
Woman	9.92 . 10^{-4}	21
Children	5.83 . 10^{-4}	32

The deposition size fraction is calculated in three different regions of the respiratory system - the head airway (HD), the tracheobronchial region (TB) and the alveolar region (AL) according to Hinds [20]. They are calculated as follows:

$$DF_{HD} = IF\left(\frac{1}{1+\exp(6.84+1.183\ln d_p)} + \frac{1}{1+\exp(0.924+1.885\ln d_p)}\right) \quad (2)$$

DF_{HD} is the deposition fraction for the head airways, d_p is the particle size, IF is the inhalable fraction, which is calculated as follows:

$$IF = 1 - 0{,}5\left(1 - \frac{1}{(1+0.00076d_p^{2,8})}\right) \quad (3)$$

DF_{TB} denotes the deposition fractions in the tracheobronchial region and is given by the following equation:

$$DF_{TB} = \left(\frac{0.00352}{d_p}\right)\left[\exp\left(-0.234(\ln d_p + 3.40)^2\right) + 63.9\exp\left(-0.819(\ln d_p - 1.61)^2\right)\right] \quad (4)$$

Deposition fraction for the alveolar region DF_{AL} is calculated using the equation:

$$DF_{AL} = \left(\frac{0.0155}{d_p}\right)\left[\exp\left(-0.416(\ln d_p + 2.84)^2\right) + 19.11\exp\left(-0.482(\ln d_p - 1.362)^2\right)\right] \quad (5)$$

The particle mass concentration was used for the calculation, which corresponds to the increase in particles due to wind-induced resuspension. Consequently, the calculated RDD value corresponds to the value that a person would breathe while walking in the city with wind blowing at the speed at which the increase in particle concentration was identified.

3 Results and discussion

Based on the explained methodology, the data were processed and evaluated. The data are divided into four seasons in five locations. For each location, the concentration of PM_{10}, $PM_{10\text{-}2,5}$, $PM_{2.5\text{-}1}$, PM_1 ($PM_{2.5}$ fraction is also evaluated for illustration) is evaluated depending on the wind speed. The data were fitted with a polynomial in a way that best fits the nature of the data. An example of the polynomial translation of the data is shown in the following figure.

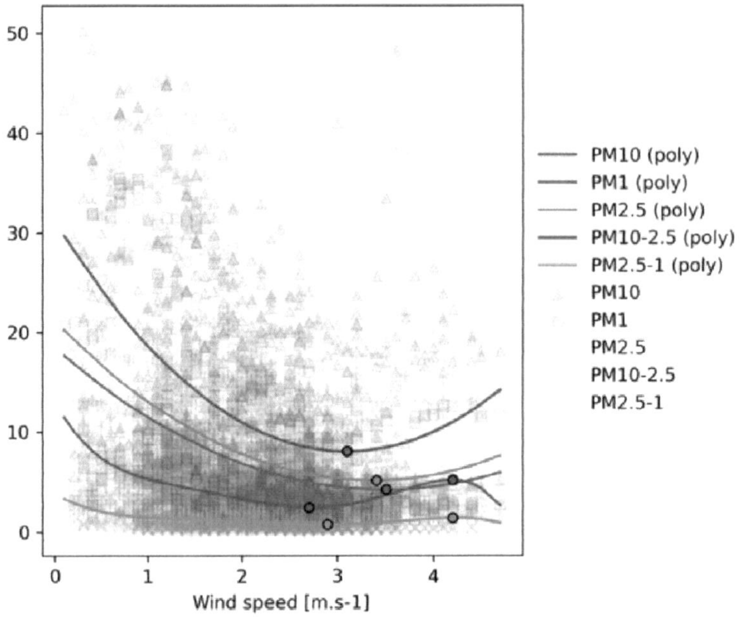

Fig. 3. Best fit polynomial to the evaluated data for location Výstavište in winter period.

For each season and location, it was determined at which wind speeds PM_{10} particles increase. Within a specified range of wind speeds, the contribution of each size fraction to the increase in PM_{10} was determined. The concentration increase was calculated as the difference in the concentration maxima and minima of the polynomials for each particle size fraction over the observed wind speed range. The particle size fraction dominantly responsible for the increase in PM_{10} concentration due to wind is summarized in the following table.

Table 2. Fraction of particles dominantly responsible for the increase in total concentration due to particle resuspension.

Location	Arboretum	Lány	Výstavište	Svatoplukova	Zvonařka
Spring	-	$PM_{10\text{-}2,5}$	$PM_{10\text{-}2,5}$	PM_1	$PM_{10\text{-}2,5}$
Summer	$PM_{10\text{-}2,5}$	$PM_{10\text{-}2,5}$	$PM_{10\text{-}2,5}$	$PM_{2,5\text{-}1}$	$PM_{10\text{-}2,5}$
Autumn	$PM_{10\text{-}2,5}$	$PM_{10\text{-}2,5}$	PM_1	$PM_{10\text{-}2,5}$	$PM_{10\text{-}2,5}$
Winter	-	-	$PM_{10\text{-}2,5}$	PM_1	$PM_{10\text{-}2,5}$

The table shows that $PM_{10\text{-}2,5}$ is a fraction, which is predominantly resuspended by wind, confirming the assumption of the studies [4,8]. However, a further assumption of a decrease in $PM_{2.5}$ concentrations with increasing wind speeds could not be confirmed. In the vast majority of cases, there were increasing concentrations of both $PM_{2.5\text{-}1}$ and PM_1 fractions due to wind. This phenomenon can be caused by fitting the data with a polynomial. Previous studies used a second-degree polynomial, which could suppress small changes in concentration with a wind speed. The behaviour of the size fractions on the wind speed interval where the increase in PM_{10} occurred is evident in Tab. 2. The locations where there was an increase in concentration, a decrease in concentration, or the concentration was constant are marked.

Table 3. Concentration behaviour of the particle fraction due to an increase in wind speed.

Season	Fraction	Arboretum	Lány	Výstavište	Svatoplukova	Zvonařka
Spring	$PM_{10-2,5}$	-	Increase	Increase	Decrease	Increase
	$PM_{2,5-1}$	-	In	Increase	Constant	Increase
	PM_1	-	Increase	Increase	Increase	Increase
Summer	$PM_{10-2,5}$	Increase	Increase	Increase	Increase	Increase
	$PM_{2,5-1}$	Increase	Increase	Increase	Increase	Increase
	PM_1	Increase	Increase	Increase	Increase	Increase
Autumn	$PM_{10-2,5}$	Increase	Increase	Increase	Increase	Increase
	$PM_{2,5-1}$	Constant	Constant	Constant	Increase	Constant
	PM_1	Decrease	Increase	Increase	Decrease	Increase
Winter	$PM_{10-2,5}$	-	-	Increase	Increase	Increase
	$PM_{2,5-1}$	-	-	Increase	Constant	Increase
	PM_1	-	-	Increase	Increase	Increase

It can be seen from Tab. 3 that the increase in concentration was evident in almost every season and location. Based on Tab. 2, it is known that the $PM_{10-2,5}$ fraction is dominantly responsible for the increase in PM_{10} due to wind. However, Tab. 3 shows that the concentration increase is also recorded for other fractions. From the point of view of health risks, it is important to know how concentrations increase for all fractions and seasons.

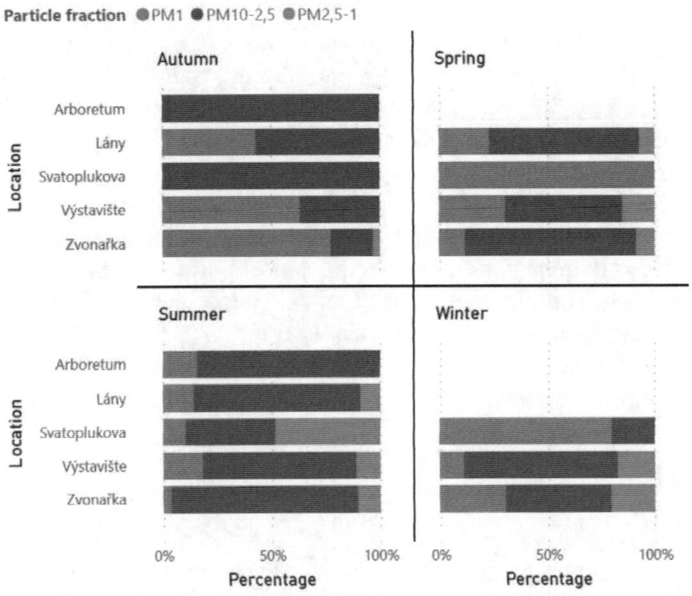

Fig. 4. The percentage increase for each fraction against the total increase.

Almost every location is dominated by an increase in the $PM_{10-2,5}$ fraction. Sites that showed similar behaviour across seasons were Výstavište, Zvonařka and Lány. Here, the contribution of the fine fraction PM_1 to the overall concentration increase decreased during the spring and summer months. The increase in the $PM_{2.5-1}$ fraction remained approximately constant or did not change at all during the year. During the autumn and winter months, the increase in the PM_1 fraction increased.

However, the overall increase in the concentration of individual fractions does not indicate a problem related to health risks. Particle size is key with respect to health risks, as smaller particles settle deeper in the respiratory system. Although the increase in the $PM_{10\text{-}2.5}$ fraction is dominant, the increase in PM_1 concentration has a more negative impact on human health. The following graph shows the number of resuspended particles that can settle in different parts of the respiratory system.

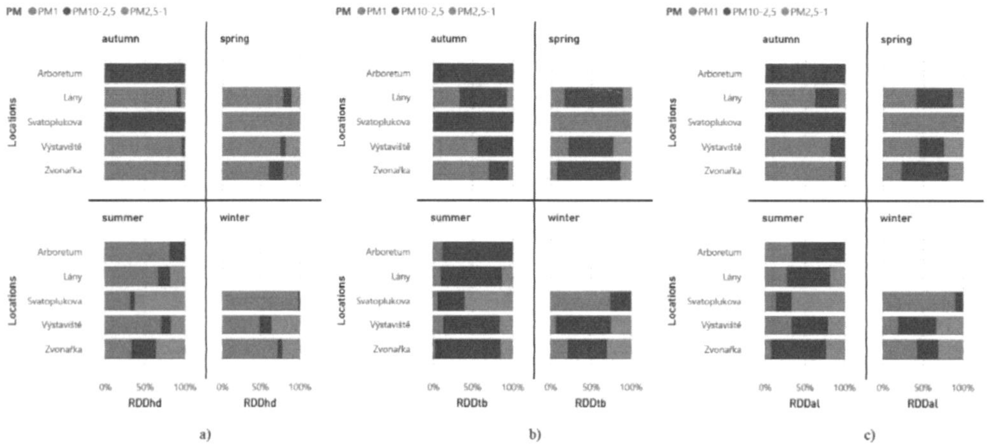

Fig. 5. The percentage of particle fraction deposided in respiratory sytem against topal deposited mass of particles. a) the head airway (HD), b) the tracheobronchial region (TB), c) the alveolar region (AL).

Comparing Fig. 4 and Fig. 5, it is evident that although the increase in concentration is dominated by the $PM_{10\text{-}2,5}$ fraction, more PM_1 particles are trapped in the human body. The PM_1 fraction is a minority fraction in the concentration increase in most cases. Considering the total amount of trapped particles in the human body, the fraction of PM_1 is not negligible, especially in head airways region. Similarly, the $PM_{2.5\text{-}1}$ fraction is too significant due to its concentration increase. However, due to health risks, the proportion of trapped particles has increased, especially in the alveolar region. In tracheobronchial region, the fraction of $PM_{10\text{-}2,5}$ particles captured is dominant, as in the overall concentration increase in Fig. 5. Thus, the results show that PM_1 particles are trapped in the head airways despite the low total concentration. $PM_{10\text{-}2,5}$ particles are predominantly trapped in the tracheobronchial region. The alveolar region is dominated by PM_1 capture, and it should be noted that $PM_{2.5\text{-}1}$ particles also make a non-negligible contribution in this region. Gender and age don't influence the results.

4 Conclusion

The present study focuses on the identification of the wind-induced resuspension process in cities. Emission monitoring data are used. The data are divided into four annual periods. Times when the resuspension process is suppressed and humidity reaches values above 60 % are not evaluated [10]. The remaining data were separated into size fractions $PM_{10\text{-}2,5}$, $PM_{2.5\text{-}1}$ and PM_1 and expressed as a mathematical dependence on wind speed. From the resulting functions, it was determined which fraction was responsible for the increase in PM_{10} concentration due to increasing wind speed. It was found that particles with the size fraction $PM_{10\text{-}2,5}$ resuspended at a dominant rate, which is in agreement with previous studies. The original assumption of a decrease in $PM_{2.5}$ concentration with increasing wind speed was not evident. This may be due to the fact that the $PM_{2.5}$ fraction has split into $PM_{2.5\text{-}1}$ and PM_1. The methodology used identified a concentration increase in the $PM_{2.5\text{-}1}$ and PM_1 fractions.

However, this increase cannot be attributed with certainty to resuspension alone. The source of these particles may be traffic or other sources [4–8], which the wind carries. A more detailed data analysis would be needed to confirm or disprove this. It would be necessary to link the particle concentration information to the NO_x concentration.

With increasing wind speed, the concentration of all size fractions increases in each season. Whatever the source of the particles, this increase has an impact on human health. Although the increase in $PM_{10\text{-}2,5}$ fraction has been dominant, the increase in PM_1 concentration is more serious for health. These particles are inhaled more by humans compared to $PM_{10\text{-}2,5}$. Within the respiratory system, different fractions are trapped in different parts. The PM_1 fraction is trapped in the head airways. The $PM_{10\text{-}2,5}$ fraction in the tracheobronchial region and the $PM_{2.5\text{-}1}$ fraction in the alveolar region.

This paper has been supported by the EU project Computer Simulations for Effective Low-Emission Energy Engineering funded as project No. CZ.02.1.01/0.0/0.0/16_026/0008392 by Czech Republic Operational Programme Research, Development and Education, Priority 1: Strengthening capacity for high-quality research and the collaboration. This research was further funded by projects of Brno University of Technology: FSI-S-20-6280, FAST/FSI-J-22-7905.

References

1. Karagulian, F.; Belis, C.A.; Dora, C.F.C.; Prüss-Ustün, A.M.; Bonjour, S.; Adair-Rohani, H.; Amann, M. Contributions to Cities' Ambient Particulate Matter (PM): A Systematic Review of Local Source Contributions at Global Level. *Atmos. Environ.* 2015, *120*.
2. Harrison, R.M.; Allan, J.; Carruthers, D.; Heal, M.R.; Lewis, A.C.; Marner, B.; Murrells, T.; Williams, A. Non-Exhaust Vehicle Emissions of Particulate Matter and VOC from Road Traffic: A Review. *Atmos. Environ.* 2021, *262*.
3. Casotti Rienda, I.; Alves, C.A. Road Dust Resuspension: A Review. *Atmos. Res.* 2021, *261*.
4. Harrison, R.M.; Yin, J.; Mark, D.; Stedman, J.; Appleby, R.S.; Booker, J.; Moorcroft, S. Studies of the Coarse Particle (2.5-10μm) Component in UK Urban Atmospheres. *Atmos. Environ.* **2001**, *35*, doi:10.1016/S1352-2310(00)00526-4.
5. Thorpe, A.J.; Harrison, R.M.; Boulter, P.G.; McCrae, I.S. Estimation of Particle Resuspension Source Strength on a Major London Road. *Atmos. Environ.* **2007**, *41*, 8007–8020, doi:10.1016/J.ATMOSENV.2007.07.006.
6. Pospíšil, J.; Jícha, M. Mezní Rychlost Resuspenze Částic PM10 v Prostředí Městské Zástavby.; Česká aerosolová společnost: Prague, 2007; pp. 33–34.
7. Grundström, M.; Hak, C.; Chen, D.; Hallquist, M.; Pleijel, H. Variation and Co-Variation of PM10, Particle Number Concentration, NOx and NO2 in the Urban Air - Relationships with Wind Speed, Vertical Temperature Gradient and Weather Type. *Atmos. Environ.* **2015**, *120*, doi:10.1016/j.atmosenv.2015.08.057.
8. Kassomenos, P.; Vardoulakis, S.; Chaloulakou, A.; Grivas, G.; Borge, R.; Lumbreras, J. Levels, Sources and Seasonality of Coarse Particles (PM 10-PM 2.5) in Three European Capitals - Implications for Particulate Pollution Control. *Atmos. Environ.* **2012**, *54*, doi:10.1016/j.atmosenv.2012.02.051.
9. Wang, Y.; Zhao, Y.; Yao, J. Large Eddy Simulation of Particle Deposition and Resuspension in Turbulent Duct Flows. *Adv. Powder Technol.* **2019**, *30*, 656–671, doi:10.1016/J.APT.2019.01.012.
10. Gotoh, K.; Takebe, S.; Masuda, H.; Banba, Y. The Effect of Humidity on the Removal of Fine Particles on a Solid Surface Using High-Speed Air Jet. *KONA Powder Part. J.* **1995**, *13*, doi:10.14356/kona.1995024.

11. Amato, F.; Pandolfi, M.; Moreno, T.; Furger, M.; Pey, J.; Alastuey, A.; Bukowiecki, N.; Prevot, A.S.H.; Baltensperger, U.; Querol, X. Sources and Variability of Inhalable Road Dust Particles in Three European Cities. *Atmos. Environ.* **2011**, *45*, doi:10.1016/j.atmosenv.2011.06.003.
12. Mei, M.; Song, H.; Chen, L.; Hu, B.; Bai, R.; Xu, D.; Liu, Y.; Zhao, Y.; Chen, C. Early-Life Exposure to Three Size-Fractionated Ultrafine and Fine Atmospheric Particulates in Beijing Exacerbates Asthma Development in Mature Mice. *Part. Fibre Toxicol.* **2018**, *15*, doi:10.1186/s12989-018-0249-1.
13. Seaton, A.; Tran, L.; Aitken, R.; Donaldson, K. Nanoparticles, Human Health Hazard and Regulation. *J. R. Soc. Interface* 2010, *7*.
14. Yang, M.; Chu, C.; Bloom, M.S.; Li, S.; Chen, G.; Heinrich, J.; Markevych, I.; Knibbs, L.D.; Bowatte, G.; Dharmage, S.C.; et al. Is Smaller Worse? New Insights about Associations of PM1 and Respiratory Health in Children and Adolescents. *Environ. Int.* **2018**, *120*, doi:10.1016/j.envint.2018.08.027.
15. Pirsaheb, M.; Sharafi, K.; Hopke, P.K.; Hadei, M.; Shahsavani, A. Carcinogenic Risks of Particulate Matter during Middle Eastern Dust Events and Normal Days. *Atmos. Pollut. Res.* **2020**, *11*, doi:10.1016/j.apr.2020.06.012.
16. Ziou, M.; Tham, R.; Wheeler, A.J.; Zosky, G.R.; Stephens, N.; Johnston, F.H. Outdoor Particulate Matter Exposure and Upper Respiratory Tract Infections in Children and Adolescents: A Systematic Review and Meta-Analysis. *Environ. Res.* 2022, *210*.
17. Pope, C.A.; Dockery, D.W. Health Effects of Fine Particulate Air Pollution: Lines That Connect. *J. Air Waste Manag. Assoc.* **2006**, *56*, doi:10.1080/10473289.2006.10464485.
18. Yang, M.; Guo, Y.M.; Bloom, M.S.; Dharmagee, S.C.; Morawska, L.; Heinrich, J.; Jalaludin, B.; Markevychd, I.; Knibbsf, L.D.; Lin, S.; et al. Is PM1 Similar to PM2.5? A New Insight into the Association of PM1 and PM2.5 with Children's Lung Function. *Environ. Int.* **2020**, *145*, doi:10.1016/j.envint.2020.106092.
19. Mei, L.; Yan, S.; Li, Y.; Jin, X.; Sun, X.; Wu, Y.; Liang, Y.; Wei, Q.; Yi, W.; Pan, R.; et al. Association between Short-Term PM1 Exposure and Cardiorespiratory Diseases: Evidence from a Systematic Review and Meta-Analysis. *Atmos. Pollut. Res.* **2022**, *13*, doi:10.1016/j.apr.2021.101254.
20. 'William, C.H. *Aerosol Technology: Properties, Behavior, and Measurement of Airborne Particles, 2nd Edition*; 2nd ed.; Wiley-Interscience Publication: Canada, 1999;
21. Darquenne, C. Deposition Mechanisms. *J. Aerosol Med. Pulm. Drug Deliv.* **2020**, *33*, 181–185, doi:10.1089/jamp.2020.29029.cd.
22. 'Trevor, H.; 'Robert, T.; 'Jerome, F. *The Elements of Statistical Learning: Data Mining, Inference, and Prediction*; 2nd ed.; Springer New York, NY: New York, 2009;
23. Azarmi, F.; Kumar, P. Ambient Exposure to Coarse and Fine Particle Emissions from Building Demolition. *Atmos. Environ.* **2016**, *137*, doi:10.1016/j.atmosenv.2016.04.029.
24. Prabhu, V.; Gupta, S.K.; Madhwal, S.; Shridhar, V. Exposure to Atmospheric Particulates and Associated Respirable Deposition Dose to Street Vendors at the Residential and Commercial Sites in Dehradun City. *Saf. Health Work* **2019**, *10*, doi:10.1016/j.shaw.2019.01.005.
25. Gupta, S.K.; Elumalai, S.P. Exposure to Traffic-Related Particulate Matter and Deposition Dose to Auto Rickshaw Driver in Dhanbad, India. *Atmos. Pollut. Res.* **2019**, *10*, doi:10.1016/j.apr.2019.01.018.

Identification of Cavitation by Noise

Jana Jablonská[1,*], *Milada* Kozubková[1], *Miroslav* Mahdal[2], *Radek* Štramberský[2], *Tomáš* Blejchař[1], and *Marian* Bojko[1]

[1]VŠB TU Ostrava, Dept Hydromechanics and Hydraulic Equipment, Ostrava, Czech Republic
[2]VŠB TU Ostrava, Department Control Systems and Instrumentation, Ostrava, Czech Republic

Abstract. The identification of cavitation is very important in technical practice for operational and especially economic reasons. The article deals with the use of another way to measure noise during cavitation. The current approach of measuring noise with an intensity probe is used in practice for identification, but it does not immediately address the position of the cavitation source for a given frequency range. Measurement by an acoustic camera is not entirely common in practice, but it allows to determine the location of the noise source for a given frequency range. To test the acoustic camera, the authors focused on the cavitating flow in a hydraulic circuit with three previously tested nozzles. Noise was measured for these nozzles using an acoustic intensity probe with two microphones. The results were evaluated by statistical methods and compared with measurements using an acoustic camera. The aim of the article is to point out the advantages of using this approach for accurate area identification of the problem.

Research background: The work is focused on the issue of cavitation and its identification in the hydraulic circuit. For cavitation research, a variant of cavitation identification by noise was chosen. However, this measurement brings problems that are only revealed through more sophisticated and accurate measurements.
Purpose of the article: The purpose of the article is to point out other possibilities of measuring cavitation noise using modern technologies and subsequently verify the results.
Methods: Metody: A common way of measuring noise is to measure it with a suitably located acoustic intensity probe. A more modern approach is area noise measurement. Measurement methodology and benefits are described.
Findings & Value added: The commonly used way of measuring noise using an acoustic intensity probe has proved to be insufficient, as it is not possible to distinguish the location of sources in the case of complex measurements. When using an acoustic camera, there are more sources of noise in a given circuit and they are detected according to the required frequencies in different places than expected. The article points out the specific identification of noise sources using the frequency spectrum of noise for selected elements.
Keywords: *cavitation, noise, acoustic camera, acoustic intensity probe.*

* Corresponding author: jana.jablonska@vsb.cz

1 Cavitation and noise measurement

Cavitation is a long-term problem in technical practice. With increasing demands on the performance of hydraulic equipment (pumps, turbines, valves, cooling systems, etc.), it occurs in almost all hydraulic elements. The critical places for the formation of cavitation from the point of view of hydraulic systems are especially the places where the velocity increases, resp. the pressure of the working fluid reduces. Cavitation occurs mainly in the narrowing of the flow cross-section, when closing or opening valves, in the impeller, in the inlet part of the hydrodynamic pump, etc. [1, 3] Cavitation is distinguished as vapor and gaseous. [4, 5]

Vapor cavitation is described as the content of bubbles is mostly vapor (water). Gaseous cavitation is defined if the bubble content is formed from a non-condensed gas such as air. [4, 5]

If at some point in the hydraulic area the pressure drops to the value of the cavitation pressure p_W (saturated vapor pressure) for the given liquid, the connection of the liquid is broken, and a cavitation bubble is formed. This is then considered to be the place of cavitation (vapor) and the creation of the cavitation area. If the pressure value decreases or remains the same, the vapor bubble will increase. In the flowing liquid, the vapor bubble moves to a region of higher pressure (than the saturated vapor pressure p_W), the vapors contained in it condense violently and the bubble will shrink quickly - it disappears, i.e., it implodes. The gases partially diffuse into the liquid. The liquid thus enters the space thus freed and the rest of the gases is compressed. [1, 2, 9] The formation and development of cavitation can be affected by temperature.

The real liquid may contain bubbles of undissolved gas (air), so-called cavitation nuclei. Cavitation bubbles arise from cavitation nuclei. [1, 2, 9] Under certain fluid conditions (pressure and temperature), the cavitation nuclei reach a critical size, after which it becomes visible to the naked eye like a cavitation bubble. When the cavitation pressure decreases, the air bubbles in the liquid increase, and therefore the expansion of the bubbles can occur even at a higher pressure (critical pressure p_{cri}) than the cavitation pressure p_W. This facilitates the formation of gas cavitation. Assuming that the cavitation nuclei have a spherical shape, the equilibrium condition is valid for the pressure inside the bubble p_B:

$$p_B = p_K + \frac{2\sigma}{R} \tag{1}$$

where p_K [Pa] is the fluid pressure around the bubble, σ [N·m^{-1}] is the surface tension and R [m] is the radius of the bubble. [1, 6, 7] Water has a relatively high surface tension, so it can be a very effective cavitation medium. [3] The residual air in the bubble is the reason why the bubble disappears in an oscillating way (the air acts as a spring when compressed). Similarly, the gases dissolved in the liquid reduce its strength and promote cavitation.

According to the Rayleigh-Plesset theory, the change in the bubble radius can be expressed, assuming that the temperature of the flowing liquid is constant, and the liquid is incompressible, by the equation for bubble dynamics as

$$\frac{p_B(t) - p_\infty(t)}{\rho_L} = R\frac{d^2R}{dt^2} + \frac{3}{2}\left(\frac{dR}{dt}\right)^2 + \frac{4\nu_L}{R}\frac{dR}{dt} + \frac{2\sigma}{\rho_L R} \tag{2}$$

where ρ_L [kg·m^{-3}] is density of liquid (water), ν_L [m^2·s^{-1}] is kinematic viscosity of liquid, [1, 6, 7]

Cavitation noise is created by the formation, growth, pulsations of cavitation bubbles, and especially by their extinction. The pressure waves propagate from the centre of the bubble to the surrounding fluid and interact with each other. Noise can be transmitted by liquid or piping to the system. [3] Thus, noise is related to turbulent flow and the existence of cavitation and can be used as a tool to identify them.

Turbulence is generally evaluated by statistical methods that give exactly defined results only for random processes in the space of homogeneous turbulence. The energy spectrum is used for evaluation (basically power spectral density of velocity fluctuations), so PSD is a basic statistical characteristic of random nature signals, which contain information about the size of structures flow. [12] It should be noted that the turbulent scale is dependent on the dissipation rate at turbulent flow for high Re numbers, and it is not dependent on viscosity. This area is called the inertial subregion and it is characterized by its slope (-5/3) (in the logarithmic coordinate system). In the case of turbulent flow with cavitation, it is necessary to use a milder requirement, i.e., the stationarity of the process in terms of statistical quantities.

Separate microphones are used to measure dynamic changes in acoustic pressure levels. These changes are evaluated using frequency spectra. To locate a noise source in an environment where it is difficult to locate individual noise sources, it is advantageous to use a microphone array, thanks to which a unit called Pressure Contribution can be evaluated for a selected point in space. This value indicates how the location affects the measured sound pressure level. [18, 19]

1.1 Circuit description and measurement

To investigate the properties of cavitation (especially noise), three similar hydraulic elements typical for the formation of hydraulic cavitation were chosen:

I - converging - diverging nozzle of circular cross section [14],
II - converging - diverging nozzle of rectangular cross-section [15]
III - Venturi tube [16]

All elements are transparent and have the same inlet flow area corresponding to an inlet diameter of 50 mm, resp. rectangular cross section (50 x 12) mm. The width at the narrowest cross-section is 14.8 mm. The length of the narrow part of the Venturi tube is 30 mm. All geometries are placed vertically in the hydraulic circuit. The authors also use a similar geometry for cavitation research [8, 10, 11].

Hydraulic parameters (pressures, flow), physical properties of water (temperature, amount of oxygen in the water tank) and noise were measured during the flow through the elements. The size of the cavitation area was evaluated visually. The measuring range is chosen so that the cavitation development is the same for all variants, i.e.

A - flow without cavitation,
B - flow with initial cavitation,
C - fully developed cavitation.

The measurement scheme is shown in fig. 1. The liquid (water) is sucked from the tank by a five-stage Grundfos pump, which is controlled by the frequency inverter YASKAWA VS mini J7. The volume flow [$l \cdot s^{-1}$] is measured by the electromagnetic flow meter ELIS PLZEŇ. The measured element is located behind the flow meter in the vertical position (see fig. 1). Hydrotechnik pressure sensors [Pa] are connected in front of and behind the measured element and records are stored in the Hydrotechnical analyser. The dissolved oxygen transmitter ENDRESS + HAUSER LIQUISYS-M is placed in the tank.

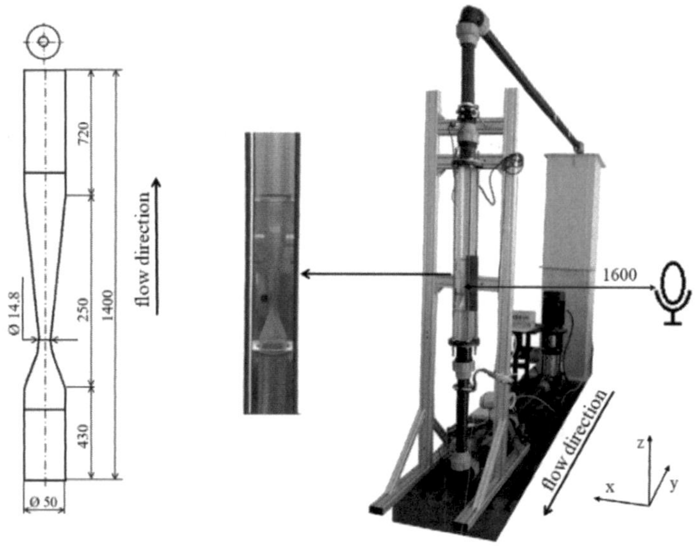

Fig. 1. Measuring circuit, location and orientation of sensors. Dimensions are in mm.

The acoustic camera Bruel & Kjaer WA1764-W-001 was used for experimental noise measurements, it is located 1.6 m in front of the examined element. If known noise sources are measured, a sound pressure level spectrum with suppressed ambient sources can be obtained. Often, this pressure contribution is calculated not for a point, but for M x N points on an area that corresponds to the area of the visual recording from the camera. A graphical representation of the noise sources can be obtained by combining these two visualizations, where the places with the highest pressure contribution are shown in colour. [17] The frequency range and quality of the localization is strongly dependent on the distribution of the microphones and the angle between the different noise sources and the microphone array.

The converging - diverging nozzle (CDN) of circular cross section was chosen for area noise measurement. The onset of cavitation was detected visually and by noise.

1.2 Evaluation of basic hydraulic parameters

Measurements that have been published in the literature [13, 14, 15, 16] are used for evaluation. For each measurement variant, pressures and volume flow rates were recorded with a time step of 0.001 s and subsequently the average values were evaluated. At the same time, the noise intensity was recorded with a given sampling frequency 32.768 kHz.

The dependence of the pressure drop on the volume flow rate is shown in fig. 2 (a). The evaluation of the dependence of the cavitation number on the Reynolds number at the measured element inlet is in fig. 2 (b). The Reynolds number is determined as

$$Re = \frac{v_{in} d_h}{\nu_L} \tag{3}$$

where v_{in} [m·s^{-1}] is the velocity at the measured element inlet, d_h [m] is the hydraulic diameter, ν_L [m^2·s^{-1}] is kinematic viscosity of liquid (water).

The cavitation number is determined as

$$Ca = \frac{2(p_{out} - p_w)}{\rho_L v_{max}^2} \tag{4}$$

where p_{out} [Pa] is the mean value of the pressure at the outlet of the measured element, p_w [Pa] je saturated pressure at given conditions, ρ_L [kg·m^{-3}] is density of water, v_{max} [m·s^{-1}] is the maximum velocity of water flow i.e., the velocity at the narrowest cross-section.

Figure 2 (a) shows that with increasing volume flow rate the pressure drop increases for all variants. The characteristic of element II is different due to the larger flow area in the narrowest cross-section, therefore the cavitation in this element was created arose at the latest. The flow was turbulent in all cases, see fig. 2 (b). Cavitation for various geometries originated at different cavitation numbers. The critical cavitation number was around 1 for the Venturi tube, for other geometries it was less than 1. The initial cavitation number also increases with air content.

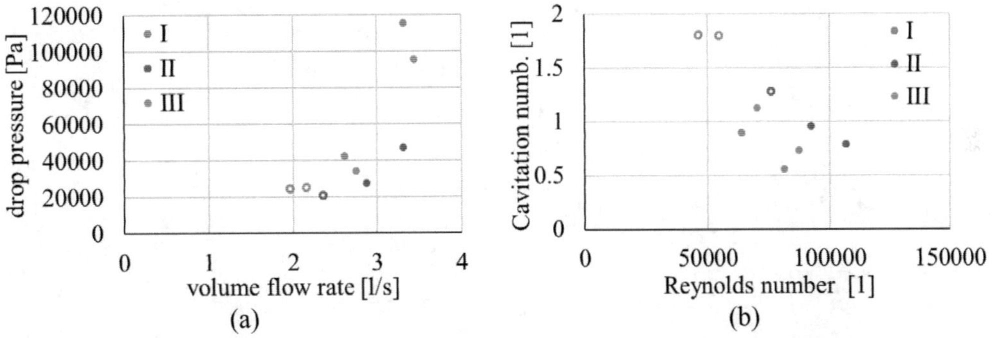

Fig. 2. Dependence of pressure drop on volume flow rate (a), dependence of cavitation number on Reynolds number (b) Unfilled points represent flow without cavitation, filled points represent flow with cavitation.

2 Evaluation by noise intensity probe

Noise measurement by a noise intensity probe with two microphones was performed using the above geometries (see chapter 1.1). The MiniSPL probe was placed 0.4 m in front of the measured element. Other noise sources have been partially shielded.

The acoustic intensity probe evaluates the acoustic intensity I [W·m^{-2}], which can be converted to the acoustic intensity level *La* [dB] as

$$La = 10\, log_{10}\left(\frac{I}{I_0}\right) \tag{5}$$

where I [W·m^{-2}] is acoustic intensity, $I_0 = 1 \cdot 10^{-12}$ W·m^{-2} is reference values of acoustic intensity.

Figure 3 shows the frequency spectra obtained from the noise measurements. For all three nozzles, it is possible to evaluate the same spectrum trend for flow without cavitation. Spectrum slope (-5/3) for frequency <1000; 10,000> is typical for turbulent flow. In cases with initial and developed cavitation, the character changes significantly. For all variants, it is clear that the noise frequencies are significant in the range of 2000 Hz to 8000 Hz.

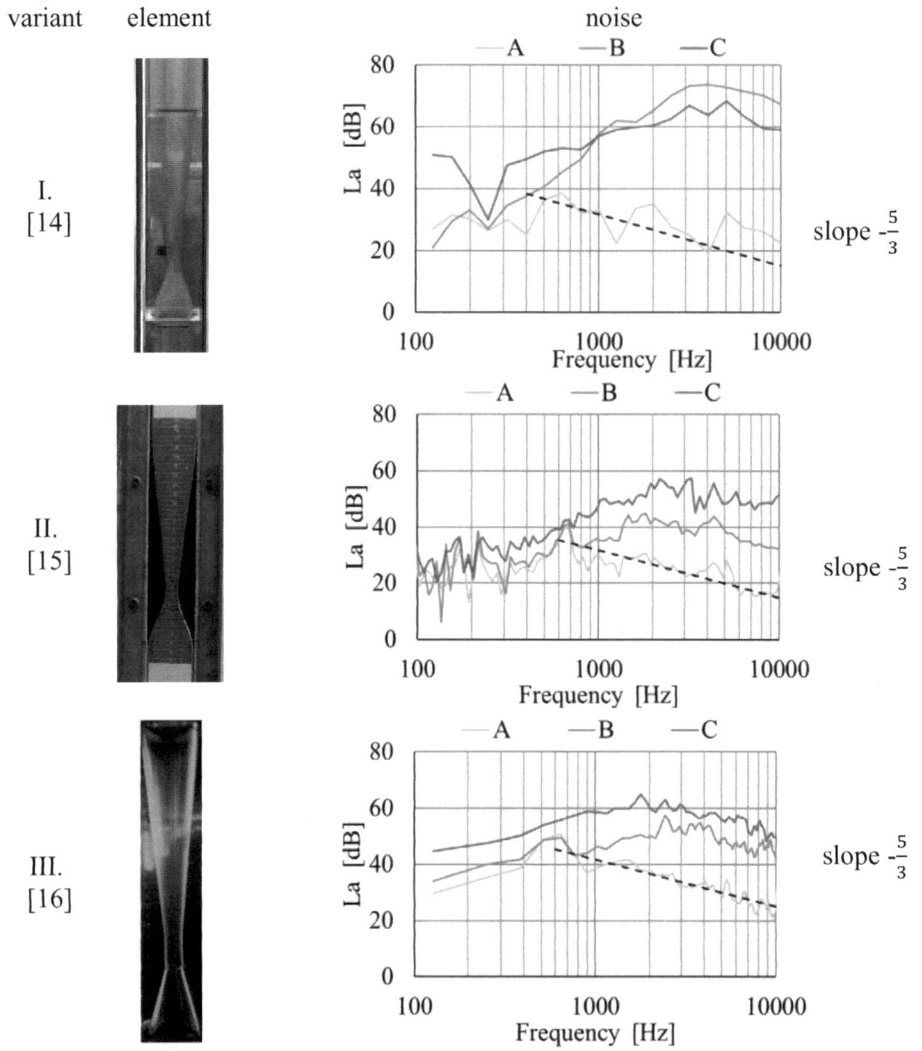

Fig. 3. Evaluation of noise measurements for defined geometries.

3 Evaluation by acoustic camera (microphone array)

Noise measurement using an acoustic camera allows you to evaluate the total noise and also the noise on the element. In this noise measurement, it is not appropriate to evaluate frequencies less than 1500 Hz, which is related to the microphone distribution of the camera used. By evaluating the measurement, it is possible to determine the total noise and subsequently the noise of selected sources.

From the frequency spectrum of the noise for the measured element it is evident that a significant increase of the acoustic noise level for the flow with cavitation (B, C) in the frequency range 2000 Hz to 10000 in comparison with the flow without cavitation (A), see fig. 4. The same was confirmed by measurement with an acoustic intensity probe, see fig. 3.

In the case of converging - diverging nozzle of circular cross section, the noise measurement by the intensity probe and the acoustic camera confirmed that the initial cavitation is noisier than the developed cavitation. Curve B is in the range of expected cavitation frequencies above curve C. The literature [3] describes that the onset of cavitation is often accompanied by a distinct hissing sound.

variant	element	noise
I.	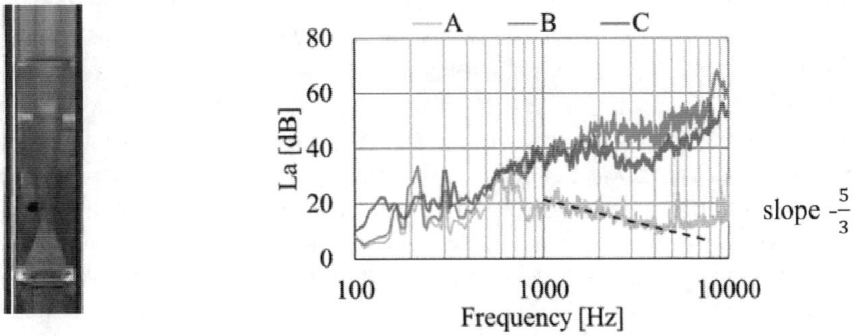	

Fig. 4. Evaluation of noise measurements by area noise for variant I.

When measuring noise in the entire monitored area, the place of its occurrence can be evaluated for a given significant frequency. Figure 5 shows the frequency spectrum of the total noise for converging - diverging nozzle of circular cross section (I) and variant C, i.e., fully developed cavitation. An increase in the PSD value can be observed in the frequency range from 5000 Hz to 6000 Hz and from 9000 Hz to 10000 Hz, which is caused by cavitation in the circuit. The PSD increase around the 8000 Hz frequency value is due to the operation of the pump's electric motor, i.e., it is not caused by cavitation.

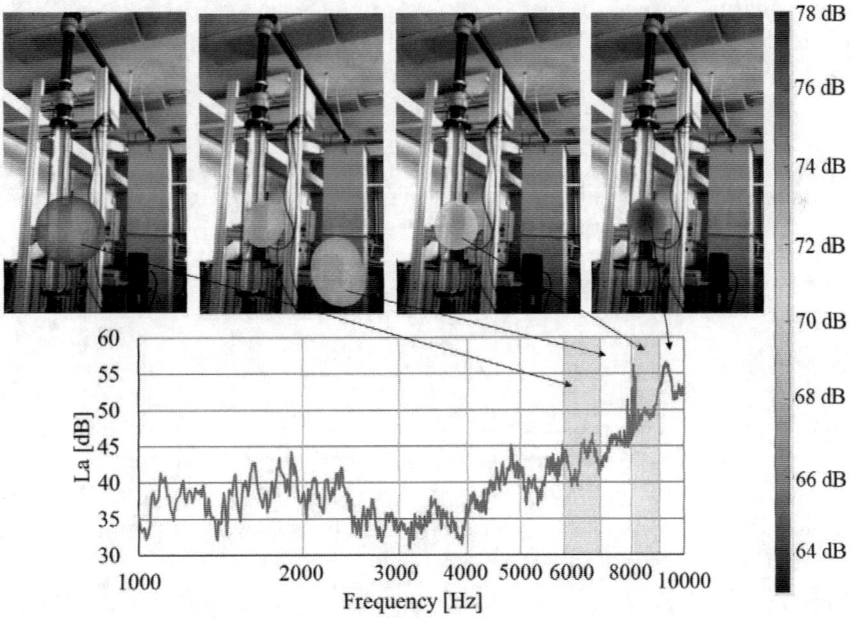

Fig. 5. Evaluation of the frequency spectrum of noise and definition of the problem in the circuit for converging - diverging nozzle of circular cross section, variant C - fully developed cavitation.

4 Conclusion

Cavitation in hydraulic systems results in noise that alerts you to possible circuit problems during normal operation. Cavitation noise may not be dominant in a given area, so better identification methods must be used.

The classic method is the measurement of noise by a noise intensity probe, where the evaluated spectrum shows an increase in intensity at frequencies in the order of <1000;

10000> Hz. A more accurate variant is the use of area noise measurement, where it is also possible to identify the location of noise sources in the room. It can be concluded that the frequency spectrum of the noise measured by the intensity probe (Fig. 3, variant I) and the acoustic camera (Fig. 4) gives identical results. However, when using an acoustic camera, it is possible to precisely define the location of the noise (Fig. 5) and the corresponding frequency. Therefore, acoustic camera recordings better identify noise sources. It is obvious that another significant source of noise is the electric motor of the hydrodynamic pump.

In the case of non-cavitating flow (variant A), it can be seen that the slope of the frequency spectrum curves in Fig. 3 and Fig. 4 corresponds to the slope at turbulent flow. The frequency spectrum of noise obtained during flow with initial or developed cavitation deviates significantly from this curve, which is a signal to identify cavitation. In addition, with increasing flow, the noise of the system increases, but in the case of converging - diverging nozzle of circular cross section, the initial cavitation (formation) is noisier than the developed cavitation.

Cavitation detection using area noise measurement helps in practice to identify the place in the circuit where cavitation with lower noise occurs. Subsequently, the problematic element or circuit can be optimized in terms of cavitation suppression using numerical simulation.

The aim of the authors in the future is to apply the numerical method of multiphase cavitation flow consisting of liquid, steam and air for water and especially for oil, where air is a major component and steam cavitation is essentially non-existent. The frequency response of cavitation noise is obtained by applying the acoustics model in the ANSYS Fluent software and the mathematical model is verified with experimental noise measurements. Numerically, it will then be possible to optimize the hydraulic element in terms of suppressing cavitation and reducing the noise of the hydraulic equipment.

This work was supported by the European Regional Development Fund in the Research Centre of Advanced Mechatronic Systems project, project number CZ.02.1.01/0.0/0.0/16_019/0000867 within the Operational Programme Research, Development and Education.
The work presented in this paper was supported by a grant SGS „Numerical modeling of transient fluid flow problems with the support of experimental research" SP2022/32".

References

1. Noskievič, J. *Kavitace v hydraulických strojích*. SNTL—Nakladatelství Technické Literatury: Praha, Czech Republic, 1989; 333p, ISBN 80-03-00206-0.
2. Zhang, Y., Qian, Z., Wu, D., Wang, G., Wu, Y., Li, S., & Peng, G. (2017). Fundamentals of cavitation and bubble dynamics with engineering applications. *Advances in Mechanical Engineering*, 9(3), 1687814017698321.
3. Raichel, D. R. (2006). *The science and applications of acoustics*. Springer Science. 663 p. ISBN-10: 0-387-26062-5.
4. Yasui, K. (2018). *Acoustic Cavitation and Bubble Dynamics*. Springer Cham. 124 p. ISBN 978-3-319-68236-5. 10.1007/978-3-319-68237-2.
5. Yasui, K. (2015). *Sonochemistry and the Acoustic Bubble*. Chapter 3 - Dynamics of Acoustic Bubbles. ISBN 9780128015308. DOI: 10.1016/B978-0-12-801530-8.00003-7.
6. Brennen, Ch. (1995). *Cavitation and Bubble Dynamics*. 44. 10.1017/CBO9781107338760.
7. Plesset, M.; Prosperetti, A. (2003). Bubble Dynamics and Cavitation. *Annual Review of Fluid Mechanics*. 9. 145-185. 10.1146/annurev.fl.09.010177.001045.

8. Maršálek, B.; Maršálková, E.; Odehnalová, K.; Pochylý, F.; Rudolf, P.; Stahel, P.; Rahel, J.; Čech, J.; Fialová, S.; Zezulka, Š. (2019). Removal of Microcystis aeruginosa through the Combined Effect of Plasma Discharge and Hydrodynamic Cavitation. *Water*. https://doi.org/10.3390/w12010008.
9. Reinke, P.; Ahlrichs, J.; Beckmann, T.; Schmidt, M. (2022) High-Speed Digital Photography of Gaseous Cavitation in a Narrow Gap Flow. *Fluids*. 7, 159. https://doi.org/10.3390/fluids7050159
10. Zhang, L.; Zhang, G.; Ge, M. (2020). Coutier-Delgosha, O. Experimental Study of Pressure and Velocity Fluctuations Induced by Cavitation in a Small Venturi Channel. *Energies*. 13, 6478. https://doi.org/10.3390/en13246478
11. Shi, H.; Nikrityuk, P. (2020). The Influence of Inflow Swirl on Cavitating and Mixing Processes in a Venturi Tube. *Fluids*. 5, 170. https://doi.org/10.3390/fluids5040170
12. Uruba, V. (2009) *Turbulence*. ČVUT v Praze. 141 s.
13. Jablonská, J.; Kozubková, M.; Bojko. M. (2021). Flow of Oil and Water through the Nozzle and Cavitation. *Processes*. 9, no. 11: 1936. https://doi.org/10.3390/pr9111936
14. Kozubkova, M.; Bojko, M.; Jablonska, J.; Homa, D.; Tůma, J. (2016). Experimental research of multiphase flow with cavitation in the nozzle. *EPJ Web of Conferences*. 114. 02058. 10.1051/epjconf/201611402058.
15. Jablonská, J.; Mahdal, M.; Kozubková, M.; (2017). Spectral Analysis of Pressure, Noise and Vibration Velocity Measurement in Cavitation. *Measurement Science Review*. 17. 10.1515/msr-2017-0030.
16. Jablonská, J.; Kozubková, M.; Mahdal, M.; Marcalík, P.; Tůma, J.; Bojko, M.; Hružík, L. (2021). Spectral analysis of gaseous cavitation in water through multiphase mathematical and acoustic methods. *Physics of Fluids*. 33. 085128. 10.1063/5.0058757.
17. Bruel & Kjær. "PRODUCT DATA. BK Connect Acoustic Camera Type ̈ 9712-W-FEN". Nærum, Denmark, BP 2534 – 18 2020-06.
18. Hald, J. (2004). *Beamforming, Bruel Kjaer Technical Review*, nr.1.
19. Tuma, J., Janecka, P., Vala, M., Richter, L. (2012). Sound source localization. In *Proceedings of the 13th International Carpathian Control Conference* (ICCC) (pp. 740-743). IEEE.

https://doi.org/10.1051/matecconf/202236902011

Braid Angle and its Influence on the Hydraulic Hoses Behaviour Under Pressure Loading

Filip Dýrr[1,*], *Adam* Bureček[1], *Lumír* Hružík[1], and *Tomáš* Polášek[1]

[1]VSB – Technical University of Ostrava, Faculty of Mechanical Engineering, Department of Hydromechanics and Hydraulic Equipment, 708 00 Ostrava, Czech Republic

Abstract. This paper deals with the experimental analysis of the effect of the braid angle on the behaviour of a hydraulic hose under pressure loading. The main structural elements of the hose are a rubber tube and a steel braid or spiral. The material of the braid and the number of braids influence the magnitude of the pressure load on the hydraulic hose. The angle of the braid affects the deformation of the hydraulic hose when the inner wall is loaded with fluid pressure. For hydraulic hoses, it is a requirement to maintain a neutral braid angle during manufacture to ensure a balance between axial and circumferential stresses under pressure loading. An experimental device was set up to measure the hoses. This device can be used to measure the diameter of the outer braid, the length and the tensile force of the hose in the axial direction and, where appropriate, to evaluate the angle of the outer braid of the hose as a function of the fluid pressure loading on the inner wall of the hose. The object of the research is to determine the effect of the braid angle on the behaviour of the hydraulic hose under pressure loading. Based on the measurements, the dependencies of the tensile force, the braid angle and the change in length of hydraulic hoses on the internal pressure were determined. In order to compare the measurement results, hydraulic hoses with different inner diameters and different numbers of braids or spirals were selected.

Research background: Description of neutral braid angle, hydraulic hose geometry, tensile force of hose.

Purpose of the article: Influence of braid angle on the hydraulic hoses behaviour under pressure loading.
Methods: Experimental measurements.
Findings & Value added: Determining of the change of length dependencies on the pressure, determining of the tensile force dependencies on the pressure.

Keywords: *hydraulic hose, braid angle, tensile force, pressure loading*

* Corresponding author: filip.dyrr@vsb.cz

1 Introduction

Hydraulic hoses are an important part of hydraulic systems. Hydraulic hoses enable the connection of individual components when their relative positions are changed. This is due to the design of the hydraulic hoses. The construction of a high-pressure hydraulic hose consists of three parts, namely a rubber tube, a steel braid or spiral and a rubber sleeve. However, for special applications, there are other materials for the manufacture of hydraulic hoses, such as thermoplastic polymers [6]. The braid material and the number of braids affect the pressure load capacity of the hydraulic hose. In addition to the steel wire braid, there are hoses where the braid is made of aramid fibres [2]. Two types of high pressure hydraulic hoses can be distinguished according to the braiding method. In the first case, the steel wires are braided together to form a so-called hose braid. Both single-braid and double-braid variants are commonly available on the market [4]. The integration of two steel braids in the manufacture of a hydraulic hose increases its pressure load capacity. In this case, a thin intermediate layer of rubber is placed between the braids. In the case of the second variant, the hose manufacturing technology consists in winding the individual steel wires side by side along the helix. The individual spirals again have a thin layer of rubber between them. For high working pressures, a hydraulic hose variant with four layers of winding spirals is commonly available on the market. The robustness of this construction obviously increases the pressure load capacity of the hose, but at the same time reduces its flexibility in contrast to the steel braided construction.

The angle at which the braid is wound is also important for the design of the hydraulic hose. This angle has a significant effect on the change in geometric dimensions of the hose when the inner wall is loaded with fluid pressure [1, 9, 10]. The lifetime of the hose also depends on this angle [3]. Failure of a hydraulic hose can cause fatal problems, for example, in the braking system of automobiles [7]. Therefore, emphasis is placed on maintaining a neutral braiding angle during the actual production of hydraulic hoses, where the optimum longitudinal feed and braiding speed of the production machine must be maintained [5]. If the neutral angle is maintained, there is a balance between axial and circumferential stresses in the hydraulic hose [12]. The design of a hydraulic hose is very complex, for this reason, it is not easy to clearly determine the material properties of hoses, such as their modulus of elasticity [8]. Hydraulic hoses significantly affect the dynamic properties of hydraulic systems, mainly due to their capacitance or viscoelastic properties [13, 14].

2 Theoretical background

The working pressure of the hydraulic oil causes axial and circumferential stresses in the hydraulic hose, which stresses the hose in both directions. The design of the hydraulic hose is such that there is a balance between the applied stresses. The balance is ensured by the neutral braid angle of the hydraulic hose $\varphi_n = 54.7356°$. Failure to maintain the neutral braid angle during manufacturing results in a change in the geometry of the hydraulic hose under the influence of the working pressure. The hose is more stressed, which can lead to a shortened service life [1]. Table 1 summarizes the geometry change when the braid angle of the hydraulic hose is changed. If the braid angle is less than neutral, the length of the hydraulic hose is shortened. In this variant, a tensile force F_T is generated in the axial direction. This phenomenon is exploited in the case of pneumatic fluidic muscles [15]. The resulting tensile force F_T can put additional stress on the hydraulic hose ends and thus shorten the life of the hydraulic hose. On the other hand, if the braid angle is greater than neutral, the hose length is extended and a compressive force is generated [11].

Table 1. Table of change in geometric properties of a hydraulic hose when the working pressure is applied to the inner wall [11].

Angle change	Hose geometry changes		
	Length	Diameter	Volume
Smaller than neutral	Decreases	Increases	Increases
Greater than neutral	Increases	Decreases	Increases

To determine the neutral angle of the braid, it is possible to consider the hydraulic hose as a cylinder closed at both ends with a radius r, and wall thickness s. Internal pressure in the hydraulic hose p generates axial stress σ_A (see Fig. 1) and circumferential stress σ_O (see Fig. 2) in the walls of the hydraulic hose [12].

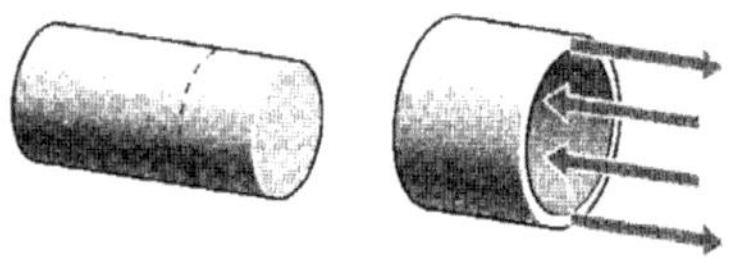

Fig. 1. Axial stress in a closed hydraulic hose [12].

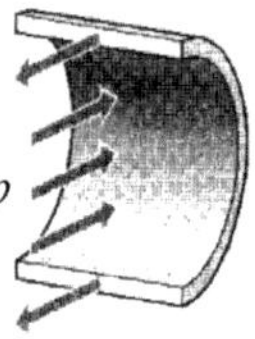

Fig. 2. Circumferential stress in a closed hydraulic hose [12].

The generated axial force F_A is the product of the internal pressure p and the internal area in the hydraulic hose [12]:

$$F_A = \pi \cdot r^2 \cdot p. \tag{1}$$

The axial stress in the wall of the hydraulic hose σ_A is equivalently given by the ratio of the axial force F_A and the area of the hydraulic hose in the cross section, assuming $r >> s$ [12]:

$$\sigma_A = \frac{F_A}{2 \cdot \pi \cdot r \cdot s}. \tag{2}$$

Expressing the axial force F_A from the equations (1) and (2) [12]:

$$F_A = \pi \cdot r^2 \cdot p = \sigma_A \cdot 2 \cdot \pi \cdot r \cdot s, \tag{3}$$

the axial stress in the hydraulic hose σ_A can be determined by the relation (4) [12]:

$$\sigma_A = \frac{p \cdot r}{2 \cdot s}. \tag{4}$$

The circumferential force F_O for a unit length of hydraulic hose can be expressed in the same way [12]:

$$F_O = 1 \cdot 2 \cdot r \cdot p. \tag{5}$$

The circumferential stress expressed for a unit length of hydraulic hose is given by equation (6) [12]:

$$\sigma_O = \frac{F_O}{1 \cdot s} \cdot \frac{1}{2}. \tag{6}$$

Expressing the circumferential force F_O from the equations (5) and (6) [12]:

$$F_O = 1 \cdot 2 \cdot r \cdot p = \sigma_O \cdot 2 \cdot 1 \cdot s, \tag{7}$$

the circumferential stress σ_O in the hydraulic hose can be determined according to the equation [12]:

$$\sigma_O = \frac{p \cdot r}{s}. \tag{8}$$

From equations (4) and (8), it can be seen that the circumferential stress σ_O is twice the axial stress σ_A. Fig. 3 shows a segment of the hydraulic hose and the forces acting on the braid wires. The braid wire makes a braid angle φ with the longitudinal axis of the hydraulic hose [12].

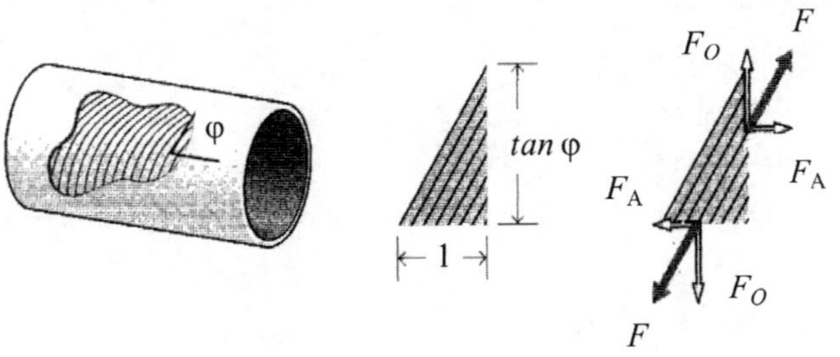

Fig. 3. Forces acting on the hydraulic hose braid [12].

After releasing an element of unit length, the height of the element equals *tan* φ. In order to maintain the equilibrium between circumferential and axial stresses when the internal pressure p is applied in the hydraulic hose, equation (9) for the circumferential load and equation (10) for the axial load must be valid [12]:

$$F_O = F \cdot \sin\varphi = \frac{p \cdot r}{s} \cdot 1 \cdot s, \tag{9}$$

$$F_A = F \cdot \cos\varphi = \frac{p \cdot r}{2 \cdot s} \tan\varphi \cdot s. \tag{10}$$

The braid angle φ can be expressed using the goniometric function [12]:

$$\tan\varphi = \frac{F_O}{F_A}, \tag{11}$$

where the axial force F_A and the circumferential force F_O can be divided into the product of the individual stresses and the surfaces on which they act [12]:

$$\tan\varphi = \frac{F_O}{F_A} = \frac{\sigma_O}{\sigma_A} \cdot \frac{1 \cdot s}{\tan\varphi \cdot s}. \tag{12}$$

After substituting equations (4) and (8) into relation (12) we get [12]:

$$\tan\varphi = \frac{2}{\tan\varphi}, \tag{13}$$

$$\tan\varphi = \sqrt{2} = 54.7356\ °. \tag{14}$$

The braid angles of the tested hydraulic hoses are determined in this work. In the case that the unloaded hose has a measured braid angle value less than the neutral angle φ_n, a tensile force F_T and a change in the geometry of the hydraulic hose is expected. For this reason, the effect of the braid angle on the build-up of the tensile force F_T as a function of the working pressure p applied to the inner wall of the hydraulic hose is investigated. The dependence of the hose geometry on the working pressure in the hydraulic hose is also investigated.

As already mentioned, the pressure load capacity is mainly influenced by the braid or spiral of the hydraulic hose. Fig. 4 shows commercially available braid variants. On the left is a close-up of a hydraulic hose with a single braid 1SN. The maximum working pressure of this hose is p_{max} = 160 bar. The middle photo shows a detail of a hydraulic hose with two braids 2SN, maximum working pressure p_{max} = 275 bar. The highest pressure p_{max} = 420 bar is achieved by the hydraulic hose with four spirals 4SP, which is shown on the right.

Fig. 4. Comparison of high pressure hydraulic hose braids with internal pressure d_{in} = 12.5 mm.

3 Experimental analysis

For the purpose of research in the field of hydraulic hoses, an experimental stand for measuring hydraulic hoses of different diameters was created (Fig. 5, Fig. 6). The support frame is made of modular aluminium profiles for easy modification. The tested hydraulic hose H is rigidly threaded to a plate which is fixed at the top of the support frame. The source of oil flow is the hydraulic power unit. The distribution of oil to the experimental stand is provided by a pressure line P and a return line T. The working pressure of oil p in the hydraulic hose is set to the required value by a proportional relief valve PRV. The working oil pressure p is measured by the pressure sensor PS. As the pressure p increases, the length Δl of the hydraulic hose H changes, which is measured by the laser distance sensor LS1. At the same time, the change in the diameter Δd of the hydraulic hose braid is measured by the optical micrometer LS2. In the case of the fixation of the hose H, as the working oil pressure p increases, the tensile force of the muscle F_T increases in its longitudinal axis, which is measured by the force sensor FS. The force sensor FS is located between the lower end of the hydraulic hose H and a plate which is rigidly connected to the frame of the structure. A CAM camera is used to take a picture of the hose H, on which the rubber cover has been removed and braid is visible. Images of the braid are taken over the full working pressure range of the hose or up to a pressure of p = 200 bar. The elements used are listed in Table 2.

Table 2. Table of elements used on the experimental device.

Symbol	Name	Type	Producer	Basic parameters
PRV	proportional relief valve	DBETBEX-1X	Rexroth	p_{max} = 315 bar
PS	pressure sensor	PR 400	Hydrotechnik	working range: (0 ÷ 250) bar
FS	force sensor	FO 200	Hydrotechnik	working range: (0 ÷ 20) kN

LS1	laser distance sensor	optoNCDT	Micro epsilon	working range: (0.5 ÷ 200) mm
LS2	optical micrometr	LS - 7070	Keyence	working range: (0.5 ÷ 65) mm
CAM	camera	FASTCAM MINI UX	Photron	frame rate: up to 2000 fps

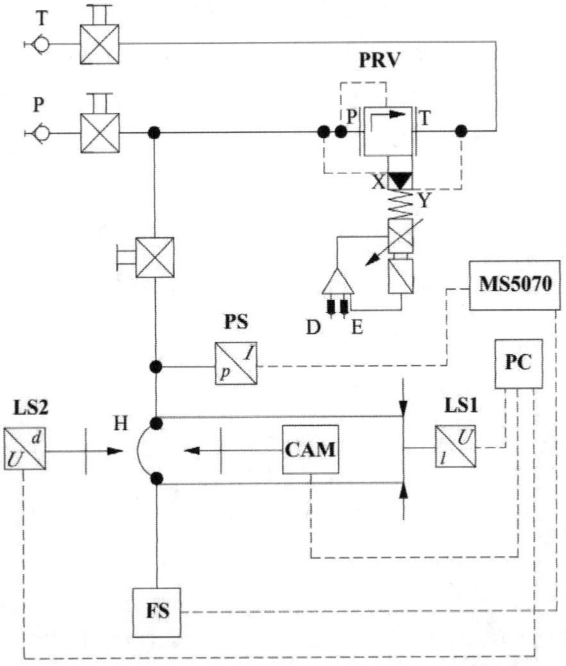

Fig. 5. Simplified scheme of experimental equipment for testing hydraulic hoses.

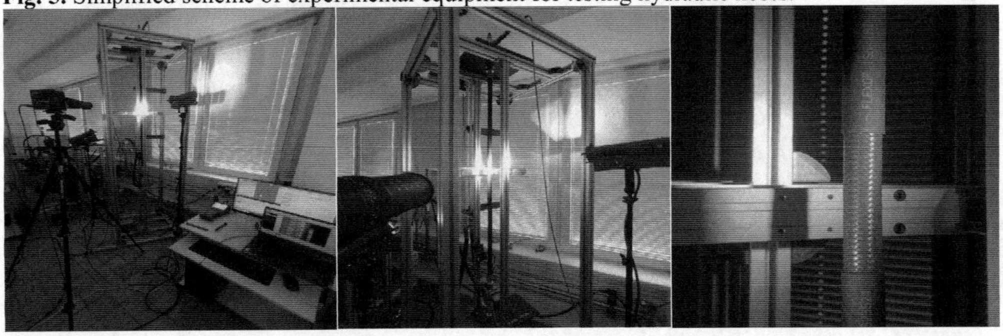

Fig. 6. Experimental equipment for testing hydraulic hoses.

Fig. 7 shows a detail photograph of the hydraulic hose braid angle φ. Using Photron FastCam Viewer software, the braid angles for the hydraulic hoses tested were evaluated over the entire working pressure range of the hose or up to a pressure of $p = 200$ bar. Furthermore, the measured data such as change in length Δl, braid diameter d, tensile force F_T over the entire working pressure range of the hose or up to pressure $p = 200$ bar were evaluated. Table 3 shows the list of hydraulic hoses tested.

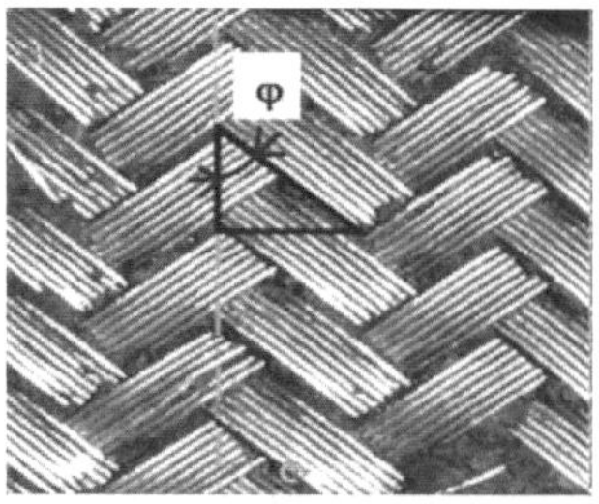

Fig. 7. Example of determining the braid angle of a hydraulic hose.

Table 3. Table of tested hydraulic hoses.

Hose	Inner diameter d_{in}	Diameter of outer braid d	Length without fittings l	Number of braids/spirals	Maximal working pressure	Initial braid angle φ_0
	[mm]	[mm]	[mm]	[-]	[bar]	[°]
DN12.5_1SN	12.5	17.8	1371	1 braid	160	54.73
DN12.5_2SN	12.5	18.8	1366	2 braids	275	52.74
DN12.5_4SP	12.5	20.5	1376	4 spirals	420	54.38
DN16_1SN	16	20.6	1350	1 braid	130	53.06
DN19_1SN	19	24.7	1345	1 braid	105	52.90

4 Results

Two variants of measurements were made. In the first case, the hose was fixed in the top plate and the lower end of the hose had one degree of freedom in the vertical direction. In this way, the change in braid angle φ and the change in hose length Δl were measured. In the second case, the hose was fixed in the top plate and the lower end was fixed into the force sensor. The tensile force F_T was measured in this way.

Fig. 8 shows the dependence of the braid angle φ on the working pressure p for hydraulic hoses with one braid and different inner diameters. The graph shows that for hose DN12.5_1SN the initial braid angle φ_0 was measured very close to the neutral braid angle φ_n. It can also be seen that for hose DN12.5_1SN the braid angle value does not change a lot with increasing working pressure p. This result confirms that at the value of the neutral braid angle, equilibrium between axial and circumferential stresses occurs and geometric changes do not occur. Hydraulic hose DN16_1SN has a braid angle value at atmospheric pressure $\varphi_0 = 53.06°$. The braid angle increases linearly with increasing pressure in the hydraulic hose. Hydraulic hose DN19_1SN has a braid angle value at atmospheric pressure $\varphi_0 = 52.90°$. Again, the braid angle increases with increasing working pressure p. The change in braid angle φ is accompanied by a change in geometry for these hoses.

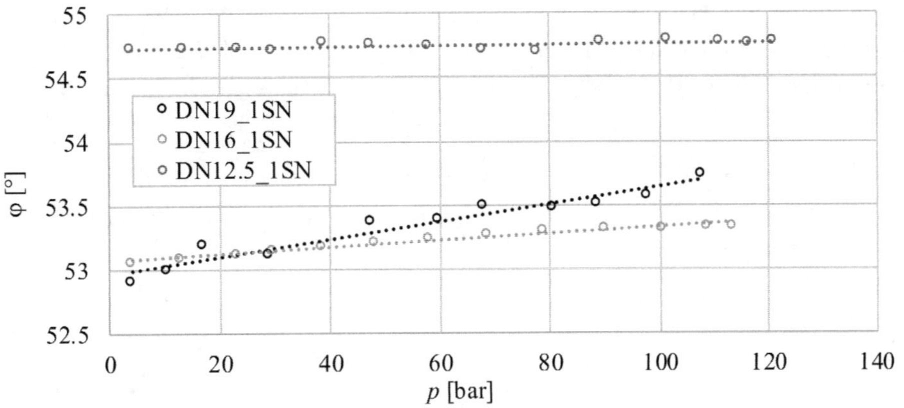

Fig. 8. Dependence of the braid angle φ on the working pressure p for one braid and different internal hose diameters.

Fig. 9 shows the dependence of the braid angle φ on the working pressure p for hoses of the same internal diameters d_{in} = 12.5 mm, but with different numbers of braids or spirals. For hose DN12.5_1SN, the values of working pressure p were not achieved as for the other hoses, due to the pressure load capacity of hose DN12.5_1SN. The graph shows that the hydraulic hose DN12.5_4SP has an outer spiral angle value at atmospheric pressure φ_0 = 54.38°. As the pressure in the hydraulic hose increases, the spiral angle increases linearly. It can be seen that the neutral angle value has been exceeded. According to the theory, the length l of the hydraulic hose should increase when the value of the neutral angle of the braid is exceeded. However, we must take into account that after removing the rubber cover we were only able to determine the spiral angle for the outer spiral. It was not possible to determine the angle for the inner spiral without destroying the hydraulic hose. The graph shows that the hydraulic hose DN12.5_2SN has a value of the angle of the outer braid at atmospheric pressure φ_0 = 52.74°. As the pressure in the hydraulic hose increases, the braid angle again increases linearly. For this hose, the braiding should result in the greatest reduction in the length l of the hydraulic hose and, in the case of fixing the hose, the greatest tensile force F_T. Again, it must be remembered that the braid angle has been determined for the outer braid only, which is visible after removal of the rubber cover.

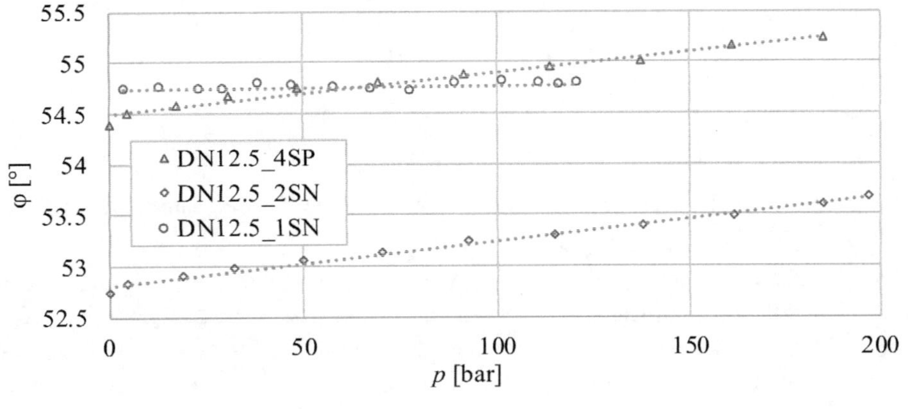

Fig. 9. Dependence of the braid angle φ on the working pressure p for the same hose inner diameter and different numbers of braids or spirals.

Fig. 10 shows the dependence of the hose length change Δl on the working pressure p for different hose inner diameters. The graph shows that the largest change in length Δl occurs for hose DN16_1SN ($\varphi_0 = 53.06°$). The smallest shortening occurs for hose DN12.5_1SN ($\varphi_0 = 54.73°$).

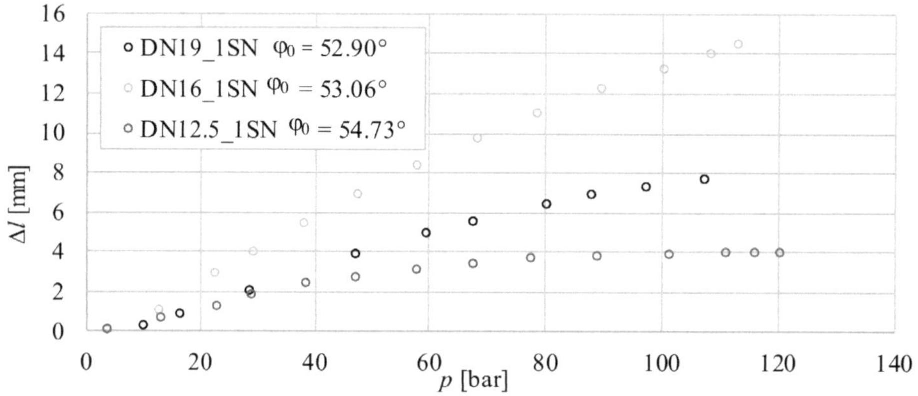

Fig. 10. Dependence of hose length change Δl on working pressure p for one braid and different hose inner diameters.

Fig. 11 shows the dependence of the hose length change Δl on the working pressure p for different numbers of hose braids or spirals. The graph shows that the greatest change in length Δl occurs for hose DN12.5_2SN ($\varphi_0 = 52.74°$). The least shortening occurs for hose DN12.5_4SP ($\varphi_0 = 54.38°$).

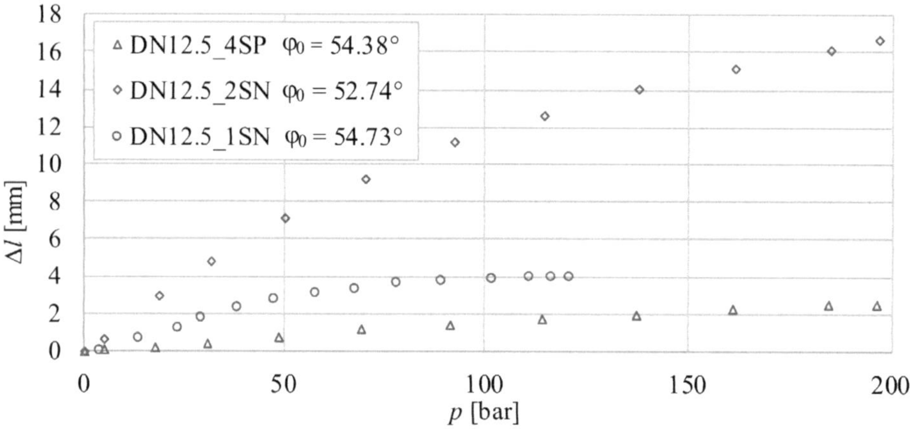

Fig. 11. Dependence of hose length change Δl on working pressure p for the same hose inner diameter and different numbers of braids or spirals.

Fig. 12 shows the dependence of the tensile force F_T on the working pressure p for different internal hose diameters. The graph shows that for hydraulic hose DN12.5_1SN ($\varphi_0 = 54.73°$) there is a very small increase in the tensile force F_T as a function of the pressure p. For hydraulic hose DN16_1SN ($\varphi_0 = 53.06°$) there is a significant increase in the tensile force F_T as a function of the pressure p, in contrast to hose DN12.5_1SN ($\varphi_0 = 54.73°$) where the increase in the tensile force F_T is negligible. The same trend of increase of the tensile force F_T as a function of pressure p applies to hose DN19_1SN ($\varphi_0 = 52.90°$). From the graph it can be determined that the initial braid angle φ_0 influences the dependence of the tensile force F_T on the working oil pressure p.

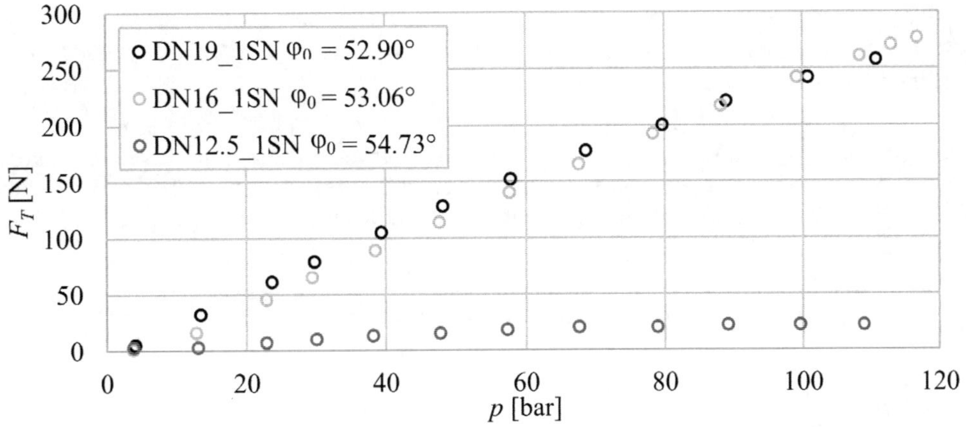

Fig. 12. Dependence of the tensile force F_T on the working pressure p for one braid and different internal hose diameters.

Fig. 13 shows the dependence of the tensile force F_T on the working pressure p for different numbers of hose braids or spirals. The graph shows that for hydraulic hose DN12.5_4SP ($\varphi_0 = 54.38°$) there is a very small increase in the tensile force F_T. The trend of the increase of the tensile force F_T as a function of the working pressure p is similar to that of the hydraulic hose DN12.5_1SN ($\varphi_0 = 54.73°$). For hydraulic hose DN12.5_2SN ($\varphi_0 = 52.74°$) there is a significant increase in the tensile force F_T. This graph again confirms the influence of the initial braid angle j$_0$ on the tensile force F_T as the working pressure p changes.

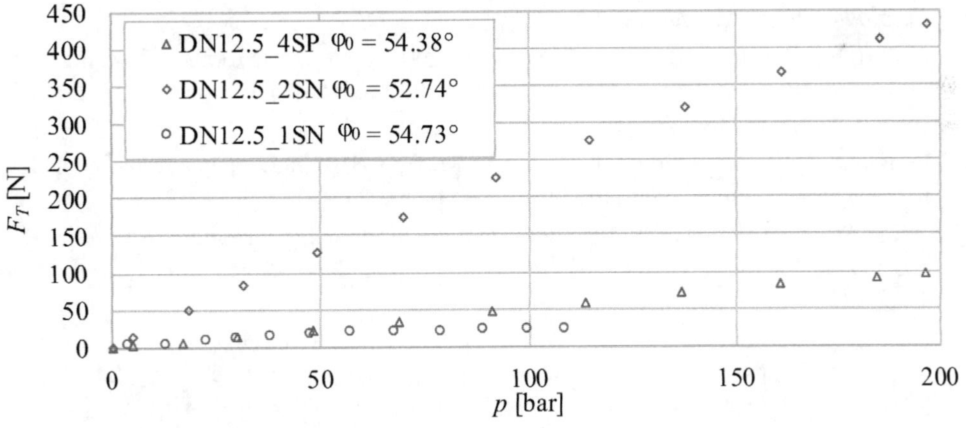

Fig. 13. Dependence of the tensile force F_T on the working pressure p for the same hose diameter and different numbers of braids or spirals.

5 Conclusions

This paper deals with the experimental analysis of the effect of the braid angle on the behaviour of hydraulic hoses under pressure loading. The dependencies of the braid angle, the change in hose length and the hose tensile force on the working oil pressure were successively measured and evaluated. Within the results, the data for the tested hydraulic hoses are compared. The result shows that the braid angle has a significant effect on the behaviour of the hydraulic hose under pressure load. The smaller the initial braid angle is compared to the neutral braid angle, the greater the change in hose length with oil working pressure and the greater the tensile force is induced. It is also observed that the characteristics determined are independent of the inner diameter or number of braids or spirals of the measured hoses.

This work was supported by the European Regional Development Fund in the Research Centre of Advanced Mechatronic Systems project, project number CZ.02.1.01/0.0/0.0/16_019/0000867 within the Operational Programme Research, Development and Education.

The work presented in this paper was supported by a grant SGS „Numerical modeling of transient fluid flow problems with the support of experimental research." SP2022/32.

References

1. Huang, D., Zhang, J. (2021). Research on the Tensile Mechanical Properties of a Braided Corrugated Hose and Its Axial Stiffness Model. *Journal of marine science and engineering*, 9, 1029.
2. Lubecki, M., Stosiak, M., Bocian, M., Urbanowicz, K. (2021). Analysis of selected dynamic properties of the composite hydraulic microhose. *Engineering failure analysis*, 125, 105431.
3. Hua, G., Changgeng, S., Jianguo, M., Guomin, X. (2020). Study on theoretical model of burst pressure of fiber reinforced arc-shaped rubber hose with good balance performance. *Polymers & polymer composites*, 29, 919-930.
4. Karpenko, M., Prentkovskis, O., Šukevičius, Š. (2022). Research on high-pressure hose with repairing fitting and influence on energy parameter of the hydraulic drive. *Eksploatacja i niezawodnosc-maintenance and reliability*, 24, 25-32.
5. Potluri, P., Rawal, A., Rivaldi, M., Porat, I. (2003). Geometrical modelling and control of a triaxial braiding machine for producing 3D preforms. *Composites part a-applied science and manufacturing*, 34, 481-492.
6. Drumond, G. P.,Pasqualino, I. P., da Costa, M. F. (2016) Study of an alternative material to manufacture layered hydraulic hoses. *Polymer testing*, 53, 29-39.
7. Kwak, S. B., Choi, N. S. (2009). Micro-damage formating of a rubber hose assembly for automotive hydraulic brakes under a durability test. *Engineering failure analysis*, 16, 1262-1269.
8. Hruzik, L., Burecek, A., Vasina, M. (2017). Mathematical simulation and measurement of expansion of hydraulic hose with oil. *Tehnicki vjesnik-technical gazette*, 24, 1905-1914.
9. Xiao, S. L., Wang, P., Soulat, D., Gao, H. (2020). An exploration of the deformability behaviours dominated by braiding angle during the forming of the triaxial carbon fibre braids. *Composites part a-applied science and manufacturing*, 133, 105890.
10. Cui, C., Dong, J. W., Mao, X. B. (2019). Effect of braiding angle on progressive failure and fracture mechanism of 3-D five-directional carbon/epoxy braided composites under impact compression. *Composites structures*, 229, 111412.
11. Hölcke, J. (2002). *Frequency response of hydraulic hoses.* Royal Institute of Technology, KTH.
12. Roylance, D. (2001). *Pressure vessels.* Department of Materials Science and Engineering Massachusetts Institute of Technology, Cambridge.

13. Feng, H., Qungui, D., Yuxian, H., Yongbin, C. (2017). Modelling Study on Stiffness Characteristics of Hydraulic Cylinder under Multi-Factors. *Strojniški vestnik – Journal of mechnaical engineering*, 63, 447-456.
14. Siwulski, T., Warzyńska, U. (2017). The advantages of a new hydraulic cylinder design with a control system. In *23rd International Conference Engineering mechnaics* (pp. 15-18).
15. Dýrr, F. Dvořák, L., Fojtášek, K., Brzezina, P., Hružík, L., Bureček, A. (2022). Experimental Analysis of Fluidic Muscles. In *Proceedings of the 25th International Conference on Hydraulics and Pneumatics.* (pp. 45-49).

Drag Estimation in the Near Wake of the NREL's Airfoils Based on Hot-Wire Data

Vitalii Yanovych[1,2,*], *Daniel* Duda[1], *Václav* Uruba[1,2], and *Pavel* Antoš[2]

[1]University of West Bohemia, Univerzitní 8, 301 00 Pilsen, Czech Republic
[2]Institute of Thermomechanics of the Czech Academy of Sciences, Dolejškova 5, 182 00 Prague, Czech Republic

Abstract. This paper presents the results of the drag coefficient estimations for different types of NREL airfoils based on the experimental data. Namely, it was S803, S807, S813, and S817 profiles with the same chord length. The investigations were conducted at three angles of attack α=0°, α=±5°, different chord-based Reynolds numbers 0.6×10^5, 1.3×10^5 and 2.6×10^5. While, measuring cross-sections were placed behind the trailing edge at $x \cdot c^{-1} \approx 0.2$, 0.4 and 1.0. Experimental data were collected using a hot-wire split fiber probe 55R55, which allowed us to estimate the characteristics of turbulent flow in stream-wise and crosswise directions. According to the obtained results, the highest and lowest *Cd* values correspond to profiles S817 and S803, respectively. Moreover, the results show that more asymmetric profiles S803 and S807 have the lowest resistance at zero angles of attack.

Research background: Application of Antonia and Rajagopalan methodology to drag assessment of various stream bodies.
Purpose of the article: Comparative evaluation of the drag coefficients of the NREL airfoils based on instantaneous velocity distribution behind.
Methods: Hot-wire anemometry with split fiber probe 55R55.
Findings & Value added: The highest drag coefficient corresponds to the S817 profile. The asymmetry of the airfoil shape has a significant impact on its drag characteristics.

Keywords: *Hot-wire, NREL's airfoils, wake topology, drag coefficient*

1 Introduction

Nowadays, investigations of the aerodynamic characteristics of the different airfoil types at relatively low chord-based Reynolds numbers less than 5×10^5 are becoming more actual from both fundamental and industrial points of view [1]. That is primarily due to the active development of innovations for small wind turbines [2], various types of drones, quadcopters, modern micro-air vehicles, and even biomechanics systems [3]. Despite their significance, the experimental data for lift and drag coefficients for low Re_c are only available for some airfoils. Usually, various aerodynamic scales are applied to determine these parameters, while

*Corresponding author: yanovych@kke.zcu.cz

for their prediction, using numerical simulations. Besides, the methodology proposed by Antonia and Rajagopalan has also become widely available to assess the drag forces of various streamlined bodies [4]. According to it, the aerodynamic forces can to determined using instantaneous velocity and its derivatives [5]. It allows obtaining drag data without surface instrumentation or external balance mechanism, which is particularly useful for different experimental techniques [6-9].

This paper aims to study by Hot-wire the behavior of the drag coefficient for different types of NREL's airfoils under various experimental conditions. These profiles family have become widespread in various fields because their geometry provides well lift-to-drag ratios. [10].

2 Experimental setup

The experiments were conducted at the Department of Power Engineering of the University of West Bohemia in the low-speed open-type wind tunnel, which has a test section 0.75 m long, 0.3 m high, and 0.2 m wide [11]. The natural level of turbulent inlet flow did not exceed 0.2 %.

Four types of NREL airfoil S803, S807, S813, and S817, were selected for the current study. They were produced of polymerized lactic acid (PLA) by 3D printing technology using the PRUSA i3 printer. In the general case, at 3D printing, the manufactured models are characterized by high surface roughness Ra ≈ 21.8 μm [12, 13]. Therefore, with the help of P2500 sandpaper with low abrasiveness, we have reduced the roughness to Ra ≈ 3~5 μm. The quality of roughness was estimated using a Taylor Hobson Surtronic Duo profilometer. All airfoil models have the same chord length and span of $c \approx 100$ mm and of $l{\cdot}c^{-1} \approx 1$, respectively. But at these dimensions, their thickness was different, namely for S803 – $b{\cdot}c^{-1} \approx 0.12$, S807 – $b{\cdot}c^{-1} \approx 0.18$, S813 – $b{\cdot}c^{-1} \approx 0.16$, and S817 – $b{\cdot}c^{-1} \approx 0.16$ (see Fig. 1 c).

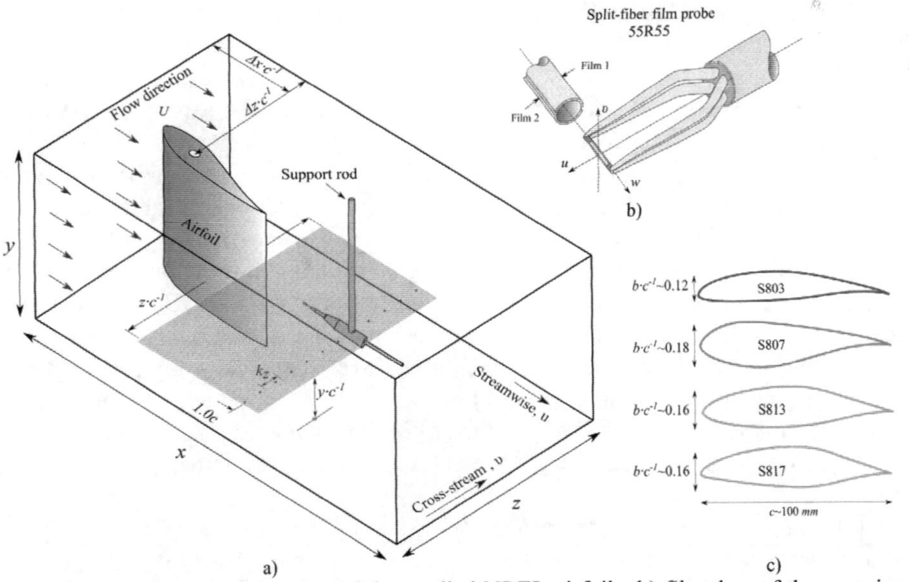

a) c)

Fig. 1. a) The geometry configuration of the studied NREL airfoils. b) Sketches of the experimental setup. c) 55R55 split-fiber film probe. The arrows in blue indicate the direction of free-stream flow U. The green field shows the Hot-wire measuring plane. Where b and c the thickness and chord length of the airfoils. While $z{\cdot}c^{-1}$, $x{\cdot}c^{-1}$ and $y{\cdot}c^{-1}$ the width, length and location height of the measuring fields, respectively. The red dots line is the position of the measuring cross-section and k_z is the distance between the measuring points. The coordinates $\Delta z{\cdot}c^{-1} \approx 1.9$ and $\Delta x{\cdot}c^{-1} \approx 2.8$ shows the position of the aerodynamic center of the airfoil.

Measurements were performed at three flow velocities $U \approx 10$, 20 and 40 $m{\cdot}s^{-1}$. Thus, the chord-based Reynolds number Re_c was around $0.6{\times}10^5$, $1.3{\times}10^5$ and $2.6{\times}10^5$. The angle of attack also had three positions, α=0° and α=±5°, that varied by a stepper motor. It should note that, accordingly to the airfoil geometry, at zero angles, the blockage ratio is equal to 2.1-2.3%, while for the positive and negative angles, it is about 3%. Respectively to the Pope and Harper, this value cannot exceed 7.5%. Otherwise, an appropriate blockage correction method must be applied to the experiments. Mainly, the wake topology was analyzed in the measuring cross-section from $z{\cdot}c^{-1} \approx -0.5$ to 0.5 at $x{\cdot}c^{-1} \approx 1.0$, and the height of $y{\cdot}c^{-1} \approx 1.8$. But for a qualitative assessment of wake evolution, some measurements were done at $x{\cdot}c^{-1} \approx 0.2$ and 0.4, also. The sketch of the airfoil model and its set-up in the test section presents in Fig. 1 a.

The 55R55 split-fiber film Hot-wire probe was applied to study the two-dimensional characteristics of turbulent flow behind airfoil (see Fig. 1 b). The general uncertainty of the measurements was approximately ± 0.5%. The Hot-wire signals for all experiments were sampled at 70 kHz. The filtering of the received signal was applied as well. For the high and low pass filter the frequency was 10 Hz, and 30 kHz respectively. The typical duration of a one-point measurement was 10 s.

3 Results and discussion

In current investigations, the average drag coefficient for each airfoil was estimated from the wake profiles of the mean velocity and Reynolds stresses, accordingly to the formula of Antonia and Rajagopalan [4]:

$$C_d = 2\int_{-\infty}^{\infty} \frac{u}{U} \cdot \left(1 - \frac{u}{U}\right) d\left(\frac{z}{c}\right) + 2\int_{-\infty}^{\infty} \left(\frac{\overline{v'^2} - \overline{u'^2}}{U^2}\right) d\left(\frac{z}{c}\right) \quad (1)$$

where u is the local mean velocity, U is the external velocity of undisturbed flow, $\overline{u'^2}$ and $\overline{v'^2}$ are the normal Reynolds stresses components in streamwise and cross-stream directions, while $z{\cdot}c^{-1}$ is a dimensionless coordinate of the measuring point in a cross-stream direction.

The above formula consists of two parts, where the first term reflects the momentum deficit and the second is the effect of turbulence fluctuation in stream-wise and crosswise directions. Generally, this dependence gives approximately constant values in the absence of backflows. That usually happens close to the bluff body due to the vortex formation behind it. One of the main issues in drag predicting by velocity profiles is at what distance from the aerodynamic profile we need to integrate to correctly evaluated the drag coefficient. Many papers are devoted to this point. For example, according to [14], if the integration operation is performed at a cross-section where the local velocity is less than $0.5U$, the resulting value of the momentum deficit will have significant inaccuracies causing a serious underestimate of the drag coefficient. Moreover, in the airfoil drag investigation, the angle of its attack also plays an important role. Because a larger angle leads to the increasing bluntness, thus promoting the intensity of vortices separation from the airfoil [15]. As a result, the integral value of the second part of equation 1, which is responsible for fluctuating forces, increases sharply.

First, using the 55R55 hot-wire probe, we estimated the patterns of the velocity deficit with increasing distance from the trailing edge of the studied airfoils. The normalized streamwise velocity distribution $u{\cdot}U^{-1}$ along the wake centerline at $Re_c \approx 0.6{\times}10^5$, depending on the incident angles α and the distance $x{\cdot}c^{-1}$ shown in Fig. 2.

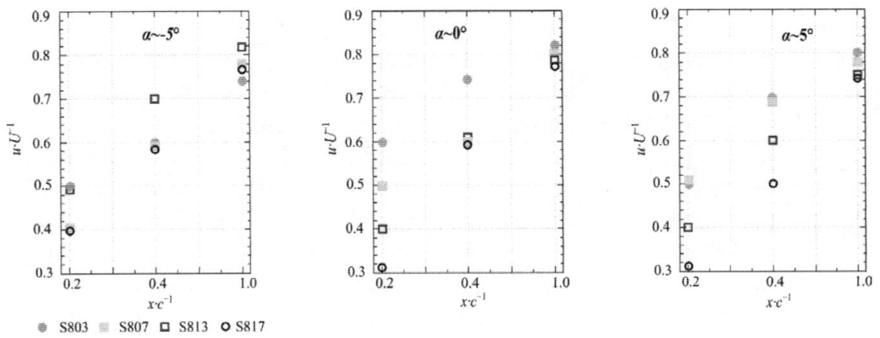

Fig. 2. Normalized streamwise velocity distribution $u \cdot U^{-1}$ along the wake centerline at $Re_c \approx 0.6\times10^5$, depending on the incident angles α and the distance $x \cdot c^{-1}$ behind airfoils.

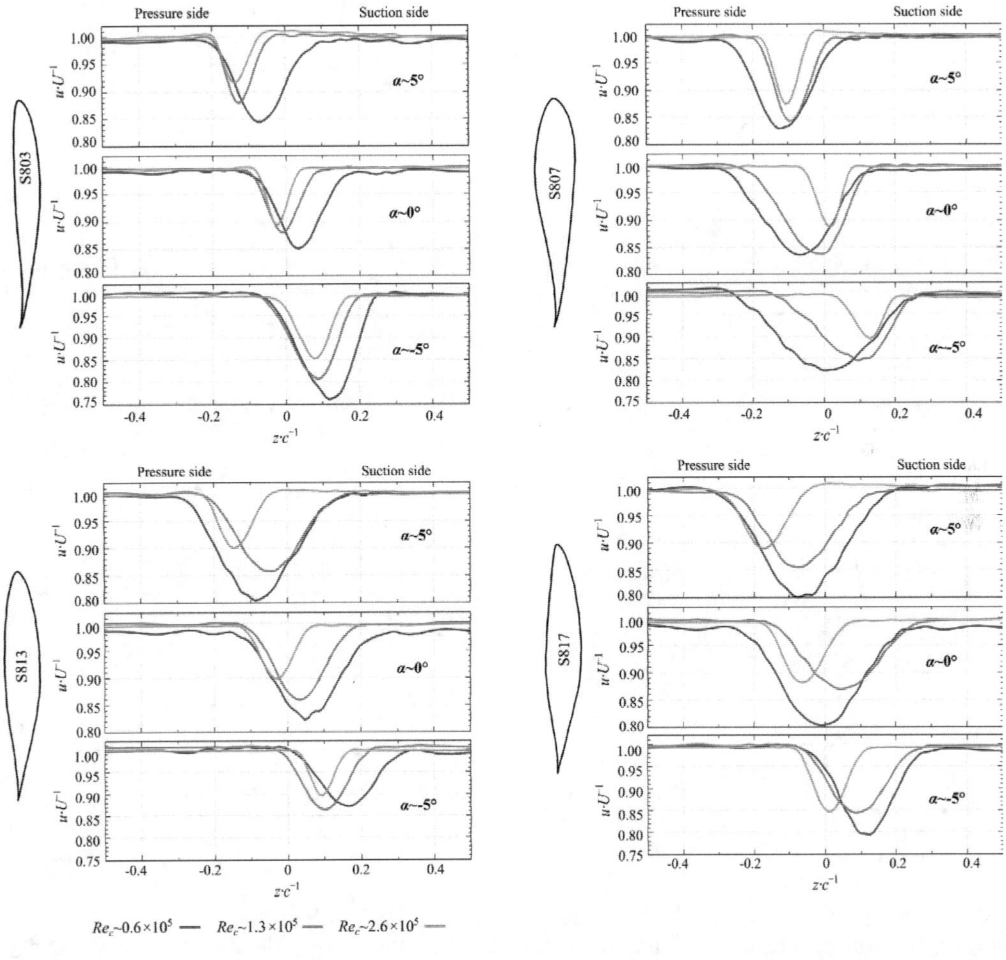

Fig. 3. Normalized streamwise velocity profiles $u \cdot U^{-1}$ at $x \cdot c^{-1} \approx 1.0$, depending on the airfoil types and incident angles α. Violet, green and yellow curves characterize the obtained data at $Re_c \approx 0.6\times10^5$, 1.3×10^5 and 2.6×10^5, respectively.

As we can see, with increasing distance up to $x \cdot c^{-1} \approx 1$, the velocity deficit decreased from 0.3–0.6 up to 0.8. Moreover, the last value of $u \cdot U^{-1}$ is approximately the same for all airfoil types, regardless of the angles of attack. Whereas, close to the trailing edge, the effect of profile geometry increases significantly. That is especially clear at $x \cdot c^{-1} \approx 0.2$ and $\alpha = 0°$, where the velocity deficit for S817 is almost twice high as for S803. This result is somewhat unexpected, as the thickest among the studied airfoils is the S807 profile (see Fig. 1 c).

Based on the above data for further analysis, we chose the cross-section at $x{\cdot}c^{-1} \approx 1$. At this distance, we evaluated the distribution of the velocity characteristics behind different airfoil types (stream-wise component only), depending on the angle of attack and various Reynolds numbers (see Fig. 3). For better visualization, the obtained data were normalized to the average freestream velocity U and placed relative to the trailing edge of airfoils at α=0°. The received results permitted us to estimate the dimensions of the wake. In general, we can observe that as the Reynolds number increases, the depth and width of the wake decrease. Accordingly, to the airfoil thickness, the largest and smallest wake widths correspond to profiles S807 and S803, respectively. This feature is especially noticeable at α =−5°, where the wake at S807 are three times thicker than at S803.

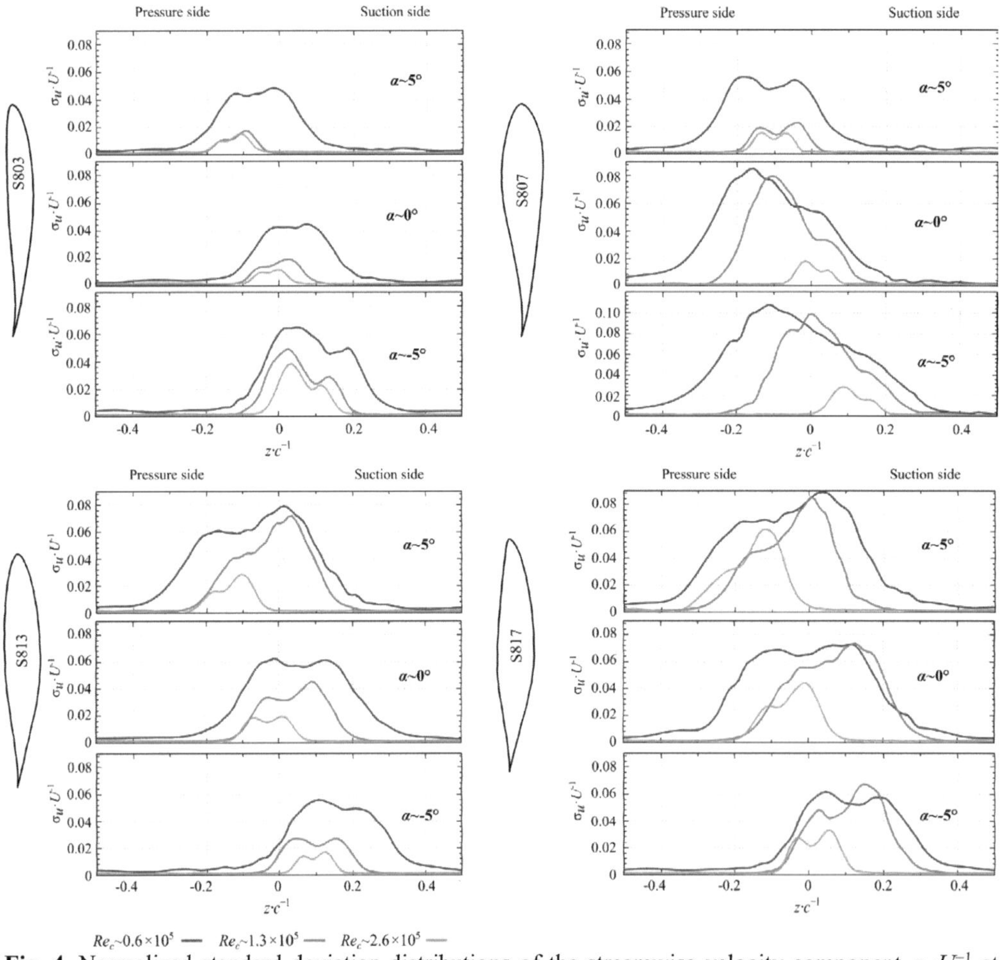

Fig. 4. Normalized standard deviation distributions of the streamwise velocity component $\sigma_u{\cdot}U^{-1}$ at $x{\cdot}c^{-1}{\approx}1.0$, depending on the airfoil types and incident angles α. Violet, green and yellow curves characterize the obtained data at $Re_c \approx 0.6{\times}10^5$, $1.3{\times}10^5$ and $2.6{\times}10^5$, respectively.

We also evaluated the patterns of the normalized standard deviation distributions in the streamwise directions at $x{\cdot}c^{-1}{\approx}1.0$ depending on the Reynolds number and angle of attack (Fig. 4). The obtained data showed that, the smallest value of $\sigma_u{\cdot}U^{-1}$ and the narrowest width of the perturbation area is inherent for the S803 at α=0°. Mainly this difference is due to the thinnest profile thickness. In general, the experimental results of the studied airfoils indicate a sharp decrease in the value of $\sigma_u \cdot U^{-1}$ with increasing Reynolds number and the presence of double peaks of the data distributions. Usually, the double peak can be observed at a symmetrical airfoil and occurs due to the interaction of vortices between the pressure and suction sides. In our case, the appearance of this phenomenon may indicate a certain shape symmetry of the studied profiles. Thus, we can see that for S813 and S817 airfoil at α=0°, the double peak is more apparent. Whereas for the most asymmetric S807 profile, the obtained curve of $\sigma_u \cdot U^{-1}$ has only one pronounced peak from the pressure side.

It should note that another relative indicator of shape asymmetry may be the anisotropy of turbulent flow. What can be estimated using the Reynolds stress tensor [16]. Usually, it has six individual components that describe the flow characteristics relative to the three spatial directions. But in our case, the split fiber 55R55 probe provides measurements only in two directions. In this case, the Reynolds stress tensor becomes statistically two-dimensional. The normalized Reynolds shear stress distributions $\overline{u'v'} \cdot U^{-2}$ at $x \cdot c^{-1} \approx 1.0$, depending on the airfoil types and incident angles, are shown in Fig. 5.

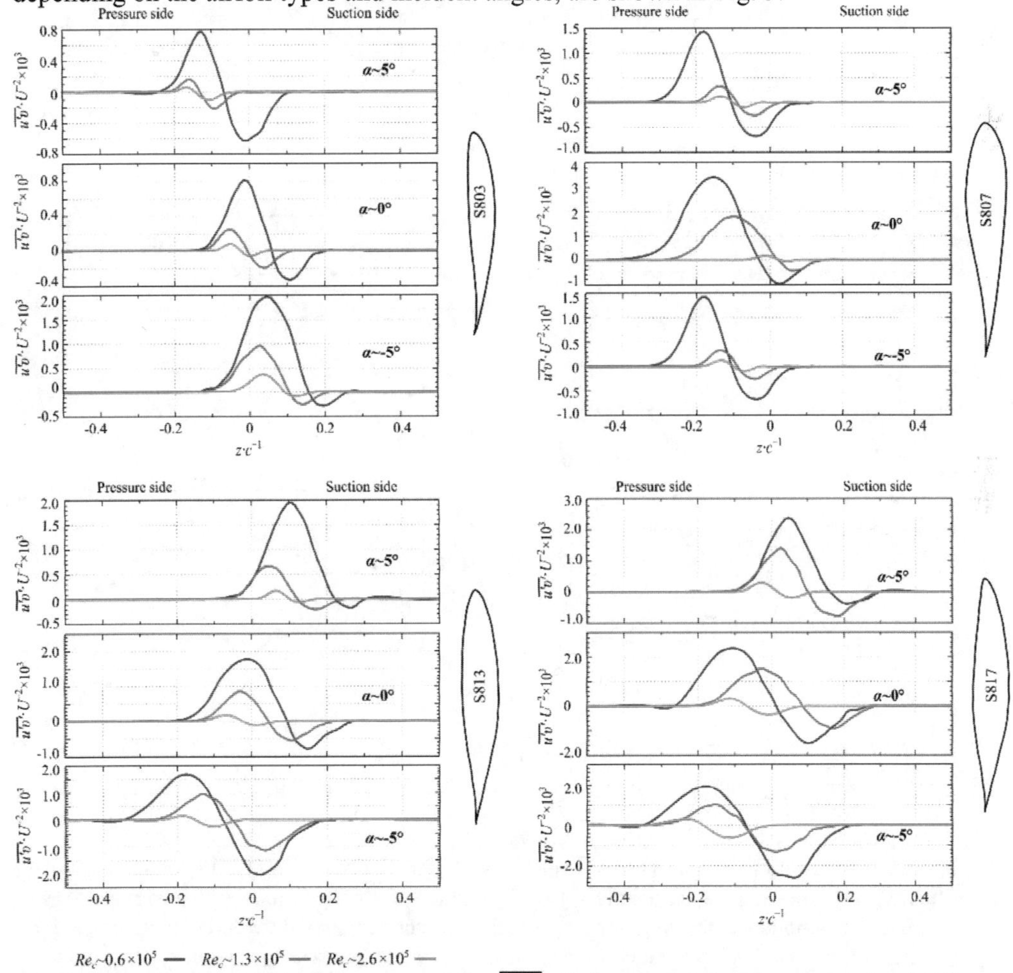

Fig. 5. Normalized Reynolds shear stress profiles $\overline{u'v'} \cdot U^{-2}$ at $x \cdot c^{-1} \approx 1.0$, depending on the airfoil types and incident angles α. Violet, green and yellow curves characterize the obtained data at $Re_c \approx 0.6\times10^5$, 1.3×10^5 and 2.6×10^5, respectively.

As we can see, the highest anisotropy is observed for S807 airfoil, while the lowest is for S803. Of course, such a pattern can be caused not only by the asymmetry of the geometry but also by the airfoil thickness.

Moreover, the characteristics of $\overline{u'v'} \cdot U^{-2}$ for the S803 strongly depends on the Reynolds number. For example, when Re_c increases four times, the maximum value of $\overline{u'v'} \cdot U^{-2}$ decreases approximately five times. Besides, the data obtained show that regardless of the airfoil type, the perturbation region of shear stress with a negative value is located directly on the suction side. This behavior is due to the growth of turbulence intensity, which increases the frictional resistance of the flow. Another interesting fact is that the greatest width of the perturbation zone corresponds to the S817 profile. That is especially noticeable at zero angle of attack. It should note that, unlike others, this airfoil type has a maximum width close to the trailing edge. Thus, this feature of the geometry can cause such a difference.

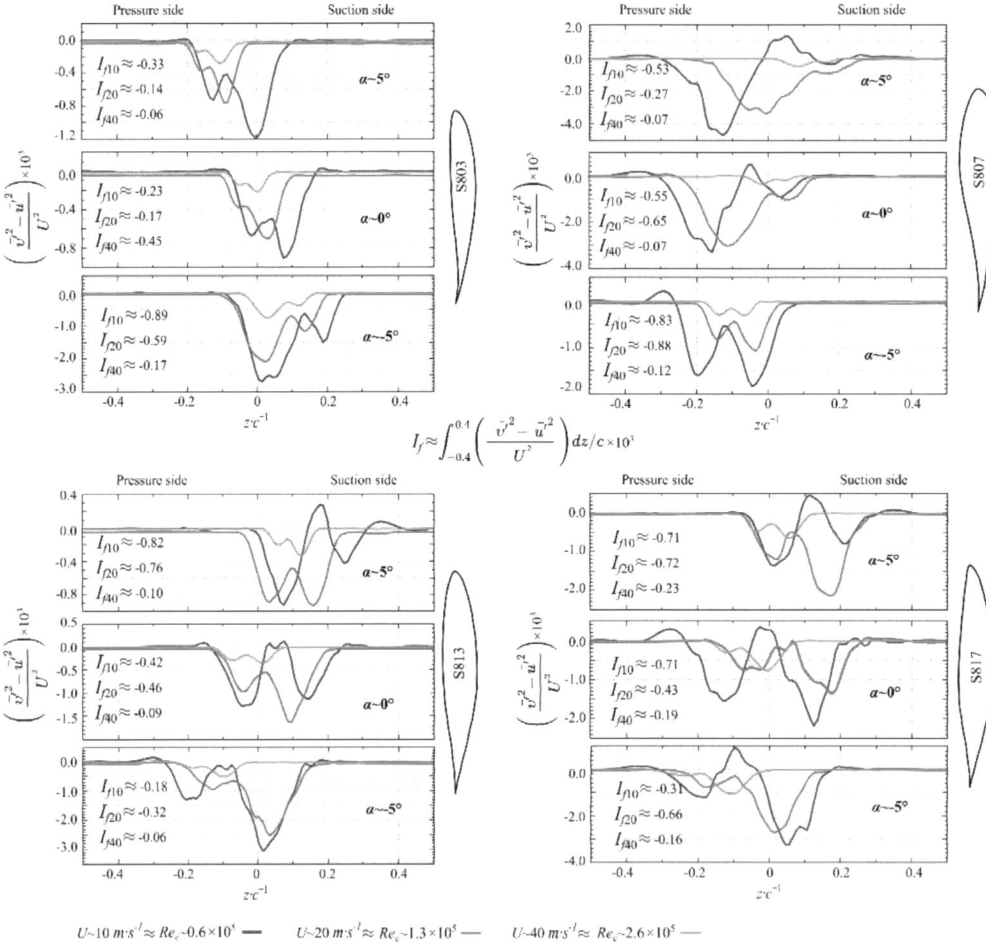

Fig. 6. Impact assessment of streamwise and cross-stream fluctuations on the drag coefficient (accordingly to Antonia and Rajagopalan [4]) for different airfoil types. Calculations were performed at $x{\cdot}c^{-1}{\approx}1.0$ and at various incident angles α, where I_f is the integral value of the turbulence contribution for each measuring cases. Violet, green and yellow curves characterize the obtained data at $Re_c \approx 0.6{\times}10^5$, $1.3{\times}10^5$ and $2.6{\times}10^5$, respectively.

The next necessary part of our investigation was the analysis of the Reynolds normal stress components characteristics. Because exactly their difference forms the part of Antonia and Rajagopalan equation, which reflects the contribution of turbulent fluctuations on the drag coefficient. In Fig. 6 shows a graphical interpretation of the normalized difference between the cross-stream and streamwise velocity fluctuations at $x{\cdot}c^{-1}{\approx}1.0$ depending on the airfoil types, angle of attack, and different Reynolds numbers. Moreover, each graph also contains the integral value of the obtained distribution I_f for each measuring case. It should note that this parameter could be also interpreted as a complex indicator of the degree of flow anisotropy. Accordingly, to the obtained data, the value of I_f decreases with increasing Reynolds number from 0.6×10^5 to 2.6×10^5. Thus, at zero angles of attack, the magnitude of I_f reduces by: for S803–2 times, for S807–8 times, for S813–5 times, and for S817 is 4 times.

Finally, using the obtained data and equation (1), we calculated the characteristics of the drag coefficient *Cd* for each airfoil types depending on the incident angle and Reynolds numbers (see Fig. 7). The obtained data, regardless of the airfoil type, at flow velocity increasing shows logical drag growing. Moreover, the highest value of *Cd* is inherent for S817, while for S803, it is the lowest. That is a somewhat unexpected result, as the S807 is the thickest of the investigated airfoils (see Fig. 1 c). Besides, we can also observe the influence of airfoil shape asymmetry on the drag coefficient characteristics. Thus, for S803 and S807, its minimum value is detected at α=0°, whereas at α=±5°, *Cd* increases sharply. While the maximum drag values for S813 and S817 are observed at α=+5° and α=−5°, respectively.

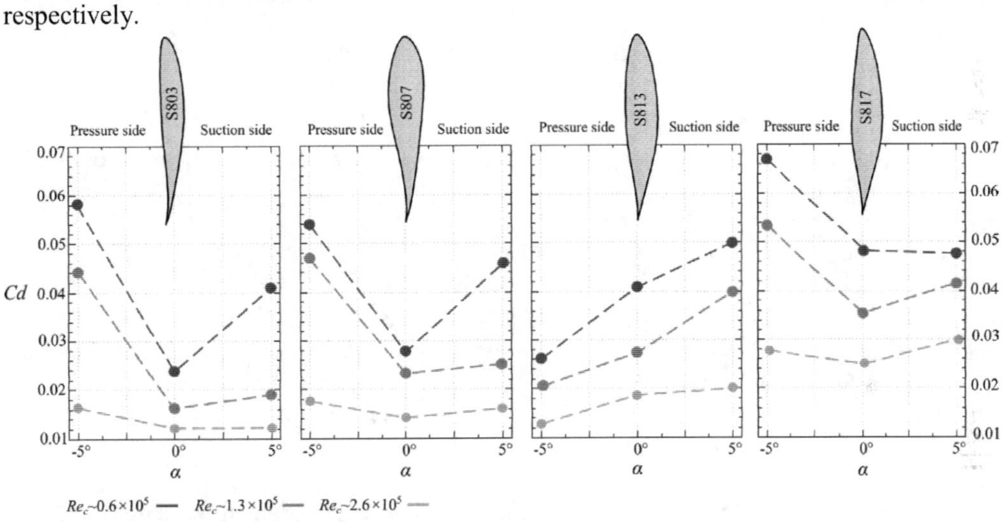

Fig. 7. Distribution of the drag coefficient *Cd* for different airfoil types depending on the incident angle and Reynolds numbers. Violet, green and yellow curves characterize the obtained data for $Re_c \approx 0.6\times10^5$, 1.3×10^5 and 2.6×10^5, respectively.

4 Conclusion

The drag coefficient in the near wake of the different NREL's airfoils was estimated. Namely, it was the profiles S803, S807, S813, and S817. Hot-wire anemometry with a split-fiber film probe 55R55 was applied to the data collection. Experimental investigations realized in several measuring cross-sections, at different Reynolds numbers 0.6×10^5, 1.3×10^5, 2.6×10^5, three angles of attack α= ±5°and 0°. As a result of the study, we obtained a series of graphical distributions for velocity characteristics, standard deviation, and Reynolds stress components. That allowed us to assess the wake topology for each measuring case. Then, using the methodology of Antonia and Rajagopalan, we evaluated the values of the drag coefficients for different airfoil types. Comparative analysis of the data showed that the highest and lowest value of *Cd* corresponds to S817 and S803 profiles, respectively. Moreover, the obtained results displayed that for S803 and S807 profiles, the minimum drag coefficient manifests at α=0°. Whereas for S813 and S817, it was detected at negative and positive angles of attack, respectively.

This work was supported by the project "Research and development of power machines and equipment" reg. No. SGS-2022-023.

References

1. Zhou, M d. Mahbub Alam, Yang, H.X., Guo, H., Wood, D.H. (2011). Fluid forces on a very low Reynolds number airfoil and their prediction. *International Journal of Heat and Fluid Flow*, 32(1), 329-339.
2. Brendel, M., Mueller, T.J. (1988). Boundary-layer measurements on an airfoil at low Reynolds number. *Aircraft*, 25, 317-612.
3. Usherhood, J.R., Ellington, C.P. (2002). The aerodynamics of revolving wings, I. Model hawkmoth wings. *Journal of Experimental Biology*, 205(11), 1547-1564.
4. Antonia, R.A., Rajagopalan, S. (1990). Determination of drag of a circular cylinder. *AIAA Journal*, 28(10), 1833-1834.
5. Son, O., Cetiner, O. (2016). Drag Prediction in the Near Wake of a Circular Cylinder based on DPIV Data. *Journal of Applied Fluid Mechanics*, 9, 1963-1968.
6. Brennan, J.C., Fairhurst, D.J., Morris R.H., McHale, G., Newton, M.I. (2014). Investigation of the drag reducing effect of hydrophobized sand on cylinders. *Journal of Physics D: Applied Physics*, 47(20), 205302.
7. Cao, S., Ozono, S., Tamura, Y., Ge, Y., Kikugawa, H. (2010). Numerical simulation of Reynolds number effects on velocity shear flow around a circular cylinder. *Journal of Fluids and Structures*, 26(5), 685-702.
8. Goodfellow, S.D., Yarusevych, S., Sullivan, P.E. (2012). Momentum coefficient as a parameter for aerodynamic flow control with synthetic jets. *AIAA Journal*, 51(3), 623-631.
9. Henderson, R.D. (1995). Details of the drag curve near the onset of vortex shedding. *Physics of Fluids*, 7(9), 2102-2104.
10. Tangler, J. Somers, D. (1995). *NREL airfoil families for HAWT's*, report of the National Renewable Energy Laboratory (January 1995, Cole Boulevard Golden, Colorado, USA).
11. Yanovych, V., Duda, D., Uruba, V. (2021). Structure turbulent flow behind a square cylinder with an angle of incidence. *European Journal of Mechanics, B/Fluids*, 85, 110-123.
12. Yanovych, V., Duda, D., Uruba, V., Tomášková, T. (2022). Hot-Wire Investigation of Turbulence Topology behind Blades at Different Shape Qualities. *Processes*, 10(3), 522.
13. Yanovych, V., Duda, D., Horáček, V., Uruba, V. (2019). *Research of a wind tunnel parameters by means of cross-section analysis of air flow profiles. AIP Conference Proceedings*, 2189, 020024.
14. Son, O., Cetiner, O. (2016). Drag Prediction in the Near Wake of a Circular Cylinder based on DPIV Data. *Journal of Applied Fluid Mechanics*, 9, 1963-1968.
15. Duda, D., Uruba, V., Yanovych, V. (2021). Wake width: Discussion of several methods how to estimate it by using measured experimental data. *Energies*, 14(15), 4712.
16. Yanovych, V., Duda, D., Uruba, V., Antoš, P. (2021). Anisotropy of turbulent flow behind an asymmetric airfoil. *SN Applied Sciences*, 3(12), 885.

Reduction of Heating Consumption in the School Building During Distance Teaching

František Urban[1,*], *František* Ridzoň[1], *František* Világi[1], *Peter* Mlynár[1], *Ladislav* Zelený[1]

[1] Faculty of Mechanical Engineering in Bratislava, Department of energy machinery, Nám. slobody 17, 812 31 Bratislava

Abstract. Thermal comfort is one of the basic human needs. There are many options for transforming primary energy sources into heat and, in our countries today, into cold. The heat/cooling supply system needs to be optimised. The criterion is to minimise the final consumption of heat, primary energy sources and emissions, including CO_2, while ensuring individual human thermal comfort.

Keywords: *heat consumption, thermal comfort, heating optimisation, distance teaching*

1 Introduction

The paper evaluates the heat consumption for the heating of the main building of the Faculty of Mechanical Engineering in Bratislava (FME) at Námestie slobody 17 in the period 2018-2021. The heat consumption was influenced by the full-time and distance form of teaching. During this period, the objective was to operate compact heat transfer stations (CHTS) and heating system circuits in such a way as to ensure the thermal comfort of staff and students while minimising heat consumption.

2 Description of the faculty of mechanical engineering building and the heat supply system

The total area of the seven-storey Main Building with an auditorium and gymnasium is 25 168 m2. The load-bearing structure of the building consists of a reinforced concrete monolithic skeleton and ceiling slabs. The building envelope consists of a system of vertical monolithic reinforced concrete columns filled with ceramic masonry made of perforated bricks. In the period from 2009 to 2015, the following measures were implemented to reduce heat consumption for heating the FME building: complete replacement of glazed structures, thermal and waterproofing of the roof, thermostatic and hydraulic regulation of the central heating system.

FME is supplied with heat from centralised heating system Bratislava - East of Bratislavská teplárenská, a.s. The original heat transfer station was replaced in 2012 by two compact heat transfer stations (CHTS) for the Main Building and the Laboratory Pavilion. In the primary circuit, the CHTS are connected to the hot water network at temperatures up to 130/70°C with a design overpressure of 2.5 MPa. In the secondary DHW circuit the heating water is heated at a temperature gradient of 90/60°C by plate heat exchangers connected in parallel. In the Main Building, the DHW distributor distribute the hot water for heating to four DHW circuits, the auditorium + gymnasium circuit, two HVAC circuits auditorium + gymnasium, and the extended lecture theatres. In both CHTS, the hot water (HW) preparation is provided by flow-through plate heat exchangers and HW storage tanks are installed.

A control system is used to manage the operation of the CHTS and their technologies, with the return water temperature being controlled equithermally. Quantitative control in the DHW circuits is provided by circulating water pumps with frequency converters, with control dependent on the air temperature in the four reference rooms.

3 Heat consumption in the main building of SJF

The heat consumption (CH) for heating Q_{CH} and the preparation of hot water Q_{HW} in the main building and the pavilion of the FME STU laboratories in the years 2018 to 2021 were analysed. BAT, Inc. provided data measured by a remote control billing metre on heat consumption at intervals of 1 hour at the FME sampling point in the

* Corresponding author: fratisek.urban@stuba.sk

periods 1.10.2018 to 31.5.2021 and 1.10.2021 to 19.1.2022 [1]. Monthly heat consumption and heat cost were also available [2]. In the CHTS installed at the FME Main Building and the Laboratory Pavilion, the daily consumption for heating and the consumption of HW are depreciated by the secondary heat metres [3].

The heat consumption of Q_{HW} for HW preparation in the evaluated period accounted for 6.0% to 7.6%. Figure 1 shows the CH in the Main Building and Laboratory Pavilion of the FME STU in each month of 2018 to 2021.

For each area j of the building envelope FME, the heat loss P_q is the heat transfer which depends on the summed heat transfer coefficient k_j, the area S_j, and the difference between the indoor air temperature t_i, and the outdoor air temperature t_e.

$$P_q = \sum_{j=1}^{j_n} k_j \,.\, S_j \,.\, (t_i - t_e) \text{ (W)}$$

Simplistically, it can be stated that the heat loss due to heat transfer varies from day to day during the heating period, depending on the temperature difference t_i - t_e. Assuming that the indoor temperature is t_i = 20 °C, the temperature difference t_i - t_e can be replaced by the number of day-degrees D_{20}. The number of D_{20} during the months January 2018 to December 2021 is shown in Figure 2. Between 2018 and 2020, the number of day-degrees was 2,856 to 2,964 K.day, 3,299 K.day in 2021. The consumption of heat for heating Q_H is directly proportional to the number of day-degrees D_{20}. The efficiency of heating control can be evaluated by the ratio Q_H / D_{20}. Figure 3 shows the Q_{CH}/D_{20} ratio during the heating periods January 2018 to December 2021.

The Q_{CH}/D_{20} ratio was related to the form of teaching in FME STU, full-time or distance learning. In the period of the COVID-19 pandemic, the activities of the FME were influenced by the guidelines of the Chief Hygienist and the orders of the Rector. From March 2020, teaching was at distance, as well as from October 2020 to May 2021. In the academic year 2021/2022, the first 7 weeks of teaching were face-to-face. Based on the worsening epidemiological situation of COVID-19, from 8 October 2021 until the end of the winter semester on 17 December 2021, teaching was carried out by distance teaching. Examinations from 3.1.2022 onwards were mostly by distance type. The staff during the semester and examination period worked mainly in the faculty.

The mode of activities at the FME during distance teaching revealed opportunities to save energy consumption. Individual heating circuits were quantitatively controlled to maintain a maximum air temperature of t_i = 22 °C in reference rooms at comfort mode during the working days. During the distance teaching and examinations, thermostatic valves were set to temper classrooms. During the attenuation modes in the economy mode, the maximum air temperature were t_i = 17 °C. The attenuation modes were set as follows in terms of start and end:

- nighttime attenuation from 19:00 h to 4:00 h,
- weekend attenuation from friday 19:00 h to monday 2:00 h,
- Christmas and New Year attenuation from 23.12.2021 12:00 to 30.12.2021 12:00, for all circuits, the air temperature was set t_i = 17 °C, from 30.12.2021 12:00 in the central heating circuit no. 3 (offices of selected employees) t_i = 21 °C, at the same time set night and weekend mode, the other circuits t_i = 17 °C until 7.1.2022 9:00 am.

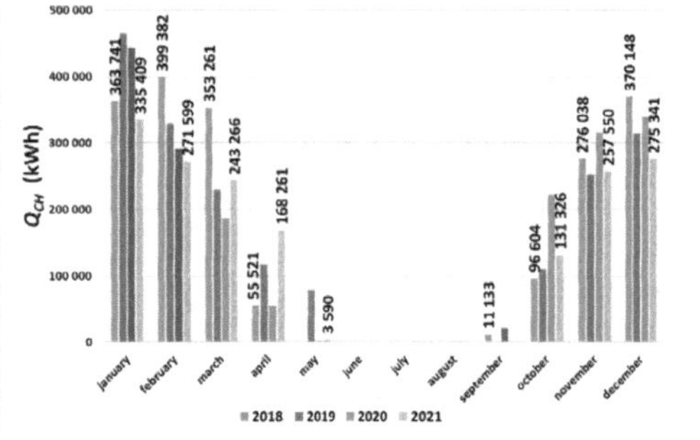

Fig. 1: Heat consumption for heating Q_{CH} during the months January 2018 to December 2021

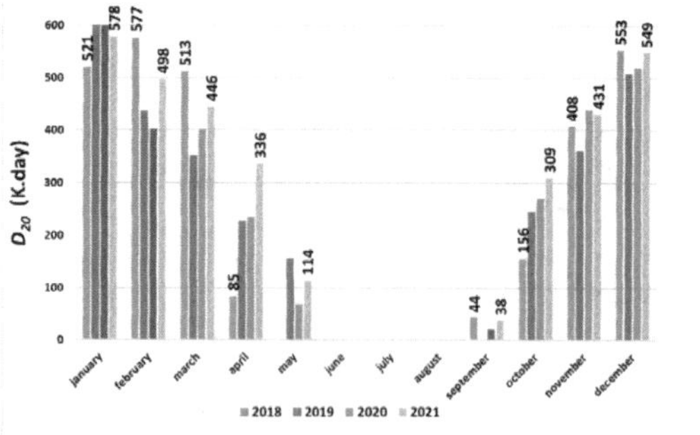

Fig. 2: Number of D_{20} day-degrees during the months January 2018 to December 2021

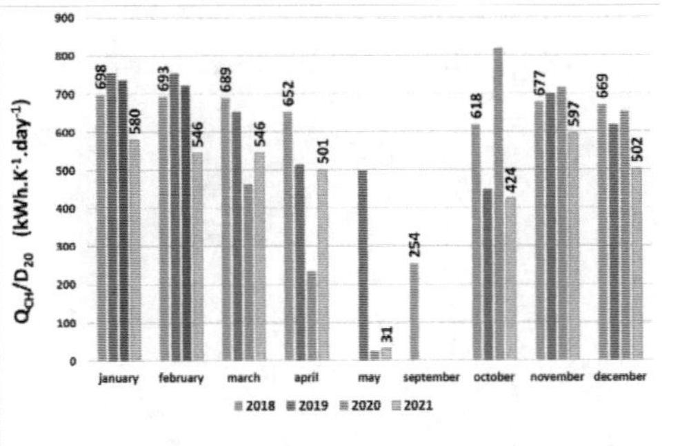

Fig. 3: Q_{CH}/D_{20} ratio during the heating periods January 2018 to December 2021

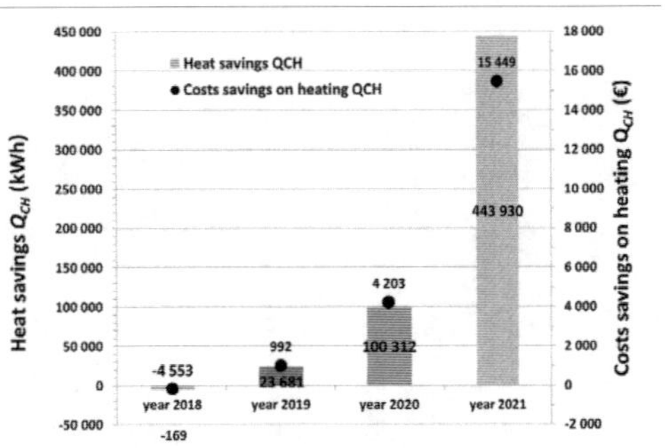

Fig. 4: Savings of heat consumption for heating Q_{CH} and variable component of heat costs without VAT during the years 2018 to 2021

It is necessary to distinguish the months of full-time teaching, i.e., January and February months in 2018, 2019 and 2020, September 2018, October, November and December in 2018 and 2019. In these months, the average values of the Q_{CH}/D_{20} ratio, denoted $\Phi(Q_{CH}/D_{20})$ [5], were calculated. In the other months of 2020 and 2021, teaching was done by distance teaching. There was an opportunity to reduce the heating consumption.
Defined were the monthly heat savings for central heating:

Heat savings: $$Q_{CH\_sav\,month} = \left(\left(\frac{Q_{CH}}{D_{20}}\right)_{month} - \emptyset\left(\frac{Q_{CH}}{D_{20}}\right)_{month}\right) D_{20\,month} \qquad (kWh)$$

Costs savings on heating: $$Q_{CH\,month} = c_{vz} \cdot Q_{CH\_sav\,month} \qquad (€)$$

Where c_{vz} (€.kWh^{-1}) is the variable component of heat costs.

From Fig. 3 it can be seen that during the distance form of teaching at the FME STU the heating consumption was reduced in the months of October, November and December in 2020 and in March 2021. In 2021 the heating consumption of the Q_{CH} was reduced by 443 930 kWh and the saving of the variable component of the heating costs excluding VAT was 15 449 € (Fig. 4). Heat costs were reduced by 9.2%.

4 Modes of heating in the main building of FME

The purpose of adjusting the heating modes in the Main Building of the FME was to ensure the thermal comfort of staff and students and at the same time to reduce the consumption of heat for heating. Thermal comfort is mainly influenced by the temperature of the outside air, the way of controlling the central heating circuits including the damping modes, and the thermal capacity of the building. To investigate thermal comfort in more detail, thermocouples - sensors for indoor air temperature t_{i232}, outdoor air temperature t_e, and wall temperature t_{w232} - were installed in reference room 232. Temperatures were sensed from 4.12.2021 to date at 15 minute intervals. Figure 5 shows the traces of the average daily air temperature values in the reference rooms t_{i232}, exterior t_e, the day-degrees D_{20}, the heat output of P_{CH} to heat the Main Building and the Laboratory Pavilion, the ratio P_{CH} / D_{20}, and the ratio of heat output to the number of day-degrees [4]. The data are shown for the period from 4.12.2021 to 19.1.2021, including working days, weekends, and the Christmas and New Year period. Noticeable are the decreases in the P_{CH} heat output and the P_{CH} / D_{20} ratio during weekend lulls from 19:00 on Friday to 02:00 on Monday, and the subsequent increase in the values of these variables during Mondays. During the weekends, the air temperature t_{i232} decreased by 0.47 ° C to 0.69 ° C in room 232, with wall temperature decreases of 0.35 °C to 0.58 °C.

In the period 23.12.2021 to 30.12.2021, all the circuits of the central heating system were set to economy mode and air temperature was t_{i232} = 17 °C. The air temperature t_{i232} decreased by 3.40 °C and the wall temperature by 3.81 °C.

The offices of the employees who started to work on 3.1.2022 are connected to the central heating circuit No. 3. Therefore, from 30.12.2021 to 7.1.2022, the regulated temperature in reference room No. 232 increased to t_{i232} = 21 °C. For the reference rooms of the other CHP circuits, the temperature t_{i232} = 17 ° C has not changed.

On Monday 10.1.2022 the working period for most of the employees started. To ensure their thermal comfort, all heating circuits switched to comfort mode from 09:00 on 7.1.2022, with a maximum regulated air temperature of t_{i232} = 22 °C in the reference rooms.

In the room 232 the average daily air temperature was changed by t_{i232} = 4.35 °C and the wall temperature by t_{w232} = 4.51 ° C during the assessment period 4.12.2021 to 17.1.2022.

In Fig. 6, the time histories of the characteristic variables from 7.1.2022 to 12.1.2022 are plotted at intervals of 1 h.

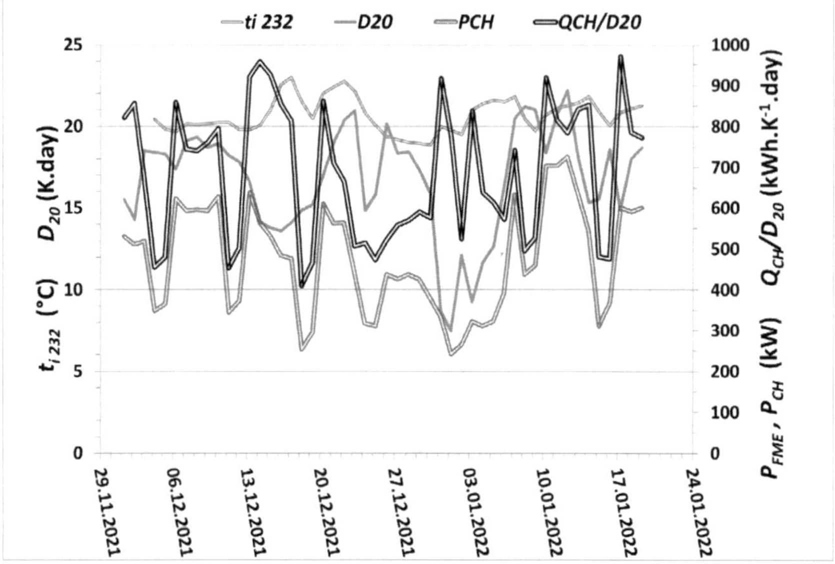

Fig. 5: Plots of the average daily characteristic values of the variables for the period 4.12.2021 to 17.1.2022 in reference room 232

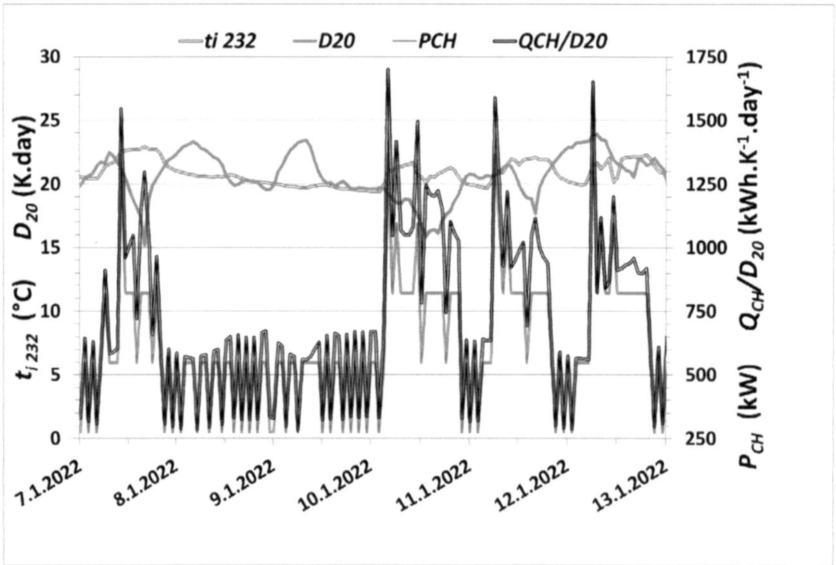

Fig. 6: Plots of the characteristic values of the variables at 1h intervals between 7.12.2021 and 12.1.2022 in reference room 232

The selected period is the period in which the Christmas and New Year lulls end and the working days begin. On 10.1.2022, the air temperature t_a in the reference rooms gradually increased from 20.7 °C to 21.6 °C. It was necessary to heat the building intensively. The thermal output of the P_{CH} varied from 273 kW to 1642 kW and the Q_{HC}/D_{20} ratio from 287 kWh.K^{-1}.day^{-1} to 1684 kWh.K^{-1}.day^{-1}. During other working days, decreases in night heating are noticeable.

5 Conclusion

During the 2021 heating season, teaching was conducted by distance. By regulating the CHTS and the heating circuits in the main building of the FME, the heating consumption was reduced by 443 930 kWh and the savings of the variable component of the heating costs excluding VAT was 15 449 €. Heat costs were reduced by 9.2%. The measurements of indoor and outdoor air temperature and wall temperature in the reference room demonstrated the thermal comfort of the FME building. The night, weekend, Christmas and New Year's damping modes were tested.

References

1. Tatarková, Z. (2022). Direct data from BAT,a.s. „Trendy MQCzima 2222240000.xlsx.“ BAT, a.s. Bratislava.
2. Nemec V. (2022). Direct data from University of Technology „teplo_radlinského_od_2018_do_2021.xls“, Slovak university of technology in Bratislava.
3. Zelený, L. (2022). Direct data from the faculty of mechanical engineering „Prevádzkové denníky KOST“, Slovak university of technology in Bratislava.
4. Jánošová, D. (2022). Direct data from ENGIE „dennostupne.xls.“ ENGIE Services a.s.
5. Masaryk, M. ml. – Bursíková, L. – Zsíros, M. (2022). The importance of shielding and solar convection for reducing the thermal load of buildings, Air Conditioning Conference 2022, Bratislava: Slovak Society for Environmental Engineering, 2022, ISBN 978 - 80 - 89878 - 92 - 5

Energy Management of a Metal Hydride Hydrogen Storage Tank Using a Loop Heat Pipe

Lucia Martvoňová[1,*], *Milan* Malcho[1], *Jozef* Jandačka[1], and *Ladislav* Ďuroška[1]

[1]University of Zilina, Faculty of mechanical engineering, Department of power engineering, Univerzitná 8215/1 01026, Zilina, Slovakia

Abstract. The article analyzes the thermal management of a metal hydride storage tank for hydrogen in the mode of filling the storage tank with hydrogen when it is necessary to cool the metal hydride filling intensively. Cooling is carried out by boiling water at low pressure and therefore also at low temperatures of around 50 °C. In the article, a heat transfer model during boiling is developed and the limits of heat transfer during boiling at low temperatures are determined.

Keywords: *metal hydride, hydrogen, loop heat pipe, energy management*

1 Introduction

Storing hydrogen in storage tanks with metal hydrides is less energy-intensive than storing hydrogen in a liquid state. Thus, the reservoirs do not have to meet the demanding conditions of low temperatures and high pressure, and minimal heat losses. The storage of hydrogen in metal hydrides, however, requires compliance with the specified temperature conditions and the need for removal or supply of heat when storing or obtaining hydrogen from the metal hydride [1].

The reservoir usually consists of individual segments with metal hydride, including a pipe with a cooling medium. Heat transfer in the hydride takes place only by conduction through the wall of the tube in which it is stored, into the volume of the hydride. On the outer wall, heat is transported by convection from or to the heat-carrying medium. More advanced methods of construction use an increase in the heat exchange surface with ribs. In this case, the metal hydride is placed between the fins, while the cooling medium flows inside the tube. The hydrogen flows into the hydride through a separate pipe that is routed under the cooling pipe. For mobile devices, the storage tank is heated by external heating during filling. Cooling is provided by forced convection while driving. There are ribs in the volume of the metal hydride, which ensure the conduction of heat from the outer shell to the inner volume [2].

In another construction solution [3], there are several separate reservoirs with metal hydride, which are connected by pipelines into one block fixed in a rigid casing, in which the pipeline for the supply or removal of heat is also placed. These separate reservoirs are subsequently also covered with lead. The main parameter that affects the possibilities of deploying such containers is the weight of the entire assembly. The disadvantage of such a solution is the low heat removal from the metal hydride only by free convection. Therefore,

* Corresponding author: lucia.martvonova@fstroj.uniza.sk

it is more appropriate to choose a design arrangement where the reservoir with the metal hydride will be cooled or heated actively since the intensity of heat transfer directly affects the degree of saturation of the hydride with hydrogen [4]. Therefore, a widespread method of cooling and heating metal hydrides is represented by reservoirs with a pair of holes, where one pair is used for the supply of hydrogen and the other pair ensures the delivery of cooling liquid by convection [5].

2 Construction of a metal hydride storage tank for hydrogen

When storing hydrogen in metal hydride reservoirs, exothermic and endothermic operations occur. For the entire system to be operationally durable, the heat exchange control devices for hydrogen storage must be designed as compact and light as possible with minimal heat loss, also in terms of thermal capacity. It is especially important for the automotive industry, where the reservoirs work in a highly non-stationary mode, which requires a flexible system of heat supply and removal. Cooling and heating systems operating on the principle of heat pipes [6], i.e., with heat transfer based on the phase transformation of the working substance without the need for forced circulation of the working substance using circulating pumps, are shown to be promising devices for the energy management of metal hydride reservoirs. Systems with a thermosyphon effect have several advantages over standard coolers. They do not need a pump to drive the heat-carrying medium, the flow is ensured by the removed heat itself [7]. The heat transfer coefficients are maximum during the phase change, the temperature of the phase transition is stable, and its value can be regulated by pressure or suitable physical properties of the working substance used. To transport heat from and to the reservoir, it is possible to design a thermosyphon heat pipe with a closed loop (Fig.1), which would ensure a homogeneous temperature field in the entire volume of the metal hydride hydrogen reservoir. The use of heat pipes for heat removal or heating is also possible when converting the chemical energy of hydrogen into electrical energy in fuel cells [8], which tend to be the most energetically efficient converters of the accumulated energy in the metal hydride reservoir in the form of hydrogen into electrical energy and when hydrogen is compressed [9].

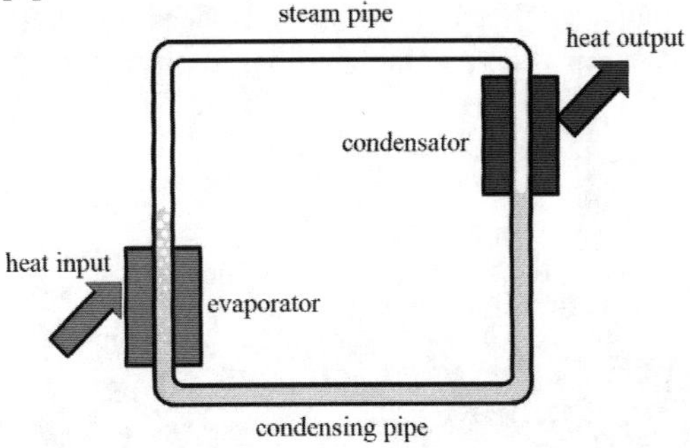

Fig. 1. Loop heat pipe.

The metal hydride storage tank for hydrogen was constructed as a loop heat pipe with an overpressure of about 0.7 MPa (Fig. 2). The metal hydride is stored in the reservoir tubes in the form of pressed briquettes of a cylindrical shape. The tubes are cooled from the side of the outer casing during hydrogen storage by heat removal into the water filling of the space between the tubes while it is heated by natural convection and boiling at a reduced pressure of saturated vapors.

Fig. 2. Realization of a metal hydride hydrogen tank.

3 Mathematical model of metal hydride reservoir cooling using boiling water

Due to the maximum ability to adsorb hydrogen into its structure, the metal hydride filling must be cooled, the recommended cooling temperature is approximately 50 °C. The evaporative part of the closed-loop heat pipe can also be used to cool the metal hydride filling during hydrogen storage. A bundle of tubes closed on one side is in the cylindrical container, which ends with an evaporator, which creates a storage space for the metal hydride (Fig. 3). The space between the tubes is filled with water, which at a lower vapor pressure boils at temperatures significantly lower than the boiling temperature at normal atmospheric pressure.

The heating of the filling due to hydrogen adsorption is simulated by the flow of warm water through a Field tube. Warm water with a temperature of 50 to 70 °C is supplied to the inter-tube space from the thermostat, which is cooled by heat removal into the water-filling space around the bundle of tubes. Under reduced pressure, the steam flows spontaneously into the steam-water heat exchanger, where it condenses and naturally flows back into the space between the tubes. The cooled water returns to the thermostat through the inner pipe. Therefore, in the mathematical model, heat transfer by convection to the outer and inner tubes was solved. Boiling of water at a defined pressure was modeled on the outer surface of the field tube, and heat transfer coefficients on all surfaces, corresponding heat flow densities and corresponding heat outputs were calculated based on the criterion equations.

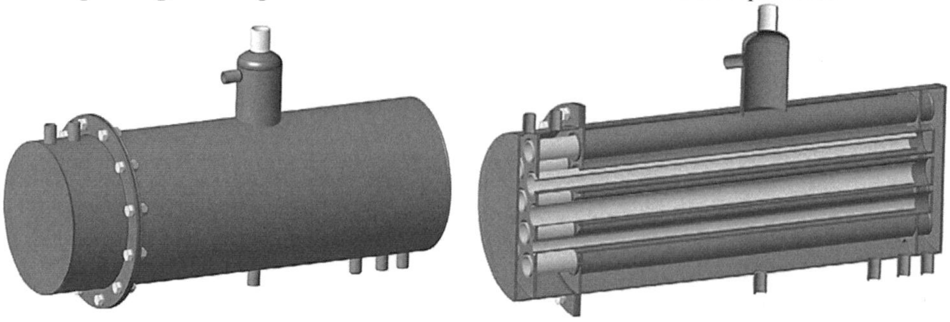

Fig. 3. Model of a metal hydride hydrogen tank.

According to the results from the mathematical model, shown in Figure 4, the heat transfer coefficient by boiling strongly depends on the temperature difference of the wall temperature at which the liquid boils and on the temperature of the water-steam phase transition. The proposed system maintains the temperature of the outer wall of the field tube at 61-64 °C at a pressure of saturated steam of about 20 kPa, and through boiling, heat output of up to 2.5 kW is transferred to the external cooling system.

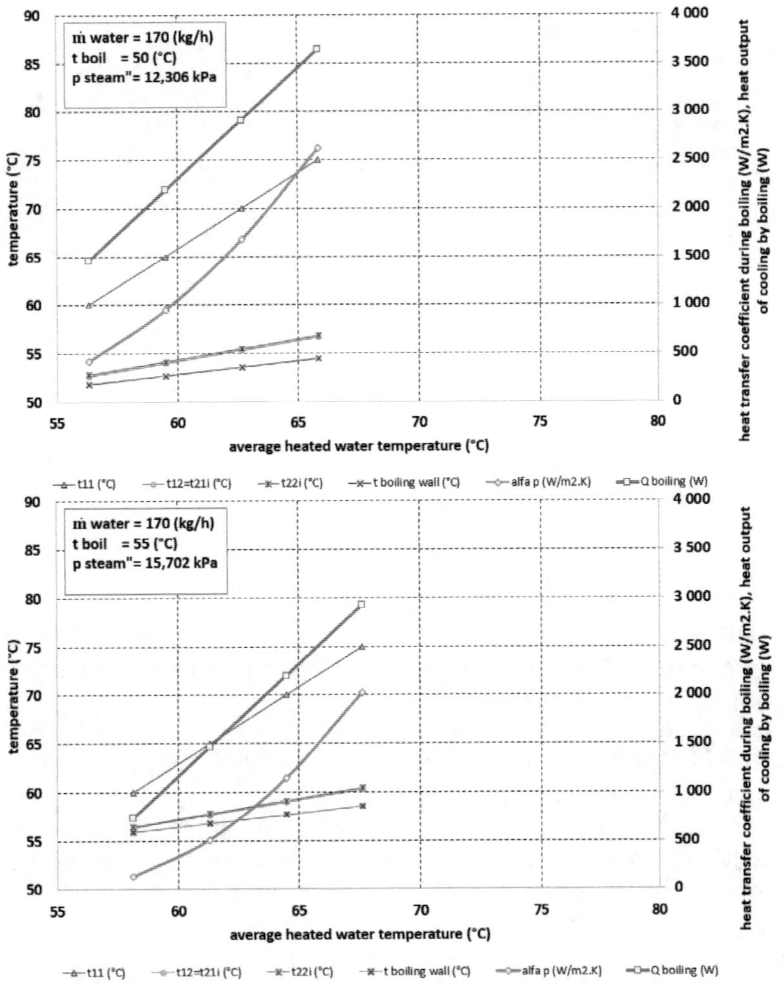

Fig. 4. Dependence of the heat output of boiling water cooling on the mean heating temperature at a pressure in the range of 12.306 – 24.947 kPa (part1).

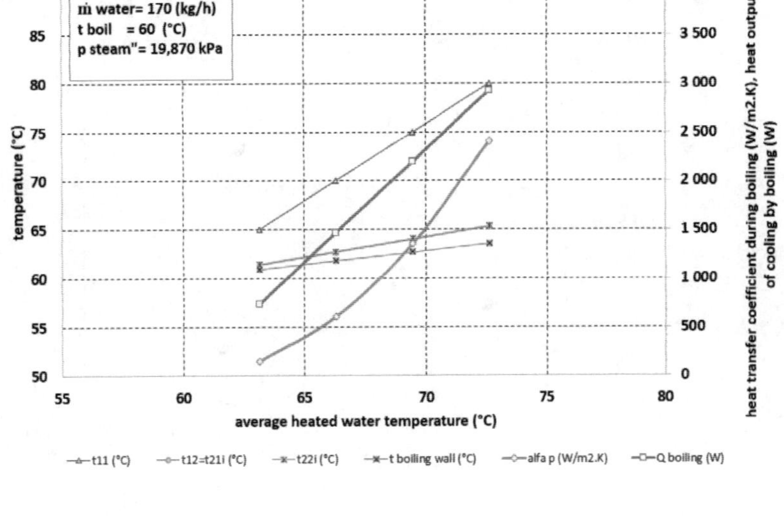

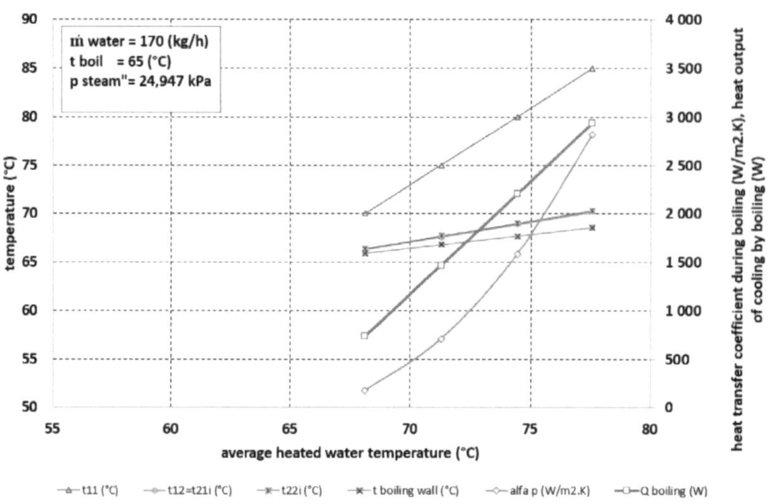

Fig. 4. Dependence of the heat output of boiling water cooling on the mean heating temperature at a pressure in the range of 12.306 – 24.947 kPa (part2).

4 Experimental measurements of thermal performance on a physical model of a metal hydride storage tank for hydrogen

The experimental measurement was carried out on a model of a metal hydride hydrogen storage tank with thermal management using a loop heat pipe (Fig. 5). The cooling mode took place in the form of cooling the tank filling by boiling water at a low pressure of saturated vapors. The reservoir functions as a loop heat pipe evaporator, the condenser of which is a plate steam-water exchanger cooled by water from the water circuit. The steam circuit is realized with a kohaflex metal hose. The heating of the metal hydride filling to the required temperature is simulated by external heating of water in a thermostat with an input of approx. 3 kW.

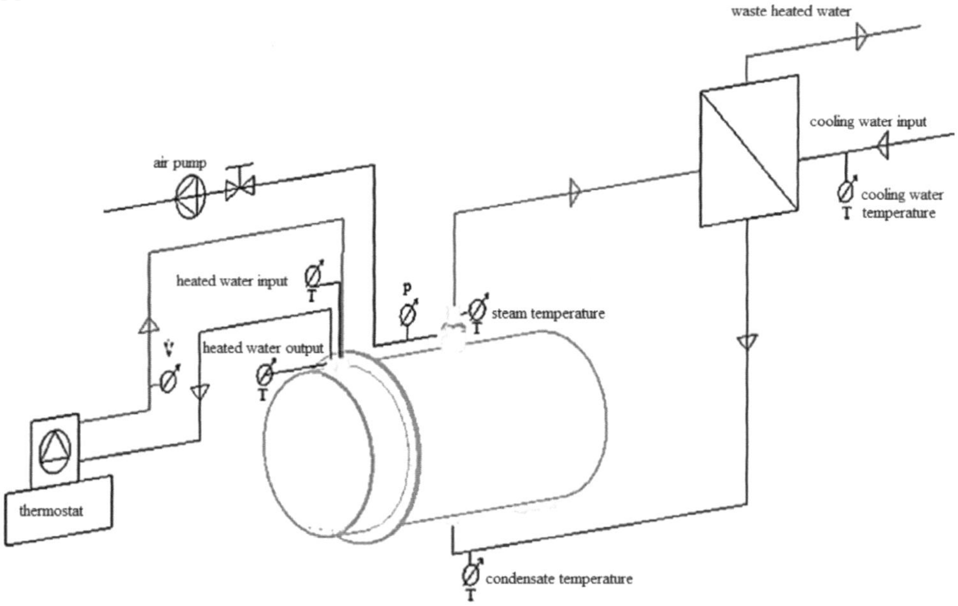

Fig. 5. Experimental measurement scheme.

When refueling with hydrogen from a metal hydride tank, the metal hydride heats up to a temperature of around 50 °C and should not exceed it within an hour. The heating of the metal hydride filling was simulated by the flow of heating water supplied from the thermostat to the space between the tubes. The circulation of heated water between the thermostat and the tank was ensured with the help of the thermostat circulation pump. The flow of heated water was measured by a magnet induction flowmeter, which limited the flow to 176.7 l/h.

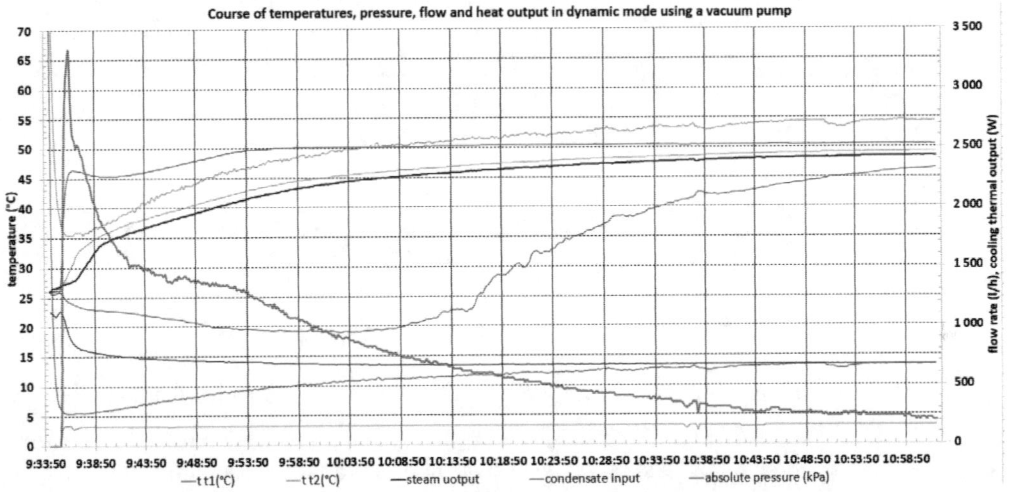

Fig. 6. Course of temperatures, pressure, flow, and heat output in the dynamic model using a vacuum pump.

In fig. 6 shows the course of the measured quantities. For one minute, the system of pipes, packaging, and filling of the reservoir took a peak heat output of up to 3.3 kW. This is due to the thermal capacity of the steel pipes, the steel container, and the water filling of the container, which had a temperature of about 25 °C at the start of the heating simulation. The water filling was heated together with the package, and after heating and intensive boiling, the warmer condensate began to return to the reservoir. The vacuum pump was no longer sufficient to remove the saturated and wet steam, and part of the steam also got into the condenser. Gradually, the boiling stabilized at a power of approx. 250 W and the heating temperatures of the tubes approached the temperature of the exhaust steam. After determining the parameters of the metal hydride alloy, it is possible to simulate the real need for cooling power on the model.

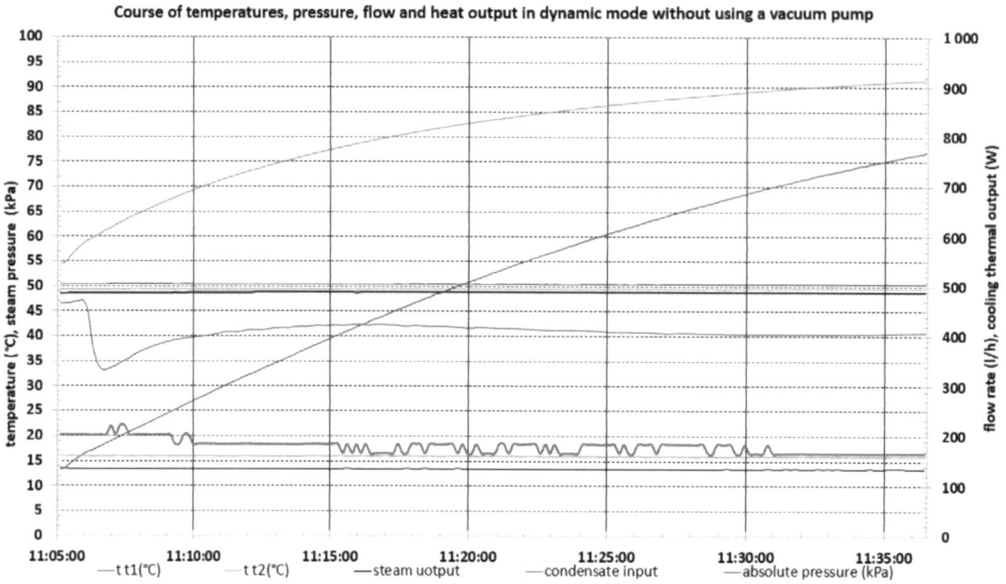

Fig. 7. Course of temperatures, pressure, flow, and heat output in a dynamic mode without using a vacuum pump.

Testing of the cooling system showed that the system is not vacuum-tight and that air, which is one of the non-condensable gases, gets into the reservoir and thus eliminates the processes of intensive heat transport on the wall during phase change.

In fig. 7 shows the course of the quantities without maintaining the pressure using a vacuum pump. The absolute pressure in the reservoir increased with time. As a result of the leak, the pressure increased from 12 kPa to 77 kPa within about 30 minutes, which resulted in an increase in the boiling point of water from 54 °C to 92 °C. The thermosiphon thus switches to Perkins tube mode and the heat output is reduced from approx. 200 W to 170 W.

5 Conclusion

The proposed cooling system has interesting properties from the point of view of heat removal dynamics. When the temperature of the metal hydride suddenly changes from the ambient temperature to a temperature of about 50 °C, the value of the cooling power at the peak is about 3.3 kW. The drop in cooling capacity at a given temperature and pressure is determined by the stable boiling temperature and heating water temperature.

With loop heat pipes, the vacuum tightness of the system is important. Leaks not only increase the pressure but also another component - air, which has diametrically different thermokinetic properties compared to water - gets into the one-component system. Therefore, in the future, it will be necessary to modify the vacuumed part of the system so that it is vacuum-tight. This problem turns out to be critical for achieving high cooling performance values of a water-filled closed-loop heat pipe system.

The experiments also confirmed that the designed simulator can test heating at different inlet temperatures and flow rates. In our case, we were limited by the maximum flow through the magnet induction flowmeter, which was about 170 l/h. In the case of a higher flow rate, more uniform heating of the outer tube would be achieved, which would correspond more closely to the real situation.

When approaching the real conditions in which heat is released from the volume of the alloy, it will be necessary to realize the heating of a similar significantly cheaper filling with the thermokinetic properties of the metal hydride alloy, apparently electrically as Joule heat.

Based on the measurement of electrical quantities, the determination of the released heat, which must be removed by the cooling system, would be significantly more accurate. Such a method of measurement would allow the measurement of surface temperatures on the surface of the metal hydride filling or even in its center. This information would make it possible to predict the accumulation dependence of the alloy more accurately on its temperature.

This work was supported by the Grant System of the University of Zilina No. 1/2021 (15898) "Verification of the principle of thermal management of a metal hydride (MH) hydrogen storage tank using a loop heat pipe", KEGA 046ŽU-4/2021 "Innovative methods of increasing the thermal efficiency of small heat sources using heat recovery through phase transformations" and KEGA 047ŽU-4/2022 "Innovative approaches to fluid dynamics modeling in energy systems".

References

1. Jinsong Zhang, Timothy S. Fisher, P. Veeraraghavan Ramachandran, Jay P. Gore: A Review of Heat Transfer Issues in Hydrogen Storage Technologies, Journal of Heat Transfer 127(12), 2005.
2. M.W.Davids et al: Metal hydride hydrogen storage tank for light fuel cell vehicle. International Journal of Hydrogen Energy, Volume 44, 2019.
3. Mykhaylo Lototskyy et al: Metal hydride hydrogen storage tank for fuel cell utility vehicles. International Journal of Hydrogen Energy, Volume 45, Issue 14, 2020.
4. S.Mellouli et al: Integration of thermal energy storage unit in a metal hydride hydrogen storage tank. Applied Thermal Engineering, Volume 102, 2016.
5. MahvashAfzal et al: Heat transfer techniques in metal hydride hydrogen storage: A review. International Journal of Hydrogen Energy, Volume 42, 2017.
6. Yuzhu Liu, Hui Wang, Ajay Prasad, Suresh G. Advani: Role of heat pipes in improving the hydrogen charging rate in a metal hydride storage tank. International Journal of Hydrogen Energy 39(20): 10552–10563 DOI:10.1016/j.ijhydene.2014.04.197
7. Gavlas, S., Jandačka, J., Malcho, M.: Zariadenie na samočinný prenos tepla so samoregulovateľnou nastaviteľnou teplotou vody. Úžitkový vzor č. SK 7672 Y1, Úrad priemyselného vlastníctva Slovenskej republiky 2016.
8. Anggito P. Tetuko, Bahman Shabani, and John Andrews: Passive Fuel Cell Heat Recovery Using Heat Pipes to Enhance Metal Hydride Canisters Hydrogen Discharge Rate: An Experimental Simulation. Energies 2018.
9. Robert C. Bowman Jr.: Metal Hydride Compressors with Gas-Gap Heat Switches: Concept, Development, Testing, and Space Flight Operation for the Planck Sorption Cryocoolers. Inorganics (ISSN 2304-6740; CODEN: INORCW) 2019.

Gas Dynamic Air Distribution for Post-Reaction Gas Afterburning in a Metallurgical Furnace

Wojciech Bialik[1,*], *Stanisław* Gil[1], and *Piotr* Mocek[1]

[1]Silesian University of Technology, Faculty of Materials Engineering, Krasińskiego 8, 40-019 Katowice, Poland

Abstract. In the paper, the course of post-reaction gas afterburning generated during the metallurgical process where the reduction of metal and semi-metal oxides is performed with the use of carbon is presented. Process waste gases can be an alternative source of energy to be used in the same process or converted to another. The participation of chemical enthalpy in the total energy stream of the post-reaction gas is 93 %, the rest is physical enthalpy. In the study, particular attention was paid to the proper mixing of fuel and oxidizer and to maintaining an appropriate ratio of excess combustion air λ_a. The dynamics of the post-reaction gas combustion processes was calculated according to several popular models but the best results were obtained using the two-step mechanism with reaction constants according to Westbrook-Dryer.

Keywords: gas dynamics, afterburning, metallurgical furnace.

1 Introduction

Due to increasing costs of energy carriers, it is economically necessary to make every effort to reduce their consumption. During chemical processes in industrial furnaces, post-reaction gases are frequently generated which demonstrate high levels of chemical and physical enthalpies [1-5]. Therefore, there is a need to utilise these properties not only for economical reasons, but also for protection of traditional fossil fuel resources and natural environment.

Management of the resulting waste energy requires its conversion to another convenient form. The first stage of this process is the usual gases afterburning, i.e. the conversion of their chemical enthalpy into a physical one. For solving similar problems of gas combustion, but in objects with smaller geometry, good compliance with the experimental data is provided by the method of stochastic fields [6], while an overview of the various modelling techniques can be found in the work [7].

A success in this field requires performing a deep analysis of the technological process, the potential to change it and availability of economical resources necessary to achieve intended goals. Another issue is the human factor, often neglected in technical considerations, which is associated with new technological solutions that require developing new skills and acquiring new knowledge.

The aim of the paper is to present the results of estimated energetic potential of high-temperature post-reaction gases during their afterburning in metallurgical furnaces.

2 Properties and components of the post-reaction gas

The mixture of reagents delivered to a metallurgical furnace, mainly composed of carbon, quartzites, iron compounds and biomass, is transformed during heating and participates in complex reduction reactions. These reactions generate large amounts of post-reaction gas. Its components are listed in Table 1. Depending on the metamorphism of the applied reducer, its humidity and the course of the technological process, the composition and amount of generated gas change. The gas components determine its chemical enthalpy which is 387.77 MJ/kmol of the gas to 390.72 MJ/kmol of the gas. On the other hand, the course of the technological process influences, among others, the temperature of gases leaving the reaction space, i.e. their physical enthalpy. This temperature, measured just above the charge surface, is 700 °C to 800 °C. The computational fraction of the physical enthalpy for 800 °C within the specific overall enthalpy (h_{ch}+h_f) of the post-reaction gas

* Corresponding author: wojciech.bialik@polsl.pl

Table 1. Molar fraction of post-reaction gas components.

Scope	Post-reaction gas components						
	CO	SiO	H_2O	C_2H_4	CH_4	N_2	H_2S
Minimum	0.674	0.036	0.148	0.111	0.025	$4.402 \cdot 10^{-3}$	$1.438 \cdot 10^{-3}$
Maximum	0.669	0.043	0.146	0.111	0.025	$4.451 \cdot 10^{-3}$	$1.526 \cdot 10^{-3}$

is 6.98 % to 7.23 %. This shows that a major part of usable energy of post-reaction gases is associated with their chemical enthalpy. Therefore, a very important factor is proper organisation of the post-reaction gas afterburning process in the restricted furnace space with controlled air distribution.

3 Organisation of the combustion process

While designing the combustion process, several assumptions were made necessary to be met in order to effectively recover a part of the energy of post-reaction gases (defined, to distinguish, as post-process or exhaust gases). The major assumptions were as follows:

- The combustion air must be sucked through the charging windows and blown in through the nozzles.
- A proper gas dynamics for mixing post-reaction gases with the air must be ensured to achieve their complete combustion for various phases of cyclic delivery of new charge portions.
- A controlled delivery of the air necessary for post-reaction gas combustion must be ensured so that the exhaust gas temperature at the hood outlet is 900 °C to 950 °C.

Meeting the presented assumptions required solving a series of technical problems associated not only with gas dynamics but also with metrology and automatics.

Another challenge for the organisation of the combustion process was the presence of gaseous SiO in the post-reaction gas (Table 1). Its afterburning generates SiO_2 dust in the exhaust gases. Although the fraction of molar SiO in the post-reaction gas is small, it carries a lot of chemical enthalpy. The comparison of the reaction enthalpy increase, taking into account phase transitions for CO and SiO with an average gas temperature of 750 °C and the pressure of 1014 hPa, is presented in equations (1) and (2) [8]:

$$CO_{(g)} + 0.5O_{2(g)} = CO_{2(g)} - 282.51\ MJ/kmol \quad (1)$$

$$SiO_{(g)} + 0.5O_{2(g)} = SiO_{2(s)} - 801.79\ MJ/kmol \quad (2)$$

The data from equations (1) and (2) indicate that with a nearly three-fold lower concentration of gaseous SiO, the achieved thermal effects are comparable to those for CO oxidation. In addition, the rate of SiO oxidation to SiO_2 is approximately four-fold higher at 800 °C than the rate of CO oxidation. Therefore, an important factor is the impact of SiO combustion on the energy balance in the discussed combustion process.

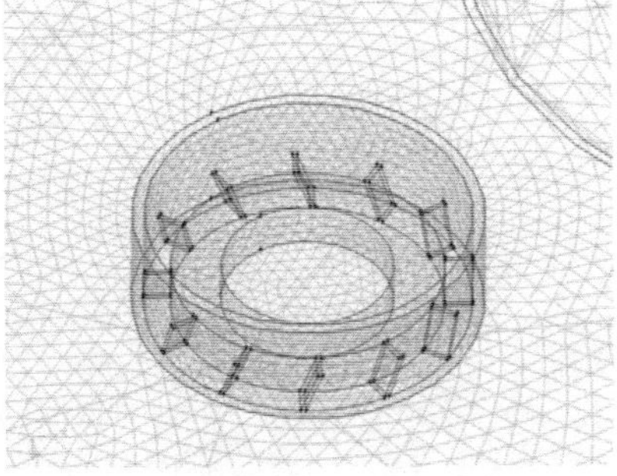

Fig. 1. Numerical division grid of the combustion air nozzle.

Fig. 2. Assembly of the ceiling nozzle.

During the earlier furnace operation, it was found that the post-reaction gas combustion might be longer and partly occur even in the exhaust gas channel depending on the degree of the charging window shutters opening. This phenomenon occurred due to insufficient mixing of the post-reaction gases and the oxidizer despite the relevant ratio of the excess air. The problem was solved by delivering the combustion air through additional ceiling nozzles. The ceiling nozzles have been designed to ensure that part of the air flow is axial and part of the stream is twirled. This division was set in half and cannot be changed without intervention in the nozzle geometry. The nozzle design is presented in Fig. 1 and its assembly in the furnace ceiling is shown in Fig.2.

The mass flow rate of the pumped air was selected based on the stoichiometric calculations of the combustion of post-reaction gases with the composition presented in Table 1. For the calculations, modified equations of substance balances for the gas components were used as follows [9]:

- C balance (kmol C/kmol of the gas)

$$n'_C = CO + CH_4 + 2C_2H_4, \tag{3}$$

- H_2 balance (kmol H_2/kmol of the gas)

$$n'_{H_2} = H_2O + 2CH_4 + 2C_2H_4 + H_2S, \tag{4}$$

- Si balance (kmol SiO/kmol of the gas)

$$n'_{Si} = SiO, \tag{5}$$

- S balance (kmol H_2S /kmol of the gas)

$$n'_S = H_2S, \tag{6}$$

- O_2 balance (kmol O_2/kmol of the gas)

$$n'_{O_2} = \frac{1}{2}CO + \frac{1}{2}SiO + \frac{1}{2}\,H_2O. \tag{7}$$

The theoretical amount of oxygen required for the complete combustion of the post-reaction gases was calculated based on the following equation:

$$n_{O_2min} = n'_C + n'_{Si} + n'_S + \frac{1}{2}n'_{H_2} - n'_{O_2}\,. \tag{8}$$

As the combustion air is the air with the oxygen molar fraction of 0.2099, the theoretical demand for dry oxygen was calculated as follows:

$$n_{a\,min} = \frac{n_{O_2min}}{0.2099}\,. \tag{9}$$

Due to imperfection of the process of mixing the oxidizer with the fuel, its excess is introduced defined as the ratio of excess air λ_a. Assuming λ_a, the real amount of the air used in the combustion process is calculated with the use of the following equation due to the organisation of the combustion process or additional technological requirements:

$$n'_a = \lambda_a\, n_{a\,min}. \tag{10}$$

For the stream of post-reaction gases generated in the furnace during the proper course of the technological process, the flow rate regarding the theoretical demand for dry air, calculated using the equation (9), is 2.66 to 2.70 kg/s. The influence of λ_a values from equation (10) on the exhaust gas temperature is illustrated in Fig. 3.

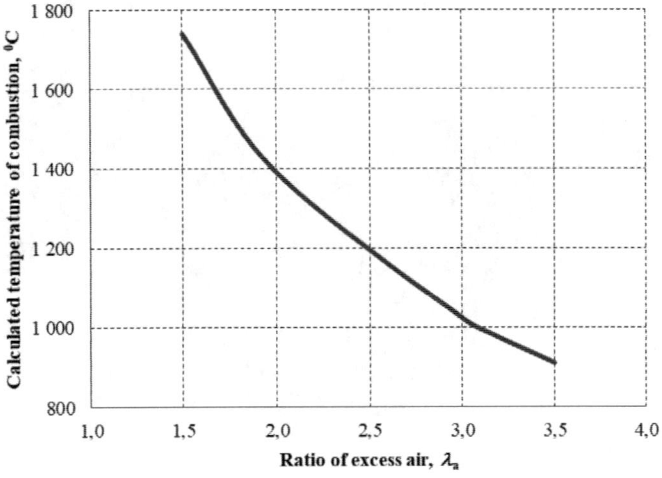

Fig. 3. Computational influence of excess combustion air on the furnace gas temperature.

The dynamics of the post-reaction gas combustion was determined according to several popular models [10-13]. The best results were achieved using the Westbrook and Dryer two-step mechanism with chemical reaction constants (see Table 2) [14,15] and the reaction rate was determined with the use of the following equation:

$$\frac{dC_j}{dt} = k\ exp\left(-\frac{E}{RT}\right) T^n \Pi_j C_j^m,\ \text{mol}/(\text{cm}^3\ \text{s}). \tag{11}$$

In the Westbrook and Dryer model [10-13], $T^n = 0$; therefore, the exponent $n = 0$. The proper selection of the equation (11), proposed by Westbrook and Dryer as the most beneficial, is confirmed not only by the computational results that are comparable to the observations, but also by the literature reports of the use of this mechanism for fuel-scarce gaseous flames.

Table 2. Chemical reaction constants of the global model [14,15].

Mechanism	Reactions		k, 1/s	E, kJ/mol	n	$\Pi_j C_j^m$ mol/cm^3
Westbrook-Dryer	1	$CH_4 + 1.5\ O_2 \rightarrow CO + 2H_2O$	$2.80 \cdot 10^9$	202.64	0	$C_{CH4}^{-0.3} C_{O2}^{1.3}$
	2	$CO + 0.5\ O_2 \leftrightarrow CO_2$	F:$2.238 \cdot 10^9$ B:$5 \cdot 10^8$	167.47	0	$C_{CO} C_{O2}^{0.25} C_{H2O}^{0.5}$

The computational effects of the model for two-step mechanism were comparable to the values obtained in the measurements. These values are presented in Table 3 and the way the measurement probe was introduced is illustrated in Fig. 4.

The composition of the exhaust gases was determined using an IMR 3000 P exhaust gas analyser equipped with the following electrochemical sensors:

- an O_2 sensor with the range of 0 - 20.9 %,
- a CO sensor with the range of 0 - 60 000 ppm,
- an SO_2 sensor with the range of 0 - 4 000 ppm,
- an NO sensor with the range of 0 - 2 000 ppm,
- an NO_2 sensor with the range of 0 - 4 000 ppm.

Due to a large range of the CO sensor in the IMR 3000 P analyser, a WAG-1 analyser was also used for the measurements, equipped with the following sensors:

- an electrochemical CO sensor with the range of 0 - 2 000 ppm,
- an electrochemical O_2 sensor with the range of 0 - 100 %,
- a FTIR CO_2 sensor with the range of 0 - 100 %,

- a FTIR CH_4 sensor with the range of 0 - 100 %.

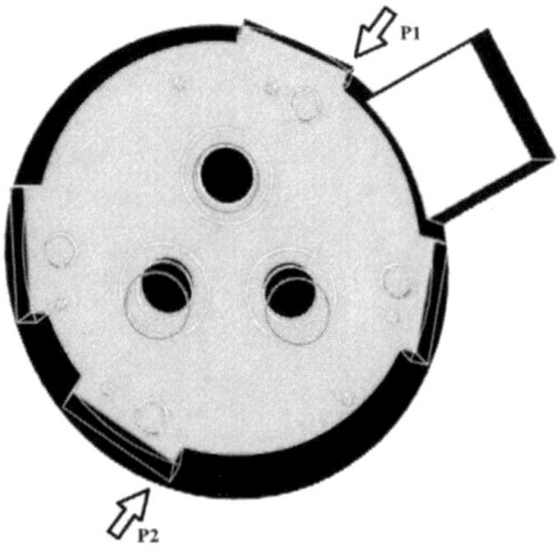

Fig. 4. Location of points where the exhaust gas analyser probe was introduced.

Temperature measurements were performed with the use of a thermocouple type K located in an aspiration probe. Comparable measurement results were obtained from a temperature probe located in the IMR 3000 P exhaust gas analyser.

The gas concentrations presented in Table 3 were converted to conventional 6 % O_2 in the exhaust gases. The measured low CO concentrations measured in unbeneficial conditions of post-reaction gas combustion (opened charging windows to perform measurements, excess amounts of the sucked combustion air, a low temperature of the gases) confirm a good design of the new furnace gas dynamics. High-temperature post-reaction gases generated in the combustion process in controlled conditions demonstrate a high usable energetic potential. Controlled, multi-step combustion air delivery ensures achievement of their high and stable temperature within the range of 700 °C to 900 °C. The measured temperatures of the exhaust gases at the gas outlet window and in the exhaust gas channel which delivers them to the recuperation system are presented in Fig. 5.

Table 3. Components of the post-reaction gases at the level of 2 metres above the edge of the furnace bath [16].

Distance from the furnace axis 2750 mm				Distance from the furnace axis 2250 mm				Distance from the furnace axis 1900 mm			
Gas temperature		P1	P2	Gas temperature		P1	P2	Gas temperature		P1	P2
		414 °C	396 °C			366 °C	220 °C			299 °C	270 °C
Component		Value		Component		Value		Component		Value	
O_2	%	20.0	20.2	O_2	%	19.9	19.7	O_2	%	20.3	19.9
CO_2	%	1.9	1.8	CO_2	%	1.5	2.4	CO_2	%	1.1	1.1
CO	ppm	50	80	CO	ppm	40	40	CO	ppm	20	20
CH_4	%	0.1	0.2	CH_4	%	0.2	0.1	CH_4	%	0.2	0.1
NO	ppm	35	82	NO	ppm	17	10	NO	ppm	16	36
SO_2	ppm	2	1	SO_2	ppm	3	1	SO_2	ppm	3	1

The thermocouple located in the furnace outlet window demonstrates higher than real temperatures of the flowing exhaust gases mainly due to the effects of the radiant flux generated during the process of post-reaction gas combustion in the furnace space.

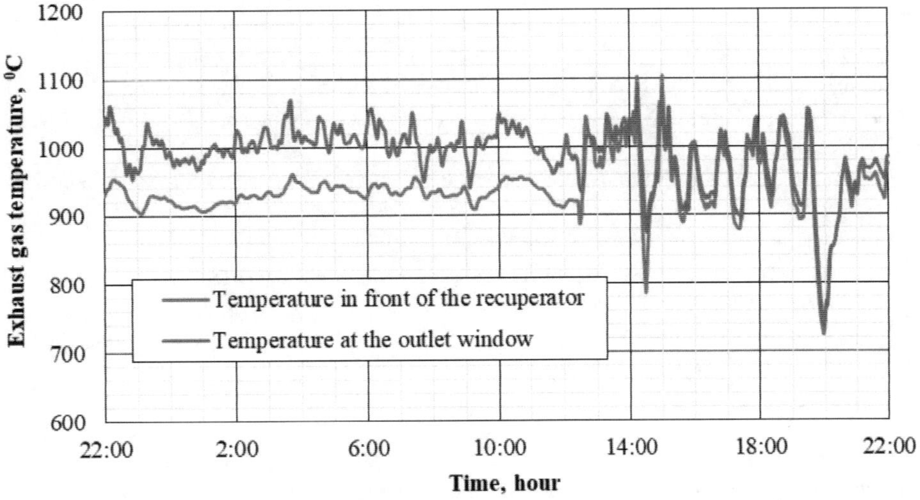

Fig. 5. Furnace gas temperatures measured in the exhaust port and in the recuperator channel.

A necessary limitation of the exhaust gas temperature is achieved by selection of the proper ratio of excess air λ_a. Its value was 3.0 to 3.2 on average during a proper course of the technological process. The measured temperatures of the exhaust gases presented in Fig. 5 are comparable to the theoretical curve values shown in Fig. 3. Required adjustments of the λ_a value to the amounts of the combusted post-reaction gas that change over time can be achieved by correcting the amounts of the air blown into the nozzle system through the fan or by opening the charging window shutters and ensuring the negative pressure in the furnace space using the outlet exhaust gas fan.

4 Conclusions

1. Waste gases from technological processes can constitute an alternative energy source to be used in the same process or after.
2. Conversion in another process.
3. The gas generated in the reduction process demonstrates a chemical enthalpy comparable to that of the coke oven gas which was a valuable fuel in metallurgy ensuring its development. The participation of chemical enthalpy in the total energy stream of the post-reaction gas is 93 %, the rest is physical enthalpy.
4. The important elements of the gas combustion process are proper mixing of the fuel and the oxidizer and maintaining the proper ratio of excess air λa.
5. The dynamics of the post-reaction gas processes was determined according to several popular models. The best results were achieved using the Westbrook and Dryer two-step mechanism with chemical reaction constants.
6. The correctness of calculations according to the Westbrook and Dryer model was verified by the measurements. Their results are comparable to the computational values.

References

1. Takla, M., Kamfjord, N.E., Tveit, H., Kjelstrup S. (2013). Energy and exergy analysis of the silicon production process. *Energy*, 58, 138-146.
2. Kizek, J., Lazić, L. (1999). Simulation method for optimization of a mixture of fuel gases. *Metalurgija*, 38(2), 109-113.
3. Gil, S., Góral, J., Ochman, J., Saternus, M., Bialik, W. (2014). An experimental study on the air and gas removal method in a model of furnace for ferroalloy production. *Metalurgija*, 53(4), 447- 450.
4. Dzurňák, R., Varga, A., Jablonský, G., ...Pástor, M., Kizek, J. (2020). Influence of air infiltration on combustion process changes in a rotary tilting furnace. *Processes*, 8(10), 1-16.

5. Els, L., Andrew, N., Noakes, S. (2013). Process Modeling and testing methods for evaluating ferroalloy furnace off-gas waste energy. *Proceedings of INFACON XIII* (pp. 889-908). Almaty, Kazakhstan.
6. Fredrich, D., Jones, W.P., Marquis A.J. (2019). The stochastic fields method applied to a partially premixed swirl flame with wall heat transfer. *Combustion and Flame*, 205, 446-456.
7. Pitsch, H. (2006). Large-eddy simulation of turbulent combustion. *Annual Review of Fluid Mechanics*, 38, 453-482.
8. Chase, M.W., Jr. (1998). *NIST-JANAF Thermochemical Tables. Fourth Edition*. American Institute of Physics, New York.
9. Szargut, J. (1985). *Termodynamika*. PWN, Warszawa.
10. Curran, H. J. (2019). Developing detailed chemical kinetic mechanisms for fuel combustion. *Proceedings of the Combustion Institute*, 37, 57-81.
11. Veynante, D., Vervisch, L. (2002). Turbulent combustion modeling. *Progress in energy and combustion science*, 28(3), 193-266.
12. Dryer, F. L., Glassman, I. (1972). High temperature oxidation of CO and CH_4. *14th Symposium (International) on Combustion. Combustion Institute*, Pittsburgh, 14(1), 987-1003.
13. Khalil, E.E., Spalding, D.B., Whitelaw, J.H. (1975) Calculation of local flow properties in two-dimensional furnaces. International Journal of Heat Mass Transfer, 18(6), 775-791.
14. Westbrook, C.K., Dryer, F.L. (1981). Simplified Reaction Mechanisms for the Oxidation of Hydrocarbon Fuels in Flames. *Combustion Science and Technology*, 27(1-2), 31- 44.
15. Westbrook, C.K., Dryer, F.L. (1984). Chemical Kinetic modeling of hydrocarbon combustion. *Progress in Energy and Combustion Science*, 10(1), 1-57.
16. Gil, S., Bialik, W., Kozłowski S. (2021). Generation of Nitrogen Oxides in Submerged Arc Furnace during the Production of Ferroalloys. *Advances in Thermal Processes and Energy Transformation*, 4, 37-41.

High Temperature Recuperators Cooperating with a Metallurgical Furnace for Heating the Air Under Pressure

Wojciech Bialik[1,*], *Stanisław* Gil[1], and *Piotr* Mocek[1]

[1]Silesian University of Technology, Faculty of Materials Engineering, Krasińskiego 8, 40-019 Katowice, Poland

Abstract. In this paper, a system which uses hot furnace gases from a metallurgical process to heat compressed air necessary for another energy process is presented. The applied construction of the recuperator ensures high temperatures of the heated air needed for its utilisation in a separate process. Their levels depend on the processes in the reactor. A limitation to the construction of the installation is creep resistance of the materials used to assemble the recuperator modules which operate under high-temperature regimes. The well-prepared gas dynamic design of the recuperator ensured low pressure loss for the flowing air. Furnace gases leaving the recuperation system still have a high energy potential which can be utilised. As it is not possible to manage such large amounts of additional energy, the problem will be solved in the future when necessary.

Keywords: *recuperators, heating air, metallurgical furnace.*

1 Introduction

Metallurgical technologies are combinations of chemical, thermodynamic and, in some cases, electrical processes which occur simultaneously within the space limited by the furnace reaction chamber. Application of theoretical analyses and mathematical models for modification of the technological process parameters or its deeper modernisation may become a factor of the technological advantage. Technology influences the competitive advantage when it plays a major role in determination of the relative cost position or product differentiation [1, 2]. The overall reduction of energy and raw material consumption in the technological process considerably impacts on the price and competitiveness of the offered product, which is frequently the aim of optimisation [3, 4].

High-temperature furnace gases are carriers of far higher amounts of usable energy. The energy balances show that their enthalpy flux constitutes approximately 40 % of the overall energy delivered to the process. Transformation of the furnace gas (fuel gas) enthalpy into another, convenient to send and use form of energy encounters many technological, constructional and functional barriers [5-8].

A waste-heat boiler which produces hot water or steam is one of the most commonly proposed and applied solutions. Water steam can be used for operating steam turbines combined with electric generators [9, 10]. Electrical energy production in a turbine set with water steam requires the use of appropriately treated water which must be systematically replenished. This generates additional costs associated with functioning of the system of circulating water preparation and its growing prices. Another inconvenience is the requirement of continuous heat reception from the water steam condensation system regardless of its periodic demand in combined recovery systems. Frequently, waste energy from technological processes can be utilised in another project following appropriate conversion in the energy recovery system. Growing prices of energy carriers make all activities aimed at their more efficient use very desirable.

The aim of this paper is to present a system which uses hot furnace gases from a metallurgical process to heat compressed air necessary for another energy process.

2 Description of the recuperation channel

Six recuperator modules combined within a serial-parallel system were placed inside the outlet furnace gas channel. Each serial air flow system is composed of three modules: 6-5-1 and 4-3-2. This solution aims to achieve comparable air heating temperatures in the serial flow sequences. The scheme of module combinations in the recuperator is presented in Fig. 1. The recuperator is located in the channel with furnace gases delivered to

* Corresponding author: wojciech.bialik@polsl.pl

the dust collection system. The air and furnace gases pass through in counterflow, which ensures optimal flow and heating characteristics of the recuperator.

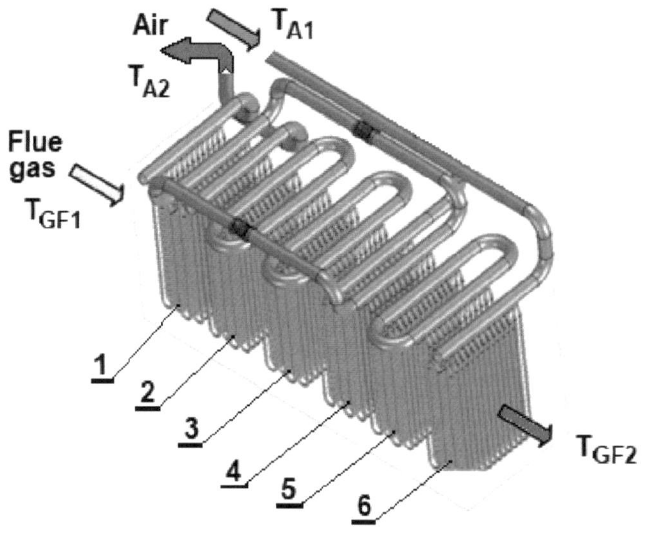

Fig. 1. Scheme of the recuperation system.

The recuperator and the furnace gas channel have been designed to meet a series of requirements being a compromise of many parameters. The major inconvenience in the project resulted from a limited available space in the process facility. Moreover, required minimisation of the furnace gas and air flow resistances in the recuperator pipes had to be ensured. The flue gas channel in the model calculations is presented in Fig. 2.

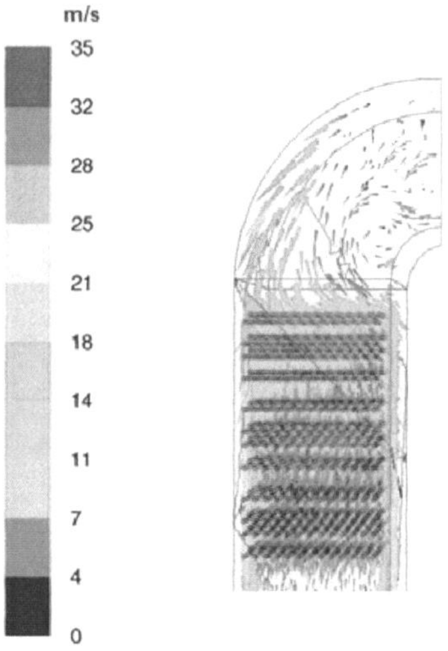

Fig. 2. Computational velocity distribution in the flue gas channel.

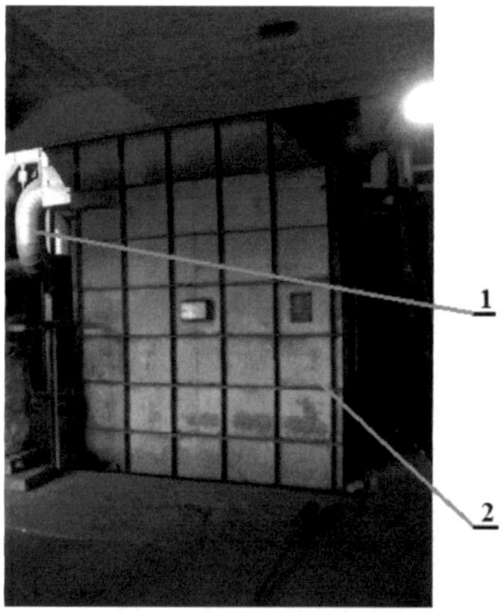

Fig. 3. Recuperation channel: 1 - air pipeline, 2 – the flue gas channel of the furnace.

The mean 24-hour temperature T_{GF1} of the afterburnt furnace gases (flue gases) flowing to the first recuperator module is 800 °C. Due to the recuperator channel size, it is measured by means of three thermocouples located within the vertical axis of the channel at three different levels (heights). The value shown in Fig. 3 presenting the T_{GF1} temperature changes versus time is the mean value of the three measurements. The mean 24-hour temperature T_{GF2} of the flue gases leaving the recuperation system is 556 °C. It is measured by means of two thermocouples located within the channel axis.

The initial measurements showed that the impact of the measurement point level (height) in the channel was considerably smaller due to the strong mixing of streams between the recuperator pipes and a lower temperature of the gases. Two thermocouples are sufficient to obtain representative results. The 24-hour T_{GF2} temperature change as the mean value based on measurements with two thermocouples is presented in Fig. 4. After reaching the assumed working parameters, temperature fluctuations of the flue gases are recorded in the channel. They result from the course of technological processes in the furnace and current conditions of the combustion of furnace gases.

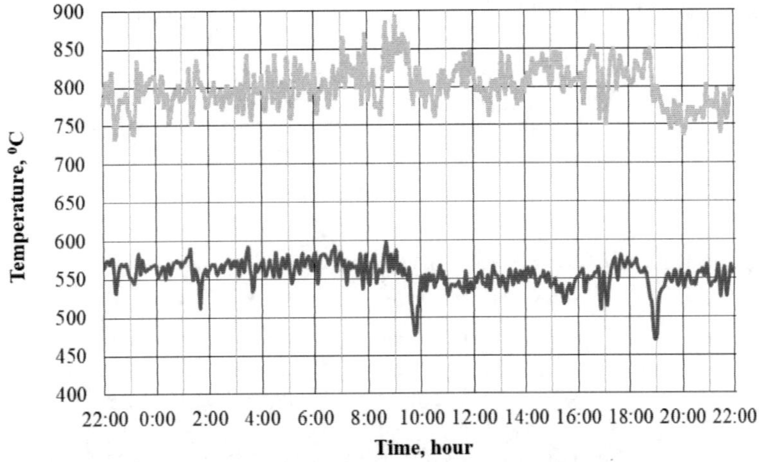

Fig. 4. Mean temperatures of the furnace gases at the inlet and the outlet of the recuperator : T_{GF1} T_{GF2}

The flues gasses in the recuperator pass their energy to the compressed air, which decreases their temperature from T_{GF1} to T_{GF2}. The modules of recuperators were designed as U-pipe structures based on many years of experience resulting from the research and implementation works carried out at the Department of Process Energy, Silesian University of Technology [11-13]. The diameter and number of the pipes in the module were selected as a compromise between the requirements of acceptable air flow rates and geometrical limitations of the flue gas channel. The final shape of a single module pipe was similar to the W letter to reduce the air pressure in the pipes and for the strength reasons. The overall pipe length in the module was selected so that it was comparable to the available commercial size, which prevented generation of unnecessary waste leading to additional costs.

The air in the compressor is heated up to approximately 385 °C and delivered to the recuperation system. The air temperature sensors were placed in the compressor outlet port as well as at the recuperation system inlet (T_{A1}) and outlet (T_{A2}) (Fig. 1). For control purposes, temperatures of two air streams at the outlets of the module sequences 6-5-1 and 4-3-2 are additionally measured. The air flowing through the two module sequences in the recuperator heats up (energy taken from the flue gases) from the mean temperature of approx. 385 °C to far above 620 °C (the mean 24-hour temperature of 650 °C), which is illustrated in Fig. 5.

The mean 24-hour temperature of the air at the outlet of the first module sequence (6-5-1) was 667 °C while the temperature for the second sequence (4-3-2) was 641 °C. The temperature difference between the module sequences was less than 4.5 % in relation to the air temperature in the first module sequences, which indicates a very good design of air flow distribution and the order of module connections. The measured 24-hour temperature courses at the outlets of the module sequences are presented in Fig. 6.

Effectiveness of the system of furnace gas energy utilisation depends not only on the achieved air heating temperature, but also on reduction of pressure generated in the recuperator and the other installation components. The level of pressure loss results from the applied technical solutions which influence gas dynamics of the system. The necessary air pressure in the pipelines is ensured by the compressor and the energy consumption associated with its operation depends on e.g. flow resistances that must be overcome.

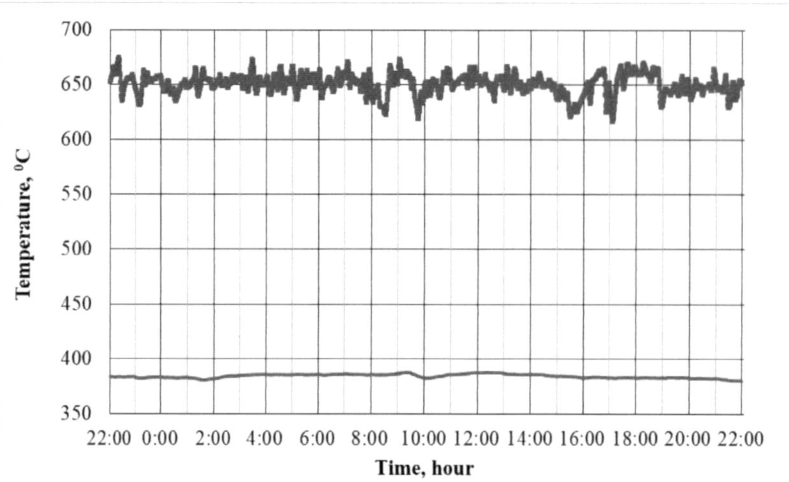

Fig. 5. Temperatures of compressed air: ▬ T_{A1} the inlet to the recuperator, ▬ T_{A2} the outlet from the recuperator.

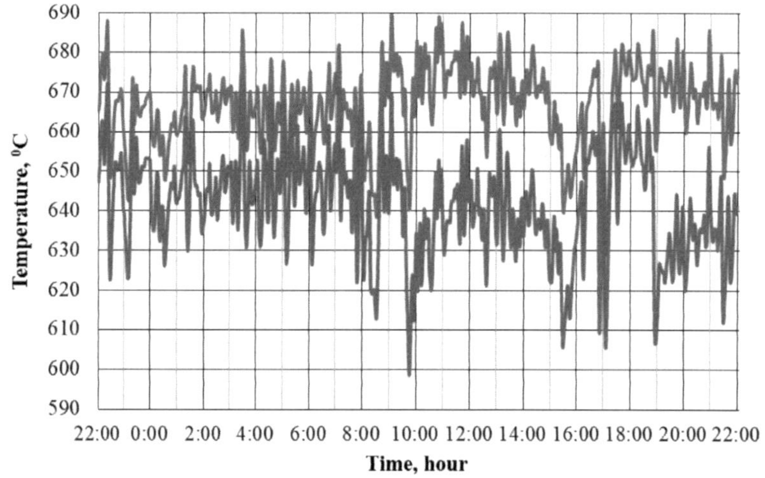

Fig. 6. Temperatures of compressed air at the outlets of the sequences 1 and 2 of the recuperator modules:
▬ sequence of modules 1 (6-5-1), ▬ sequence of modules 2 (4-3-2).

The relative pressure loss in the air heating system for the pressure measured at the outlet port of the compressor is illustrated in Fig. 7. The mean pressure loss is 1.09 % of the pressure generated in the compressor, which is an over three-fold smaller value than that estimated in the models for the initial geometrical solutions.

The pressure loss measured in the recuperator was 48 % on average of the overall value recorded during the operation of the compressed air heating system. The measurements performed during the system operation indicated that the mean absolute increase in the air enthalpy flux was 1100 to 1200 kW.

An additional effect of the application of the energy recovery system from post-process gases was the reduction of their temperature by an average of 245 °C, what is shown in Fig. 4. Reduction of the exhaust gas temperature without excessive dilution of ambient air caused the reduction of their flow rate flowing through the dusting exhaust fan. This enabled the fan to operate with changed operating characteristics.

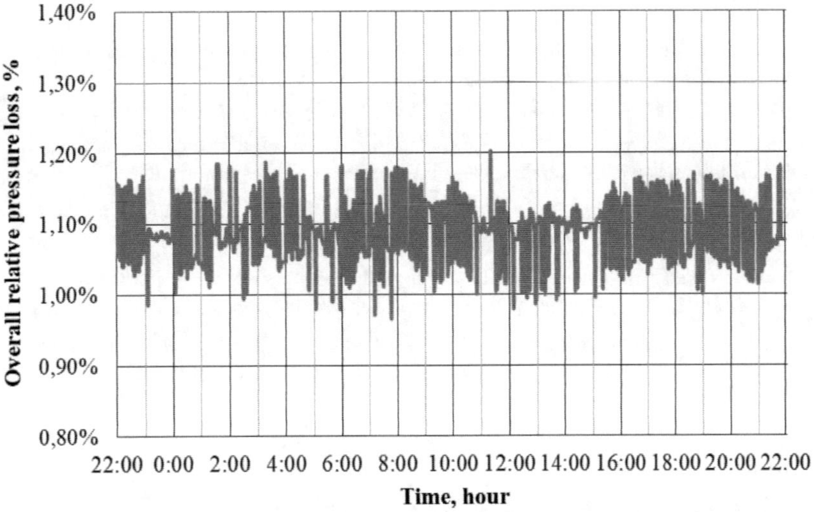

Fig. 7. Overall relative air pressure loss recorded between the compressor and the receiver.

As a result, the power of the electric motor was reduced, which was 750 kW before the modernization. Thanks to that, an additional effect of reducing the energy consumption of the technological process was achieved. The measured power of the fan motor during the recuperator's operation is shown in Fig. 8.

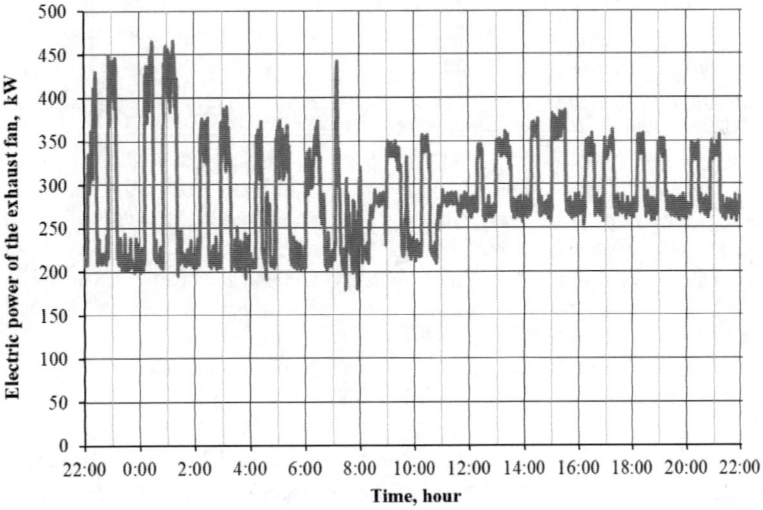

Fig. 8. Stream of electric energy supplied to the exhaust fan engine.

The system for regulating the negative pressure of post-process gases in the furnace affects the operating conditions of the fan and changes its operating characteristics along with the rotation of the rotor. The amount of the produced post-process gases changes over time and depends on the frequency of feeding new portions of batch materials to the furnace, their weight and the manner of its handling. Therefore, in Fig. 8. cyclical changes in fan operating conditions are visible. The influence of cyclical changes in the thermal parameters of the process is also observed in the other charts.

3 Conclusions

1. The applied construction of the recuperator ensures high temperatures of the heated air necessary for its utilisation in a separate process. The temperature level depends on processes in the metallurgical reactor and the resulting furnace gases of the appropriate temperature.
2. A limitation to the construction of the installation is creep-resistance of the materials used to assemble the recuperator modules which operate under high-temperature regimes.
3. A cyclic delivery of the substrates to the metallurgical reactor, every 30÷40 minutes, is observable in the profiles of the temperature courses of furnace gases and heated air.
4. The well-prepared gas dynamic design of the recuperator ensured low pressure loss for the flowing air which was only 1.1% on average of the compressor available pressure.
5. Furnace gases leaving the recuperation system still have a high energy potential which can be utilised. As it is not possible to manage such large amounts of additional energy, the problem will be solved in the future when necessary.

References

1. Porter, M.E. (1991). Towards a Dynamic Theory of Strategy. *Strategic Management Journal*, 12, 95-117.
2. Porter, M.E. (2006). *Competitive advantage. Achieving and maintaining better results*. Helion Publishing House, Gliwice.
3. Yamaki, T., Matsuda, K., Na-Ranong, D., Matsumoto H. (2018). Energy saving performance of reactive distillation process for TAME synthesis through multiple steady state conditions. *Chemical Engineering and Processing - Process Intensification*, 130, 101-109.
4. Babaie, O., Esfahany, M.N. (2020). Optimization and heat integration of hybrid R-HIDiC and pervaporation by combining GA and PSO algorithm in TAME synthesis. *Separation and Purification Technology*, 236, No. 116288, https://doi.org/10.1016/j.seppur.2019.116288.
5. Gorog, J.P., Brimacombe, J. K., Adams, T. N. (1981). Radiative Heat Transfer in Rorary Kilns. *Metallurgical Transactions B*, 12, 55-70.
6. Masuyama, F. (2001). History of power plants and progress in heat resistant steels. *SIJ International*, 41(6), 612–625, https://doi.org/10.2355/isijinternational.41.612.
7. Belyanovskaya, E., Rimár, M., Lytovchenko, R.D., Variny, M., Sukhyy, K.M., Yeromin, O.O., Sykhyy, M.P., Prokopenko, E.M., Sukha, I., Gubinskyi, M.V., Kizek, J. (2020). Performance of an adsorptive heat-moisture regenerator based on silica gel-sodium sulphate. *Sustainability*, 12(14), No. 5611, https://doi.org/10.3390/su12145611.
8. da Rosa, O.C., da Silva, A.K., Gosselin, L. (2022). Geometric and operational analysis of a heat recuperator operating with s-co2 near the pseudocritical region. *Applied Thermal Engineering*, No. 118933, https://doi.org/10.1016/j.applthermaleng.2022.118933
9. Zhang, L., Chennells, M., Xia, X. (2018). A power dispatch model for a ferrochrome plant heat recovery cogeneration system. *Applied Energy*, 227, 180-189.
10. Hjartarson, H., Palsson, H., Saevarsdottir, G. (2010). Waste heat utilization from a submerged arc furnace producing ferrosilicon. *Proceedings of INFACON XII* (pp. 739-778). Helsinki, Finland.
11. Tomeczek, J, Wiśniewski, T. (1995). Radiation characteristics of high temperature tubular heat recuperators. *Gaswarme International*, 44(10), 487-492.
12. Tomeczek, J., Bialik, W., Wiśniewski, T. (2012). Possibilities of using the energy of flue gases from high-temperature furnaces for the production of electricity. *Hutnik*, 79(3), 144-151.
13. Gil, S., Góral, J., Ochman, J., Saternus, M., Bialik, W. (2014). An experimental study on the air and gas removal method in a model of furnace for ferroalloy production. *Metalurgija*, 53(4), 447- 450.
14. Kaiser, F. (2015). Steady State Analyse of existing Compressed Air Energy Storage Plants. Proceedings of Power and Energy Student Summit (pp. S02.2.). Dortmund, Germany.
15. Kowalczyk, T. (2022). Comparative analysis of hybrid energy storage based on a gas–gas system and a conventional compressed air energy storage based on a recuperated gas turbine round trip efficiency, exergy losses, and heat exchanges start-up losses. Energy Conversion and Management, 258, No. 115467, https://doi.org/10.1016/j.enconman.2022.115467.

Numerical Predictive Combustion Model of Hydrogen Enriched Natural Gas

Branislav Zvada[1,*], *Radovan* Nosek[1], *Peter* Ďurčanský[1], *Andrej* Kapjor[1], and *Nikola* Kantová Čajová[2]

[1]Department of Power Engineering, Faculty of Mechanical Engineering, University of Zilina, 010 26 Zilina, Slovakia
[2]Research Centre, University of Zilina, 010 26 Zilina, Slovakia,

Abstract. Hydrogen was established as one of the main pillars of energy stability in the Europe Union. One of the ways how to achieve this goal is natural gas enriched with hydrogen. Due to this is very important to know the properties of this fuel and its behaviour during combustion. The main scope of the research is to provide a better understanding of the emissions, efficiency, and performance of the heat source when combusting hydrogen and gas fuel mixture. In this paper is described hydrogen characteristics, hydrogen fuel preparation, an overview of gas fuel combustion in gas appliances with the hydrogen additive, a mathematical model for the combustion process estimation. In the conclusion, multiple predictive models were compared. We can state that, based on calculations of a numerical predictive model, as hydrogen concentration raised emissions, as nitrogen, carbon dioxide, wet exhaust, and water, are decreasing.

Keywords: *hydrogen; combustion; numerical model; emissions*

1 Introduction

Due to the climate and energy crisis partially caused by the war in Ukraine and due to the effort of the European Union to reduce our demand for energy supplies from the Russian Federation. One of the possible ways how to reduce our dement on imported energy commodities and one hand reduce our impact on our climate is the combustion of enriched natural gas by green hydrogen. Green hydrogen which is produced from renewable energy sources such as electricity from wind and solar power plants, as directly produced from water by photolyze. As we can see combustion of hydrogen enriched natural gas is in these days very actual theme. In Europe is natural gas used in significant amounts in house heating systems and for their reliable and safe operation properties of this fuel is needed. Nowadays, the maximum concertation of hydrogen in natural gas varies from 0.05 % in Belgium up to 10 % in Germany. Some limitations aspect can be in specific applications like CNG pumps or in chemical processes. Based on previous research we can follow that hydrogen-enriched

* Corresponding author: branislav.zvada@fstroj.uniza.sk

natural gas can be used up to 25 % of the volume concentration of hydrogen in standard gas boilers due to his wobe index (W25 = 47.52 MJ/m^3) which is the same group of gaseous fuels as natural gas. Regards emission impact works of other authors show emissions decreasing with rising hydrogen concentrations in the fuel mixture. Same trend we can observe in results in experimental works of other researchers, which were done on domestic boilers, industrial boilers or even on basic household cooktop burners. In this article, stoichiometric combustion numerical model was made, due to specify the properties of hydrogen-enriched natural gas. Based on the volume of hydrogen mixed in natural gas properties such as flame speed, flashback properties, flame head speed, wobe index, or calorific value is changed. If we want to change fuel for our gas devices, we must know its exact properties and how this fuel changes the combustion process. Due to the possibility of variation in the concertation of hydrogen was calculations in this article made with various concentrations of hydrogen in natural gas. [1, 2, 3, 4, 5, 6]

Table 1. Comparison of lower heat values of different fuels.

Fuel	Heat value kWh/kg	Heat value kWh/m^3
Hydrogen	33.2	2.98
Natural gas	13.8	9.7
Wood	4	
Coal	8	

Table 2. Some physical properties of methane and hydrogen.

Parameter	Unit	Methane	Hydrogen
Density	kg.m^{-3}	0.648	0.0813
Viscosity	10 −5 Pa·s	1.11	0.89
Velocity of laminar flame	m.s^{-1}	0.4	2.1
Lower limit of flammability	% volume	5	4
Upper limit of flammability	% volume	15	75
Ignition energy value	10^{-5} J	33	2
Heat value	MJ.m^{-3}	34	10.2
Heat value	MJ.kg^{-1}	49.9	120.1
Gross caloric value	MJ.m^{-3}	37.8	12.5
Gross caloric value	MJ.kg^{-1}	55.5	142.1
Temperature of adiabatic flame	K	2226	2318
Wobbe index	MJ.m^{-3}	51.9	48.5

2 Hydrogen as natural resource

The hydrogen atom is the lightest and simplest from all elements. In state of the molecule as H_2 is a colourless, non-toxic, highly combustible gas with no odours and with a low density, one-fourteenth of the density of air. To condensate gas hydrogen to liquid form at atmospheric pressure, the temperature has to be reduced to 20.3 K. [14] Released hydrogen is dispersing rapidly. Compared to other fuels, hydrogen has the highest combustion energy per unit mass. Hydrogen is used in gaseous, liquid, or slosh forms. Liquid hydrogen is transparent with a little light blue tint. Slush hydrogen is a mixture of liquid and solid at the temperature of triple point. [15] The conditions of hydrogen triple point, are, temperature at 13.8 K and pressure 7 200 Pa. The highest possible temperature, at hydrogen vapours can be liquefied, is 33.145 K which is critical temperature. The corresponding critical pressure is 1 300 kPa, density in this critical point is 31.263 kg/m^3. Above this temperature is impossible to condense hydrogen into its liquid phase just by increasing the pressure. At normal conditions, temperature, pressure, etc., hydrogen is not very reactive substance, to accelerate reaction, catalysator or spark is needed, it proceeds with high rate an explosive violence. Hydrogen is an extremely flammable gas. It burns in air and oxygen to produce water. In air chlorine mixture, it can explode spontaneously just by spark, heat or even sunlight. [7, 8]

Flame visibility. There is common misconception about that hydrogen flames are not visible. Visible emissions from hydrogen flames are considerably weaker, than emissions from comparable hydrocarbon flames, however they are visible. [7]

Flammability limits. The flammability range of hydrogen is from 4 % up to 75 % by volume in air at NTP (normal temperature and pressure). Hydrogen flammability range expand with temperature, hydrogen with temperature 100 °C at NTP has its down flammable limits 3 %. [7] Hydrogen adiabatic premixed flame temperature for stoichiometric mixture in air is higher compared to other fuels and it is 2403 K. [16] Stoichiometric hydro-gen-oxygen is composed from 66.66 % of hydrogen and 33.33 % of volume of oxygen. For conditions in normal air it is 29.59 % volume of hydrogen and 70.41 % of air volume. [9,10,11,12,13]

3 Predictive numerical model of hydrogen enriched natural gas combustion

Calculations physical properties of hydrogen enriched natural gas were make in terms of EN ISO 6976:2016.

For basic calculation and verifications of accuracy or eliminations of mistakes was for first calculation used standard natural gas with compositions as in ISO 6976:2016.

For determination of basic properties for our comparison was made calculations with properties of natural gas which is used in Slovakia.

Table 3. Composition of natural gas in Slovak republic.

Country	Molar percentage in mixture (mol %)	Molar fraction in mixture (mol)
methane	94.2981	0.942981
ethane	3.1006	0.031006
propane	0.6829	0.006829
izo-butane	0.0965	0.000965
n-butane	0.1119	0.001119
izo-pentane	0.0269	0.000269
n-pentane	0.0209	0.000209
neo-pentane	0.0019	0.000019
hexane	0.0364	0.000364
carbon dioxide	0.7250	0.007250
nitrogen	0.8987	0.008987
hydrogen	0	0

Molar fracture

A molar fraction is defined as the ratio of the number of moles of a given component present in a solution to the total number of moles present in a given solution. The result of adding the molar fractions of all components is one, resp. hundred percent.

$$(x_i)_{ZP+H2} = \frac{100 - (x_{H2})_{ZP+H2}}{100} \cdot (x_i)_{ZP} \tag{1}$$

Compression factor

The compression factor, also known as the gas compressibility factor, tells us how much the actual gas differs from the ideal gas at a given pressure and temperature.

$$Z(t_2, p_2) = 1 - \left(\frac{p_2}{p_0}\right) \cdot [\sum x_i \cdot s_i\,(t_2 \cdot p_0)]^2 \tag{2}$$

Gross calorific value

Combustion heat defines the amount of heat released by the complete combustion of a unit amount of gas with a stoichiometric amount of oxygen or air at a constant pressure and temperature. All the resulting combustion products cooled to the initial temperature remain in the gaseous state except for the water which condenses.

$$(H_c)_G(t_1) = (H_c)_G^0(t_1) = \sum xi \cdot [(Hc]_G^0]_i\,(t_1) \tag{3}$$

Net calorific value

The calorific value defines the amount of heat released by the complete combustion of a unit amount of gas with a stoichiometric amount of oxygen or air, at a constant pressure and temperature, all combustion products cooled to the initial temperature being in the gaseous state.

For mass based

$$(H_v)_N(t_1; t_2; p_2) = \frac{(H_c)_N^0\,(t_1)}{V} \tag{4}$$

For molar based

$$(H_c)_N^0\,(t_1) = (H_c)_G^0\,(t_1) - \sum x_i \cdot \frac{b_i}{2} \cdot L^0(t_i) \tag{5}$$

Density

$$\rho(t_2, p_2) = \frac{\rho^0(t_2, p_2)}{Z(t_2, p_2)} \tag{6}$$

Relative density

$$G(t_2, p_2) = \frac{G^0 \cdot Zair(t_2, p_2)}{Z\,(t_2, p_2)} \tag{7}$$

Wobe index

Wobe index is used as criterium for interchangeability of gaseous fuels. Wobe index is defined as the ratio of combustion heat (or calorific value) and square root of relative density.

$$W_G(t_1; t_2, p_2) = \frac{(H_v)_G}{\sqrt{G(t_2, p_2)}} \tag{8}$$

Table 4. Calculated properties for specific percentage of hydrogen in mixture.

Parameter	Sym.	Unit	Value					
Temperature	t	°C	15					
Pressure	p	kPa	101.325					
Molar percentage of hydrogen in mixture	x_{H2}	mol %	0	5	10	15	20	25
Density	ρ	kg.m^{-3}	0.725	0.693	0.661	0.629	0.597	0.565
Relative density	G		0.592	0.565	0.539	0.5134	0.487	0.461
Gross calorific value	H_{vG}	MJ.m^{-3}	38.72	37.38	36.044	34.707	33.370	32.033
Net calorific value	H_{vN}	MJ.m^{-3}	34.91	33.67	32.430	31.197	29.918	28.740
Wobbe number	W	MJ.m^{-3}	50.32	49.69	49.069	48.496	47.116	47.850

Table 5. Stechiometric values of wet air and emissions for different mixtures.

Parameter		Value					
Molar percentage of hydrogen in mixture (mol %)		0	5	10	15	20	25
Minimal value of wet combustion air in mixture (Nm^3.Nm^{-3})		9.877	9.504	9.130	8.757	8.384	8.010
Minimal amount of exhaust in mixtures (Nm^3.Nm^{-3})	N_2	7.627	7.338	7.050	6.761	6.473	6.185
	CO_2	1.042	0.990	0.938	0.886	0.834	0.782
	H_2O	2.139	2.084	2.028	1.973	1.917	1.862
	wet exhaust	10.808	10.412	10.016	9.620	9.224	8.828

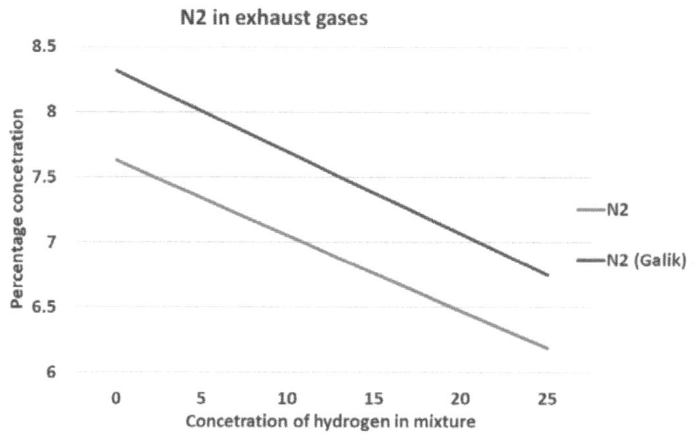

Fig. 1. Comparison N_2 emissions with another predictive model.

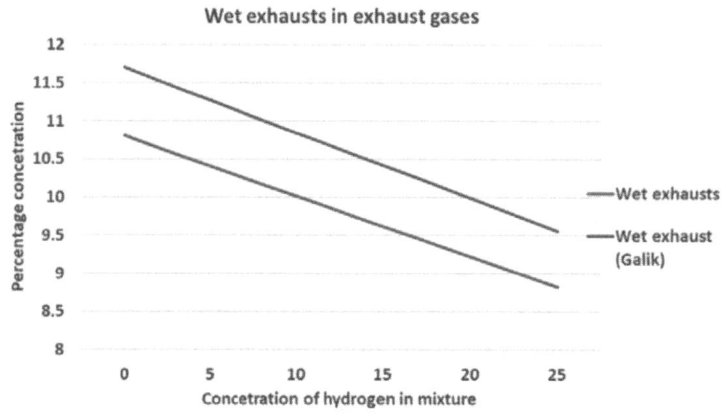

Fig. 2. Comparison wet exhausts emissions with another predictive model.

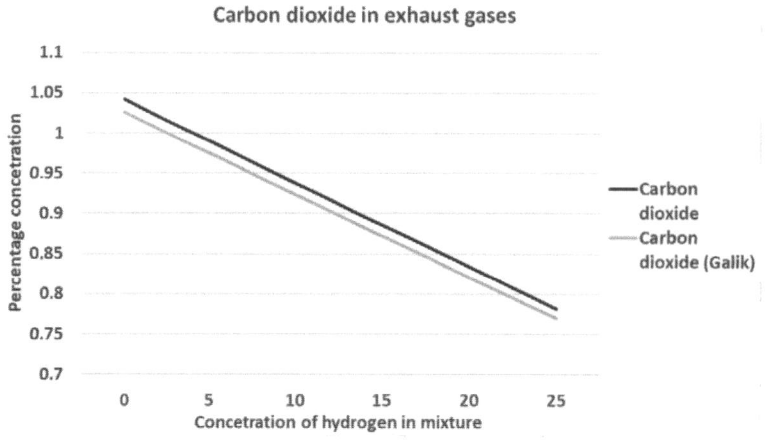

Fig. 3. Comparison carbon dioxide emissions with another predictive model.

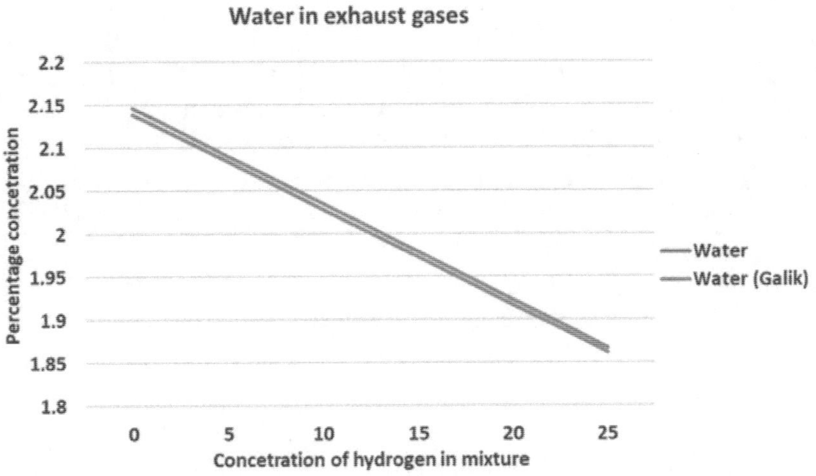

Fig. 4. Comparison water emissions with another predictive model.

4 Conclusion

One of the possibly way how to stabilize and reduce our demand from fossil fuels is blend of natural gas and hydrogen. Due to reality that natural gas grid is very complex system and likewise whole industry which depend on natural gas is very fragile on composition of natural gas is necessary to know as much detail as possible to start our trans-formation from pure natural gas to hydrogen enriched natural gas. Based on these we make numerical predictive combustion model of hydrogen enriched natural gas to better understand behaviour of these gaseous fuel which we want to use in future. In this article emission changes were evaluated and then compared with another author. As we can see on charts above data have the same trend and we can state that as hydrogen concentration raised emissions calculated by us are decreasing. In terms of emission reduction is hydrogen mix very potential fuel for our energy demands. However, for future using of hydrogen enriched natural gas is important to solve multiple problems for example flame stability or quenching distance. Even though some problems hydrogen enriched natural gas is fuel with very high potential for future.

This research is supported by the project VEGA-1/0479/19 "Influence of combustion conditions on the production of solid pollutants in small heat sources".

References

1. Zhao, Y., McDonell, V., Samuelsen, s., (2019). Influence of hydrogen addition to pipeline natural gas on the combustion performance of a cooktop burner. *International journal of hydrogen energy*, 44(4), 12239-12253.
2. Zhao, Y., McDonell, V., Samuelsen, s., (2019). Experimental assessment of the combustion performance of an oven burner operated on pipeline natural gas mixed with hydrogen. *International journal of hydrogen energy*, 44(4), 26049-26062.
3. Latosov, E., Pakere, I., Murauskaite, L., (2021). Impact of Grid Gas Requirements on Hydrogen Blending Levels. *Environmental and climate technologies*, 25(1), 688-699.
4. Cellek, MS., Pinarbasi, A., (2018). Investigations on performance and emission characteristics of an industrial low swirl burner while burning natural gas, methane,

hydrogen-enriched natural gas and hydrogen as fuels. *International journal of hydrogen energy*, 43(2), 1194-1207.

5. Boulahlib, MS., Medaerts, F., Boukhalfa, M., (2021). Experimental study of a domestic boiler using hydrogen methane blend and fuel-rich staged combustion. *International journal of hydrogen energy*, 46(75), 37628-37640.
6. Lo Basso, G., Nastasi, B., Garcia, DA., (2017). How to handle the Hydrogen enriched Natural Gas blends in combustion efficiency measurement procedure of conventional and condensing boilers. *Energy*, 123, 615-636.
7. Subramani, B. V.. (2015). *Compendium of Hydrogen Energy Volume 1: Hydrogen Production and Purification*. Cambridge: Woodhead Publishing.
8. Sorensen, B. (2005). *Hydrogen and fuel cells: Emerging technologies and applications*. Elsevier Academic Press.
9. Ying, YY., Liu, D., (2019). Detailed influences of chemical effects of hydrogen as fuel additive on methane flame *International journal of hydrogen energy*, 40(9), 3777-3788.
10. Kakoulaki, G., Kougias, I., Taylor, N., (2021). Green hydrogen in Europe - A regional assessment: Substituting existing production with electrolysis powered by renewables. *Energy conversion and management*,228, 113649.
11. Kong, M., Feng, S., Xia, Q., (2021). Investigation of Mixing Behavior of Hydrogen Blended to Natural Gas in Gas Network. *Sustainability*, 13(8), 4255
12. Kouchachvili, L., Entchev, E., (2018). Power to gas and H-2/NG blend in SMART energy networks concept. *Renewable energy*, 125, 456-464.
13. Yangaz, M., Ozdemir, MR., Sener, R., (2020) Combustion performance of hydrogen-enriched fuels in a premixed burner. *Environmental technology*, 41(1), 2-13.
14. Mayrhofer, M., et. al. (2021), Assessment of natural gas/hydrogen blends as an alternative fuel for industrial heat treatment furnaces. In: *International journal of hydrogen energy* . 8(3), 111-135
15. Santoli, L., et. al. (2013), Energy characterization of CHP (combined heat and power) fuelled with hydrogen enriched natural gas blends. In: *Energy*. [online]. 2013, [cit. 2022-02-28]. ISSN 0360-5442.
16. Schiro, F., et. al. (2020), Modelling and analyzing the impact of hydrogen enriched natural gas on domestic gas boilers in a decarbonization perspective. In: *Carbon resources conversion*. 15(3), 118-135

Utilization of the Energy Potential of Waste from the Automotive Industry

Peter Pilát[1,*], and *Marek* Patsch[1]

[1]University of Žilina, Faculty of Mechanical Engineering, Department of Power Engineering, Univerzitná 8215/1, 010 26 Žilina, Slovakia

Abstract. Human life is inextricably linked to the generation of waste, whether at the municipal or industrial sphere. Waste is an important source of secondary raw materials and stored energy, which nowadays is advantageous to use. The most suitable way of waste recovery is its recycling and reuse. The share of waste that is unsuitable for recycling for various reasons is high, it currently exceeds the processing capacity of the Slovak Republic and is therefore deposited in landfills. The energy stored mainly in chemical bonds is therefore not used. The share of energy stored in this way is considerable, and its use would help not only in the field of rational waste management, but also in the field of power industry. The simplest way of energy recovery of waste is its direct combustion, more complex technologies are its gasification and pyrolysis. All technologies are called as Waste to Energy systems. Combustion of waste releases stored chemical energy, which is converted into heat, which is used for the production of electricity and heating. The products of gasification and pyrolysis are gaseous and liquid fuels that can be directly used in heat sources for the production of electricity and heat for heating, but also as an input material for the creation of high-grade fuels and for the chemical industry. The article deals with the conceptual design of a small facility for the recovery of waste, especially plastics, from the automotive industry and sorted municipal waste. The concept discusses the technical possibilities of preparing the input material, the technical ways of using it by gasification or pyrolysis, and the possibilities of using the resulting products of processing plastic waste. The assumptions for the design of a small waste recovery facility are verified on an experimental pyrolysis waste treatment device, the article contains the results of laboratory pyrolytic waste treatment and the properties of the resulting products.

1 Introduction

In every period of human society, waste was created, and we also know from archaeological finds about garbage dumps, where society stored things that they could no longer use or did not want to use. Even today, society creates waste, which society wants to get rid of. The most rational solution is to use waste as a secondary raw material. However, waste that can be reused as a raw material must meet certain criteria that cannot be met for a large part of

[*] Corresponding author: peter.pilat@fstroj.uniza.sk

the waste, and that is mainly the emphasis on species purity and purity from the point of view of contamination by other substances, that is, a large part of the waste can be used as an input raw material it cannot be used again in production. The second option is to use the waste as energy. Even according to the guidelines of the European Commission, waste that can be recycled or recovered for energy can be landfilled to a limited extent. Thus, even the European Commission encourages us that if it is not possible to reuse the waste, and if it can be used for energy, then it is necessary to use it for energy purposes. Plastics are one such waste that has a high energy value and is highly polluted, meaning that it is not suitable for reuse. The simplest way of energy recovery is incineration, where the energy from combustion is directly used in the form of heat, e.g. in cement plants or it will be used for heating and electricity production in incinerators, but the capacities of incinerators are limited in Slovakia and their location is not exactly the most suitable, which makes logistics difficult. An ecologically and energetically more appropriate solution is the use of new plastic waste processing technologies, e.g. gasification or pyrolysis, which transform this waste back into the starting raw material – gas and liquid phase, which can be used either directly as fuel or as input raw material for the petrochemical and chemical industry. The construction of such facilities in the territory of the Slovak Republic is not yet very widespread, the reason may be the financial difficulty or the almost non-existent market where the products of decomposition of plastic waste could be used. Existing operations, e.g. for the gasification and pyrolysis processing of plastic waste are rather pilot projects in Slovakia, they have limited capacity, they are unevenly distributed, which also implies the need to transport suitable waste over long distances from a wide area. Transporting waste over long distances can hardly be said to be economical and ecological. Large gasification and pyrolysis processing facilities are available in the world and on the market, which are investment-intensive and unsuitable for the conditions of the Slovak Republic. The starting point for this situation is the construction of smaller gasification and pyrolysis processing facilities, sized according to the amount of waste in the immediate vicinity, evenly distributed throughout the territory of the Slovak Republic, support for their construction by the state, e.g. appropriate legislation, subsidies, etc., and also the creation of a market for gasification or pyrolysis products - synthesis gas, oil, etc. Such smaller processing plants of non-recyclable plastic industrial and municipal waste could be located near larger sources of plastic waste, e.g. near processing facilities for discarded cars, or for associations of municipalities, or in areas for sorting municipal waste. Such an operation could be a modular system, e.g. placed in containers and could be adapted to specific requirements, types and quantities of processed waste. [1,2,3,6]

Gasification or pyrolysis processing of single-type sorted plastic waste (e.g. PET bottles, HDPE, LDPE, PP) is technologically manageable, the technology of processing mixed plastic waste, which can be contaminated with various impurities, such as waste from the automotive industry, is problematic. The disparity of this input material for energy recovery increases the technological complexity of the entire process. This waste is specific mainly due to its state (solid and liquid), species composition and various fractions (depending on the method of previous processing). Plastics, rubber, operating fluids (fuels, engine and transmission oils, cooling and brake fluids,...), leather, synthetic textiles, upholstery, etc. can be considered energetically usable waste from the automotive industry and car recycling. The largest share of energy-usable waste from scrapped cars is made up of plastics, which can generally be divided into polypropylene 35% of the total weight of plastics in a car wreck, polyurethane 20%, PVC 11%, ABS 10%, polyamide 9%, polyethylene 5% and other unspecified 10% . In the category of plastic waste, it is also possible to include the remains of cable scrap, they are made up of cable insulation, terminals, connectors, etc., they are mainly made of PVC and ABS and are no longer recycled, they are intended primarily for energy recovery. The biggest problems in the energy recovery of plastics are with PVC, whose presence in the fuel is limited due to the resulting emissions (if PVC in the fuel is above 1%, an increased temperature in the combustion chamber and an increase in the residence time are necessary). The composition of the input fuel for the process of energy recovery of plastics is given by the technology of processing the car wreck. The crushed plastic portion after shredding is already very difficult to separate and this plastic portion is primarily intended for energy recovery, the input fuel has unspecified properties and composition. The most accurate chemical and physical properties and composition of the input fuel are achieved when the car wreck is completely dismantled. [2,3]

2 Pyrolysis

Pyrolysis is the heat treatment of waste substances in a pyrolysis furnace or reactor at a temperature of 250 to 1650 °C without air access, or with limited air access and at reduced atmospheric pressure. The result of pyrolysis decomposition is liquid substances (pyrolysis oil) and gaseous substances (pyrolysis gas).
The principle of this method of energy recovery of plastic waste is polymer degradation, which can take place with the participation of a catalyst, the reaction takes place in a closed reactor under the influence of heat. The reaction takes place either in the presence of a small amount of air or without the presence of an oxidizing agent, which is most often replaced by an inert gas, most often nitrogen and most often at normal atmospheric pressure. The batched mixture of plastic waste is depolymerized in the reactor, split into lower hydrocarbon chains (decomposition of long chains of plastic materials into a fluid mixture of saturated hydrocarbons). The waste plastic material is converted into gas, and by its subsequent cooling, it turns into oil fractions, the raw material from which it was originally made, and synthesis gas, which contains non-condensable molecules. By cooling the resulting fluid mixture of saturated hydrocarbons, the process of fractional distillation begins, i.e. the separation of individual oil fractions, when the plastic waste is converted back into the basic raw material. The resulting product of the process, which is called catalytic cracking, is a mixture of petroleum fractions, containing diesel, gasoline and oil components, paraffin, wax and other parts. [4,5]

Depolymerization most often takes place in the presence of a catalyst. When adding a catalyst, plastic waste acquires elasticity already at low temperatures (around 200 °C). The advantage of the catalytic depolymerization process at lower working temperatures is also the lower fuel consumption needed to heat the reactor. The catalyst is continuously added to the reactor, which ensures its continuous and stable activity and thus balanced working i.e. temperature conditions with a favorable effect on uniform product quality.

3 Input raw materials and output products

With the described technologies, it is possible to process various types of waste, municipal and industrial, biomass, plastics, old tires, etc. From plastic waste, waste polyolefins (polyalkenes) such as HDPE, LDPE and LLDPE are most often processed using the given technologies, i.e. polyethylenes of different specific weights and PP – polypropylene. The suitability of these polymers results from the fact that they are composed exclusively of carbon and hydrogen and their basic structural units are ethylene CH2 = CH2 and propylene CH3–CH=CH2.

Depolymerization using the methods described above produces synthesis gas from the input materials, which can be divided into condensable and non-condensable molecules by cooling (Table 1 shows an example of the composition of the synthesis gas depending on the input raw material). Depending on the input raw material, if it is plastic, after cooling it contains mostly liquid depolymerization products - crackers, which, depending on the conditions of the decomposition process, and especially depending on the height of the decomposition temperature, have an oily, waxy, or diesel character. These emerging products have a similar fractional composition to those produced primarily from petroleum. The remaining part after depolymerization is water and solid residue in the form of foreign residues from the original waste plastic raw material, in the form of e.g. metal residues, glass, stones, earth and other mechanical impurities. The resulting liquid and gaseous fractions do not contain a significant amount of halogens, sulfur, nitrogen or metals, as the original raw material does not contain them either. Therefore, from the point of view of chemical composition, there are practically only C1 to C5 hydrocarbons in the gas portion, and the output product of the entire technology is a mixture of liquid carbon and hydrogen, in a mass ratio of 86:14%, which are structurally predominantly unsaturated hydrocarbons without a significant content of aromatic compounds. The output product corresponds to the fractional composition of light heating oil, diesel fuel and wax, that is, raw materials that can be further processed in the petrochemical industry as a valuable raw material or can be directly used for the production of heat and electricity.

Application of the given technologies to other types of plastic waste, e.g. from the automotive industry is particularly problematic due to the diversity of the input material, which has a fundamental impact on the choice of depolymerization temperature, the length of the depolymerization period, the use of catalysts, the final product and residues after depolymerization.

Table 1. Example of synthesis gas composition depending on the type of input raw material

	H2 [%]	CH4 [%]	C2-C4 [%]	CO [%]	CO2 [%]	N2 [%]	Density [kg.Nm-3]	Calorific value [MJ.Nm-3]	Calorific value per ton of input raw material [kW/ton input]
biomass	15	26	3	35	17	4	1.10	17.10	2591
plastics	25	38	18	9	5	5	0.80	28.00	7778
tires	19	40	28	3.5	6.5	3	0.90	36.00	3333

4 Gasification and pyrolysis device

For gasification and pyrolysis, reactors are used, most often cylindrical in shape, depending on the amount of processed waste, they are constructed either as continuous or discontinuous. Both technological processes require the supply of a certain amount of heat, its amount depends on the working temperature, amount and type of processed waste. Heat to the process is supplied either by burning gaseous and liquid fuels or by electric heating. Heating with electric energy is realized directly, through an electric spiral, which can be placed in the feed screw. For safety reasons, heating by burning liquid or gaseous fuels in burners is most often realized through an intermediate medium located in the double wall of the reactor. The reactor has thermal insulation due to the economy of operation. Reactor capacities range from several hundred kilograms to several tons of processed waste per hour. The retention time of the batch in the reactor and the operating parameters depend on the reactor design and the type of waste being processed, an example of operating parameters is shown in tab.2. The solid residue after the technological process is removed through the bottom of the reactor by a cooled screw conveyor. The amount of solid residue depends on the type of processed waste, technology and operating parameters.

Table 2. Examples of operating parameters of pyrolysis reactors

	Biomass[1]	Polymer plastics[2]	Tires
Working temperature	250 – 700 °C	650 – 800 °C	650 – 800 °C
Retention time	5 – 15 min	15 – 25 min	15 – 25 min
Amount of synthesis gas	15 – 80 %	60 – 95 %	30 – 60 %
Amount of synthetic oil (after cooling)	5 – 40 %	5 – 30 %	20 – 40 %
Solid residue3	15 – 90 %	2 – 30 %	35 – 45 %

[1] wood chip, agricultural dendromass, sawdust,...
[2] plastics, energetically significant components of municipal and industrial waste,
[3] changes depending on operating conditions (torrification / pyrolysis / high-temperature pyrolysis)

5 Preparation of raw material

The input raw material for the pyrolysis process was plastics that were obtained from a local scrap yard, where they were dismantled from car wrecks. Plastics were not selected according to purity, so some were with a surface treatment, e.g. varnishing and some without surface treatment. The material was sorted by type and was a mixture of plastics from the interior, bumpers, fenders and various covers from the engine compartment. These were mainly ABS, PP, PE and their approximate weight ratio, in which they are found in the car, was preserved. Plastics were selectively crushed (fig. 2) on a knife crusher (fig. 1). Subsequently, a mixture of crumbs was created according to mass representation. The crumb contained all the fractions that were created by crushing plastics in a knife crusher, i.e. from small particles of irregular shape to strips several centimeters long.

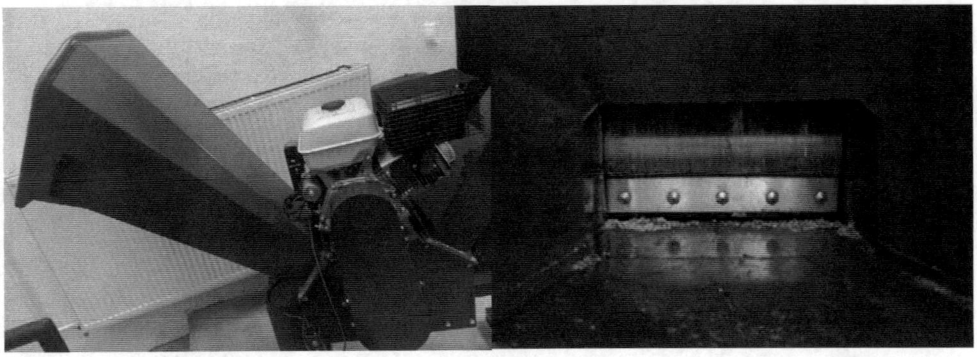

Fig. 1. Rotary grinding device

Fig. 2. The created mixture of plastic waste

6 Pyrolysis reactor

A small experimental pyrolysis reactor fig. 3, the main parts of which are a removable pyrolysis chamber, which is heated by a gas block burner and a condenser, where vapors generated in the pyrolysis chamber condense into pyrolysis oil. The non-condensed vapors coming out of the condenser are currently burned at the torch and in the future will be returned to the combustion process to heat the pyrolysis chamber, thereby reducing the energy requirements of the entire device. The proposed pyrolysis reactor works discontinuously, that is, after the end of the pyrolysis process, the chamber must be cleaned and filled again with a new charge. This reactor was designed so that we could carry out experimental measurements with different materials. During the tests, the entire pyrolysis process and also the entire reactor will be optimized to make the process as economical and efficient as possible considering the conditions.

Fig. 3. Pyrolysis reactor (condenser is separate)

7 Experimental measurement

The laboratory discontinuous pyrolysis reactor (Fig. 3) was filled with a mixture of prepared plastic waste, one working batch weighed 3.5 to 4 kg. The reactor was filled in a cold state (at ambient temperature), the condensation exchanger was cooled to an average temperature of 13.5 °C. Different working temperatures in the range of 300 to 600 °C were chosen for pyrolysis. The course of working temperatures during the selected cycle of pyrolysis is shown in fig. 4. The pyrolysis process took place without added catalyst. In fig. 4, four phases of the pyrolysis process can be observed:

• Preheating phase – the working space is filled with an inert atmosphere, the temperature of the pyrolysis reactor and the charge rise from the ambient temperature, the phase is completed when the first vapors are released.

• Phase of vapor formation – the first vapors began to be released at a temperature of 62 °C, the supply of inert atmosphere from the pressure bottle was stopped, the start of the pyrolysis process no longer allows the entry of atmospheric oxygen, the vapors flow through the condensation exchanger but are non-condensable.

• Condensate formation phase – condensate began to form at a working temperature of 240 °C, at temperatures below 350 °C the condensate has an oily character, after complete cooling it has a waxy character with high viscosity, at higher working temperatures the viscosity of the product decreased.

• Cooling – after the end of the pyrolysis process (minimum amount of condensate formed), the reactor is cooled to ambient temperature so that it can be opened and solid residues can be removed.

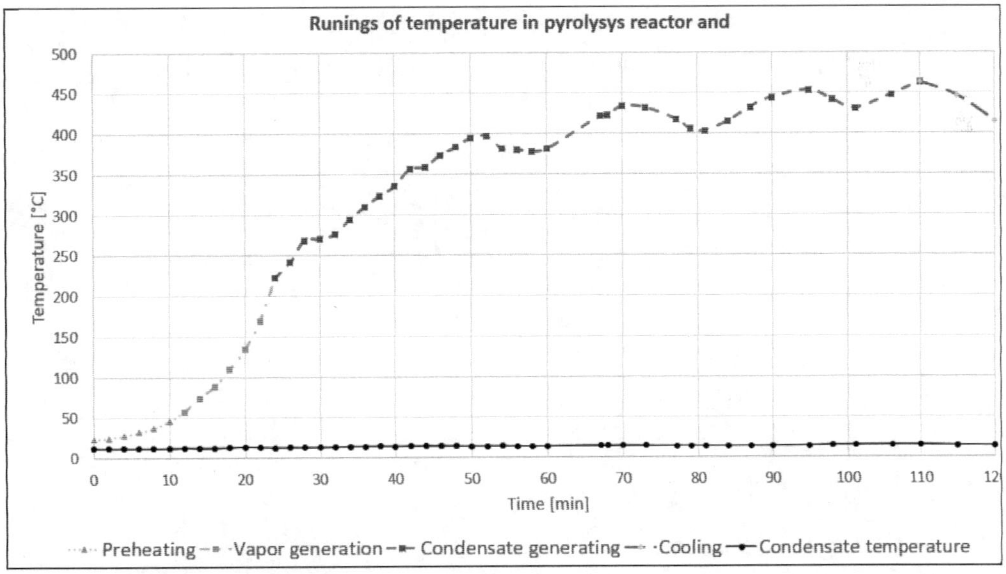

Fig. 4. Working temperatures during pyrolysis

The resulting products of the pyrolysis processing of plastics were pyrolysis oil and gas, the by-product was a solid residue. Table 3 shows their representation in selected pyrolysis cycles. The average mass fraction of pyrolysis oil was 60.44%, pyrolysis gas 29.74% and solid residue 9.81%. Crude pyrolysis oil was made up of all condensable hydrocarbons that were formed during pyrolysis, it was represented by fractions with low viscosity (consisting of gasoline and diesel fractions, Fig. 5) to high viscosity fractions (oily and waxy, resembling fuel oil, Fig. 6). The pyrolysis oil was dark brown to black in color, with a strong odor. The low-viscosity fractions were filtered (Fig. 7), the treated samples were subsequently weighed again. An example of the results of the sample processing after the pyrolysis process is shown in Table 4. The filtering took place at room temperature, which had an impact on the permeability of the sample through the filter, all the part of the sample that did not pass through the filter was included as waste from the process. The light fraction constituted about 40% of the weight of the obtained pyrolysis oil, the heavy about 33% and about 26.5% of the weight was removed by filtration. This amount of waste could be reduced if the filtering was carried out in a heated state of pyrolysis oil, the waste separated in this way is still highly flammable, which was confirmed by a combustion test.

After opening the pyrolysis reactor, the solid residue was removed, which, in addition to carbon, probably contains other components (metal pieces from the input material, heavy metals from surface coatings and adhesives, etc.). The solid residue can be used in the following pyrolysis cycle as a catalyst because it contains a high proportion of active carbon.

Table 3. Weights and percentage by weight in selected samples

	1. sample		2. sample		3. sample		4. sample	
	Weight [g]	%	Weight [g]	%	Weight [g]	%	Weight [g]	%
Pyrolysis oil	2140.0	59.3	2459.0	61.5	2118.4	60.1	2218.7	57.4
Pyrolysis gas	1139.8	31.6	1186.8	29.7	998.2	28.3	1196.6	31.0
Solid residue	329.1	9.1	345.2	8.9	406.2	11.5	446.8	11.60
batch	3608.9	100.0	4000.0	100.0	3522.8	100.0	3859.1	100.0

	5. sample		Average	
	Weight [g]	%	Weight [g]	%
Pyrolysis oil	2534.8	63.9	2534.8	63.9
Pyrolysis gas	1115.2	28.1	1115.2	28.1
Solid residue	317.0	8.0	317.0	8.0
batch	3967.0	100.0	3967.0	100.0

Fig. 5. Low-viscosity fractions after pyrolysis

Fig. 6. High-viscosity fractions after pyrolysis

Fig.7. Filtering samples

Results of partial chemical analysis:
After a partial chemical analysis, the following were mainly represented in the pyrolysis oil: ethylbenzene, hexene, styrene, heptene, benzene, cyclohexane, cyclooctatetraene and other chemicals.

8 Conclusion

In the future, pyrolytic processing of waste will play an important role in the use of waste, not only for direct energy use, but also as a source of input material for the production of noble fuels and lubricants, and also for the chemical industry for the re-production of new materials. Prospective research in this area shows that it is also possible to process mixed multi-species plastic waste, with an average gain of approximately 90% of the original mass usable for energy and other purposes (liquid and gaseous phase of pyrolysis products). The solid residue is suitable as a catalyst for the next cycle of pyrolysis, because of its increased content of harmful substances it is then suitable to be deposited in a landfill.

Such waste treatment requires further research in the field of the influence of the operating temperature of the pyrolysis reactor, the length of the residence time, the use of various catalysts to improve the economy of the operation of the pyrolysis reactor, and also in the field of pyrolysis oil and gas processing.

The article was created within the work on the task solved by the UNIVNET association no. 0201/0004/20 and it was also supported from the funds of the Ministry of Education, Youth and Sports of the Slovak Republic.

References

1. A. Beukens, Incineration Technologies. Springer, 2013
2. M. Čarnogurská and M. Lazár, *Plazmové spracovanie a zhodnocovanie odpadu.* SjF TUKE Košice, 2013
3. J. Jandačka, *Energetické využitie komunálneho odpadu.* UNIZA-EDIS, 2014
4. J. Najser, O. Němček, M. Mikeska, *Characterization of syngas produced from biomass gasification.* MATEC Web of Conferences, vol.168, 2018
5. J. Najser, V. Peer, M. Vantúch, *Biomass gasification for liquid fuel production.* AIP Conference proceedings, vol. 1608, p. 71-75, 2014
6. Information from EU regulations

Pyrolysis Treatment of Municipal Solid Waste and Automotive Waste with Study of Each Component Energy Potential

Michal Holubčík[1,*], *Andrej* Klačko[1], *Jozef* Jandačka[1], and *Juraj* Drga[1]

[1]Department of Power Engineering, Faculty of Mechanical Engineering, University of Zilina, Univerzitná 1, 010 26 Zilina, Slovakia

Abstract. Many research is aimed at improving municipal solid waste disposal and producing usable energy. Pyrolysis technology not only decreases the volume of municipal solid waste, but it also produces pyrolysis oil, pyrolysis gas, and carbon, all of which have a high calorific value and are widely used in industrial activities. This article focuses on the plant-based pyrolysis of municipal solid waste and automotive plastic trash. The research also discusses the pyrolysis energy potential of municipal solid waste components such as plastics, biomass, rubber, and plastics. The energy potential of plastic waste from vehicle components was also investigated in this study, which used thermogravimetric and elemental analysers. According to the findings of the examination of the most common plastic waste from automobiles, it is possible to determine the potential treatment of this waste by pyrolysis. By analyse municipal solid waste, it was discovered that waste treatment by pyrolysis might lower the environmental load in Slovakia in the future, ensuring a greater quality of life, inexpensive and sustainable energy for humanity, and strengthening waste treatment innovation.

Keywords: *municipal waste, pyrolysis, waste recovery, automotive industry, energy*

Acronyms

ABS	acrylonitrile butadiene styrene
AD	anaerobic digestion
BMP	biochemical methane potential
MSW	municipal solid waste
NK	nomenclature
PE	polyethylene
PET	polyethylene terephthalate
PP	polypropylene
PU	polyurethane foam
PVC	polyvinyl chloride

[*]Corresponding author: michal.holubcik@fstroj.uniza.sk

TAP solid alternative fuel

1 Introduction

As a result of increased human consumption because of economic expansion and urbanization, is seeing an increase in the creation of municipal solid waste (MSW). According to the World Bank, global MSW production will reach 3.4 billion tonnes by 2050 [1]. China is one of the developing countries with the most MSW production. Human health depends on air quality. Emissions, especially particulate matter (PM), contained in the air can penetrate the lungs and blood circulation, which can cause various respiratory and cardiovascular diseases [2]. According to the National Statistical Office, MSW output in 2019 was 242 million tonnes, up 6.16 percent from the previous year [3]. MSW not only takes up precious ground, but it also pollutes the air and water, endangering public health and the ecosystem.

MSW can be disposed of in several ways. Many countries use landfills, which can hold a lot of MSW [3]. Leachate from landfills, on the other hand, can damage groundwater [4]. The volume of MSW can be reduced to a larger extent by combustion than by heat treatment. It does not require a larger plot area than landfill treatment. Combustion heat can be used to generate electricity or heat [5]. The ecology is harmed by emissions of fly ash, slag, and pollutant gases like SO_2, NO_X, and dioxins. Scientists have been focusing on converting garbage into useable resources in recent years.

Approximately 15 to 16 million cars are sold in Europe each year. The amount of incoming market and about the same is also leaving. The oldest vehicles are scrapped after 20 or 30 years of operation, generating 15 million tonnes of waste. In Slovakia, it is about 50-60 thousand cars a year [6]. The vehicle consists of metals, plastics, fluids, rubber, glass, textiles. For some types of waste, the tip can be set. If sampling lines are created, you can, for example, go straight to metallurgical primary production with separated metals, figuratively speaking. [7]

The processing of old used vehicles is an important factor in environmental protection. To be able to reuse the materials used in the construction of the car and not cause a lot of waste, it is necessary to raise the issue of their quality dismantling while preserving the quality of the secondary raw materials that can be returned for reprocessing. The technologies currently used in the processing of old vehicles are shredding (US approach), selective dismantling (French approach) and complete dismantling (German approach) [7].

It is significantly easier to process only the components while disassembling a car. Pyrolysis is particularly well suited to the processing of automotive polymers. Plastics that can no longer be recycled can only be pyrolysed, are polluted, and hence are unsuitable for alternative recycling methods. [8]

Because there is no burning, the method does not produce hazardous pollutants or have an environmental impact, as is the case with trash incineration. This is because during combustion, or oxidation, the flame comes into direct contact with the reaction mixture. The waste is isolated from the reactor and furnace, which use propane-butane (or natural gas) as fuel and solely serve to achieve the appropriate plant temperature. The polymers are gasified rather than burned [8].

2 Physical composition of MSW

Living patterns, lifestyles, and tourism are all changing fast because of growing urbanization and increased urban migration. MSW has a different composition and characteristics according on the area and season. This mixture is critical for obtaining high-quality pyrolysis fuel.

In 2020, Slovak homes produced more municipal garbage, but the good news is that the percentage of recyclable waste grew to 44 %. Less than half of municipal rubbish ended up in landfills for the first time in history (48 %). In 2020, the population of Slovakia produced an average of 446 kg of municipal waste. This is 53 kg more than the average annual value of waste produced in the previous five years (2016 - 2020). In total, the households of the Slovak Republic produced 2.43 mil. tones of municipal waste. This follows from data on municipal waste in 2020, which were currently published by the Statistical Office of the Slovak Republic.

From a regional point of view, most waste - more than 500 kg per capita per year - was produced in three regions of western Slovakia, namely Trnava (584), Bratislava (531) and Nitra (521). On the contrary, the least waste was generated by the inhabitants of the Prešov, Košice and Banská Bystrica regions. In these regions, the amount of waste per capita did not exceed 400 kg. In the Prešov Region, one person produces on average two thirds less waste than in the Trnava Region.

MSW is classified into four groups according on its composition: recyclable, kitchen, hazardous, and other garbage. Waste plastics and rubber, kitchen waste, paper, textiles, biomass waste, ceramics, glass, and metal make up the majority of MSW [9].
In Table 1 is stated municipal waste quantity by type of waste in tons and then municipal waste quantity by waste subgroups in tons in Table 1.

Table 1. Municipal waste quantity by type of waste in tons [9].

Part of the region	2020	2019	2018	2017
Slovakia	2 434 039	2 369 725	2 325 178	2 136 952
Bratislava Region	357 774	336 551	338 169	324 714
Western Slovakia	932 476	920 299	903 764	818 118
Trnava region	330 111	323 490	313 247	280 807
Trenčín region	251 835	245 845	247 929	222 573
Nitra region	350 530	350 964	342 588	314 739
Middle Slovakia	564 713	557 486	548 324	499 987
Žilina region	312 432	311 036	299 601	274 936
Banská Bystrica Region	252 281	246 450	248 723	225 051
Eastern Slovakia	579 076	555 390	534 920	494 133
Prešov Region	288 350	283 339	271 709	238 309
Košice region	290 727	272 051	263 211	255 824

Table 2. Municipal waste quantity by waste subgroups in tons [9].

Waste component	2020	2019
Acids	2.62	1.89
Batteries and accumulators other than those mentioned in 20 01 33	59.46	204.04
Batteries and accumulators specified in 16 06 01, 16 06 02 or 16 06 03 and unsorted batteries and accumulators containing these batteries	7 734.57	6 846.04
Biodegradable kitchen and restaurant waste	2 579.45	4 156.34
Biodegradable waste	298 348.71	247 673.43
Bulky waste	191 728.02	211 326.59
Cardboard-based multilayer composite materials (cardboard-based composites)	3 408.60	2 471.81
Clothing	4 166.95	4 503.78
Cytotoxic and cytostatic drugs	0.01	0.09
Detergents containing dangerous substances	6.61	8.53
Detergents other than those mentioned in 20 01 29	0.58	0.82

Discarded electrical and electronic equipment other than those mentioned in 20 01 21 and 20 01 23 containing dangerous parts	4 234.17	3 658.61
Discarded electrical and electronic equipment other than those mentioned in 20 01 21, 20 01 23 and 20 01 35	14 866.52	10 645.25
Discarded equipment containing chlorofluorocarbons	4 463.39	4 262.68
Dissolvent	21.96	34.99
Edible oils and fats	529.96	425.23
Fluorescent lamps and other wastes containing mercury	53.67	49.4
Glass	78 777.65	71 524.09
Chimney sweeping wastes	-	30.86
Inks, printing inks, adhesives and resins containing dangerous substances	625.58	515.39
Inks, printing inks, adhesives, and resins other than those mentioned in 20 01 27	70.34	49.91
Market waste	355.43	984.61
Medicinal products other than those mentioned in 20 01 31	0.38	0.4
Metal packaging	11 271.72	10 567.75
Metals	371 266.75	336 195.26
Mixed municipal waste	1 144 885.77	1 166 418.83
Municipal wastes not otherwise specified	6.19	12.55
Oils and fats other than those mentioned in 20 01 25	169.83	184.81
Other non - biodegradable wastes	2 785.99	3 167.15
Packaging containing residues of dangerous substances, contaminated with dangerous substances, including empty pressure vessels	125.42	110.54
Paper and cardboard	97 720.29	106 961.28
Pesticides	10.44	9.46
Photochemical substances	0.27	0.1
Plastics	68 536.31	57 922.52
Principles	0.45	0.19
Small construction waste	68 417.73	70 864.95
Soil and aggregates	3 055.47	2 808.70
Textiles	487.59	420.1
Together	2 434 039.46	2 369 725.16
Wastes from sewage treatment	1 455.35	1 828.15
Wastes from street cleaning	20 092.60	22 664.99
Wastes not otherwise specified	875.96	29.77
Wood containing dangerous substances	60.17	86.76
Wood other than those mentioned in 20 01 37	30 780.51	20 096.54

3 Physical plastic composition of automotive waste

Because plastics are formable, lightweight, and typically seen as highly recyclable materials, they are becoming a more popular material in the design and production of complicated consumer items such as vehicles. However, the actual recycling of heterogeneous plastics employed in such long-lasting items is difficult, and different scenarios for how basic products, such as water bottles, are retrieved through curb or container recycling operations provide quite different possibilities. Although there is equipment for recycling plastics, technological and economic restrictions limit their use in high-level consumer or industrial applications. Plastics are more likely to wind up in energy recovery or as shredder residues, and eventually in landfills, for these reasons [9].

The current recovery and recycling procedure for a discarded automobile includes dismantling, de-pollution, and shredding, as well as physical and mechanical treatment of shredder waste and energy recovery. Presently, plastics and composites contribute to shredder residue, and as the use of these materials increases, so does the amount of shredder residue generate. The percent plastics by mass in an average vehicle has gone from 6 % in 1970 up to 16 % in 2010 and is expected to reach 18 % in 2021 [10]. The automotive industry accounts for a significant percentage of plastics demand with estimates ranging up to 30% and always rising [10].

Table 3. Examples of parts assemblies having significant non-metallic materials content [7].

Part Assembly	**Average**		**Range**	
	Metals (% Weight of part)	**Non-Metals (% Weight of part)**	**Metals (% Weight of part)**	**Non-Metals (% Weight of part)**
Front door	70 %	30 %	66 % – 72 %	28 % – 34 %
Rear door	69 %	31 %	66 % – 74 %	26 % – 34 %
Steering wheel	60 %	40 %	55 % – 65 %	35 % – 45 %
Steering column	80 %	20 %	77 % – 83 %	17 % – 23 %
Dashboard	44 %	56 %	28 % – 54 %	46 % – 72 %
Seat	66 %	34 %	21 % – 81 %	19 % – 79 %

Given European regulations requiring 95 percent of a car to be processed for reuse and recovery and 85 percent to be processed for reuse and recycling by January 1, 2015, identifying feasible improvements and alternatives in the plastics recycling chain is critical. [8]

4 Materials and methods

Polyurethane foam (PU), polypropylene (PP), polyethylene terephthalate (PET), polyethylene (PE), polyvinyl chloride (PVC), and acrylonitrile butadiene styrene are some of the most often used plastics in automotive applications (ABS).
According to Table 5, we performed an analysis of waste plastics from the following car components. We have selected components that contain the most plastics such as polypropylene (PP), polyethylene (PE) and acrylonitrile butadiene styrene (ABS). Among these vehicle components, we have narrowed the selection to the following: front door upholstery, dashboard, and painted plastic front bumper [11].

The component samples were crushed with a mechanical crusher to prepare plastic samples suitable for analysis on instruments such as a thermogravimetric analyzer LECO TGA 701 for the determination of moisture, volatile, fixed carbon, and ash content; an elemental analyzer LECO CHN628 for the determination of carbon, hydrogen, nitrogen content; a separate module LECO 628S for sulfur content. The particle size was mainly in the range of 0.5 mm to 2 mm for all samples. The results of these analyzes are shown in Tables 4-5. In the elemental analyzer were pre-weighed samples combusted in the absence of atmospheric gases. The thermogravimetric analyzer measured the weight loss of individual samples during increasing the temperatures. Individual samples were heated at 107 °C (air atmosphere), then at 900 °C (nitrogen atmosphere), and finally cooled at 550 °C (air atmosphere). There were used standard methods for solid biofuels by the relevant standards STN EN ISO 18122 and STN EN ISO 18123.

5 Results and discussion

The results of these analyses are shown in Tables 4 and Table 5 on dry basis. Results values in the tables are average values from three repetitions of measurements. These achieved results are consistent with the other results in works [12, 13, 14] stated in Table 4 for comparison. The achieved results of the elemental analysis confirm the high potential for obtaining elements such as mainly carbon and hydrogen. The values of the carbon content compared with others differ by up to 10 % and the hydrogen content differ up to 6 %, which could be influenced by the production quality of the preparation of samples, used devices, atmosphere, and method. Based on the results, investigated materials are still attractive for pyrolysis processing in terms of the content of the examined elements.

Table 4. Results of carbon, hydrogen, nitrogen, and sulfur content (dry basis).

Sample	Nitrogen (%)	Carbon (%)	Hydrogen (%)	Sulfur %
PP	1.94	72.38	8.14	0.12
PP [12]	0.14	84.80	14.55	0.23
PP [13]	0.00	85.97	14.03	0.00
PE	3.13	75.54	7.65	0.61
PE [13]	0.00	85.43	14.57	0.00
ABS	0.00	83.63	13.68	0.15
ABS [12]	0.00	75.5	8.20*	8.20*
ABS [14]	0.00	84.3	0.00	0.068

Table 5. Results of thermogravimetric analysis.

Sample	Moisture Wet (%)	Fixed Carbon Dry (%)	Volatile Dry (%)	Ash Dry (%)
PP	0.00	1.26	98.54	0.20
PE	0.00	2.16	97.21	0.64
ABS	0.00	0.46	98.13	1.41
Mixed	0.00	1.25	98.50	0.25

By analysing the obtained results, we can conclude that the processing of MSW as well as selected waste from the automotive industry is a promising solution to waste pol-icy not only in the Slovak Republic but also globally in the world. The initial expenditure for the technology's creation and, finally, the portfolio of end customers may be the challenge with obtaining fuel from garbage. They are in insufficient numbers to build big pyrolysis facilities. The goal of this study was not to compare the efficacy of pyrolysis waste treatment technologies, but to show that pyrolysis is a potential option to dispose of both current and landfill garbage. Simultaneously, began to investigate pyrolysis plants for smaller companies and community facilities during the study.

6 Conclusion

Analysing the results of the above assessments, it can be assessed that addressing waste policy, whether it is municipal waste or waste from the automotive industry, can be taken to help us combat climate change. At the same time, it was found that due to the high content of hydrocarbons, pyrolysis treatment of MSW or waste from the automotive industry has prospects in the future how to reduce the amount of waste produced and turn it into further use in the form of fuels in industry. Since there will be lower existing landfills as well as reduced production of new ones. The Government of the Slovak Republic will adopt in 2023 a law banning the landfill of any waste. At the same time, it supports the processing of waste and its recovery for other forms of energy, such as pyrolysis treatment. Preparations for the wording of the Act on the Prohibition of Landfilling of Waste are thus looking for alternative methods of waste management, apart from the conventional separation and recycling method for waste treatment by pyrolysis. Studies are also being conducted on the production of TAP (solid alternative fuel) for cement plants.

The department deals with the pyrolysis treatment of MSW and wastes from the automotive industry. This work was also focused on plastic waste from car components, which was tested on thermogravimetric and elemental analysers. No nitrogen was measured for the ABS sample. The highest content of nitrogen was measured for the Mixed sample. The measured carbon content was in the range of 72.38 % to 83.63 %. The lowest value was for the PP sample and the highest value for the ABS sample. The lowest content of hydrogen was for the PE sample and the highest value for the ABS sample. The sulfur content was in the range of 0.12 % (PP sample) to 0.61 % (PE sample). The values of individual volatile matter were similar in the range from 97.21 % to 98.54 %. The highest ash content was measured for the ABS sample with a value of 1.41 %. Other samples had ash con-tent lower than 0.64 %.

Finally, with innovative waste management solutions, it is possible to meet the goals of sustainable development and thus achieve responsible consumption while adhering to an appropriate waste management scheme, such as: the separation of plastics for recycling and pyrolysis treatment. In addition, reducing emissions from waste treatment can pre-vent climate change, such as rising global temperature.

This research is supported by the project APVV-21-0452 Impact of using small electrostatic precipitators to reduce particulate pollutants in domestic fuel combustion, KEGA 032ŽU-4/2022 Implementation of knowledge about modern ways of reducing, environmental burden in the energy use of solid fuels and waste into the pedagogical process, VEGA-1/0233/19 - Design modification of a burner for burning solid fuels in small heat sources, and by the UNIVNET association no. 0201/0004/20 and it was also supported from the funds of the Ministry of Education, Youth and Sports of the Slovak Republic.

References

1. S. Kaza, L. Yao, P. Bhada-Tata, F. Van Woerden, What a Waste 2.0: A Global Snapshot of Solid Waste Management to 2050 The World Bank (2018).
2. Čajová Kantová, N., Čaja, A., Patsch, M., Holubčík, M., Ďurčanský, P.: Dependence of the Flue Gas Flow on the Setting of the Separation Baffle in the Flue Gas Tract, in: APPLIED SCIENCES-BASEL, Volume 11, Issue 7, Article Number 2961, DOI 10.3390/app11072961, Published: 2021.
3. NBSC China Statistical Yearbook, in National Bureau of Statistics of China (2020).
4. J. Piekutin Monitoring of Groundwater in the Area of a Reclaimed Municipal Waste Landfill J. Ecol. Eng., 20 (2019), pp. 262-268.
5. E. Preziosi, E. Frollini, A. Zoppini, S. Ghergo, M. Melita, D. Parrone, D. Rossi, S. Amalfitano Disentangling natural and anthropogenic impacts on groundwater by hydrogeochemical, isotopic and microbiological data: Hints from a municipal solid waste landfill Waste Manag., 84 (2019), pp. 245-255.
6. Z. Fu, S. Zhang, X. Li, J. Shao, K. Wang, H. Chen MSW oxy-enriched incineration technology applied in China: Combustion temperature, flue gas loss and economic considerations Waste Manag., 38 (2015), pp. 149-156.
7. UNIVENT, Stav a vízie zhodnocovania odpadov z automobilového priemyslu SR, Bratislava: Slovenská technická univerzita v Bratislave vo Vydavateľstve SPEKTRUM STU, ISBN: 978-89-227-5039-4.

8. Lu, J.S.; Chang, Y.; Poon, C.S.; Lee, D.J. Slow pyrolysis of municipal solid waste (MSW): A review. Bioresour. Technol. 2020, 312, 123615.
9. Lindsay Miller *, Katie Soulliere, Susan Sawyer-Beaulieu, Simon Tseng, and Edwin Tam: Challenges and Alternatives to Plastics Recycling in the Automotive Sector, Materials 2014, 7, 5883-5902.
10. http://datacube.statistics.sk/#!/view/sk/VBD_SK_WIN/zp3001rr/v_zp3001rr_00_00_00_sk.
11. M. Wang, Y. Yang, S. Jin, L. Gu, H. Zhang Social and cultural factors that influence residential location choice of urban senior citizens in China – the case of Chengdu city Habit. Int., 53 (2016), pp. 55-65.
12. Su ShiungLamadAlan D.RussellaChern LeingLeebSu KiLamcHoward A.Chasea, Production of hydrogen and light hydrocarbons as a potential gaseous fuel from microwave-heated pyrolysis of waste automotive engine oil (2012).
13. UNIVENT, Progresívne technológie zhodnocovania odpadov v automobilovom priemysle, Bratislava: Slovenská technická univerzita v Bratislave vo Vydavateľstve SPEKTRUM STU, ISBN: 978-80-553-3867-5, 2021.
14. G. Gałko, M. Rejdak, D. Tercki, M. Bogacka, M. Sajdak Evaluation of the applicability of polymeric materials to BTEX and fine product transformation by catalytic and non-catalytic pyrolysis as a part of the closed loop material economy (2021).
15. A. Batuera, J.Longb, H.Dub, D. Chena, Multi-products oriented co-pyrolysis of papers, plastics, and textiles in MSW and the synergistic effects.

Analysis of the Influence of Boundary Condition on Simulation Accuracy of Solidification Thermokinetics Model

František Kavička[1,*], *Jaroslav* Katolický[1], *Tomáš* Mauder[1], *Lubomír* Klimeš[1], and *Josef* Štětina[1]

[1]Brno University of Technology, Faculty of Mechanical Engineering, Czech Republic

Abstract. For a sample steel casting of a cylindrical shape that was cast in a metallic cylindrical mould, the analysis of influence and importance of the main boundary condition on the simulation accuracy of the temperature field in the system of casting-mould-environment was performed. For this system, the influence of the boundary condition on the frame of the mould, on the bottom of the mould, on the top of the casting, and on the casting-mould interface was analysed. As a comparing quantity for accuracy, the time-dependent temperature field (by means of isotherms) of the casting and of the mould and the total solidification time were selected. When the simulation considers the boundary conditions of one of these boundaries of the system, the heat transfer coefficient on the remaining three boundaries of the system is equal zero. The resultant heat flow along the axis is zero always. The conclusions can be used for fine-tuning of solidification models as well as for other applications; e.g. analysis and solution of inverse heat transfer problems.

1 Mathematical model

The solidification and cooling of a classically cast (i.e. gravitationally poured) casting (ingot) and simultaneous heating of the mould and chills is, from the viewpoint of thermokinetics, a very complicated case of 3D transient heat and mass transfer [1-5]. In systems comprising the casting (ingot), the mould, chills and environment, all three kinds of heat transfer take place. Since these problems cannot be solved analytically, even with the second-order partial differential Fourier equation (1) (where mass transfer is neglected and conduction is considered as the most important of the three kinds of heat transfer), it is necessary to engage numerical methods. The Fourier equation (1) for a casting (ingot) must be adapted to describe the temperature field of a casting in all its three phases: in the melt, in mushy zone and in the solid phase.

$$\frac{\partial(h\rho)}{\partial\tau} = \nabla\cdot\left(k_{eff}\nabla T\right) \quad (1)$$

Here it is necessary to introduce the enthalpy, which is

* Corresponding author: kavicka@fme.vutbr.cz

$$h = \int_0^T \left(c(\zeta) - \Delta H \frac{\partial f_s}{\partial \zeta} \right) \mathrm{d}\zeta \tag{2}$$

where ΔH is the latent heat (J/kg) and f_s is the solid fraction (-). In equation (1), T [K] is temperature, τ [s] is time, k_{eff} [W/m.K] is the effective heat conductivity and ρ [kg.m^{-3}] is the density.

The original model of the transient temperature field of a system including the casting (ingot)-mould (chills)-ambient was developed [5], which is based on the first and second Fourier's laws on transient heat conduction, and the first and second law of thermodynamics. Further, the numerical method of finite differences with explicit formula for the unknown temperature of the mesh node (*i, j, k*) in the next time step is applied, which is a function of temperatures of the same node and six adjacent nodes in the Cartesian coordinate system from the previous time step. The model considers non-linearity of the task; it means the dependence of thermo-physical properties of all materials of the systems on the temperature [6-7] and the dependence of heat transfer coefficients (boundary conditions) on the temperature of all external surfaces. The model is equipped with and interactive graphical environment for automatic generation of a mesh and for evaluation of results, so-called pre-processing and post-processing.

2 Influence of boundary conditions

For a sample steel casting of a cylindrical shape that was cast in a cast iron cylindrical mould, the analysis of influence of boundary conditions on simulation accuracy of temperature field in the system casting-mould-environment was performed (Fig. 1). The case of solidification in a non-metal mould was not conducted because the boundary conditions here have a negligible effect [7-9]. The analysis was conducted with respect to the heat transfer coefficient on the surface of the casting and mould, the bottom of the mould, the outer surface of the frame of the mould, as well as the casting-mould interface (Fig. 2). When the calculation considers the boundary conditions of one of these boundaries of the system, the heat transfer coefficient on the remaining three boundaries of the system is equal to zero. The resultant heat flow along the heat axis is zero, $q = 0$.

Each of the four coefficients of total heat transfer h is given by the sum of conventional h_c and h_r component. The conventional component is determined using the similarity theory from the general criteria equation $Nu = C(Pr\ Gr)^n$, which applies to natural convection in limited and unlimited space. The Nu criterion (the dimensionless convective coefficient), according to this relationship, is a function of dimensionless properties of the flowing liquid (the Pr criterion) and the uplift (the Gr criterion). The numerical evaluation of h_c is performed in the general way. The same applies to the radiation component, which is transformed into the reduced heat transfer coefficient.

The heat transfer outside the frame of the mould h_f does not affect the total time of solidification of the analysed cylinder. If h_f changes from 0 to 10^6 W/m^{2}K (the real values range close to 10^1 W/m^{2}K), then not only the total time of solidification is constant but the temperature field along the axial section of cylinder, illustrated using isotherms, is the same. Even the temperature field through the section of the mould does not differ significantly. The course of the isotherms (Fig. 3), for example 180 s after casting, for $h_f = 0$ W/m^{2}K and $h_f = 10^6$ W/m^{2}K is unexpectedly very close.

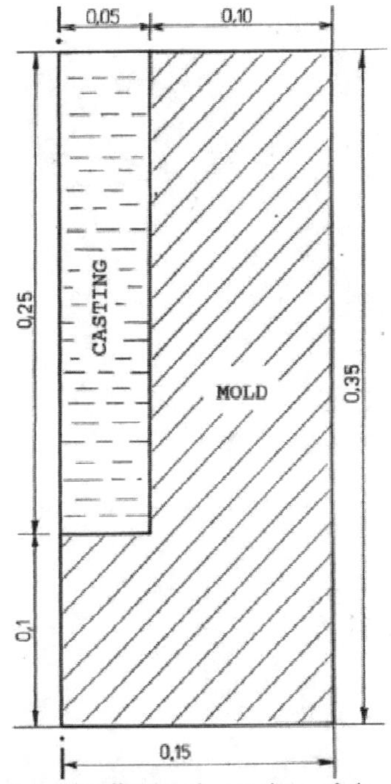

Fig. 1. Longitudinal axis section of the system casting-mould-environment.

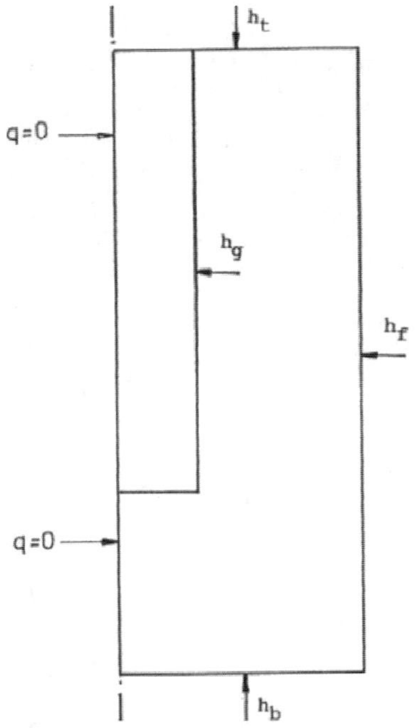

Fig. 2. The boundary conditions.

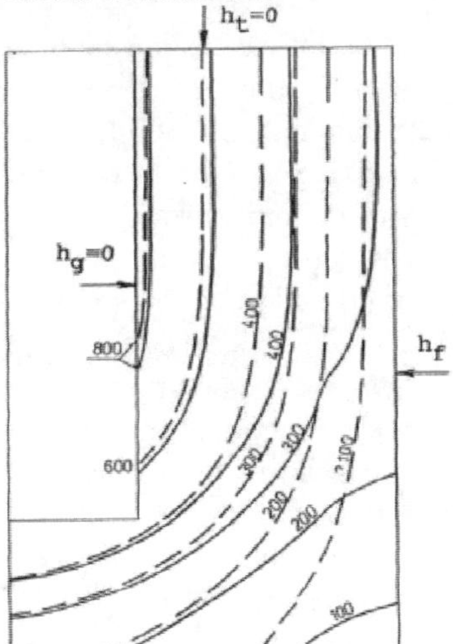

Fig. 3. The analysis of the boundary condition condition on the frame of the mould. Time: 180 s, $h_f = 0$, $h_f = 10^6$.

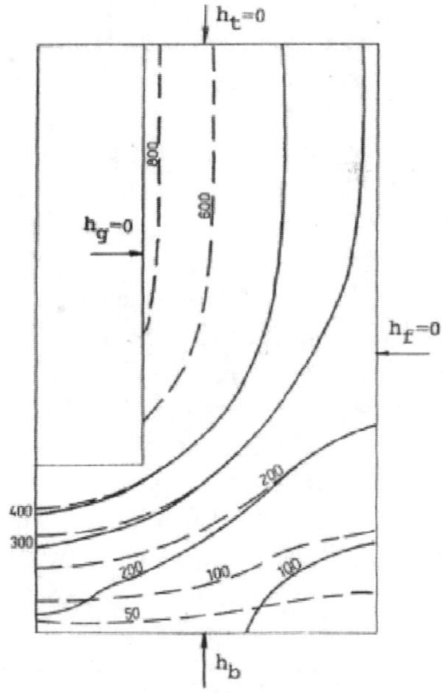

Fig. 4. The analysis of the boundary on the bottom of the mold. Time: 180 s, $h_b = 0$, $h_b = 10^6$.

The heat transfer from the bottom of the mould h_b also has no effect on the temperature field of the casting during solidification. The effect on the mutual position of the isotherms inside the mould, for both extreme values of h_b, (Fig. 4) is even smaller than in Fig. 3. For example, the 600 - 800 °C isotherms coincide. Even the range of values of the heat transfer coefficient for the top of the casting and mould h_t in a wide range from 0 to 10^6 W/m^{2}K simply does not affect the total time of solidification, even though it does affect the dynamics of the temperature field (Fig. 5), however not very significantly.

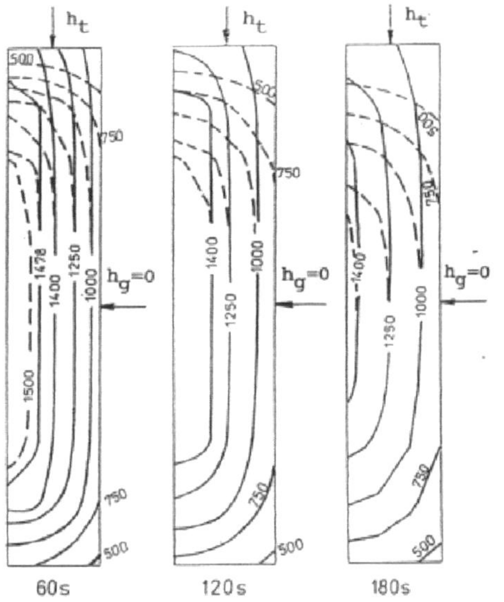

Fig. 5. The analysis of the boundary condition on the top of the casting and mould Time: 60, 120, 180 s, h_t = 0 W/m^{2}K, h_t = 10^6 W/m^{2}K.

The most difficult task is to define the boundary condition for the casting-mould interface (contact surface) and in the gap. A discussion covering the derivation of this is not the subject of this paper. Furthermore, it has significant effect on the total time of solidification, as well as on the formation of the temperature field within the casting and the mould. It could be stated that the majority of recognized authors dealing with numerical solidification models in sand (non-metal) moulds assume an ideal physical contact between these two surfaces. This simplification is fully justifiable. During the solidification in metal moulds, this assumption is ruled out. In most cases, it is necessary to consider a change in the temperature resistance of the interface, i.e. gaps as reciprocal values h_g for the contracting casting and expanding mould. In order to simplify the analysis, it is presumed that the selected size of the gap between the casting and the mould is constant during solidification. It should be noted that here too, h_g is the sum of the conventional and radiation components. Fig. 6 illustrates the influence of the size of the gap on the total time of solidification and the relative extension of the total time of solidification in percentage. If the gap widens from 10^{-4} m to 10^{-2} m, then the total solidification time moves from 342 s to 1035 s, i.e. to 202.6 %.

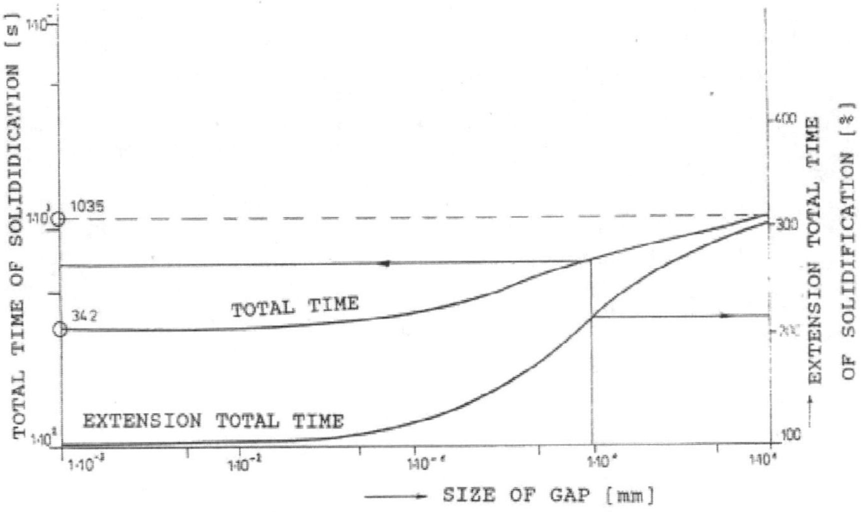

Fig. 6. The analysis of the boundary condition on the casting-mould interface (in the gap).

3 Conclusion

The main boundary conditions in the solution of heat transfer in a casting-nonmetalic mould-environment system have significant influence on the actual casting process of gravitationally cast steels, and on the accuracy of numerical simulation and optimization. It is anyway necessary to reduce energy and material consumption in foundry and metallurgical production using the numerical thermokinetics models. An original numerical model of the 3D transient temperature field was used for the influence analysis of boundary conditions on the accuracy of simulated temperature field and its numerical optimization. The influence of the boundary condition on the frame of the mould, on the bottom of the mould, on the top of the casting and on the casting-mould interface was analysed.

This work was supported by the project Adaptive Soft Computing Framework for Inverse Heat Transfer Problems with Phase Change, reg. no. 22-31173S funded by the Czech Science Foundation and by the project Computer Simulations for Effective Low-Emission Energy Engineering, reg. no. CZ.02.1.01/0.0/0.0/16_026/0008392 funded by Operational Programme Research, Development and Education, Czech Republic.

References

1. Miettinen, J., Louhenkilpi, S. & Laine, J. (1996). Solidification analysis package IDS. *Proceeding of General COST 512 Workshop on Modelling in Materials Science and Processing*, M.Rappaz and M. Kedro eds., ECSC-EC-EAEC, Brussels.

2. Kavicka, F., et al. (2016). *Numerical Models of Solidification and their Application in Metal and Ceramic Technology.* Litera Brno.

3. Kavička, F., et al. (2009). Numerical optimization of the method of cooling of a massive casting of ductile cast-iron. *Materiali in tehnologie/Materials and technology*, 43(2), 73–78.

4. Molínek, J., et al. (2007). Optimization of technological parameters of the gravity-cast rolls for rolling. *Final research report of the project GAČR GA106/04/1334*, Ostrava, Czech Republic.

5. Stetina, J., et al. (2004). Optimization of a concasting technology via a dynamic solidification model of a slab caster. *Proceedings* of *The 5th pacific rim international conference on advanced materials and processing (PRICM-5)*. (pp. 3831-3834). China: Beijing.

6. Stetina, J., et al. (2002). The importance of thermophysical properties for the accuracy of the numerical simulation of a concasting process. *Book of abstracts of The sixteenth european conference on thermophysical properties.* (pp. 319–320). UK: London.
7. Kavicka, F., et al. (1996). Analysis of the effect of thermodynamic properties and boundary conditions on the calculation accuracy of solidification thermokinetics. *Proceedings of The 62nd World Foundry Congress.* (pp. 2-10). USA: Philadelphia.
8. Hlousek, J. (1993). A coherence between a casting shape and its total solidification time. *Proceedings of the Second International Conference on computational modelling of free and moving boundary problems.* /pp. 15-21). Italy: Milano
9. Kavicka, F., et al. (2012). Definition of boundary conditions for the numerical model of transfer temperature field of a concast steel slab. *Communications,* 4A, 13-17.

Radial Force Loads on the Impeller of a Spiral Casing Pump – Comparison of Experiment and CFD Simulation

David Bleho[1,*], *Robert* Olšiak[1], *Branislav* Knížat[1], and *Peter* Tóth[1]

[1]Faculty of Mechanical Engineering STU in Bratislava, Nám. slobody 17, 81231 Bratislava, Slovak Republic

Abstract. Computational fluid dynamics (CFD) is becoming a widely used tool in the industrial fields. Experimental methods are expensive and time-consuming and CFD is a suitable substitute. In order to investigate the accuracy between the two methods, the radial force on a 6-bladed impeller was measured. The results reveal that the steady-steady analysis method fails, and higher accuracy is achieved using transient simulation. Fairly good accuracy is achieved when simulating the power parameters, where it is sufficient to use the steady-state method as well to save time. On the other hand, for force load detection, where the magnitude of the radial force varies with the position of the impeller, the transient method should be used. The distribution of the non-stationary radial force vectors is symmetric around the origin and lies mainly in 6 regions, which are the same as the number of impeller blades.

Keywords: *Radial force; CFD; Experiment; Centrifugal pump*

* Corresponding author: david.bleho@stuba.sk

1 Introduction

When designing the mechanical parts of the pump, it is necessary for the designer to determine the magnitude of the anticipated load on the shaft and bearings that may occur during operation. This load may be of different origin (mechanical or hydraulic) and therefore the resulting value is usually the sum of contributions from several sources. The hydrodynamic forces acting on the impeller contribute the most to the load on the rotor parts. These forces are the result of the interaction of rotor parts (impeller) and the stator parts (volute). Their vectors are usually decomposed into axial and radial directions, as implied by the principle of load trapping in the bearings. The magnitude of the hydrodynamic forces affects the sizing of the basic structural elements, which include the shaft (determination of its diameter) and the bearings (type and size). Therefore, it has an impact on the commercial parameters of the pump such as price and overall lifetime [6]. It is clear that the ability to determine these forces is a necessary part for the designer if he wants to optimize the design of the pump. The nature of the hydrodynamic forces is determined by the interaction of the impeller-volute. Since the flow at the impeller-volute interface is non-stationary, the resulting forces also have a pulsating character, as a result of which the spreading vibrations are accompanied by acoustic phenomena [10], [5]. Hydrodynamic pumps are machines that have undergone a rather intensive development over the last hundred years and exist in dozens of basic designs, differing in impeller type, distributor type, number of stages, number of blades, etc. This is because there is no uniform procedure for calculating radial and axial loads, but it is necessary to consider each pump separately. Dealing with this issue, we can find in the literature presentations of many results of either experimental or numerical research. Some of the fundamental works in the field of radial forces research are [1], [2], [12] and [13]. The researchers' attention is mainly devoted to spiral hydrodynamic pumps. The spiral as a diffuser has an asymmetrical shape and causes asymmetries of pressure and velocity fields in non-optimal modes.

This creates radial components of hydrodynamic forces of considerable magnitude acting on the impeller. Csanady [2] gives a theoretical analysis of radial forces caused by a flow in the spiral. His approach is restricted to a spiral of logarithmic shape, and he proposes a procedure for calculating the magnitudes of these forces. The results of the calculations are compared with the experimental results of [13]. The article [13] by Iversen et al. is devoted to the experimental investigation of radial forces in a spiral pump. In addition to the results of force measurements, a model is developed to determine the pressure distribution at the inlet of the volute at non-optimal modes. The asymmetry of the pressure distribution outside the optimum mode is the cause of the radial force. A procedure for calculating the radial force is also given and compared with experimental data. In the article [1], Biheller presents experimental results for three different types of spirals. We can also find semi-empirical and empirical formulas approximating the measured data there. Adkins and Brennen [12] in their article deal with experimental and theoretical analysis of the causes of radial forces in a spiral pump. The authors compiled a mathematical model of flow in the impeller – spiral system, which is used to calculate the pressure fields behind the impeller and the associated radial forces. An experimental device made for the measurement of radial forces and pressure fields behind the impeller is presented. An extensive comparison of the results, measured and calculated pressure fields and radial forces, is presented. An example of reducing radial loads using a double spiral is presented in [8]. A pump with a long shaft used for fluid transport is described. The hydraulic load causes deflection of the shaft, which leads to a reduction in operational stability. By CFD methods, it is shown in this work that these undesirable phenomena can be eliminated by using a double spiral (by symmetrizing the flow fields). Similarly [11], describes the development of a multichannel spiral, which aims to significantly reduce the radial load on the impeller caused by hydrodynamic forces. The authors reached the final solution using computer simulations of the flow in the pump. In general, the work of determining the radial force on the impeller is either purely computational (using e.g., CFD methods) [4] or a combination of experiment and calculation (based on CFD methods or the flow hypothesis itself). In some publications, the experiment is technologically simpler when the pressure force is calculated from the measured pressure fields, or only the measured pressure fields themselves are presented [3], [7]. A more technologically demanding experiment is represented by work where the radial force is directly measured using strain gauge force or stress sensors. This group includes articles [9], [12], [1], resp. [13]. In [4], the first method of experimental determination of force is referred to as indirect and the second method is referred to as direct.

2 Radial force analysis

The determination of the hydrodynamic radial force is an essential step in the design of spiral hydrodynamic pumps. The designer has several options available today – for example, he can use empirical or semi-empirical equations of various authors (e.g., Stepanoff or Biheller [1]) which are now considered classical in the literature. These relations have been derived under certain general assumptions, that may not be satisfied. Their accuracy is therefore questionable; they can only be used for preliminary analysis. Another possibility is to determine the force for a particular impeller shape. In Fig. 1 [1] shows the silhouette of the radial impeller. The radial force acting on an impeller of this shape is a combination of the pressure action on its outer surface and the hydrodynamic effect of the flowing fluid due to a change in its momentum. The components of the radial force F_x and F_y can be expressed by relations (1), (2) respectively.

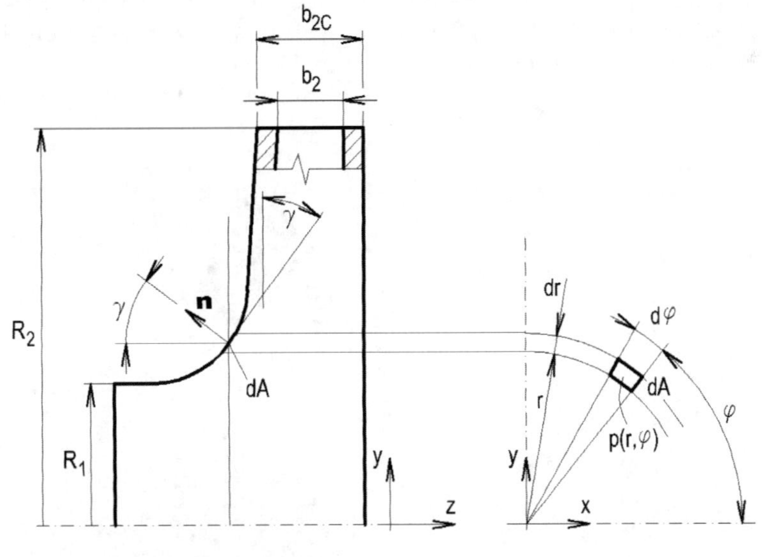

Fig. 1. Contour of the radial impeller [1].

$$\boldsymbol{F_x} = -b_{2C}\, R_2 \int_0^{2\pi} p_2(\varphi)\, \cos(\varphi)\, \mathrm{d}\varphi \tag{1}$$

$$\boldsymbol{F_y} = -b_{2C}\, R_2 \int_0^{2\pi} p_2(\varphi)\, \sin(\varphi)\, \mathrm{d}\varphi \tag{2}$$

where b_{2C}, R_2 and φ are geometric parameters evident from Fig. 1.; p_2 is the pressure (index 2 means the impeller output).

These fields need to be specified – i.e., by calculation or measurement. The designer is most satisfied if he has an efficient calculation procedure. In [13] a one-dimensional spiral flow model is proposed for this purpose. The flow is described by an ordinary differential equation whose solution gives the pressure distribution at the outlet of the impeller as a function of flow. The idea of flow is based on the fact that the pressure distribution is strongly influenced by the mixing process and frictional losses in the volute. However, this approach does not provide satisfactory results because the experiment and calculation do not match quantitatively low models of this type. Based on some hypothesis it will always be incomplete and will not sufficiently reflect the real phenomena taking place in the pump. Therefore, CFD methods have become indispensable in refining the calculation [10], [4], [3], [8], [9]. The advantage of the CFD approach is the possibility of obtaining such parameters, the measurement of which would be very technically demanding, such as shaft deflection (article [8]) or non-stationary pressures and radial forces [8], [4]. On the other hand, the CFD computation is dependent on the choice and setting of the mesh, turbulence model, boundary conditions, etc. Therefore, the results must be compared with the measurements at least for selected cases in order to generalize the procedure. Such a comparison is provided e.g., by article [9].

3 Experiment and measurement method

3.1 Model pump

The impeller selected for the experiment was developed and manufactured at the Department of Hydraulic Machines. The characteristic dimensions of the meridional section and the shape of the impeller are evident from Fig. 2 and the spiral tapered diffuser in Fig. 3. The operating parameters, along with the speed-acceleration, are evident from Tab. 1. The impeller is made of a combination of materials steel (rear disc), plexiglass (blades) and dentacryl (front disc). The spiral is made by wax core casting technology using dentacryl material.

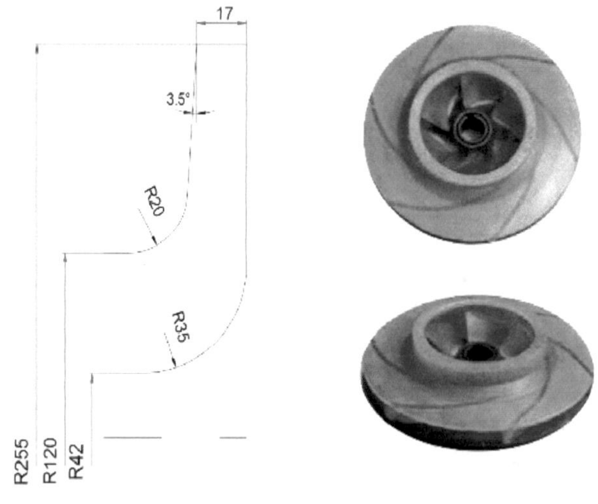

Fig. 2. Meridian section and impeller.

https://doi.org/10.1051/matecconf/202236904002

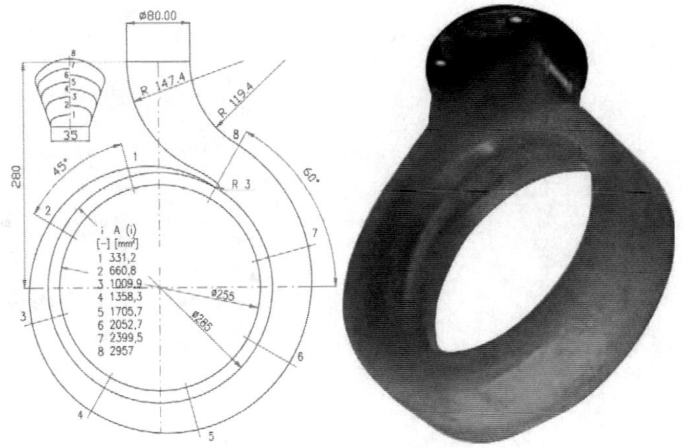

Fig. 3. Hydraulic spiral design and overall view.

Table 1. Main pump parameters.

Parameter	Sign	Value
Flow rate	Q (m^3/h)	20.31
Head	H (m)	13.78
Rotational speed	n (r/min)	1200
Specific speed	n_b	0.072
Efficiency	η (%)	77.13
Blade number of guide vane	Z_V	6

3.2 Radial force measurement method

The chosen method of radial force measurement consists in releasing the shaft from its mechanical and hydraulic parts so that it can perform a spherical motion around the rear bearing and then transferring the applied force induced in the front bearing outside the pump body to the three strain gauge force transducers.

The bearing is housed in a sleeve that is released from the bearing housing with sufficient radial clearance. Three thrust rods are fixed in the sleeve at an angle of 120° to each other, which are in point contact with the tapered tip of the strain gauges by means of limiting screws. The transfer of radial force to the strain gauges is affected only by the pull of the prestressed tension rods without contact with the body of the bearing housing.

Fig. 4. Drawing and fixing the sensors on the model pump.

3.3 Experimental circuit

For the purpose of this experiment, a closed test circuit was used. The circuit starts in a tank of fluid whose level is above the shaft axis. The liquid from the tank flows through the pipe *Pt1* to the slide on the suction *P1*, where it is fed through the pipe *Pt2* to the pump inlet by means of a flexible hose *Phs*. A pressure tap is located in the *Pt2* pipe for pressure measurement, which is led to a manometer for measuring the suction pressure (p_s). A *Pt3* pipe is connected to the flange of the volute with a pressure tap connected to a pressure gauge for measuring the pump discharge pressure (p_v). The pump flow is controlled by means of a shut-off slide *P2* attached to the *Pt3* pipe. The speed control of the pump during its operation is carried out by changing the speed of the electric motor by means of the thyristor control *TR*. The circuit from the pump discharge continues through a flexible hose *Phv* to three parallel apertures with pipe diameters ϕ80, ϕ100 and ϕ150. In each aperture, a standardized orifice for flow measurement with the corresponding diameter is incorporated in the pipe. By selecting an aperture with a suitable pipe diameter, it is possible to reduce losses in the system and thus ensure that the performance of any pump alternative is measurable beyond the optimum point. The circuit is closed by a *Pt4* pipe which returns the liquid back to the collection tank.

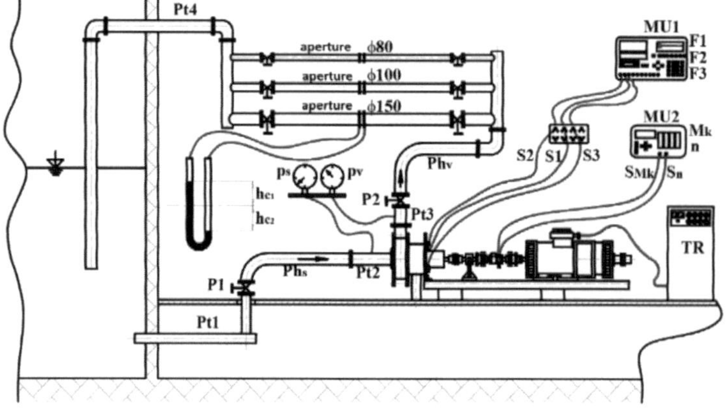

Fig. 5. Experimental circuit.

Mass flow measurement

The mass flow rate was measured using the pressure differential upstream and downstream of the orifice on a given pipe. The pressure differential was determined from the deflection (h_{C1}, h_{C2}) of the mercury U-gauge and the corresponding orifice constant (K_C).

$$Q_m = K_c \cdot \sqrt{h_{C1} + h_{C2}} \quad \left[\frac{m^3}{s}\right] \tag{3}$$

Specific energy measurement

The specific energy is defined by the pressure, kinetic and positional energy between the inlet and outlet of the pump.

$$Y_m = \frac{p_v - p_s}{\rho} + \frac{v_v^2 - v_s^2}{2} + (h_v - h_s) \cdot g \quad \left[\frac{J}{kg}\right] \tag{4}$$

Power measurement

The transfer of torque to the pump shaft is carried out by means of an elastic coupling with a rubber strain member, which allows possible small misalignment of the pump shaft due to its loading by radial force.

$$P_m = 2 \cdot \pi \cdot n \cdot M_k \quad [W] \tag{5}$$

Determine the pump efficiency using the equation:

$$\eta = \frac{\rho \cdot Q \cdot Y}{P} \quad [-] \tag{6}$$

4 CFD prediction of radial force

Numerical analysis using CFD method was performed to find the radial hydraulic force on the impeller with six blades. The three-dimensional model for the fluid body of the centrifugal pump is built with CATIA V5. Ansys CFX commercial software was used to generate the mesh, perform the numerical calculations and create the computational model. First, a model of the impeller with the spiral was generated in CAD and then discretized into a mesh using Ansys Meshing. The straight section of the suction pipe must be long enough for the correct definition of the boundary conditions, because in some cases there are intense secondary phenomena on the suction side of the radial pump [18] and the fluid velocity at the inlet becomes stable if the length of the inlet pipe is more than twice the diameter of the inlet [19]. In this case, the length of the straight section of suction pipe was equal to five times the inlet impeller diameter.

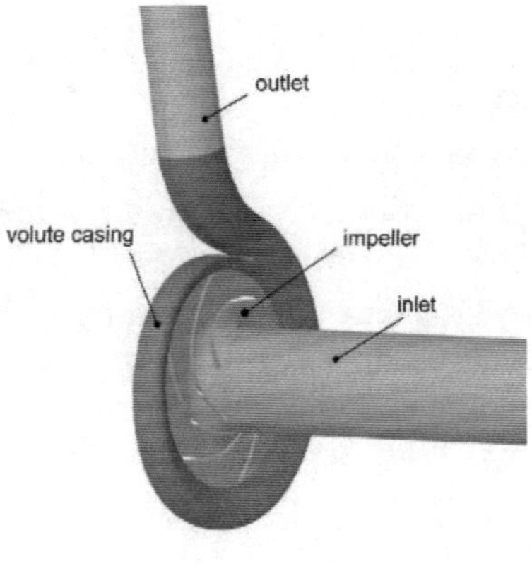

Fig. 6. Assembly depiction.

The mesh was generated as unstructured and consisted of tetrahedral elements – impeller and volute, and hexahedral elements – suction and discharge pipes Fig.7. The total number of grid elements was 5.9 M (suction pipe: 0.4 M, impeller: 2.5 M, spiral with discharge pipe: 3 M), and the element size was set to achieve a balance between calculation accuracy and calculation time, because the rapid increase in the number of grid elements puts a burden on the computational device. As can be seen on the Fig. 7, areas close to the walls and in the gaps between the rotor blade tip and the pump body, the computational mesh was significantly densified.

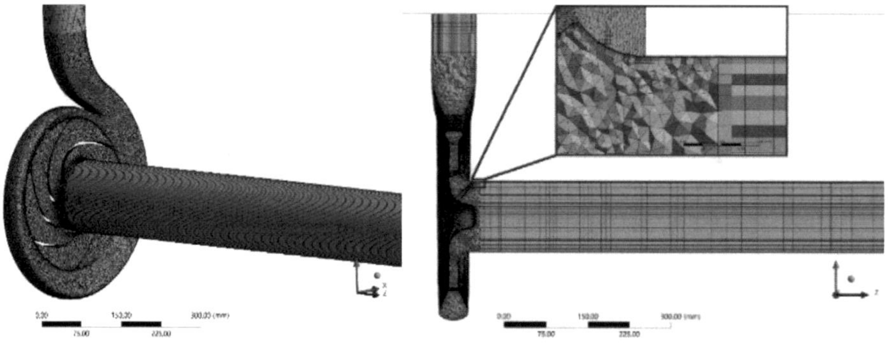

Fig. 7. Mesh of different parts.

The simulation model was divided into three domains: A – suction pipe, B – impeller and C – spiral casing with discharge pipe. By studying the strong interaction between the impeller blade and the spiral tongue [16], a non-stationary numerical analysis (transient) was performed. This eliminated the negative clocking effect [14] at the stator-rotor interface, especially at the narrowest point between the impeller and the spiral nose. The flow model was based on the solution of the RANS equations for the flow of an incompressible fluid. In this work, Menter's turbulent shear stress transport (SST) model was implemented. According to [15] Mentor's SST model appeared to be a suitable choice for flow in a multi-blade centrifugal pump. The interfaces between the rotor and stator parts of the computational domain were of the Transient Rotor Stator type.

The choice of the time step Δt for the transient simulation was set to 4.16667×10^{-4} s. One time step corresponded to a 3° rotation of the impeller, thus 120 transient results are included to calculate one revolution of the impeller. At least 8 full impeller revolutions were required for each operating point to steady the flow, which represents 960 iterations. During the calculations, the radial forces in the x and y directions are monitored and recorded at each time step. The radial force F_{rad} value was calculated based on the procedures implemented in the CFX software or CFD Post applied in the CFD simulation. General Gird Interface (GGI) is used to interface the dynamic and static parts during CFX pre-processing.

A steady-state calculation with the same turbulent model and boundary conditions was also performed to compare the time-steady state with the time-unsteady state.

No-slip boundary conditions are enforced on all solid walls. The following boundary conditions were applied: mass flow rate and direction of the absolute velocity vector at the inlet to the computational domain, static pressure at the outlet of the computational domain equal to 1 atm (101 325 Pa), rotor speed 1200 rpm. The working medium was water.

5 Results and discussion

The characteristics of the head and efficiency of the centrifugal pump obtained from the experiment are shown in Fig. 8, and these were compared with those obtained from the unsteady numerical simulation. It can be observed in Fig. 8 that the simulated head and efficiency of the centrifugal pump were slightly higher than the experimental values, but both have the same trend. The experimental and simulation values of the specific energy and efficiency at the design flow rate are $Y_e = 133.913$ J/kg and $\eta_e = 77.10$ % and $Y_s = 132.2$ J/kg and $\eta_s = 80.60$ %, respectively. The measured efficiency includes mechanical losses, which were not considered in the simulation. This accounts for some of the observed differences between experimental and simulation results.

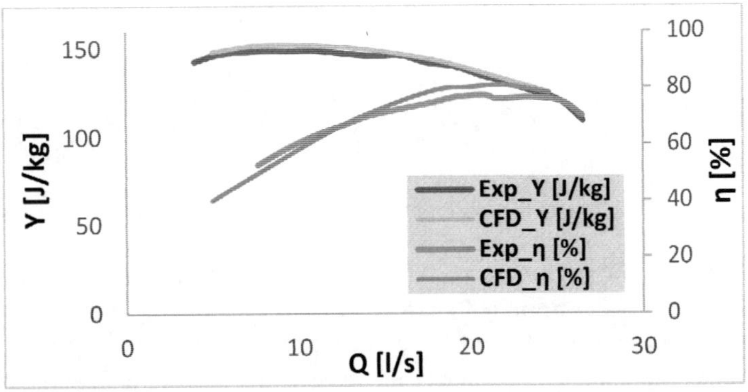

Fig. 8. Characteristics of the centrifugal pump.

The radial force on the impeller is a vector with magnitude and direction and it changes with the rotation of impeller. In order to compare the radial force under different conditions, the radial force vector's modulus at every time-step in the last calculation cycle is extracted, and the average values of all the radial force modulus are calculated based on these data. The average values of the radial force under different conditions are displayed in Fig. 9. It is obvious that a unsteady simulation has achieved more accuracy with experimental data, this can cause a large dependence of the impeller position to the stator position.

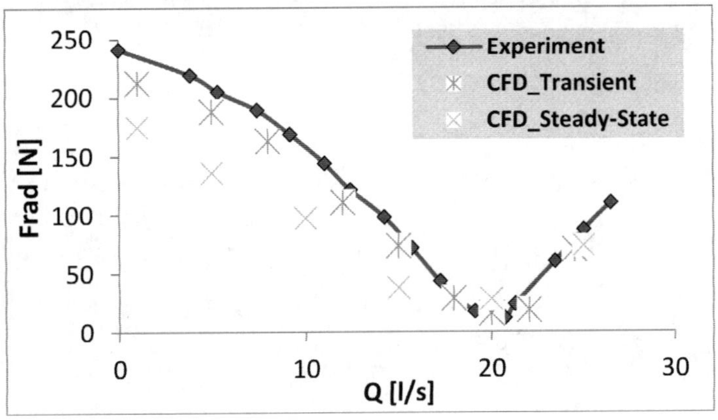

Fig. 9. Radial force magnitude.

Fig. 10 gives the distribution of radial force vectors in the last rotation cycle. Every point in the graph represents the radial force of the impeller at one instant. It can be seen that the direction and magnitude of the radial force vector change dramatically with time. The vector distribution of radial force under different flow rate in a cycle is regular and (0,0) is the symmetric centre of the vector distribution for all the cases. The radial force vectors are mainly in 6 regions, and the number is same as the number of blades of the impeller. This indicates that most radial forces are greater by non-operating conditions.

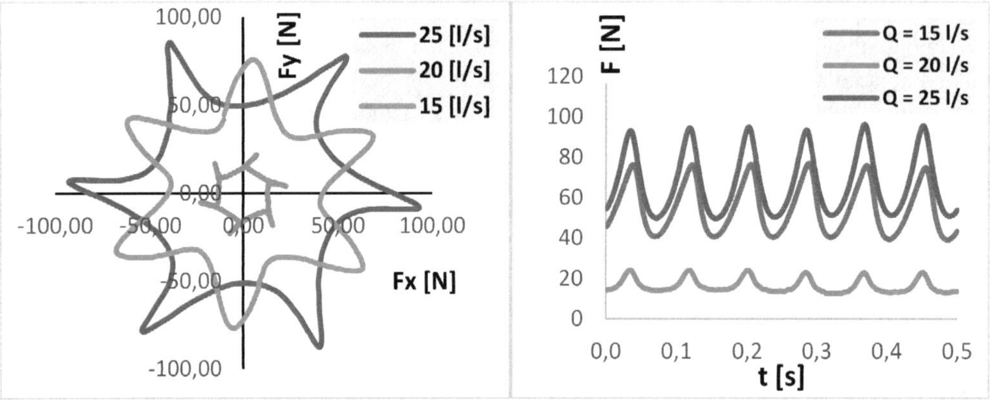

Fig. 10. Distribution of radial force vectors and time domain curves of radial force pulsation.

Fig. 11 shows the time history of instantaneous unsteady pressure contours at the boundary surface near the impeller outlet. The instantaneous non-stationary pressure contours for one revolution τ of one of the six pump impeller blades are compared. This rotation is divided into six steps to clarify changes in flow structure with time during one revolution of the impeller. In the reference design, high-pressure zones occur widely on the boundary surface near the impeller outlet, as shown in Fig. 11(e), and a high-pressure zone caused by impeller-volute interactions becomes gradually larger. Consequently, this results in the unbalancing phenomena, along with the fluid-induced vibrations caused by unsteady radial forces, throughout the annulus passage area of the pump.

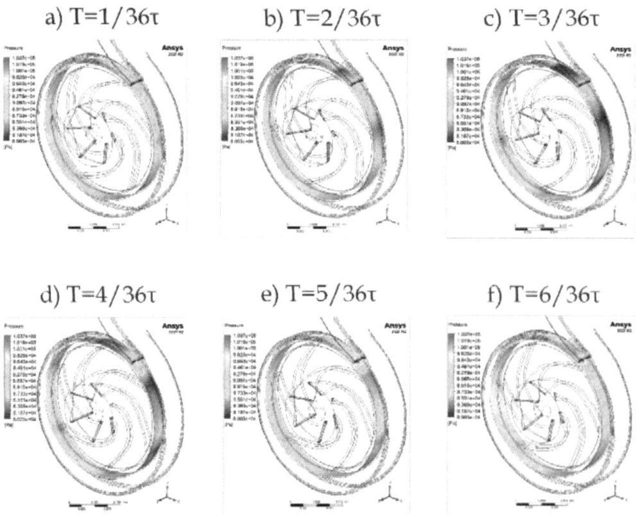

Fig. 11. Unsteady pressure contours during one revolution of the impeller.

The procedure of analysis of measured data in the frequency domain is based on a group of mathematical algorithms of fast Fourier transformation (FFT). The frequency domains of radial force pulsation of the impeller are shown in Fig. 12. The amplitudes of the peaks at 20 l/s are less than at 25 l/s and also at 15 l/s. The amplitudes of the main peaks appear by the first blade passing frequency (BPF) [20], which is 120 Hz. The pulsation frequency for the highest amplitudes, which indicates the main factor of the radial force fluctuations is the interaction between impeller and volute. This is mainly due to the small gap between rotary impeller and stationary volute, the gap is just 14 mm.

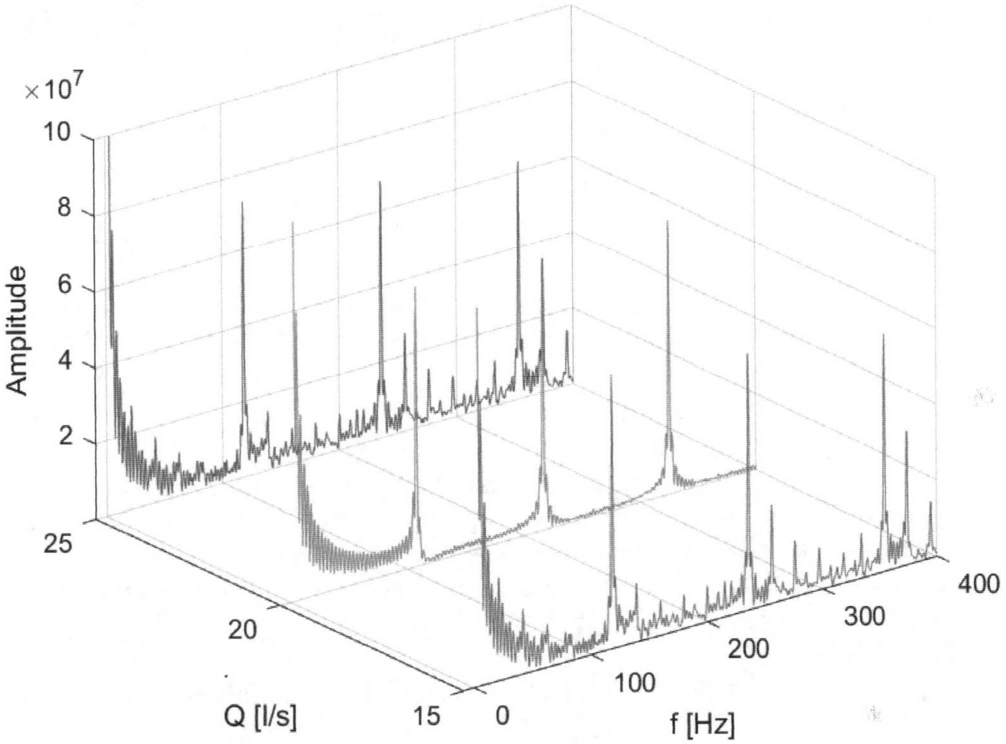

Fig. 12. Frequency domains of the impeller.

6 Conclusion

In this article, the behaviour of the hydrodynamic force inside a centrifugal pump under different operating conditions was investigated. The investigation of the radial force of the pump was carried out experimentally with the support of CFD methods. The analysis of the results enables to drawn the following conclusions.

1. Unfortunately, steady-state calculations cannot capture all the interaction effects with complete fidelity. Therefore, transient simulation was used for the calculations, the accuracy of which is satisfactory for the radial force investigation process, as the CFD results showed (Fig. 9) sufficient agreement with the experimental measurements. However, the calculated values are lower than the measured ones. The reason for this phenomenon may be due to the assumption that the simulation does not took into account all the processes occurring during real fluid flow in a multi-blade pump (e.g., backflow in the gap between the blade and the front disc). The pump performance parameters show good agreement with the measured curves (Fig. 8) but the efficiency curve is slightly higher. This may result in the CFD not considering mechanical losses. The measurement confirmed the fact that the maximum values of the pressure amplitude in the impeller are when the pump is operated outside the design conditions.

2. The radial force and vibration displacement of the impeller were lowest at the design flow rate and increased at off-design flow rates. The variation trend of the radial

displacements in the X- and Y-directions was consistent with that of the radial forces. The radial force vectors are in a centre of symmetric distribution, which mainly lie in 6 regions, corresponding to the number of blades of the impeller. The frequency for the largest amplitude is always in steps of 120 Hz, which is due to the periodic motion of pump impeller rotation.

References

1. Biheller H. J.: Radial Force on the Impeller of Centrifugal Pumps With Volute, Semivolute, and Fully Concentric Casings, Journal of Engineering for Power, July 1965, 319-322
2. Csanady G. T.: Radial Forces in a Pump Impeller Caused by a Volute Casing, Journal of Engineering for Power, October 1962, 337-340
3. Barrio R., Fernández J., Blanco E., Parrondo J.: Estimation of radial load in centrifugal pumps using computational fluid dynamics, European Journal of Mechanics B/Fluids 30 (2011) 316–324
4. Zou Z., Wanga F., Yao Z., Tao R., Xiao R., Li H.: Impeller radial force evolution in a large double-suction centrifugal pump during startup at the shut-off condition, Nuclear Engineering and Design 310 (2016) 410–417
5. Yu-qin W., Ze-wen D.: Influence of blade number on flow-induced noise of centrifugal pump based on CFD/CA, Vacuum 172 (2020) 109058
6. Godbole V., Patil R., Gavade S.S.: Axial Thrust in Centrifugal Pumps – Experimental Analysis, 15th International Conference on Experimental Mechanics, Porto/Portugal, 22-27 July 2012
7. Meng D., Jiang T., Deng H, and Hou G.: Numerical Simulation Research on Radial Force of Centrifugal Pump with Guide Vanes, Hindawi Shock and Vibration Volume 2021, Article ID 6638123, 10 pages, https://doi.org/10.1155/2021/6638123
8. Kang C., Zhua Y., Lia Q.: Effects of hydraulic loads and structure on operational stability of the rotor of a molten-salt pump, Engineering Failure Analysis 117 (2020) 104821
9. Karaskiewicz K., Szlaga M.: Experimental And Numerical Investigation Of Radial Forces Acting On Centrifugal Pump Impeller, Archive of Mechanical Engineering Vol. LXI 2014 Number 3 10.2478/meceng-2014-0025
10. Cui B., Li J., Zhang C., and Zhang Y.: Analysis of Radial Force and Vibration Energy in a Centrifugal Pump, Mathematical Problems in Engineering Volume 2020, Article ID 6080942, 12 pages https://doi.org/10.1155/2020/6080942
11. Alemi H., Nourbakhsh S. A., Raisee M. and Najafi A. F.: Development of new "multivolute casing" geometries for radial force reduction in centrifugal pumps, Engineering Applications of Computational Fluid Mechanics, 2015 Vol. 9, No. 1, 1–11, http://dx.doi.org/10.1080/19942060.2015.1004787
12. Adkins D. R., Brennen C. E.: Analyses of Hydrodynamic Radial Forces on Centrifugal Pump Impellers, Transactions of the ASME, Vol. 110, MARCH 1988, p. 20-28
13. Iversen H. W., Rolling R. E., Carlson J. J.: Volute Pressure Distribution, Radial Force on the Impeller, and Volute Mixing Losses of a Radial Flow Centrifugal Pump, Transactions of the ASME, April 1960, p. 136-143
14. Chu W., Wang Y., Tan L.,Xu Y., Numerical investigation of clocking effect on a centrifugal pump with inlet guide vanes, Emaraldinsight, 33 471 (2016)

15. Wang Y., Ding Z., Influence of blade number on flow -included noise of centrifugal pump based on CFD/CA, Elsevier, (2020), p. 1-8
16. Chalghoum I., Elaoud S., Kanfoudi H., Akrout M., The effects of the rotor-stator interaction on unsteady pressure pulsation and radial force in a centrifugal pump, Springer (2018), p. 672-681
17. Karaśkiewicz K., Slaga M., Experimental and numerical investigation od radial forces acting on centrifugal pump impeller, Manuscript (2013), p. 445-454
18. Zou Z., Wang F., Yao Z., Tao R., Xiao R., Li H., Impeller radial force evolution in a large double-suction centrifugal pump during startup at the shut-off condition, Elsevier (2016), p. 410-417
19. Wei-dong C., Xun D., Qi-xiang H., Effect of impeller reflux balance holes on pressure and axial force of centrifugal pump, Springer (2015), p. 1695-1706 1699
20. Kim, J. H., Choi, Y. S., State-of-the-Art Design Technique of a Single-Channel Pump for Wastewater Treatment, IntechOpen (2018)

This work was supported by the Slovak grant agency KEGA, project No 016STU-4/2022.

https://doi.org/10.1051/matecconf/202236904003

Vortex Identification in a Free-Surface Flow Problem Solved Using SPH

Petr Jančík[1,*], and *Tomáš* Hyhlík[1]

[1]Department of Fluid Mechanics and Thermodynamics, Faculty of Mechanical Engineering, Czech Technical University in Prague, Technická 4, 160 00 Prague 6, Czechia

Abstract. This article focuses on vortex structure identification methods and their implementation into smoothed particle hydrodynamics (SPH) framework. The most common criteria based on the local velocity gradient tensor analysis like Q-criterion, Δ-criterion, and λ_2-criterion are introduced together with their implementation for SPH data. A two-dimensional 'double dam break' problem was chosen as a test case because it results in a violent transient free-surface flow with emerging and vanishing eddies of various sizes and intensities. Q-criterion and Δ-criterion results were virtually identical, whereas λ_2-criterion appeared to be the most restrictive in vortex identification, so it was the best in suppressing incorrect findings due to the numerical solution imperfection. The analysis of the problem was therefore conducted using the λ_2-criterion.

Keywords: *CFD; vortex; SPH; free surface*

1 Introduction

Vortices are characteristic structures that are present in almost every flow. They significantly influence the transport of momentum and other quantities in fluids. Even though a vortex seems like a relatively simple structure, there is not even a clear definition. One of the most intuitive definitions was given by Lugt [1]: a vortex is the rotating motion of a multitude of material particles around a common center.

However, this definition is vague and not directly applicable to flow analysis. Therefore, some objective criteria have been proposed. The most obvious thing to do is the evaluation of vorticity in the flow field, but it did not turn out to be precise because it cannot distinguish between rotation and shear deformation [2]. More elaborate methods usually work with the velocity gradient tensor. The most widely used are Q-criterion [3], Δ-criterion [4], and λ_2-criterion [5]. Other methods have been proposed as well, for example, Γ_2-criterion suitable for PIV data analysis [6], residual vorticity originating from the triple decomposition [7], or relatively recently Ω-criterion [8].

This paper aims to present the implementation of the picked vortex identification criteria (Q-criterion, Δ-criterion, and λ_2-criterion) for 2D SPH data and find the one which gives the best results. A subdomain of the solution of 'two collapsing columns of liquid' served for the criteria comparison. Then, the problem of two collapsing liquid columns was analyzed using

* Corresponding author: petr.jancik@fs.cvut.cz

the best performing criterion (λ_2). The origin, development, and final dissipation of the most noticeable vortices are presented.

2 Methods

2.1 Smoothed particle hydrodynamics

Smoothed particle hydrodynamics (SPH) is a mesh-free particle numerical method suitable for transient free-surface problems. Many interacting particles of constant mass represent the fluid. A system of ordinary differential equations determines the evolution of particle properties and their motion. The continuity equation and the momentum equation used in this work are the following:

$$\frac{\mathrm{D}\rho_i}{\mathrm{D}t} = \rho_i \sum_j \frac{m_j}{\rho_j}(\mathbf{v}_i - \mathbf{v}_j) \cdot \nabla W_{ij} \tag{1}$$

$$\frac{\mathrm{D}\mathbf{v}_i}{\mathrm{D}t} = -\sum_j m_j \left(\frac{p_i}{\rho_i^2} + \frac{p_j}{\rho_j^2} + \Pi_{ij}\right) \nabla W_{ij} + \mathbf{g} \tag{2}$$

where indices i and j denote different particles, ρ is density, p is pressure, m is mass, $\mathbf{v}$ is velocity vector, and $\mathbf{g}$ is gravity acceleration vector. The function W_{ij} is the weight function that mediates the interactions between particles. In this work, a truncated Gaussian smoothing function was used. The fluid is considered inviscid, but there is the artificial viscosity term Π_{ij} that serves for numerical stabilization [9].

In SPH, a common approach is to consider incompressible fluid as compressible one. Because of this, an equation of state is necessary to close the system of equations. In this work, we use

$$p = c^2(\rho - \rho_0) \tag{3}$$

where c is the numerical speed of sound and ρ_0 is the fluid density at zero gauge pressure. A too-small value of c would lead to very compressible fluid, and a too-large value would mean the necessity of prohibitively small integration steps. Usually, a good choice of c is when the variation in fluid density remains below 1 %.

Wall boundary conditions are modeled as free-slip using virtual particles according to [10]. The authors of this paper used a virtually identical solution method in their previous work [11]. For more details about the method, see, e.g., [12].

2.2 Vortex identification methods

In this section, the used vortex identification criteria are defined and their implementation for a 2D flow field. To do so, let us define the velocity gradient tensor

$$\nabla\mathbf{v} = \left[\frac{\partial}{\partial x}, \frac{\partial}{\partial y}\right]^T \cdot [u, v] = \begin{bmatrix} u_x & u_y \\ v_x & v_y \end{bmatrix}. \tag{4}$$

To evaluate the elements of the velocity gradient tensor in the SPH framework, a rather standard SPH procedure was used. For example, the result for the first element is

$$(u_x)_i = \sum_j \frac{m_j}{\rho_j}(u_i - u_j)\frac{\partial W_{ij}}{\partial x}. \tag{5}$$

The remaining elements can be found in the same way.

2.2.1 Q-criterion

In Q-criterion [3], vortices are defined by a positive value of the second invariant of $\nabla\mathbf{v}$. It can be written

$$Q = \frac{1}{2}[\mathrm{tr}(\mathbf{\Omega}\mathbf{\Omega}^T) - \mathrm{tr}(\mathbf{S}\mathbf{S}^T)]\,, \tag{6}$$

where $\mathbf{\Omega}$ is the anti-symmetric part of $\nabla\mathbf{v}$ and $\mathbf{S}$ is the symmetric part of $\nabla\mathbf{v}$.It can be interpreted as the excess of rotational motion over shear strain rate.

In two dimensions, it can be written as

$$Q = u_x v_y - u_y v_x - \frac{1}{2}\left(u_x + v_y\right)^2. \tag{7}$$

In this form, Q can be evaluated using formula 5. A vortex is identified where $Q > 0$.

2.2.2 Δ-criterion

According to Δ-criterion [4], vortices are present there, where the eigenvalues of $\nabla\mathbf{v}$ are complex numbers. In a 3D case, this condition is equivalent to

$$\Delta = \left(\frac{Q}{3}\right)^3, + \left(\frac{R}{2}\right)^2 > 0, \tag{8}$$

where Q is defined in equation 6 and $R = \det(\nabla\mathbf{v})$.

In a 2D case, the characteristic equation of $\nabla\mathbf{v}$ is

$$\lambda^2 - \lambda\left(u_x + v_y\right) - u_y v_x + u_x v_y = 0\,, \tag{9}$$

and the eigenvalues are complex when the discriminant of the characteristic equation is negative. Therefore, the criterium stands

$$\Delta = 4(u_y v_x - u_x v_y) + \left(u_x + v_y\right)^2 < 0. \tag{10}$$

2.2.3 λ$_2$-criterion

The λ_2-criterion [5] define vortex there, where the tensor $\mathbf{\Omega}\mathbf{\Omega}^T + \mathbf{S}\mathbf{S}^T$ has at least two negative eigenvalues. If the eigenvalues are ordered $\lambda_1 \geq \lambda_2 \geq \lambda_3$, then it is equivalent to $\lambda_2 < 0$.

In a 2D case, λ_2-criterion can be expressed [13]

$$\lambda_2 = u_y v_x - u_x v_y + \frac{1}{2}\left(u_x + v_y\right)^2 + \frac{1}{2}\left|u_x + v_y\right|\sqrt{\left(u_x - v_y\right)^2 + \left(u_y + v_x\right)^2}. \tag{11}$$

3 Results and discussion

3.1 Analyzed problem

The vortex identification methods were applied to the flow field resulting from the collapse of two liquid columns in the gravity field. The initial configuration is depicted in Fig. 1. The total number of particles was 60 000, and 3 seconds or the simulation are presented. The problem was solved as a free-surface problem, so the surrounding gaseous phase is omitted.

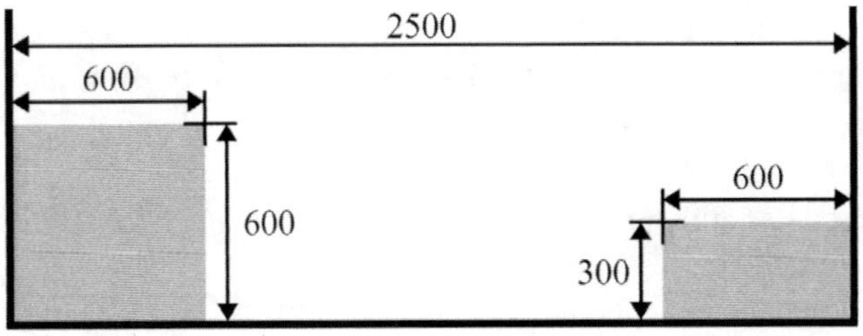

Fig. 1. The initial configutation of the analyzed problem. Dimensions in millimeters.

3.2 Method comparison

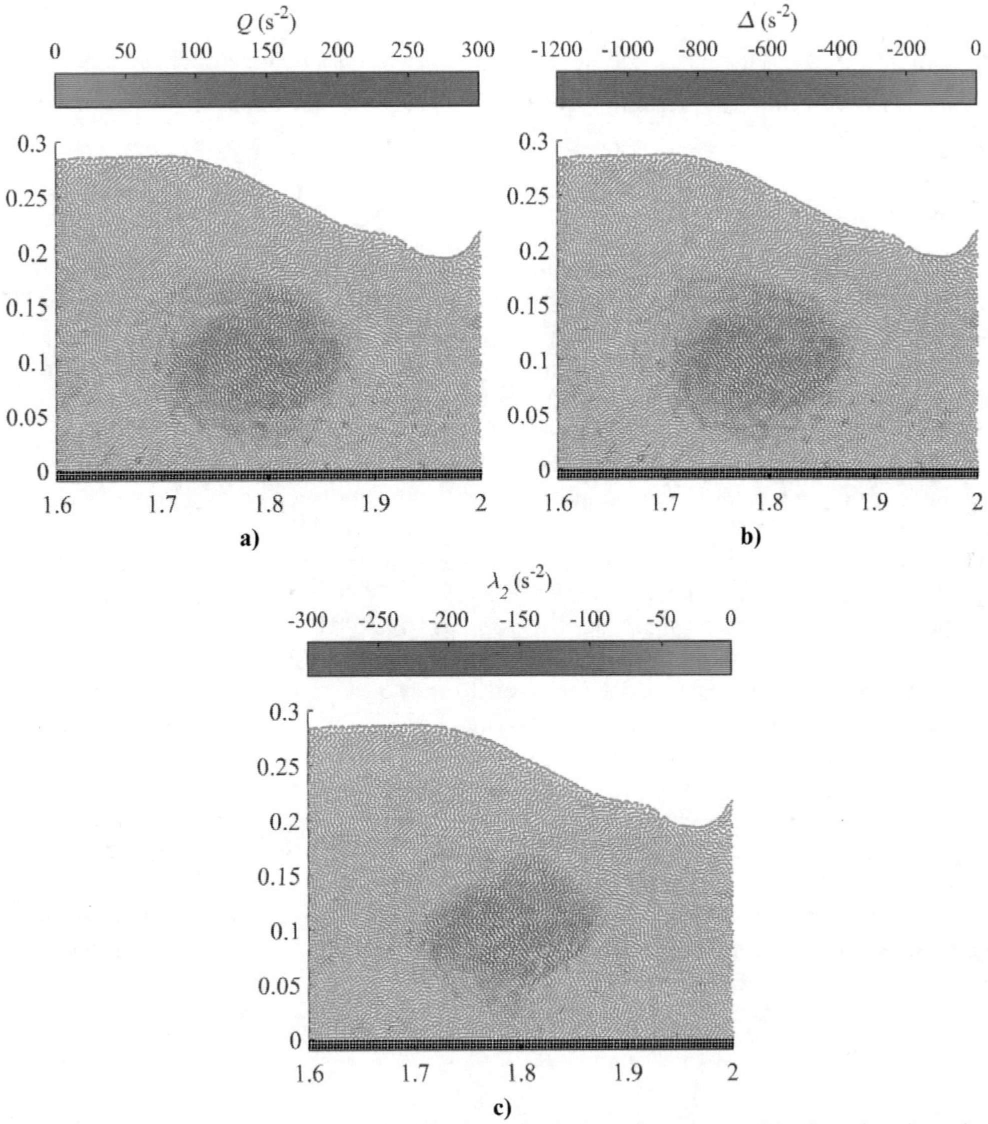

Fig. 2. The comparison of the vortex identification criteria for a chosen part of the flow domain at time 2 s. a) Q-criterion, b) Δ-criterion, c) λ_2-criterion. Dimensions in meters.

When analyzing equations 7, 9, and 11, it is noticeable that they are equivalent if the fluid is incompressible because the terms containing $(u_x + v_y)$ are zero (the continuity equation). All the discussed criteria thus reduce to

$$u_x v_y - u_y v_x > 0. \tag{12}$$

However, the used computational model considers a compressible fluid (very weakly), so the results may slightly differ. Fig. 2 illustrates the comparison of the criteria. The Q-criterion and Δ-criterion yield virtually the same results. Since they differ in the relative weight of the term $(u_x + v_y)^2$, the conclusion is that this term makes a negligible difference. Apart from the vortex identified in the middle (Fig. 2), there are small areas identified as vortices as well. However, these are numerical artifacts rather than actual vortices. The results by the λ_2-criterion are slightly different. This criterium tends to be more restrictive and identifies smaller areas than the previous, as it suppresses the numerical artifacts better. Even though the term $\frac{1}{2}|u_x + v_y|\sqrt{(u_x - v_y)^2 + (u_y + v_x)^2}$ is close to zero, it affects the result. From the compared criteria, the λ_2-criterion seems to yield the best results, so it was used in the following analysis.

3.3 Problem analysis

After the collapse of the two liquid columns, the surges met approximately in the middle of the tank. That resulted in a violent flow with rolling waves. The impacts of these waves were the general source of vortex structures that were identified in the flow field. Since the walls were a free-slip boundary condition, there were no vortices originating from the turbulent boundary layer. Chosen instances of the flow at the various time are presented in Fig. 3, where the most distinct vortices are marked for easier orientation.

At $t = 1$ s, there are no distinct vortical structures. There are some small areas of high λ_2, but they are only a result of the violent impact of the rolling wave on the surface, and they disappear rather quickly.

At $t = 1.5$ s, there appeared two vortices in the right part of the tank (Ia and Ib). In the middle of the tank, there were two cavities as a result of rolling wave impact, and the fluid around them was going to form another two vortices (II and III).

At $t = 2$ s, vortices Ia and Ib merged into one larger vortex, but the intensity decreased due to dissipation. Vortices II and III evolved from the cavities, and they were relatively strong and well defined. There were also two more cavities on the left (IV) and on the right (V) that became vortices as well.

At $t = 2.5$ s, vortex I was still there and well defined, while vortex III became considerably weaker, and vortex II vanished completely. Vortices are noticeable now but not very strong or well defined. Once again, another vortex was emerging as a result of a rolling wave impact (VI).

At $t = 3.0$ s, all the followed vortices became relatively weak due to dissipation, and only the most recently created vortex (VI) remained relatively strong. Overall, the occurrence and intensity of vortices decrease from this point as the flow becomes calmer.

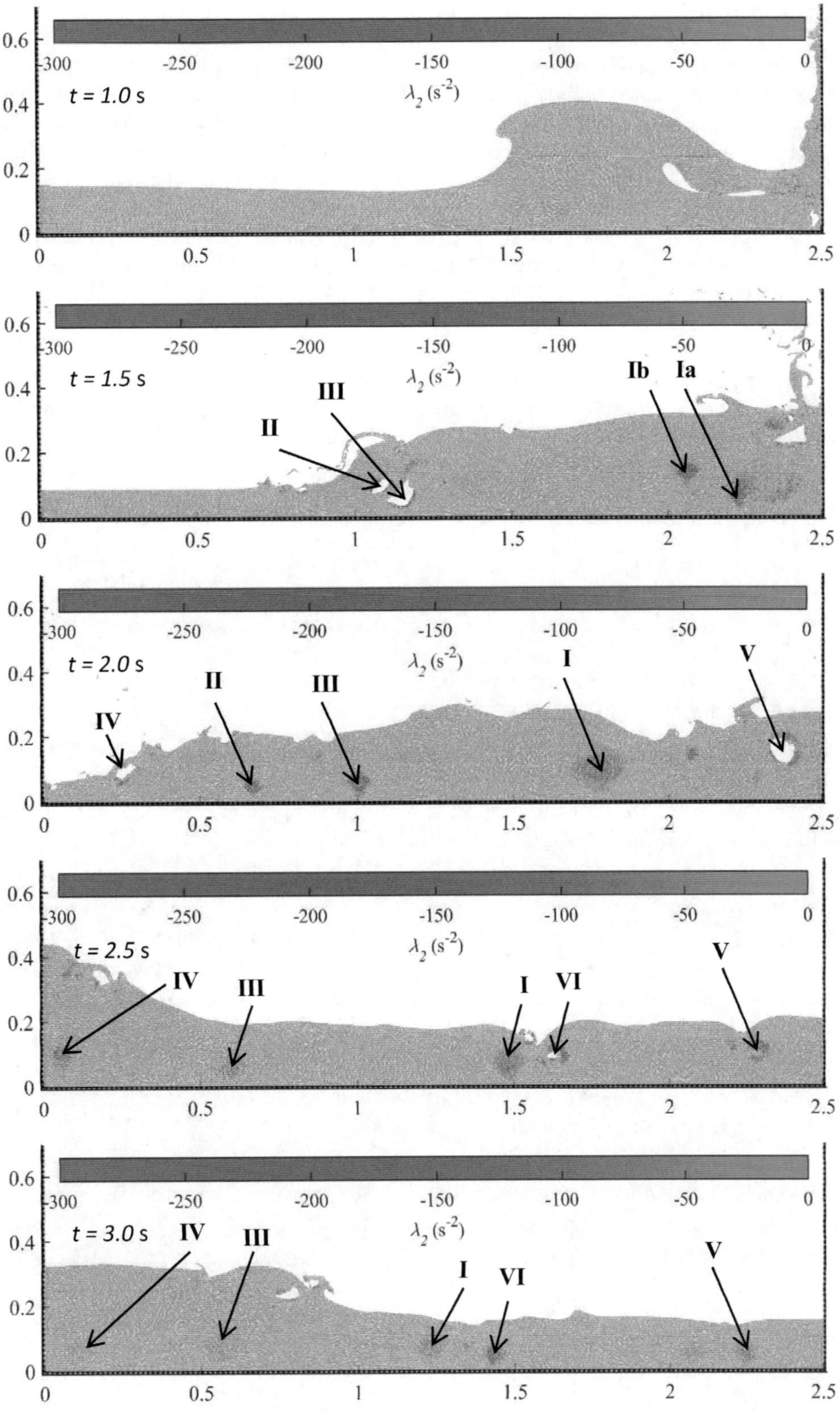

Fig. 3. The results of λ_2-criterion at various time instances with marked vortices. Dimensions in meters.

4 Conclusion

The most widely used vortex identification criteria (Q-criterion, Δ-criterion, and λ_2-criterion) were introduced together with their possible implementation for 2D SPH data. Quite expectably, they yielded almost identical results. However, the λ_2-criterion proved to be the most suitable because it managed to suppress errors most likely originating from the data imperfection.

The problem of two liquid column collapse was analyzed by the λ_2-criterion. It became apparent that the most significant vortices originate from the impact of rolling waves on the liquid surface. The boundary layer was not modeled, so no significant vortices were identified near the walls.

In the future, the criteria could be further refined to give better results, or some other criteria could be implemented as well. Not only methods based on velocity gradient tensor are available. A combination of these criteria with, for example, previously tested the Finite-time Lyapunov exponent (FTLE) for identification of Lagrangian coherent structures [14] may give remarkable results.

This work was supported by the Grant Agency of the Czech Technical University in Prague, grant No. SGS20/112/OHK2/2T/12.

References

1. Lugt, H. J. (1983). *Vortex flow in nature and technology*. Wiley.
2. Robinson, S. K. (1991). Coherent motion in turbulent boundary layer. *Annual Review of Fluid Mechanics*, 23, 601–639.
3. Hunt, J. C. R., Wray, A. A., Moin, P. (1988). Eddies, streams, and convergence zones in turbulent flows. *Center for Turbulence Research Report CTR-S88,* 193–208.
4. Chong, M. S., Perry, A. E., Cantwell, B. J. (1990). A general classification of three-dimensional flow fields. *Physics of Fluids A: Fluid Dynamics*, 2(5), 765-777.
5. Jeong, J., Hussain, F. (1995). On the identification of a vortex. *Journal of Fluid Mechanics*, 285, 69-94.
6. Graftieaux, L., Michard, M., Grosjean, N. (2001). Combining PIV, POD and vortex identification algorithms for the study of unsteady turbulent swirling flows. *Measurement Science and Technology*, 12(9), 1422-1429.
7. Kolář, V. (2007). Vortex identification: New requirements and limitations. *International Journal of Heat and Fluid Flow*, 28, 638–652.
8. Liu, C, Wang, Y.,Yang, Y., Duan, Z. (2016). New omega vortex identification method. *Science China Physics, Mechanics & Astronomy*, 59, 684711.
9. Monaghan, J. J., Gingold, R. A. (1983). Shock simulation by the particle method SPH. *Journal of Computational Physics*, 52(2), 374-389.
10. Adami, S., Hu, X. Y., Adams, N. A. (2012). A generalized wall boundary condition for smoothed particle hydrodynamics. *Journal of Computational Physics*, 231(21), 7057-7075.
11. Jančík, P., Hyhlík, T. (2019). Forces on vertical walls after dam break using weakly compressible SPH method. *AIP Conference Proceedings*, 2118(1), 30018.
12. Liu, G. R., Liu, M. B. (2003). *Smoothed particle hydrodynamics - A meshfree particle method.* World Scientific.

13. Chen, Q., Zhong, Q., Qi, M., Wang, X. (2015). Comparison of vortex identification criteria or planar velocity fields in wall turbulence. *Phisics of Fluids,* 27, 085101.
14. Jančík, P., Hyhlík, T. (2021). Finite-time Lyapunov Exponent analysis used on a free-surface flow problem solved by Smoothed Particle Hydrodynamics. *2nd International Conference on Fluid Flow and Thermal Science*, 108.

Adaptive Shading System as a Building Heat Load Reduction

Michal Masaryk[1,*]

[1]Institute of Energy Machinery, Faculty of Mechanical Engineering, Slovak University of Technology, Námestie slobody 2910/17 812 31 Bratislava, Slovakia

Abstract. This article deals with conceptual design and experimental implementation of active shading system to reduce building heat gains. First, the problem of building overheating in the summer period and the requirement of lighting in the building interior is explained. Then, the concept of the automated shading system is presented as an efficient yet simple solution to this problem. Along with that, the initial phase of experimental verification, namely the design and additives used, is described. In conclusion, the results of the initial phase are discussed and the future stages of the project along with its significance for the industry are explored.

Research background: The issue of overheating of utility buildings in the summer period and the requirement of sufficient lighting of their interior. Measurements in the respective literature and on-site in industrial buildings.
Purpose of the article: Explaining the conceptual design of an automated shading system as a remedy to the overheating problem and also to raise awareness of this widespread issue.
Methods: Direct measurements of the respective quantities in the industrial sites were conducted. Laboratory experiments along with measurements of several used shading solutions were conducted.
Findings & Value added: Some substances showed promising results in creating a translucent heat barrier for sunlight. Further solutions are planned to be tested.

Keywords: *shading; skylights; heat load; utility buildings; roofs;*

* Corresponding author: michal_masaryk@stuba.sk

1 Introduction

By large utility buildings, one big thermodynamic issue occurs on a regular basis. These natural, yet problematic phenomena are heat gains of buildings in the warm period of the year - from April to September in Slovakia. In this period of the year, solar radiation increases by so much that significant heat gains occur in the interior of large buildings. The buildings affected are mostly of industrial nature but other utility buildings, such as warehouses and workshops, are also severely affected. [1] [2]

The resulting overheating of buildings is mostly caused by construction materials and architectonic elements used. These do, unfortunately, allow for extensive penetration of solar heat into the interior of the building. The aforementioned problematic architectonic elements are first and foremost skylights on the roof. Skylights usually take up 10-15 % of the overall roof area yet they contribute at least 60 % to the overall solar heat gains. As for the construction materials, the traditional layout used is ill-suited for thermal insulation.[8] In fact, some of the best heat accumulators are commonly used by roof building, including asphalt and corrugated metal panels [2], [9]. While these provide great protection against the elements of weather, they have a disastrous effect in terms of the heat load reduction of the specific building. While all these facts hold true, it is virtually impossible to omit elements such as skylights when designing an utility building. (In addition to the redesign of existing buildings not being a financially viable option). [1]

The simple reason for the continued inclusion of glass surfaces in the design of buildings is lighting. Natural light and workplace illumination is a key parameter for optimal working conditions and long-term health of the employees. [6] The optimal value of illuminance (measured in lux) in the workplace is between 500 and 2000 [1]. These levels of workplace illumination can only be reached by using artificial lighting or, with much reduced expenses, by using natural lighting, e.g. from skylights. With skylights and other glass surfaces being the source of the vast majority of the heat gains in the warm months, the real challenge lies in designing a system that would allow for absorbing or reflecting the heat-bearing part of the solar radiation spectrum, the infrared radiation, while letting enough visible light into the interior of the building [11]. [3]

2 Adaptive shading system

Automated adaptive shading system, the layout and experimental design of which is the subject of the Institute of Energy Machinery at Faculty of Mechanical Engineering research team's work. Its purpose is to ensure the reduction of heat gains generated through glass surfaces of buildings, specifically skylights. This effect will be achieved through an effective combination of a specific physical principle, a simple design and an electronic automating system.

The operating principle will be based on creating a layer of fluid on the transparent surface, with an appropriate additive, that would ensure the filtering of the infrared spectrum of solar radiation while allowing for maximum permeability of visible light [7]. A liquid layer of only water cannot be used as its IR filtration capacity is severely limited (as shown in figure 1). While several approaches were considered, such as opaque fluid "strips" with variable width or using a specialized chemical compound in a very specific range of concentrations, eventually the most suitable approach was chosen - water-based shading solution with variable concentration of the additive.

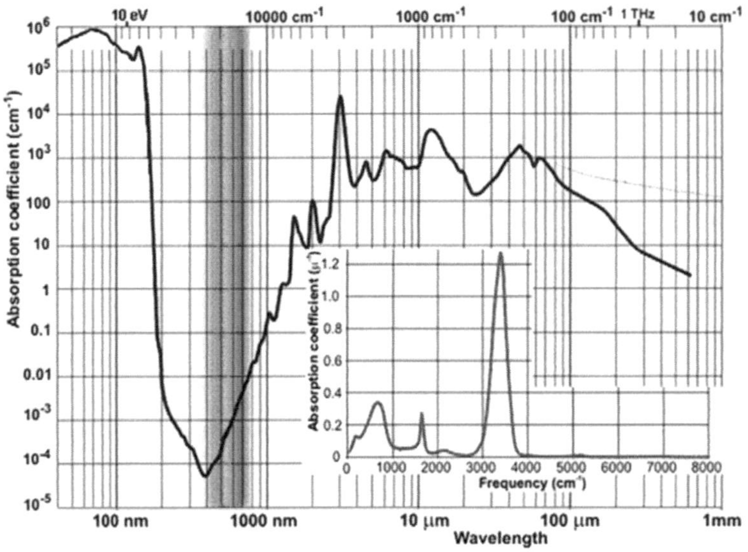

Fig. 1. Absorption of wavelengths in water [12].

2.1 System limitations

Before describing the experimental implementation of the aforementioned system, let us have a look at the practical limitations, in the realm of which the system must function in order for its usage to be financially and technically viable. There are three conditions that define how the system should operate.

The first is the requirement of simplicity and minimal maintenance. The reason for this is the capability to replace standard exterior shading systems - mostly electromechanical systems. [5] Next, a sufficient IR shading/visible transparency efficiency is required. In order for the adaptive shading system to be able to compete with hi-tech systems, such as active-layer windows [4], it has to be able to reach at least intermediate levels of performance. Finally, the system has to be innovative and resource-efficient. [4] For the shading system to be appealing to potential users, it has to have some element of technological innovation, traditionally coupled with optimal use of resources [8]. This requirement will be fulfilled by the electronic automation part of the system.

2.2 Experimental design and testing

In order to determine the appropriate additive to create an effective shading solution, several experiments with various chemical compounds and coloring agents were conducted. The chosen concept of the system was for it to use a sufficiently available additive in a water-based solution with variable concentration. Due to this, commercially accessible dyes and other coloring chemical compounds were used and tested on the prototype system. It is important to distinguish between variable concentration of the additive as a function of the shading system (electronically controlled, using separate tanks and pumps) and varying concentrations of the additive by the experiments (added manually and their effects being observed).

The experiments were conducted on a large, 130 cm x 160 cm window and on a smaller, 52 cm x 50 cm glass plate. During the winter months, a 65 W IR lamp and a regular 60 W table lamp were used as light sources for measurements. From March onwards, measurements were made on direct sunlight, as it was already available. In the case of each additive, solar irradiance E_e [W/m^2] and illuminance E_v [lux] were measured - at first without the added chemical compound, then with it. 3D model of the testing stand is shown on figure 2.

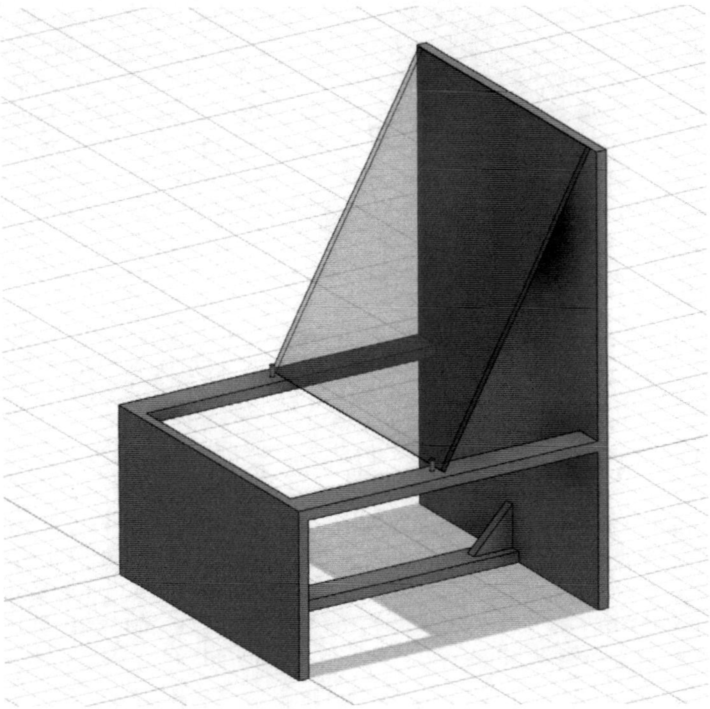

Fig. 2. 3D model of experimental stand.

2.2.1 Tested additives

Silver Paint

The first tested chemical compound added to the water solution was silver paint. The rationale behind its use was the presence of aluminum particles on which the paint is based. Similar to other metals, aluminum has high reflectivity in the IR spectrum [3] and having its particles in a jellied form was suitable for creating a water solution. Various concentrations were tested and the results are shown in table 1.

Table 1. Silver paint tests.

Substance	Irradiance E_e [W/m^2]	Illuminance E_e [lux]
Water 20 l	96	144
Silver paint 190 g	76	123
Silver paint 290 g	66	104
Silver paint 390 g	51	93

In general, the silver paint has reached satisfying efficiency in filtration of IR radiation. However, its use presented a whole set of other issues, which made it unsuitable to be used. The first being its toxicity - this trait makes it unsuitable due to health and environmental concerns. Next, it left a notable residue after use, even after washing the system with clean water, which is a significant technological obstacle.

Liquid Foil

The next tested substance added to the water solution was a specific type of paint "liquid foil". This substance was originally intended to be used as paint, applied in several layers. Experiments were conducted with both a painted dried glass surface as well as a surface with a liquid layer being poured onto it. Although the manufacturer did not state the exact formula of the substance, it was a playmaker-based compound that was non-toxic and soluble in water. Results of the measurements are shown in table 2.

Table 2. Liquid foil test.

Substance	Irradiance E_e [W/m²]	Illuminance E_e [lux]
Water 3l	498	4350
Liquid foil 20g	315	2603

After the initial experiment, the water solution with this additive was deemed unsuitable for general use. This was due to extensive residue that was hardly removable just a few minutes after the shutdown of the system. On the other hand, the painted surface, which was used as a comparison, did show notable improvement in filtering IR radiation and was kept as an efficiency benchmark.

Acrylic paint

Acrylic paint was chosen as an inexpensive and widely available alternative to other tested additives. Experiments were conducted with black, white, metallic orange and metallic yellow acrylic paints. Several concentrations were tested, while in some cases - for example after a very weak performance, it was redundant to continue the process with lower concentrations and the testing did not continue. Test results of acrylic paints are shown in table 3.

Table 3. Acrylic paints tests.

Substance	Irradiance E_e [W/m²]	Illuminance E_e [lux]
Water 3 l	114	129
Black acrylic 13 g	47	26
White acrylic 50 g	86	105
Substance	**Irradiance E_e [W/m²]**	**Illuminance E_e [lux]**
Water 3 l	152	210
Orange acrylic 60 g	134	191
Yellow acrylic 50 g	146	196

The measured shading efficiency of acrylic paints was mixed. It was heavily dependent on the color shade used. The metallic colors had the poorest results and were deemed unsuitable due to very weak performance. On the other hand, white acrylic paint yielded significantly better results, though a higher volume was needed. Finally, black acrylic paint yielded the best results, reaching the performance of white paint while requiring ⅓ of its volume. Black acrylic paint was deemed effective and its performance and effects will be further tested in future experiments.

3 Conclusions and further project continuation

Of the conducted experiments, a clear conclusion can be drawn - there is a potential for a recycled liquid-based shading system that will use simple yet effective water solution in order to provide shading of glass surfaces of buildings, without reducing the input of visible light. It is important to note, that the aforementioned experimental project is in its initial phase and the second round of experiments is being conducted.

Of the tested additives and the resulting water solutions, some have been shown to have little suitability and value for the project, mostly due to their poor performance or the capability to filter visible light instead of IR light - a property inverse to the desired one. On the other hand, by some substances it was demonstrated that they have a satisfying capacity to achieve the desired IR filtration without significantly reducing the illuminance of the building interior.

As was mentioned in the description of the designed system, its significant part is an automation and control system that would ensure its innovative and effective function. This electronic automation part is yet to be designed and its function and specifics yet to be determined. However, just like the underlying operating principle of the shading system - simplicity, the automation system is likewise projected to be built on an existing widely available control platform, such as Arduino.

The potential of the concept and its capacity to supplement or replace existing exterior shading systems are vast [7], to say at least. With the extensive use and fast-paced implementation of automation coupled with an increased focus on environmentalism and energy-efficient systems [10], projects such as these are becoming the center of attention of business and scientific communities alike.

References

1. Hraška, J. (2009). *Tieniaca Technika Budov.* Slovenská Technická Univerzita v Bratislave
2. Huang, H. (2015, September 14). *Solvothermal synthesis of Sb:SnO2 nanoparticles and IR shielding coating for smart window* ScienceDirect. https://www.sciencedirect.com/science/article/abs/pii/S0264127515304160
3. Chiang, F. B. Y. (2016, November 10). *Tungsten doped VO2/microgels hybrid thermochromic material and its smart window application* Royal Society of Chemistry. https://pubs.rsc.org/en/content/articlehtml/2017/ra/c6ra24686a
4. Nundy, S. (2021, April 12). *Electrically actuated visible and near-infrared regulating switchable smart window for energy positive building: A review* ScienceDirect. https://www.sciencedirect.com/science/article/abs/pii/S0959652621010738
5. de Vries, S. B. (2021, June 2). *Multi-state vertical-blinds solar shading – Performance assessment and recommended development directions* ScienceDirect. https://www.sciencedirect.com/science/article/pii/S235271022100601X

6. Zhiyong, L. (2020, February 3). *Simulation study on light environment performance and heat gain of applying a bimetal automatic shading device to rooms* ScienceDirect. https://www.sciencedirect.com/science/article/abs/pii/S0378778819328890
7. Mendis, T. (2020, January 11). *Economic potential analysis of photovoltaic integrated shading strategies on commercial building facades in urban blocks: A case study of Colombo, Sri Lanka* ScienceDirect. https://www.sciencedirect.com/science/article/abs/pii/S0360544220300153
8. Danis, J. (2022, March 1). *Direct heat flux sensing for window shading control in passive cooling systems* ScienceDirect. https://www.sciencedirect.com/science/article/abs/pii/S0378778822001219
9. Jorgensen, J. D. (2022, February 5). *Thermal resistance of framed windows: Experimental study on the influence of frame shading width* ScienceDirect. https://www.sciencedirect.com/science/article/abs/pii/S0925753522000236
10. Goetz-Koehler, M. (2022, January 12). *Switchable photovoltaic window for on-demand shading and electricity generation* ScienceDirect. https://www.sciencedirect.com/science/article/abs/pii/S0038092X21011208
11. Hassan, M. M. (2022, June 2). *Energy saving potential of photovoltaic windows: Impact of shading, geography and climate* ScienceDirect. https://www.sciencedirect.com/science/article/abs/pii/S0038092X22003656
12. Beć, K. (2019, February 1) *Breakthrough Potential in Near-Infrared Spectroscopy: Spectra Simulation. A Review of Recent Developments* ResearchGate. https://www.researchgate.net/figure/The-spectrum-of-absorption-coefficient-of-liquid-water-2_fig1_331285672

The Possibilities of Dynamic Phenomena Calculation in the Penstock of Hydropower Plant via Different Methods

Peter Hlbočan[1,*], *Peter* Ďurčanský[2], *Zoltán* Csuka[1], *Roman* Kiš[1], and *Zuzana* Kolková[1]

[1]ZTS – VÝSKUM A VÝVOJ, a.s., Lieskovec 598/75, 018 41 Dubnica nad Váhom, Slovak Republic
[2]University of Žilina, Faculty of Mechanical Engineering, Department of Power Engineering, Univerzitná 8215/1, 010 26 Žilina, Slovak Republic

Abstract. When designing a hydropower plant with a long pressure penstock, special attention must be paid to the pressure conditions in the penstock, especially during emergency flow shutdown. In this paper, the main principles of the calculation of static pressure rise in the pressure pipeline during rapid flow shutdown are presented. The results obtained by analytical methods (calculation of direct and indirect water hammer) and numerical methods (mathematical modelling in Matlab software environment) are compared on specific examples. The influence of some input parameters of the numerical model on the results of the numerical calculation is analysed separately.

Keywords: *water hammer; hydropower plant penstock; emergency shutdown of flow; calculation methods, Matlab modelling*

1 Introduction

When designing hydropower plants with penstocks, special attention must be paid to the analysis of dynamic phenomena. This issue is quite complex and a large number of papers have already been published on this topic, e.g. [1], [2], [3], [4], [5], [6], [7].

Among these phenomena we can also include the pressure increase in the penstock caused by the rapid shut-off of the flow at the end of the penstock. The reason for the rapid flow closing may be e.g. an emergency condition on the turbine or a sudden de-phasing of the generator accompanied by a subsequent increase in the turbine speed up to the value of runaway speed. In these circumstances, the water supply to the turbine must be stopped urgently (by closing the distributor or the shut-off valve in front of the turbine). In this case, however, there is (theoretically) a risk of a water hammer occurring at very rapid closure, which would have destructive effects on the penstock or the other equipment. It is therefore essential to set the emergency shutdown time of the equipment appropriately so that the maximum pressure rise in the penstock or upstream of the turbine does not pose a safety risk. Appropriate closing time setting requires calculating the pressure rise for several alternative closing times and selecting a value that will not cause a dangerous pressure rise and at the same time will not cause an unnecessarily long duration of the turbine runaway speed.

* Corresponding author: peter.hlbocan@ztsvav.eu

The above problem can be solved by several different methods. One possibility is to calculate the pressure increase with the application of classical theory and analytical (explicit) equations applying specific idealizations of the real system. The other option is to use mathematical modelling and numerical mathematics methods that are now common in commercial software such as Matlab & Simulink.

The aim of this paper is to compare the calculation of the pressure rise during penstock closure using analytical methods (idealized explicit equations) and numerical methods. At the same time, the influence of the time step size on the results obtained using numerical calculation methods is analyzed. The comparison is made for a simple piping system (constant diameter along the length, homogeneous material and constant wall thickness) which can form the penstock of a hydropower plant.

The phenomenon of water hammer became an important research topic in the 19th century for the field of water turbines and pumps. The experimental work of Zhukovsky (1900) defined the foundations of the current theory of water hammer. His research enabled the formulation of the fundamental equation in the theory of water hammer, which relates the change in pressure to the change in velocity at instantaneous valve closure [5].

Allievi (1902) developed a mathematical approach that analytically solved the piping system and found pressure values for different time steps and locations in a one-dimensional model of a thin-walled pipe for any linear flow variation. In 1913, Allievi presented a general theory of water hammer based on his previous mathematical approach. The applicability of the equations to various model problems was achieved by introducing a characteristic parameter ρ_A , which describes a certain element of the pipeline [3].

Jaeger (1933) developed the general theory of Allievi for specific cases. He developed analytical solution methods for systems with more than one pipe, thick-walled pipes, concrete-covered pipes, and systems with equalization chambers. The analytical methods based on Allievi's theory provide a general understanding of the water hammer phenomenon, but the formulas derived from it are limited to only a few model cases. The methods neglect friction and the second-order terms contained in the system of hyperbolic partial differential equations. They also do not allow the solution of problems with multiple characteristic parameters ρ_A. They provide a solution only for the pressure field and treat the flow drop during valve closure as linear [3].

Due to the existence of certain limitations of the analytical solution methods, so-called graphical solution methods of the hydraulic order have been introduced. Schnyder's (1936) and Bergeron's (1935) graphical methods gave results for the pressure and velocity fields for each time step at any location in a one-dimensional model of a thin-walled pipe for any type of shut-off valve. Parmakian (1963) presented a modified Schnyder-Bergeron graphical method that is applicable to systems with multiple characteristic parameters ρ_A, with consideration of hydraulic losses and with composite piping. The modified graphical method provides an accurate and fast solution for simple piping systems. However, for complex piping systems containing many hydraulic elements and branches, the graphical method did not prove successful and was replaced in a short time. Numerical methods [3], [6] solved the above problem.

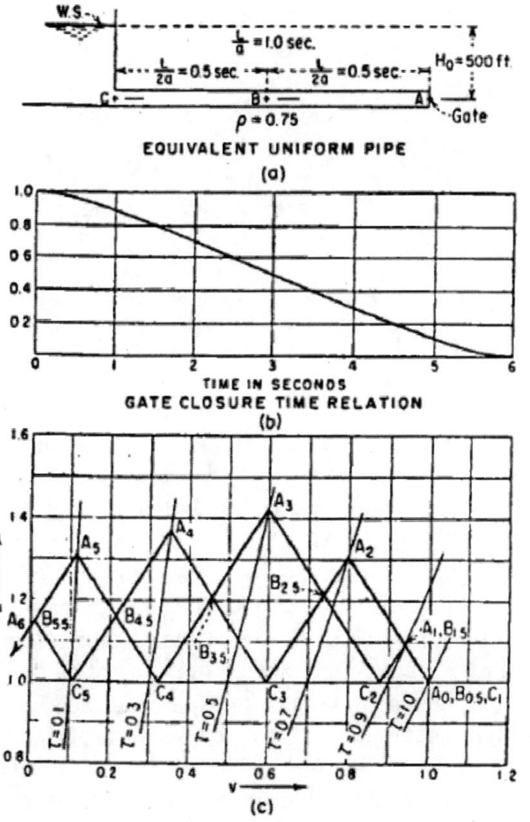

Fig. 1. Example of a graphical method for solving the hydraulic schock.

2 Numerical solution methods

The development of computer technology has made it possible to solve hyperbolic partial differential equations by numerical methods. Gray (1953) used the Eulerian method of characteristics (MOC). The method allows one to determine the order of approximation while neglecting the effects of friction, kinetic energy, and pressure wave shape change. Lister (1960) compared two methods of MOC discretization: the method of specified time intervals also called the fixed grid method (Fig. 2) and the so-called characteristic grid method. He found that the fixed grid method is much simpler in terms of computation and also in terms of constructing the computational grid. Because of its accuracy, simplicity, numerical efficiency and programming inexpensiveness, the MOC discretized using the fixed grid scheme became the most widely used method for solving the water hammer in its time. Perkins et al. (1964) established a numerical stability condition: $\Delta t/\Delta x \leq 1/a_c$ For the characteristic fixed-grid scheme, which is also referred to as the Courant stability condition. Lai (1961) in his work included a nonlinear hydraulic loss term in the MOC and implemented it in a computer code [6].

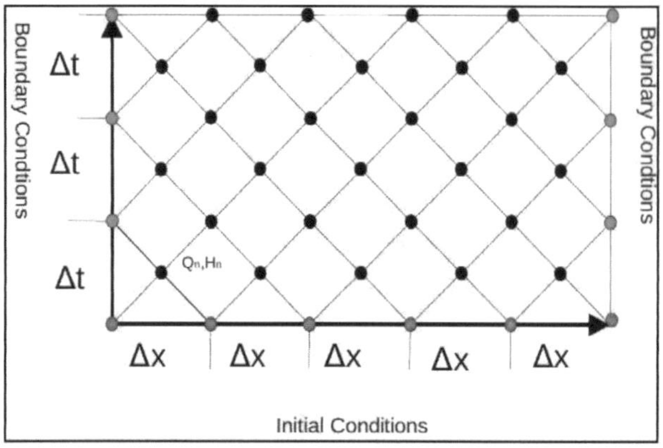

Fig. 2. Example of a fixed network for the method of characteristics (MOC).

The Lagrangian method, also known as the Wave Plan Method, was introduced by Wood et al. (Wood later compared WPM with MOC and found that for tasks such as large pipelined networks, the wave plan method requires less CPU time than MOC. Chaudhry (1985) proved that the second order finite difference method (FDM) gives better results than the first order MOC. However, he found that this method increases computation time and data storage size requirements. The first-order finite volume method (FVM) of Gudonov type was first used to solve water hammer problems by Guinot (2002). Zhao (2004) developed an explicit second-order Gudonov type method and compared its results with MOC with spatial line interpolation. He found that the second-order Gudonov type explicit finite volume method achieves a given level of accuracy with significantly less CPU time than MOC with spatial line interpolation [8].

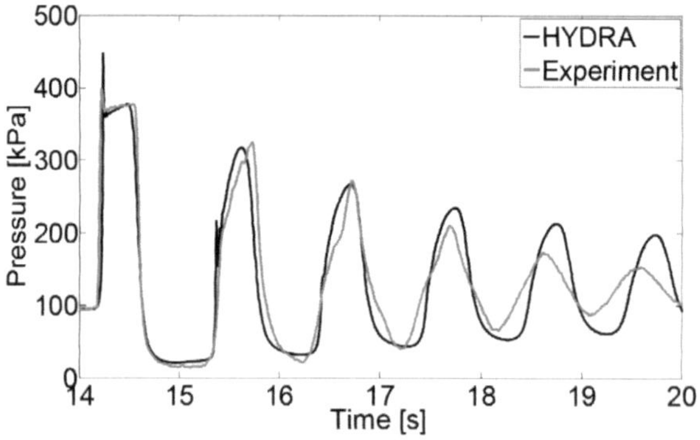

Fig. 3. Sample of results: the Lax-Wendroff method versus experiment [4].

In order to predict the pressure change caused by valve closure in a gravity pipe with air present, a numerical model based on the Lax-Wendroff method was developed. Compared to traditional methods, this study provides a more advanced approach by taking into account the influence of the flow velocity in the pipe on the propagation of the pressure wave and on the simulation of transient processes. In practice, there are many different water pipeline systems operated with the presence of a certain air content, therefore the Lax-Wendroff Method (LWM) is significant for its relatively accurate prediction of water hammer processes, especially under these conditions [7]. Fig. 3 compares the processed results from the simulation of water hammer by HYDRA software which uses the explicit Lax-Wendroff method against the results from the experiment. The results show that the method gives very accurate results especially for a few initial pressure pulses [4].

3 Analytical calculation of the pressure rise in the penstock

Analytical calculation methods apply equations to solve for the impact pressure in the penstock. A more detailed derivation and explanation of the calculation procedure is given, e.g., in [2]. For a penstock with a given length, constant diameter and wall thickness, equation (1) holds for the reflection time of the shock wave.

$$T_r = \frac{2 \cdot l}{v_v} \tag{1}$$

Where:
T_r - reflection time of the pressure wave (water hammer phase),
l - length of pipe (penstock),
v_v - the velocity of propagation of the pressure wave in the liquid, taking into account the compressibility of the liquid and the expansion of the pipe

$$v_v = \frac{\sqrt{\frac{K}{\rho}}}{\sqrt{1+\frac{K \cdot d}{E \cdot s}}} \tag{2}$$

Where:
K - the bulk modulus of elasticity of the liquid,
ρ - liquid density,
d - internal diameter of the pipe,
E - modulus of elasticity of the pipe wall material,
s - pipe wall thickness.

In the event of sudden closure of the pipe (closure time is less than the reflection time of the shock wave), a direct water hammer occurs. Then (3) applies.

$$T_f \leq T_r \tag{3}$$

Where:
T_f - the actual pipe closing time.

The increase in static pressure after sudden closure of the pipe (in direct water hammer) can be determined according to equation (4) or the increase in pressure can be expressed in metres of water column (5).

$$\Delta p = \rho \cdot v_0 \cdot v_v \tag{4}$$

Where:
v_0 - steady flow velocity of the liquid in the pipe (before closing or before the start of closing).

$$\Delta H = \frac{v_0 \cdot v_v}{g} \tag{5}$$

Where:
g - gravity acceleration.

In the case of a gradual closure of the pipe, where the closure time is longer than the reflection time of the shock wave (6), an indirect water hammer occurs. The increase in static pressure, expressed in metres of water column, can be determined by equation (7), which applies to a 'loose' pipe.

$$T_f \gg T_r \tag{6}$$

$$\Delta H = \frac{v_0 \cdot v_v}{g} \cdot \frac{2 \cdot l}{v_v \cdot T_f} \tag{7}$$

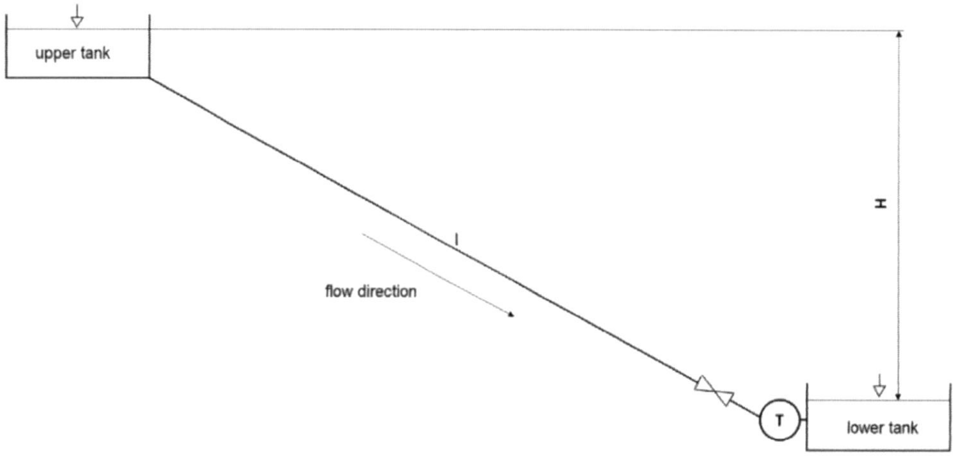

Fig. 4. Schematic diagram of the model penstock of the hydroelectric power plant.

The above calculation procedure can be applied to calculate the pressure rise in the model penstock according to the scheme in Fig. 4. The pre-closure flow rate, gradient, penstock length, pipe diameter, wall thickness and other parameters required to calculate the pressure rise are given in Table 1. For the case of a direct water hammer, an almost instantaneous pipe closure (0.05 s) is considered, which is only a hypothetical value that is in fact technically difficult to achieve, and for an indirect water hammer, a closure time of 12.3 s is considered. The calculated pressure surges for the closure times considered are also given in Table 1.

Table 1. Calculation of direct and indirect water hammer on the model system.

Flow in the penstock before closure	Q	**m³ /s**	8.02
Net power plant gradient	H	**m**	7.5
Length of penstock pipe	l	**m**	40
Nominal diameter of the penstock	DN	**m**	2
Penstock pipe wall thickness	s	**mm**	20
Outer diameter of the penstock	d_o	**m**	2.032
Inner diameter of the penstock	d	**m**	1.99
Water flow velocity before closing	v_0	**m/s**	2.6
Gravitational acceleration	g	**m/s²**	9.81
Water density	ρ	**kg/m³**	1000
Modulus of volumetric elasticity of water	K	**Pa**	2 030 000 000
Modulus of elasticity of the pipe wall	E	**Pa**	2.2E+11
Pressure wave propagation velocity in the penstock	v_v	**m/s**	1028.51
Pressure wave reflection time	T_r	**s**	0.078
Closing time - direct hydraulic impact	T_f	**s**	0.05

Pressure rise - direct water hammer	**p**	**kPa**	***2646.7***
	H	**m**	***269.80***
Closing time - indirect hydraulic impact	T_f	s	12.31
Pressure rise - indirect water hammer	**p**	**kPa**	***16.7***
	H	**m**	***1.70***

4 Numerical simulation of water flow

The analytical models allow the fast calculation of the parameters needed to determine the main design variables of a given work. However, it is more appropriate to use numerical methods and simulation programs to calculate all the variables, or to calculate them with consideration of time changes. A suitable tool is, for example, the commercial software Matlab or its open-source version SciLab. These tools allow 0D/1D and 2D analysis of physical problems and tracking of selected physical quantities as a function of changing boundary conditions and parameters.

For the purpose of the given simulation, Matlab software with Simulink superstructure was used, which simplifies the calculation by representing the steps and dependencies of the computational simulation by function blocks.

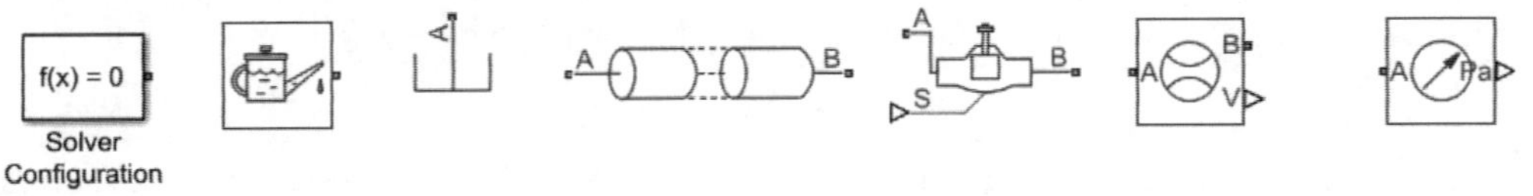

Fig. 5. Function blocks used - calculation setup, liquid properties, ideal reservoir, piping, valve, flow sensor, pressure sensor.

Fig. 5 shows the blocks used to set the boundary conditions. The "computation setup" block defines the basic parameters of the computation, i.e. mainly the solver type, the number of iterations and the time step of the computation. The "fluid properties" block defines the type of fluid and the properties to be considered by the calculation. The ideal reservoir block defines the reservoir type and the surface pressure. The calculation blocks themselves represent real parts of the technical equipment in question, i.e. the penstock pipe and the shut-off fitting, the valve. For simulation purposes, a pressure sensor and a flowmeter are still included in the network to record the actual state of these quantities.

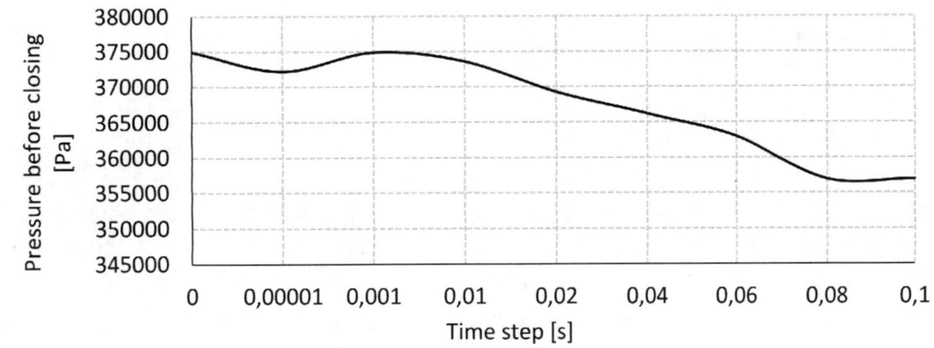

Fig. 6. Effect of time step on the calculated pressure before closure.

Before the final simulation calculations were performed, an analysis of the effect of the time step size on the results obtained was performed. The results of that analysis are shown in Fig. 6. From Fig. 6 it can be seen at which values of time step approximately the same simulation results are obtained.

The actual results of the numerical calculation of pressure rise performed for the closing time of 12.31 s and comparison with the result of the analytical calculation is shown here in Table 2.

Table 2. Comparison of analytically and numerically calculated pressure rise values.

	Closing time [s]	Δp - analytically [kPa]	Δp - numerically [kPa]
Indirect impact	12.31	16.7	18.8

5 Conclusion

The paper presents the results of analytical and numerical calculations of the pressure rise in the penstock of a hydropower plant caused by emergency shutdown the flow. The results of the analysis of the influence of the time step on the obtained results of the numerical calculation are presented. The comparison of the analytical and numerical calculation results showed that the result of the numerical calculation of the pressure rise in the penstock is more than 10 % different. For a more detailed evaluation of the individual calculation methods accuracy, it would be necessary to compare the obtained results with the results of operational measurements on the penstock. It should also be noted that there is still a relatively large potential for further improvement of the mathematical model applied in the numerical calculations.

Authors would like to thank to project: Regenerácia použitých batérií z elektromobilov / Regeneration of used batteries from electric vehicles ITMS2014+: 313012BUN5

References

1. Vintr, M. (2018).Výpočet přechodových stavů u MVE Skhalta, *Hydroturbo 2018, 24. medzinárodná konferencia o vodnej energetike 25. – 26. september 2018*, Bratislava, SR, ISBN 978-80-227-4829-2.
2. Gančo, M. (1990). Mechanika tekutín 2. vydanie 240 strán, Slovenská vysoká škola technická v Bratislave, Strojnícka fakulta, *Alfa – vydavateľstvo technickej a ekonomickej literatúry Bratislava*, skriptá, ISBN 80 – 05 – 00549 – 0.
3. Mambretti, S. (2014). Water hammer simulations, Universidade Estadual de Campinas, Brazil Ashurst Lodge, Ashurst, Southampton *SO40 7AA*, UK, ISBN: 978-1-84564-680-6.
4. Himr, D. (2013). Numerical simulation of water hammer in low pressurized pipe: comparison of SimHydraulics and Lax-Wendroff method with experiment, *EPJ Web of Conferences 45 01037*. DOI: 10.1051/34501037.
5. Provenzano, P.G, Baroni, F., Aguerre, R. J. (2011). The closing function in the waterhammer modeling, *Latin American Applied Research 41:43-47*, ISSN 0327-0793.
6. Mahdy, M. (2019). Analysis of water hammer using method of characteristics for different pipes material, *International Journal of Latest Technology in Engineering, Management & Applied Science (IJLTEMAS) Volume VIII, Issue I*, ISSN 2278-2540.
7. Zhang, Boran. Wan, Wuyi. Shi, Mengshan. (2018) Experimental and numerical simulation ofwater hammer in gravity pipe flow with continuous air entrainment, *Water 2018, 10, 928*; DOI:10.3390/w10070928.

8. Ghidaoui, Mohamed S. Zhao, Ming. McInnis, Duncan A. Axworthy, David H. (2018) A review of water hammer theory and practice, *Applied Mechanics Reviews Copyright © 2005 by Asme, Vol. 58*, DOI: 10.1115/1.1828050.

Analysis of the Efficiency of the Experimental Design of an OFF-GRID Photovoltaic System for Local Purposes of Electricity Generation

Michal Šmajda[1], *Nikolas* Polivka[1,*],and *Dmytro* Shmyhol[1]

[1]Technical university of Košice, Faculty of manufacturing technologies, Bayerová 1, 080 01 Prešov, Slovakia

Abstract

Research background: Increasingly the implementation of vehicles with alternative propulsion such as battery or hydrogen vehicles there are higher demands on electricity production. The main factor affecting the environment such vehicles is mainly a source of electricity, which recharges the batteries of vehicles or hydrogen produced as fuel for fuel cells.
Purpose of the article: Presented analysis examines the effectiveness of the experimental design of a local OFF-GRID photovoltaic system intended to support of alternative automotive drives.
Methods: As a basis for determining the efficiency of the proposed device is the analysis of a run-in photovoltaic system is used installed in the same geographical conditions with data validation with database PV-GIS.
Findings and added value: From the analysis of the investigated design of the photovoltaic system shows that electricity production during the year it is directly dependent on the intensity of global radiation. The proposed solution makes it possible to partially cover the energy requirements in transport with active use RES.

Keywords: *photovoltaic system; global radiation; solar energy; electricity*

1. Introduction

With the constant progress in the field of transport and increasing emission requirements the boundaries of internal combustion engines are increasingly shifting up to the efficiency limit. A possible solution is implementation of vehicles with alternative propulsion. The most promising method of transportation is currently applied electric transport in the form of battery vehicles. The main current competitor of this technology is implementation of fuel cells in electric cars which are powered by hydrogen [1].

Battery vehicles are a strong competitor to internal combustion engines. The technology used to drive it is considered an emission-free mode of transport. The main advantage of these vehicles is the increasingly strong infrastructure of charging stations [5, 10].

* Corresponding author: nikolas.polivka@tuke.sk

However, from an ecological point of view, the whole cycle of energy from production up to the wheel must be taken into account, it is a so-called well-to-wheel cycle. From this point of view, it can be concluded that the degree of ecology of electric cars also depends on the method of electricity production. In the case of the use of conventional methods of energy production, such as a coal-fired power plant where the energy source is fossil fuels in the form of coal, the cleanliness of battery cars is debatable. With the ever-increasing number of battery cars, the need for energy supply is also increasing, which significantly increases the carbon footprint. However, the introduction of electric vehicles will not achieve significant reductions in CO2 emissions, especially in countries with a high emission factor for electricity generation [11, 13].

In the case of hydrogen vehicles with fuel cell technology, hydrogen is the primary energy source. Despite the fact that these vehicles emit no local emissions and the energy carrier is hydrogen, which is one of the cleanest and best energy carriers, an important factor influencing the environmental friendliness of this technology is the origin of hydrogen fuel. At present, steam reforming of natural gas is the most widespread hydrogen production technology in Europe. The problem is that natural gas is a fossil fuel which affects the efficiency of hydrogen vehicles. One of the best and purest ways to produce hydrogen is water electrolysis. This method of hydrogen production, despite its environmental friendliness, requires high energy requirements, which limits its wider application for the needs of the transport sector. To expand the use of water electrolysis, it is necessary to reduce the energy consumption, cost and maintenance of current electrolyzers and, in turn, increase their efficiency, durability and safety [3, 7, 14].

Battery and hydrogen vehicles are currently the most promising path to emission-free transport, but it is necessary to take into account not only their real-time operation but also their way of obtaining energy for propulsion. With the ever-increasing number of such vehicles, the need for electricity production is also growing, which significantly affects the stability of the electricity grid. A suitable solution to this discrepancy between the production and the need for electricity is, in particular, the ever-increasing implementation of renewable sources. Renewable energy technologies hold the promise of clean, abundant energy from self-renewing sources such as hydro, wind and solar power [6, 12, 15].

From the point of view of efficiency and low geographical intensity, the most promising way of generating electricity is solar energy. Such a source in terms of system efficiency can be placed directly in places of electricity consumption such as. on the roof of gas stations, where it can serve as a source of battery charging or a source of power to electrolysis to produce hydrogen. With the correct dimensioning of RES in the transport sector, it is possible to effectively and safely cover the growing need for energy resources. [2, 8, 9, 10].

2 Design of an experimental OFF-GRID photovoltaic system

Photovoltaic panels are the basic element for converting solar energy into electricity. It is this element that provides the greatest possible degree of optimization as technology advances. For the purpose of designing this system, we have therefore considered panels that have the highest possible performance and are commonly available on the market. The panel system consists of 120 monocrystalline panels with a power output of 550 Wp for a total system output of 66 kWp.

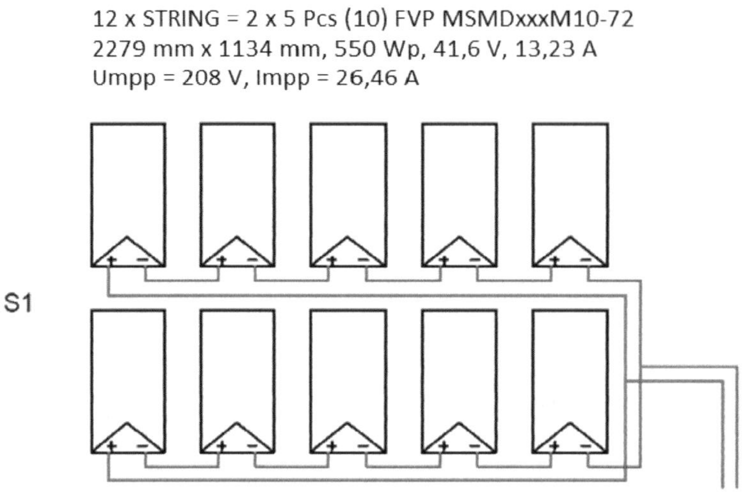

Fig.1 Photovoltaic string wiring diagram

For the design of our system, we have chosen a series-parallel circuit. Each string consists of 10 panels and they are connected in series in groups of 5 and these 2 groups are interconnected in parallel. In total, the whole system consists of 12 strings, which are connected in parallel into a single unit by means of separate regulators. The calculation of voltage, current and power of a single string is as follows:

$$Umpp = 5 \times 41{,}6\ V = 208\ V \tag{1}$$

$$Impp = 2 \times 13{,}23\ A = 26{,}46\ A \tag{2}$$

$$Pmax_{string} = 208\ V \times 26{,}46\ A = 5503{,}68\ W = 5{,}5\ kWp \tag{3}$$

$$Pmax_{celk.} = 5{,}5\ kWp \times 12 = 66\ kWp \tag{4}$$

One of the main elements of the photovoltaic system, which ensures the correct operation of the entire system and adjusts the voltage from the panels, is the voltage regulator or their system. The main task of the charge/voltage regulator is to adjust the voltage, which is then routed to the batteries or directly to the inverters and then to the grid. For larger applications, a 48 V system is best suited, which is what we have used in this design, so the regulators must also be dimensioned for a 48 V system.

In the case of our system, we have chosen the regulator with the designation Victron MPPT 250V/100A-Tr VE.Can. The basic condition in the dimensioning of the panel array is mainly the compliance with the limits of the regulator and for this reason the PV strings are dimensioned in terms of voltage, current and power levels.

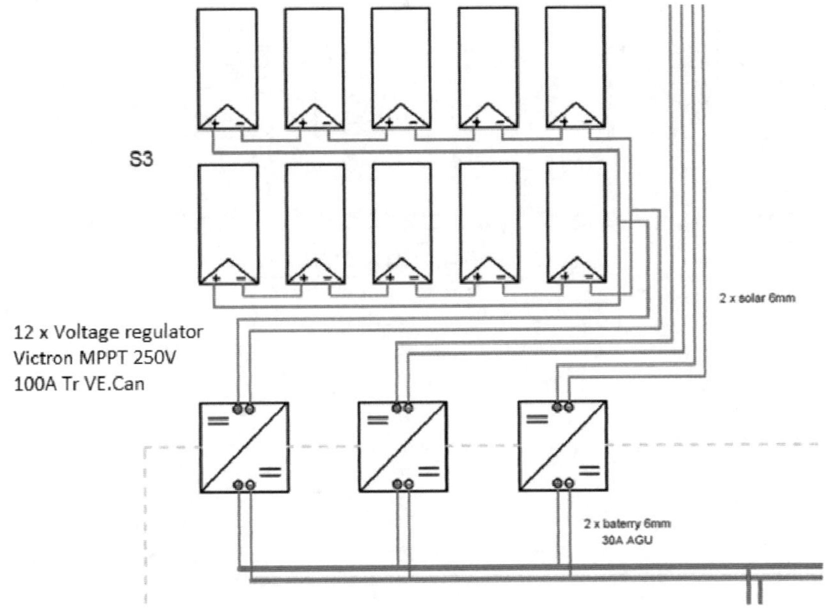

Fig. 2 Solar regulators wiring diagram

Because each PV string requires its own solar regulator, so 12 regulators were also used in the design. Figure 2 shows the principle of connecting the string to the corresponding regulator as well as their parallel interconnection.

The principle of functionality or operation of the regulators is based on the intensity of solar radiation. The battery charging process starts when the panel voltage exceeds the battery voltage (48 V) by 5 V, otherwise the charging process stops when the panel voltage drops to 1 V higher than the battery voltage.

Each of the regulators provides a charging current of 26.46 A, which means that when twelve units are connected in parallel, the total charging current of the batteries is 317.52 A. The parallel connection of the regulators is ensured by a double CYA battery cable with a cross section of 6 mm2 and AGU 30 A fusing, flowing into a double connection rail with a conductor diameter of 100 mm2.

For the purposes of our design, BYD LVS 16.0 batteries with LiFePO4 technology were used. The main advantage of these batteries is that they are lithium-iron-phosphate batteries of the latest generation with the possibility of ON-GRID but also OFF-GRID connection, the possibility of up to 100% discharge but also the possibility of assembling larger units according to the required capacity.

Depending on the calculation of the required capacity, a total of 8 BYD LVS 16.0 batteries with a total capacity of 128 kWh were used for the design of the photovoltaic system. Each battery consists of four battery packs with a capacity of 4kWh. In total, the capacity can be increased up to 256 kWh if required, which represents 16 of these batteries or 64 battery packs.

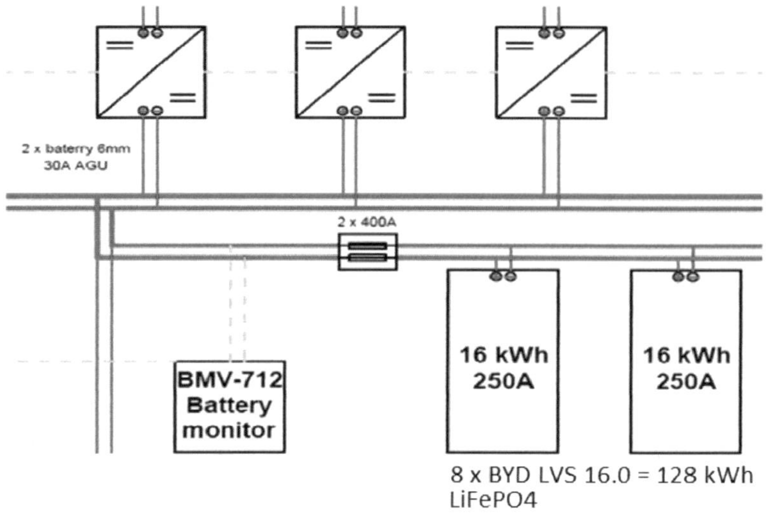

Fig.3 Solar battery wiring diagram

Figure 3 shows the schematic connection of the regulators to the batteries and also to the so-called battery tracker. The batteries are connected in parallel and the fusing between the charge regulators as well as the voltage converters is provided by two knife fuses with a rated breaking current of 400 A on both the positive and negative conductors.

The final component of the whole photovoltaic system is the voltage converter. The inverter covers the mismatch between the current required for the appliance system and the current coming from the solar panels. As the general whole PV assembly operates with direct current, it is necessary to invert the direct current to alternating current for common grid sized appliances or devices [4]. In this particular case, it is a change from a DC system with a nominal voltage of 48 V to a single-phase AC system with a voltage of 230 V. For the purpose of our photovoltaic system design, considering the output power, a combined inverter with purely sinusoidal output voltage was chosen as the most suitable, with the designation Victron Quattro 48V 15000 VA/200 A - 100 A/100 A. Figure 4 shows a schematic representation of the wiring of a group of three inverters in relation to the batteries and regulators and also to the power distribution board.

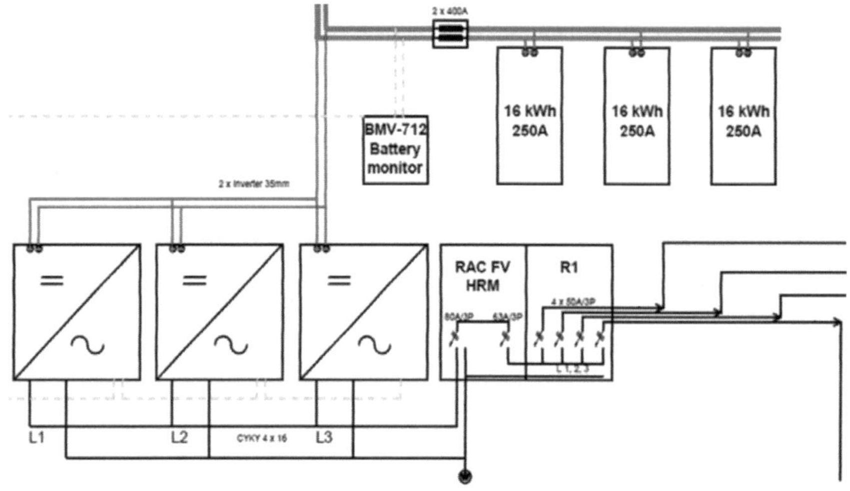

Fig. 4 Voltage converters wiring diagram

All three inverters shown in Figure 4 are connected in parallel at the input from the regulator and battery pack using a cable designed for a CYA type inverter with a cross-sectional area of 35 mm2. In addition to the three-phase output, this circuit also provides the possibility, in the case of an alternative reverse power source (grid, generator,...), to charge the batteries with a current of up to 600 A for a specific type of inverter [4].

The output of the inverters is a three-phase alternating current with a voltage of 400 V and a frequency of 50 Hz, which flows through a conductor marked CYKY 4 × 16 into the photovoltaic switchboard to the main switching point (HRM) with a fusing of 80 A.

3 Measuring the efficiency of the proposed experimental photovoltaic system

Considering the measurements made on existing PV systems with the same geographical conditions and validations of the data with the PV-GIS database, we can determine with high accuracy what amount of electricity we can produce. The main difference of the measured PV systems is first of all the dimensionality of the assembly but also the way of handling the surplus energy. Since the proposed PV system is characterized as an OFF-GRID circuit, it is not possible to distribute the excess energy to the utility grid. Therefore, this energy can be characterised as untapped. In terms of the technologies used in the design, the loss of performance of the panels in terms of their age can be excluded, and the total loss of the components used with losses on the conductors is roughly 10%.

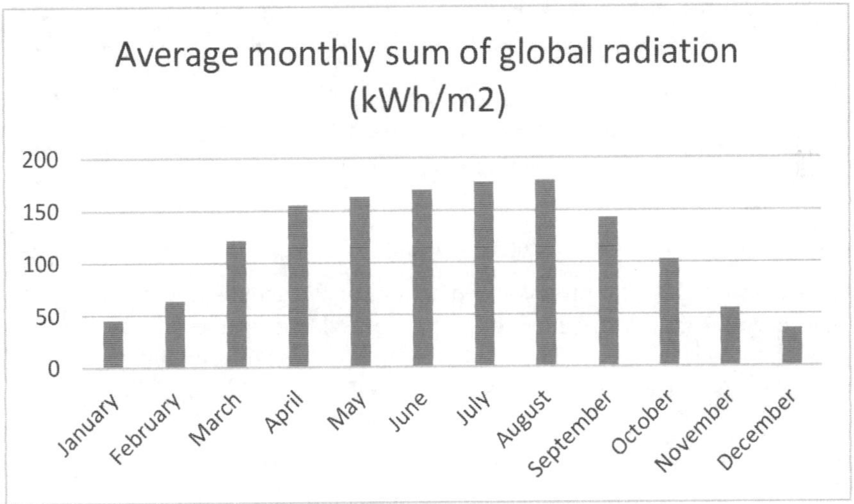

Fig. 5 Rate of global radiation incident on PV panels

Due to the correct calculation of the efficiency of the PV system, it is necessary to know the solar energy intensity. Since the measured system includes a global radiation sensor that collects real-time data using an AHLBORN transducer called ALMEMO 2490, we can determine the rate of incident radiation on the PV panels. Although the average irradiation rate for the city of Prešov is about 1100 kWh/m2 per year, we can see from the measured data that when the panels are angled at 35°, the incident energy rate is 1410 kWh/m2 per year. Figure 5 shows the sum of the global radiation rates of the panels for each month during the year. The graph shows that the highest radiation exposure with a value of 178.5 kWh/m2 is in the month of August and the lowest exposure is in the month of December when the sum of global radiation reaches the level of 36.5 kWh/m2.

Based on measured data in real operation and global radiation data compared with PV-GIS, the expected system production can be determined. Figure 6 shows the potential energy production on an average day and also the fraction of energy that we are unable to capture. One of the reasons is e.g. to reach the maximum consumption of the powered appliances and energy surpluses during the production peak, when the PV system has to go into a slight decay. The graph shows that the highest average value of usable energy can be achieved in the month of June at 197 kWh, while the lowest value with 50 kWh on average per day is expected for the month of December. Although June is the most efficient month in terms of energy production, the system has the highest average losses in August, averaging 63 kWh per day. Due to the low efficiency of energy production in the winter months, it can be concluded that the energy loss is negligible. This is mainly due to the persistent condition where the full capacity of the batteries cannot be reached.

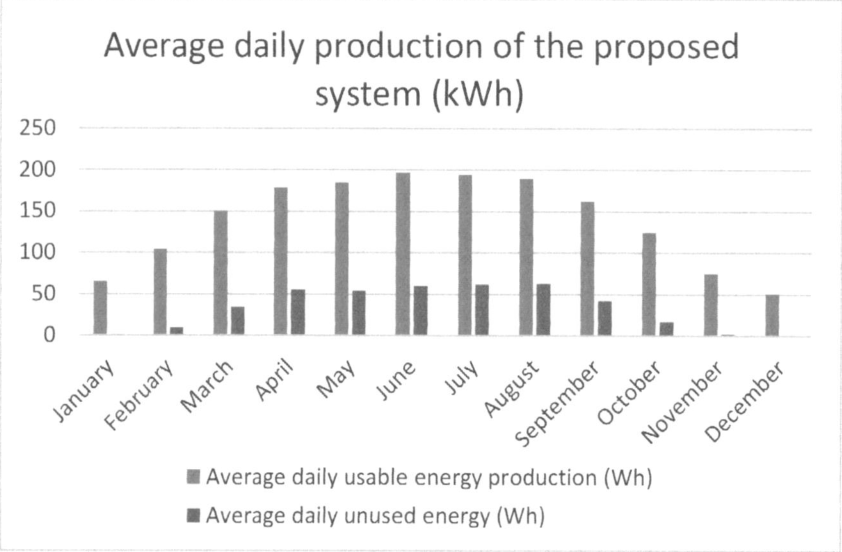

Fig. 6 Average daily production of the proposed system

When calculating the total production potential of the proposed system, it is necessary to take into account the energy that we can recover but also the energy that we cannot capture due to the surplus. It is apparent that the highest level of potential production is in the months of June to August where the value exceeded 250 kWh per day.

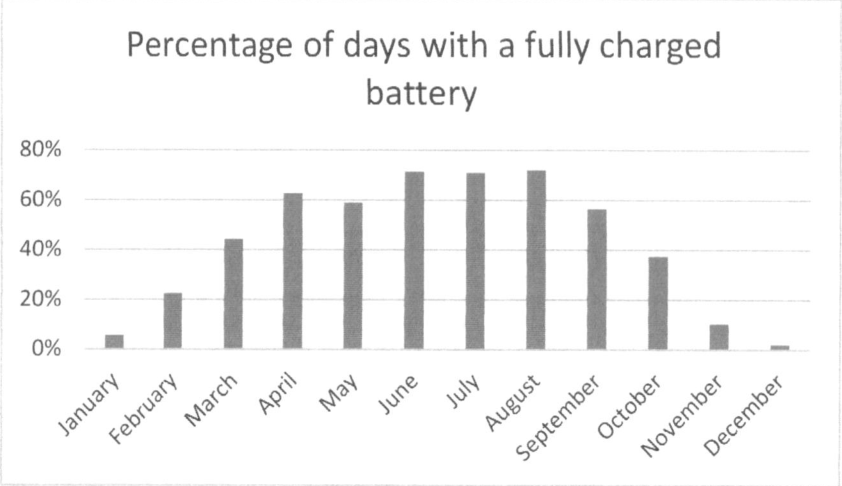

Fig. 7 Percentage of days with a fully charged battery

In Figure 7 it can be seen to what extent it is realistic to achieve a full battery charge within a year. The chart shows that the highest rate of full battery capacity is in the summer months (June-August), with the standard rate of full charge over 50% of the days in a month being between April and September. In the case of winter operation, reaching full capacity is minimal. This is mainly due to the short days of low global radiation. The system supplies power to the load as a priority, which greatly affects the battery charging process.

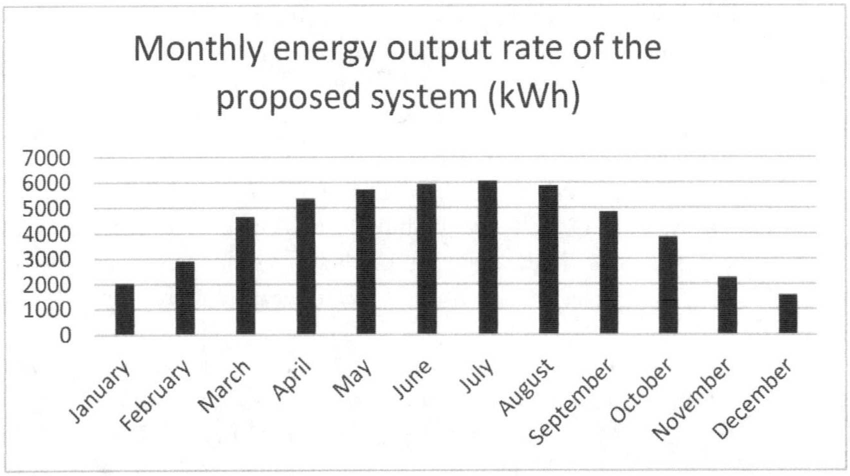

Fig. 8 Total monthly energy production

Based on the previous analysis, it is possible to determine the total share of energy produced per calendar year. The graph in Figure 8 shows that the highest level of electricity generated is in the month of July with a value of 6045 kWh. Throughout the year, 51,047 MWh of energy can be generated using the proposed system.

4 Conclusion

The design of the photovoltaic system resulted in an independent device capable of operating without additional support from the grid. The presented analysis shows that it is currently not possible to ensure constant electricity production during the whole year, however, the presented proposal has significant benefits for the application of RES in transport infrastructure. Considering the large scale of the entire PV system, it can be stated that the installation requirements are minimal. Devices of this type can be installed on the roofs of petrol stations, thus eliminating the degradation of farmland and increasing the economic use of buildings. Despite the variations in electricity production from month to month throughout the year, this proposal is a good basis for further development of zero-emission transport.

This article was supported by the state grant agency for supporting research work and co-financing the project KEGA 023TUKE-4/2021.

References

1. Wen, Y., Chen, Y., Wang, P., Rassol, A., Xu, S. (2022). Photovoltaic–electric vehicles participating in bidding model of power grid that considers carbon emissions. *Energy Reports*, *8*, Pages 3847-3855.
2. Hu, J., Li, P. (2022). Energy-sharing method of smart buildings with distributed photovoltaic systems in area. *Energy Reports, 8(5*), Pages 622-627.
3. Joubi, Abdulrahman, Akimoto, Y., Okajima, K. (2022). A Production and Delivery Model of Hydrogen from Solar Thermal Energy in the United Arab Emirates. *Energies, 15(11)*, Page 4000.
4. Lunardi, A., Normandia Lourenço, L.F., Munkhchuluun, E., Meegahapola, L., Sguarezi Filho A.J. (2022). Grid-Connected Power Converters: An Overview of Control Strategies for Renewable Energy. *Energies, 15(11),* Page 4151.
5. Alabri, W., Jayaweera, D. (2022). Optimal coordination of unbalanced power distribution systems with integrated photovoltaic systems and semi-fast electric vehicles charging stations. *IET Generation Transmission & Distribution, 16(12)*, Pages 2399-2415.
6. Winkler, J.K., Grahle, A., Syré, A.M., Martins-Turner, K., Gohlich, D. (2022). Fuel cell drive for urban freight transport in comparison to diesel and battery electric drives: a case study of the food retailing industry in Berlin. *European Transport Research Review, 14(2)*, Page 2.
7. Henry, A., McCallum, C., McStay, D., Rooney, D., Robertson, P., Foley, A. (2022). Analysis of wind to hydrogen production and carbon capture utilisation and storage systems for novel production of chemical energy carriers. *Journal of Cleaner Production, 354*, Page 31695.
8. Cheekatamarla, P. (2022). Role of On-Site Generation in Carbon Emissions and Utility Bill Savings under Different Electric Grid Scenarios. *Energies, 15(10)*, Page 3477.
9. Geng, S., Zou, R., Zhang, S., Guo, D. (2022). Research on site combination optimization framework of distributed photovoltaic power station from dual perspectives. *Energy Reports, 8*, Pages 4401-4415.
10. Ahmad, F., Iqbal, A., Ashraf, I., Marzband, M., Khan, I. (2022). Optimal location of electric vehicle charging station and its impact on distribution network: A review. *Energy Reports, 8*, Pages 2314-2333.
11. Tang, Y., Cockerill, T., Pimm, A., Yuan, X. (2021). Reducing the life cycle environmental impact of electric vehicles through emissions-responsive charging. *iScience, 24(12),* Page 3499.
12. Muhindo, S.M., Malhame, R.P., Joos, G. (2021). *Energies, 14(24*), Page 8517.
13. Petrović, D.T., Pešic, D.R., Petrović, M.M., Mijailović, R.M. (2020). Electric cars: Are they solution to reduce CO2 emission. *Therm. Sci.*, vol. 24, no. 5, pp. 2879-2889.
14. Santos, Diogo M. F., Sequeira, César A. C., Figueiredo, José L. (2013). Hydrogen production by alkaline water electrolysis. *Química Nova* [online]. 2013, v. 36, n. 8 [Accessed 30 June 2022] , pp. 1176-1193. Available from: <https://doi.org/10.1590/S0100-40422013000800017>. Epub 04 Oct 2013. ISSN 1678-7064.
15. Bull, S. R. (2001). Renewable energy today and tomorrow. *Proceedings of the IEEE*, vol. 89, no. 8, pp. 1216-1226.

Comparison of Performance Parameters of Four Types of Refrigerants

Alexander Čaja[1,*], *Nikola* Čajová Kantová[2], *Andrej* Kapjor[1], and *Martin* Vantúch[1]

[1]Department of Power Engineering, Faculty of Mechanical Engineering, University of Žilina, Univerzitna 1, 010 26 Žilina, Slovakia
[2]Research centre, University of Žilina, Univerzitna 1, 010 26 Žilina, Slovakia

Abstract. The use of refrigerants requires to meet ecology and legislation requirements. The aim of the legislation is also to reduce the global warming potential (GWP). Natural refrigerants such as carbon dioxide, ammonia, and hydrocarbons such as propane, isobutene, and propylene are used mainly because of their environmental benefits compared to synthetic refrigerants. This work compares two natural refrigerants: R-744 and R-290 and two synthetic refrigerants: R-134a and R-32. All these refrigerants are used mainly in heat pumps. In this work were calculated and compared the following performance parameters: heat rejected in the condenser, cooling power, compressor power, and the coefficient of performance (COP). The most important parameter is the coefficient of performance because when it decreases, electric energy consumption increases. It requires reaching a high value of COP. Further, this work deals with the economic analyses of selected refrigerants. The last aim of this article is their ecologic analysis based on the GWP factor.

Keywords: *refrigerants, heat pumps, global warming potential coefficient of performance*

1 Introduction

Nowadays, the use of refrigerants requires to meet ecology and legislation requirements. Legislative regulations set the use of refrigerants in electrical equipment, heating, and cooling. The use of refrigerants was governed by technical standards, although hazardous refrigerants such as sulphur dioxide, ammonia, and methyl chloride were used. There are already regulations describing the detailed use of individual types of refrigerants. The aim of the legislation is also to reduce the global warming potential (GWP) for all refrigerants. While the GWP in Europe averaged 2000 in 2015, it is currently 900 and is expected to fall to 400 by 2030 [1, 2].

GWP represents the ability to capture heat in the atmosphere or to bounce it back to the earth's surface. Its value determines how many times a given gas contributes more to the greenhouse effect than carbon dioxide. The equation (1) is used to calculate its quantity [3].

* Corresponding author: alexander.caja@fstroj.uniza.sk

$$GWP_x = \frac{\int_0^{TH} ax.[x(t)]}{\int_0^{TH} ar.[r(t)]} \quad (1)$$

TH is the time horizon over; *ax* is the radiative efficiency, *ar* is the radiative efficiency of the reference sample; *x(t)* is the time-dependent decay in abundance of the substance following an instantaneous release of it at time t=0 and *r(t)* is the time-dependent decay of the reference sample [2, 3].

Refrigerants based mainly on fully halogenated hydrocarbons (CFCs), later partially halogenated hydrocarbons (HCFCs) have been used in the past. Both groups of refrigerants are already banned for environmental protection by reducing emissions. Fluorocarbons and their mixtures are currently used (HFCs). These HFCs have been developed as a replacement for ozone-depleting refrigerants [4]. However, the Europe regulation EU 517/2014 calls for a massive reduction in the consumption of HFCs in Europe. This reduction scheme is known as HFC phase-down [5].

Natural refrigerants such as carbon dioxide, ammonia, and hydrocarbons (propane, isobutene, and propylene) are used mainly because of their environmental benefits. With zero ozone depletion potential (ODP) and very low to zero GWP factor, they are considered a great alternative to synthetic refrigerants because they pollute the atmosphere less [6,7].

In this work were calculated and compared the following parameters of refrigerants used in heat pumps: heat rejected in condenser, cooling power, compressor power and the coefficient of performance. Heat pumps are alternative devices for the production of thermal energy compared to conventional production by the combustion of fossil fuels. The principle of their function is based on the thermodynamic circulation of mechanical cooling equipment [7].

Further, this work deals with the economic analyses of selected refrigerants. The last aim of this article is their ecologic analysis based on the GWP factor.

2 Material and method

As refrigerants were chosen two natural refrigerants (R-744, R-290) and two synthetic refrigerants (R-134a, R-32) described in Table 1. Nowadays, all these refrigerants are used mainly in heat pumps.

Table 1. Refrigerants.

Code	Chemical	Name
R-744	CO_2	Carbon dioxide
R-290	C_3H_8	Propane
R-134a	CH_2FCF_3	1,1,1,2-Tetrafluoroethane
R-32	CH_2F_2	Difluoromethane

Heat rejected in condenser (*Qk*) was calculated according to equation (2), where *Qo* is cooling power and *P* is compressor power [8].

$$Qk = Qo + P \quad (2)$$

Cooling power (*Qo*) was calculated according to equation (3), where *m* is mass flow rate and *qo* is refrigeration duty. Mass flow rate (*m*) is further calculated according to equation (4), where *h1* and *h5* is enthalpy [8].

$$Qo = m.qo \tag{3}$$

$$m = \frac{Qo}{qo} = \frac{Qo}{h1-h5} \tag{4}$$

Compressor power (*P*) was calculated according to equation (5), where *a* means a difference between enthalpy *h2* and *h1* [8].

$$P = m.a \tag{5}$$

$$a = h2 - h1 \tag{6}$$

The coefficient of performance (*COP*) was calculated according to equation (7) and it is the ratio of cooling power and compressor power [8].

$$COP = \frac{Qo}{P} \tag{7}$$

It is contemplated that all of the selected refrigerants can be used in a single-stage steam circuit shown in Figure 1.

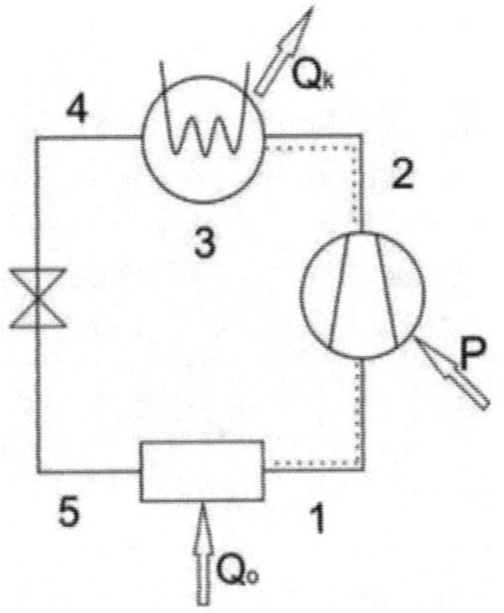

Fig. 1. Single-stage steam circuit [8].

As input parameters were the same temperatures for all calculated refrigerants: evaporation temperature of -6 °C and condensation temperature of 40 °C.

3 Results

Results from the calculation of individual performance parameters of chosen refrigerants are stated in Table 2. The most important calculated parameter is the coefficient of performance with its high value. When COP decreases, electric energy consumption increases. Of the selected refrigerants, R-744 has the lowest value of COP, and R-32 has the highest value of COP.

Table 2. Performance parameters of individual refrigerants.

Code of refrigerants	Qk [kW]	P[kW]	COP[-]
R-744	14 169	4 169	2.40
R-290	12 523	2 523	3.96
R-134a	12 474	2 474	4.04

R-32	11 299	1 299	7.70

The comparison of prices for selected refrigerants is shown in Table 3. This comparison was realized for 1 kilogram of refrigerants commonly available in Slovakia. The most economical of the selected refrigerants is the R-744 because the price is around 5 € per 1 kg of refrigerant. R-290 has the second-best price with the price of 10 € per 1 kg of refrigerant. R-134a and R-32 were half as expensive as the previous two refrigerants.

Table 3. The comparison of prices for individual refrigerants.

Code of refrigerants	The amount [kg]	Price [€]
R-744	1	5
R-290	1	10
R-134a	1	20
R-32	1	20

Based on the work of Krainer and Duda, the GWP for selected refrigerants is stated in Table 4. According to an ecological point of view, the most environmentally friendly refrigerant is R-744, because its GWP factor has a value of 1. The second most environmentally friendly refrigerant is R-290, where its GWP reaches 3. R-32 has a significantly higher GWP value of 675. R-134a has the highest GWP value of 1430 from all selected refrigerants.

Table 4. The comparison of GWP for individual refrigerants [4].

Code of refrigerants	GWP [-]
R-744	1
R-290	3
R-134a	1430
R-32	675

4 Conclusion

The world's effort is to find a refrigerant with a low value of GWP factor and also ODP factor. Natural refrigerants could be a suitable solution. However, they have a lower cooling factor and thus consume more electricity in the cooling circuits. The calculation compared two natural refrigerants and two synthetic refrigerants. The most ideal refrigerant in terms of price and ecology would be carbon dioxide R-744. However, it has the lowest refrigeration factor of the selected refrigerants. Which means it has the highest energy consumption. Of the selected refrigerants, the suitable refrigerant is propane R-290. Propane is a natural refrigerant with better properties than R-32 and R-134a, and also R-290 has a low GWP factor.

This publication has been produced with the support of the project KEGA No. 021ŽU-4/2021: Primary energy conversion into heat/cold using thermodynamic cycles and compressor cycle with working substance (refrigerant) CO2 and co-financed within the work on the task solved by the UNIVNET association no. 0201/0004/20 supported from the funds of the Ministry of Education, Youth and Sports of the Slovak Republic.

References

1. Rojko (2021) *Aká je budúcnosť chladív v tepelných čerpadlách?*ENERGIE-PORTAL. Retrieved from : *https://www.energie-portal.sk/Dokument/aka-je-buducnost-chladiv-v-tepelnych-cerpadlach-107265.aspx*
2. C.H. de Paula, W.M. Duarte and T.T.M. Rocha et al. (2020) Optimal design and environmental, energy and exergy analysis of a vapor compression refrigeration system using R290, R1234yf, and R744 as alternatives to replace R134a. *International Journal of Refrigeration*, Volume 113, 10-20.
3. Sedlář, J. (2022) *Refrigerants - introduction, definition, history*. TZBINFO. Retrieved from : *https://vetrani.tzb-info.cz/klimatizace-a-chlazeni/13626-chladiva-uvod-definice-historie*
4. Krainer, R., Duda, J. (2015) *Chladiva používaná v tepelných čerpadlech*. TZBINFO. Retrieved from : *https://vytapeni.tzb-info.cz/tepelna-cerpadla/12647-chladiva-pouzivana-v-tepelnych-cerpadlech*
5. Regulation (EU) No 517/2014 of the European Parliament and of the Council. (2014)
6. Kosco, J., Kudelas, D., Taus, P. (2016) Aplication of 1 MW heat pump york 0955ha and analysys of problems in audit of procedure. *International Multidisciplinary Scientific GeoConference-SGEM,* ISSN 1314-2704, 139-145.
7. Havelský, V. (2013) *Čo treba vedieť o tepelných čerpadlách*. Slovenský zväz pre chladiacu a klimatičnú techniku. Retrieved from: *https://szchkt.org/a/docs/news/232*
8. Čaja A., Kapjor, A., Tuhovčák, J., Hejčík, J. (2020) *Chladiace obehy a tepelné čerpadlá* - Vyd. 1 - [Košice]: Equilibria, s.r.o., 2020. - 139 p., fotografie, grafy, ilustrácie, schémy. - ISBN 978-80-8143-262-0.

Calculation of the Static Pressure Rise in the Hydropower Plant Penstock During Emergency Closing of the Flow

Peter Hlbočan[1,*], *Roman* Kiš[1], *Marta* Murgašová[1], *Zoltán* Csuka[1], and *Peter* Ďurčanský[2]

[1]ZTS - VÝSKUM A VÝVOJ, a.s., Lieskovec 598/75, 018 41 Dubnica nad Váhom, Slovak Republic
[2]University of Žilina, Faculty of Mechanical Engineering, Department of Power Engineering, Univerzitná 8215/1, 010 26 Žilina, Slovak Republic

Abstract: The paper presents an approach to calculate the static pressure rise upstream of a valve in a pressure penstock of a hydropower plant (HPP), i.e. a valve upstream of a turbine, during an emergency closure of the flow. The calculated value of the pressure rise is applied for the design of the appropriate time of the valve emergency closure in the HPP penstock and for the sizing of the various parts of the HPP (structural and mechanical). The basic theoretical background of the problem, calculation relations, the range of validity of the used calculation relations are explained and then the procedure applied to the solution of a specific task from practice is explained.

Keywords: *water hammer; static pressure rise, hydropower plant penstock; water turbine; emergency closing*

1 Introduction

It is a well-known fact that the magnitude of the static pressure rise upstream of the valve in a penstock (during valve closure) depends on the time of valve closure. At the same time, the magnitude of the maximum increase in static pressure decreases with increasing closure time. The total closing time is therefore an important parameter that influences a number of other technical parameters of the hydropower plant (HPP). With very short closure times of the valve in the penstock, the pressure increase upstream of the valve is too large and represents a potential safety risk due to the possible exceeding of the allowable stresses on the penstock and the associated mechanical and structural components of the HPP. In other words, both the structural and the mechanical components of the HPP must be oversized to be able to handle the considered maximum pressure increase in the penstock. Sizing the affected components of the HPP for too large additional load (due to a large increase in the static pressure in the penstock) may be technically infeasible under standard conditions, or it may represent a significant increase in investment costs. On the other hand, for larger closure times the pressure increase is smaller. Here, however, the very purpose of the emergency closing of the valve in the HPP penstock has to be taken into account.

* Corresponding author: peter.hlbocan@ztsvav.eu

The need for emergency closing may occur in the event of a sudden machinery failure, or e.g. in the case of so-called energy-free operation of the turbine. This occurs when a grid failure becomes and the generator suddenly disconnects from the electrical grid, causing the turbine to operate without load. This results in an increase in the speed of the whole system up to the value of runaway speed, which in the case of Kaplan turbines can reach about 2.5 times the normal operating speed. Particularly in the case of high specific speed turbines with high operating speeds, the operation of the turbine at runaway speeds can cause significant loads on the entire turbine-generator assembly, particularly the bearings. In more detail, the issue of turbine runaway speed is discussed in the literature, e.g. in [7]. It is necessary to consider the maximum time of possible energy-free operation already at the project and design phase of the HPP and to ensure that all critical structural nodes are designed for the considered duration of energy-free operation at runaway speeds.

In the project, design and to some extent also in the implementation phase of the construction of a power plant, special attention must be paid to the planned safety and reliability of operation. However, the design and construction phase is not the only phase of the HPP life cycle where the safety and reliability of the HPP operation needs to be addressed. Studies investigating the actual operation of the HPP are equally important, as fault and outage-free power generation is highly important for HPP operators and, as experience has shown, problems can occur after years of use, e.g. due to operational wear of the materials.

Computational studies at specific design stages are aimed at safe and reliable operation, so all possible operational situations that may arise should be taken into account. Steady-state operation of the HPP is the safest condition because there are no changes in the hydraulic quantities in the system. However, if any of the hydraulic variables change during operation of the hydropower plant, steady-state operation is violated and the system state changes abruptly – so-called transients, which occur during the change from one steady state to another steady state. They can cause significant static pressure changes in the penstock. Excessively high pressures can lead to considerable physical damage. Damage to turbine parts, valves or the penstock can occur. This damage can in some cases cause environmental damage and also endanger human health and life.

The importance of precise handling of the pressure rise issue during emergency shutdown and subsequent strength analyses of individual structural or building nodes is documented by concrete examples from practice, where the neglect of a precise approach to this issue has led to the devastation of the HPP.

In the past, there have been more or less serious hydropower plant accidents, which have occurred for various reasons. In the accidents mentioned below, the turbine penstock was severely damaged due to the closing of the inlet valve too quickly or the neglect of the effects of ageing of the materials on the reduction of the maximum operating pressure.

In 1950, a catastrophic accident was recorded at HPP Oigawa (Japan). The penstock ruptured 14 years after commissioning due to the sudden closure of the inlet butterfly valve (Fig. 1.). The main cause of the accident described by Bonin [8] was insufficient safe throttle valve adjustment. This accident killed three employees, caused $500 million in damage and the loss of 90 GWh of electricity.

Fig. 1. Failure of the Oigawa (Japan) penstock in 1950 [9].

Another accident occurred in 1997 at the Lapino HPP after 70 years of operation. The steel penstock broke during the loading of only 50 % of the rated capacity during the acceptance tests of the new regulator (Fig. 2.). In this case, the rapid closing of the valve was set to approximately 2 s at 100 % stroke. The cause of this accident was identified by Adamkowski [9] as material fatigue and corrosion of the penstock after 70 years of operation. The fatigue and corrosion led to poor quality welded joints, which affected the MAWP (maximum allowable working pressure) parameter of the penstock. Insufficient reinforcement at points of high stress concentration also contributed to penstock failures.

Fig. 2. Failure of the penstock of the Lapino power plant (Poland) in 1997 [10].

2 Theoretical background

When closing the valve on a HPP penstock quickly, a so-called water hammer may occur. Several approaches can be taken to deal with water hammer. In practice, analytical computational procedures known from "classical" fluid mechanics are commonly used [1, 2]. The problem of water hammer is generally well known and is the subject of a relatively large number of papers, e.g. [3 – 6]. In general, a distinction is made between direct and indirect water hammer.

From the point of view of the occurrence of direct or indirect water hammer, the reflection time of the shock wave (T_r), which is determined according to (1). It depends on the length of the pipe (l) and the velocity of propagation of the pressure wave in the fluid (in the case of a HPP penstock in water v_v). The velocity v_v can be determined from (2) for the case of a flexible 'loose' pipe and water with real compressibility (elasticity). This depends on the volumetric elasticity modulus of the fluid (K), the density (ρ), the diameter of the pipe (d), the elasticity modulus of the pipe wall material (E) and the pipe wall thickness (s).

If the total closure time (T_f) is less than the shock wave reflection time (T_r), a so-called direct water hammer occurs, which causes a relatively high increase in static pressure (5) upstream of the closure and thus poses a rather large safety risk to both the structural and mechanical components of the HPP. In opposite case, if the total closure time (T_f) is longer than the reflection time of the shock wave (T_r), an indirect water hammer will occur, whereby with increasing closure time, the increase in static pressure in front of the valve (6) is smaller. The relationships (1) to (6) are explained in more detail in reference [1].

Table 1. Calculation relations used for pressure rise calculations in the penstock

Shock wave reflection time	[s]	$T_r = \frac{2 \cdot l}{v_v}$	(1)
Velocity of propagation of a pressure wave in a liquid	[m/s]	$v_v = \frac{\sqrt{\frac{K}{\rho}}}{\sqrt{1 + \frac{K \cdot d}{E \cdot s}}}$	(2)
Direct water hammer	[s]	$T_f \leq T_r$	(3)
Indirect water hammer	[s]	$T_f \gg T_r$	(4)
Direct impact pressure rise (in meters of water column)	[mH_2O]	$\Delta H = \frac{v_0 \cdot v_v}{g}$	(5)
Increase in indirect impact pressure (in meters of water column)	[mH_2O]	$\Delta H = \frac{v_0 \cdot v_v}{g} \cdot \frac{2 \cdot l}{v_v \cdot T_f}$	(6)

3 Calculation of pressure rise during emergency valve closing

In this paper, we present the calculation of the pressure rise in a penstock for a specific practical example. The calculation of the static pressure rise has been carried out for the case of closing the valve at the end of the HPP penstock (upstream of the turbine) according to the scheme in Fig. 3.

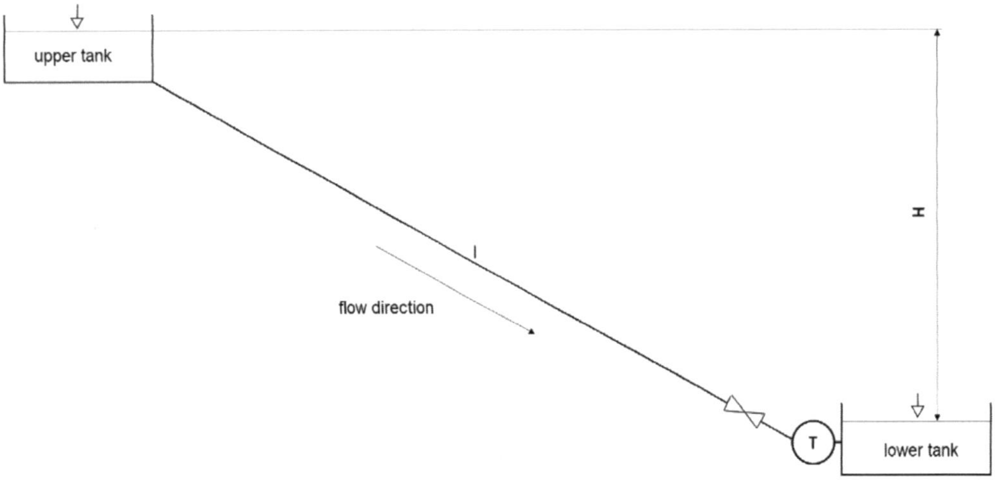

Fig. 3. Schematic of the model penstock of the hydropower plant.

The specific operational, geometrical and material parameters of the system are given in Table 2. Table 2 also lists the values of the pressure wave propagation velocity and the pressure wave reflection time, which determines the occurrence of either direct or indirect water hammer. The calculation of the pressure rise was first carried out for the case of very rapid (virtually instantaneous) closure of the valve ($T_f = 0.05\ s$), which is less than the value of the shock wave reflection time for the system ($T_r = 0.078\ s$) and hence in this case we speak of a direct water hammer. Subsequently, a series of calculations were carried out for several different closure times greater than the shock wave reflection time, corresponding to the indirect water hammer case.

The calculated pressure rise values were then compared (for the direct shock case they are in Table 3 and for the indirect shock case they are in Table 4). At the same time, the results of the indirect water hammer calculations for total closure times of 0.8 s to 30 s are processed graphically in Fig. 4. The obtained results can be used as a basis for an appropriate choice of the total valve closing time so that all the operational, technical and safety conditions mentioned in the introduction of this paper are respected.

Table 2. Parameters of the system on which the pressure rise calculation was performed.

Flow in the penstock	Q	m^3/s	8.02
Net power plant gradient	H	m	7.5
Pipe length	l	m	40
Pipe diameter	DN	m	2
Pipe wall thickness	s	mm	20
Outer diameter of the pipe	d_o	m	2.032
Inner diameter of the pipe	d	m	1.99
Water flow rate upstream of valve	v_0	m/s	2.6
Gravitational acceleration	g	m/s^2	9.81
Water density		kg/m^3	1000
Modulus of volumetric elasticity of water	K	Pa	2030000000
Modulus of elasticity of the pipe wall	E	Pa	220000000000
Pressure wave propagation speed	v_v	m/s	1028.51
Pressure wave reflection time	T_r	s	0.078

Table 3. Calculated values of static pressure rise – direct water hammer.

Closing time – direct water hammer	T_f	s	0.05
Increase in pressure - direct water hammer	Δp	kPa	2646.7
	ΔH	mH_2O	269.80

Table 4. Calculated values of static pressure rise for different valve closing times – indirect water hammer.

Closing time - indirect water hammer	T_f	s	0.8	1	2	3	4	5
Increase in pressure - indirect water hammer	Δp	kPa	257.3	205.9	102.9	68.6	51.5	41.2
	ΔH	mH_2O	26.23	20.99	10.49	7.0	5.25	4.2
Closing time – indirect water hammer	T_f	s	10	15	20	25	30	-
Increase in pressure - indirect water hammer	Δp	kPa	20.6	13.7	10.3	1	6.9	-
	ΔH	mH_2O	2.1	1.4	1.5	0.84	0.7	-

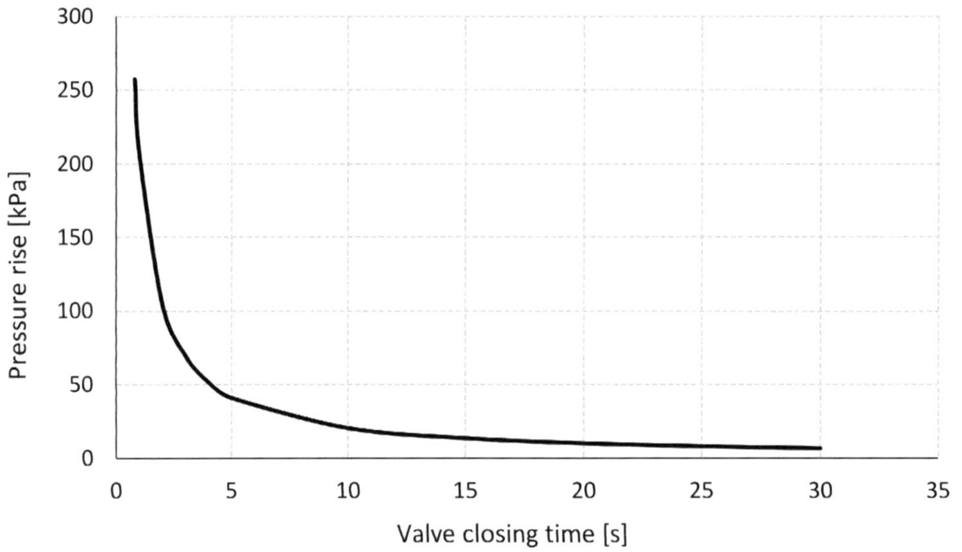

Fig. 4. Calculated dependence of static pressure rise upstream of valve as a function of closing time.

4 Conclusion

In this paper, we have explained the procedure of calculating the static pressure rise that occurs during the emergency closure of the valve at the end of the HPP penstock upstream of the turbine using a specific example. According to the results obtained, it is necessary to select a suitable value of the total time of the valve emergency closure so that the corresponding value of the static pressure increase is satisfactory from the point of view of safety, reliability of the equipment operation and from the point of view of mechanical stress on the machine parts and structural parts of the HPP. At the same time, the total closing time must not be too long so as not to prolong the duration of runaway speed in the case of energy-free operation of the turbine, which also poses a risk in terms of increased stress on the mechanical parts of the HPP (bearings, etc.). The specific value of the calculated pressure rise (for the selected shutdown time) is further used for the purposes of elasticity-strength analyses of individual parts of the structure, or for the purposes of dimensioning specific parts of the penstock and the valve upstream the turbine.

Authors would like to thank to project: Regenerácia použitých batérií z elektromobilov / Regeneration of used batteries from electric vehicles ITMS2014+: 313012BUN5
This research is supported by the project "Implementation of knowledge about modern ways of reducing the burden on the environment in the energetic use of solid fuels and waste into the pedagogical process", KEGA 032ŽU-4/2022.

References

1. Gančo, M. (1990). *Mechanika tekutín 2. vydanie.* Slovenská vysoká škola technická v Bratislave, Strojnícka fakulta, *Alfa – vydavateľstvo technickej a ekonomickej literatúry Bratislava.*
2. Janalík, J., Šťáva, P. (2002) *Mechanika tekutín.* Vysoká škola báňská – technická univerzita Ostrava, Fakulta strojní, katedra hydromechaniky a hydraulických zařízení.
3. Vintr, M. (2018).Výpočet přechodových stavů u MVE Skhalta, *Hydroturbo 2018, 24. medzinárodná konferencia o vodnej energetike 25. – 26. september 2018*, Bratislava, SR.
4. Lupa, S. – I., Gagnon, M., Muntean, S., Abdul – Nour, G. (2022) The Impact of Water Hammer on Hydraulic Power Units. *Energies, 15, 1526.*
5. Garg, R., K. (2018) Water Hammer Analysis of a Run-of-River Hydroelectric Power Plant in Lower Himalaya: A Case Study. *Water and energy international.*
6. Karadžič, U., Bergant, A., Starinac, D., Božovič, B. (2019). Water Hammer Investigation of the Shut-Down of a High-Head Hydropower Plant at Very High Reynolds Number Flows. *Strojniški vestnik - Journal of Mechanical Engineering 65(2019)7-8, 430-440. Journal of Mechanical Engineering.*
7. Nechleba, M. (1962). *Vodní Turbíny jejich konstrukce a príslušenství*, SNTL Praha.
8. Bonin, C.C.Water Hammer Damage to Oigawa power station. *J. Eng. Power. 1960, 82, 111–116.*
9. Adamkowski, A. Case Study: Lapino Powerplant Penstock Failure. *ASCE J. Hydraul. Eng. 2001, 127, 547–555.*

On Vorticity Dynamics in Turbulent Wake behind Circular Cylinder

Václav Uruba[1,2,*], and *Pavel* Procházka[2]

[1]Department of Power System Engineering, Faculty of Mechanical Engineering, University of West Bohemia, Universitní 8, 306 14 Plzeň, Czech Republic
[2]Department of Fluid Dynamics, Institute of Thermomechanics of the Czech Academy of Sciences, Dolejškova 5, 182 00 Praha, Czech Republic

Abstract. The circular cylinder turbulent wake is studied experimentally using Particle Image Velocimetry (PIV) method. The vorticity component parallel to the cylinder axis is evaluated from measured velocity field. Dynamics of the vorticity field is analysed using the Oscillation Pattern Decomposition (OPD) method. The 3 dominant modes corresponding to the 1st, 2nd and 3rd harmonics of Strouhal frequency are detected and presented. The corresponding topologies are characterized by waves of vortical systems. The results are similar to the dynamics of velocity field, however the procedure is much easier to implement, as only half number of degrees of freedom are active.

Keywords: *cylinder; wake; turbulence; vorticity; dynamics*

1 Introduction

Circular cylinder in cross-flow is a canonical case representing flow around a bluff-body. This problem is addressed in many studies of all kinds, experimental, theoretical and numerical studies could be find in literature published in last decades and even much earlier.

It is known, that for Reynolds numbers higher than about 44 the Kármán-Bénárd vortex street is generated in the wake. The two dimensionless parameters characterize the wake dynamics. The Reynolds number *Re* is the dimensionless velocity and the Strouhal number *Sr* is the dimensionless frequency:

$$Re = \frac{U \cdot D}{\nu}, \quad Sr = \frac{f \cdot D}{U}, \tag{1}$$

where U is the incoming flow velocity, D is the cylinder diameter, ν is the kinematic viscosity of the fluid and f is the shedding frequency of vortices in the wake.

In literature, extensive information resources could be found, starting from dawn of fluid mechanics until now. The first systematic work by von Kármán [1] followed by overview paper by Roshko [2], and Williamson [3] to mention the most cited. Interest of researches is not changing a lot during the time, however the technical means applied in the research, involving both experimental approach and mathematical modelling, have developed

* Corresponding author: uruba@it.cas.cz

substantially. Recently, we have appropriate means available for addressing even the most complex aspects of the phenomenon.

The flow dynamics in the wake represented by the Kármán vortex street has been simplified very often to a single-frequency, periodical process characterized by the Strouhal frequency. In reality, this approach could be accepted for the fully laminar case when the Reynolds number value is very low, bellow 150. For higher *Re*, the wake undergoes laminar-turbulent transition and the wake is characterized rather by a broad-band process with random component and it becomes fully 3D topology, see e.g. [4, 5]. The advanced methods for turbulent spectra from spatial data have been developed recently, see e.g. [6].

In the presented paper we will show the analysis of vorticity in the cylinder wake. The study uses the velocity data analysed directly in preceding studies [7, 8].

2 Experiment and data treatment

The experimental setup, used instrumentation and data analysis methods are to be presented in this chapter.

2.1 Experimental setup

The experiments were carried out in the IT ASCR Prague, a cylinder is placed in closed test section of the wind tunnel with the test section cross section 250 x 250 mm^2. The air flow in the test section was of a good quality, velocity departures were within 1 % of the set velocity across the test-section (outside boundary layers) and the intensity of turbulence was below 0.2 %.

In Fig. 1 there is the schematic view on the cylinder in the flow perpendicular to the cylinder axis. The green plane, parallel to the flow and perpendicular to the cylinder axis, represents the Plane of Measurement (PoM), (*xy*), in which the velocity measurement take place.

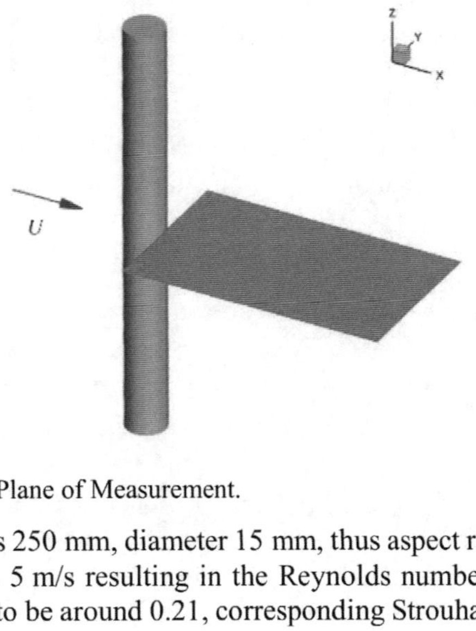

Fig. 1. Experimental setup with Plane of Measurement.

Length of the cylinder was 250 mm, diameter 15 mm, thus aspect ratio was more than 16. The flow velocity was set to 5 m/s resulting in the Reynolds number of about 4 815. The Strouhal number is expected to be around 0.21, corresponding Strouhal frequency was about 71 Hz, see [7].

2.2 Instrumentation

The Particle Image Velocimetry (PIV) technique has been applied in the PoM, see Figure 1.

The velocity vector field was measured using Particle Image Velocimetry (PIV) method. The measurement apparatus consists of laser and CMOS camera by Dantec company. The laser is New Wave Pegasus, Nd:YLF double head with wavelength of 527 nm, maximal frequency 10 kHz, shot energy is 10 mJ (for 1 kHz) and corresponding power is 10 W per one head. The camera Phantom V611 with resolution of 1 280 x 800 pixels is able to acquire double snaps with frequency up to 3000 Hz (full resolution) and it uses internal memory 8 GB. The data were acquired and post-processed in Dynamic Studio and Matlab software.

The PIV measurement was performed in the plane perpendicular to the cylinder axis and parallel to the flow direction (*xy*). The evaluated velocity fields consisting of 159 x 99 vectors were acquired in two different regimes. The data was acquired with frequency 2 kHz, the record contained 4 000 double-snapshots representing 2 s in real time.

More detailed description of the experimental setup could be found in [7, 8].

2.3 Data interpretation

The PIV method acquires instantaneous velocity field, 2 components in each measuring point. In the presented paper we will work with vorticity field, which was evaluated from the velocity field.

Vorticity $\boldsymbol{\omega}$ is a vector, mathematically it is the curl of the velocity $\mathbf{u}$:

$$\boldsymbol{\omega} = \nabla \times \mathbf{u} . \tag{2}$$

The physical interpretation of the vorticity is the local spin of a fluid particle. It generates vorticity rather than the rigid body rotation of a fluid mass, in which particles do not move relative to each other. It could be interpreted as twice the rotation rate of fluid particles. The vorticity is used to detect *vortices* very often, however the nonzero vorticity indicates presence of *shear* regions, which are natural part of any vortex, but not limited to them.

Essential feature of vorticity is its invariance to Galilean transform. Any linear transformation of a velocity field could change its topology as for the vortices presence, changing them into waves, however the vorticity field remains unchanged.

In our experiment, we acquired the in-plane velocity components u and v in the PoM (*xy*). Thus, we could evaluate the single vorticity component ω_z perpendicular to the PoM. For simplicity, we will refer to this scalar quantity as ω hereinafter:

$$\omega = \frac{\partial v}{\partial x} - \frac{\partial u}{\partial y} . \tag{3}$$

So, from the acquired velocity vector field $\mathbf{u} = [u; v]$ we obtained the scalar field ω with half number of variables.

2.4 Data analysis

The vorticity field is subjected to dynamical analysis, the spatial aspects of dynamical behaviour are to be shown next. The topology of typical structures in terms of fluctuation velocity fields are to be presented.

The Oscillation Pattern Decomposition (OPD) method is applied. The OPD method provides series of OPD modes. Each OPD mode is characterized by its topology in complex form (consisting of real and imaginary parts), frequency f and attenuation of the pseudo-periodic (oscillating) behavior. Attenuation or amplitude decay is described by so called e-folding time τ_e representing the mean time period of the mode amplitude decay by factor "e".

The other decay characteristics is the dimensionless "periodicity" p which expresses the e-folding time in multiples of periods of the corresponding OPD mode defined by its frequency:

$$p = \tau_e \times f \, . \tag{4}$$

The periodicity value could be considered as a measure of "relevance" of a given OPD mode, the higher p, the higher relevance. The frequency will be represented in dimensionless form as Strouhal number Sr, see (2).

More details on the OPD method could be found in [9, 10].

3 Results

The results are to be presented in dimensionless form, coordinates are given in multiples of the cylinder diameter D.

3.1 Statistics

The time series of vorticity fields were subjected to a standard statistical analysis.

In Fig. 2 distributions of mean vorticity field (a) and standard deviations (b) are shown, respectively

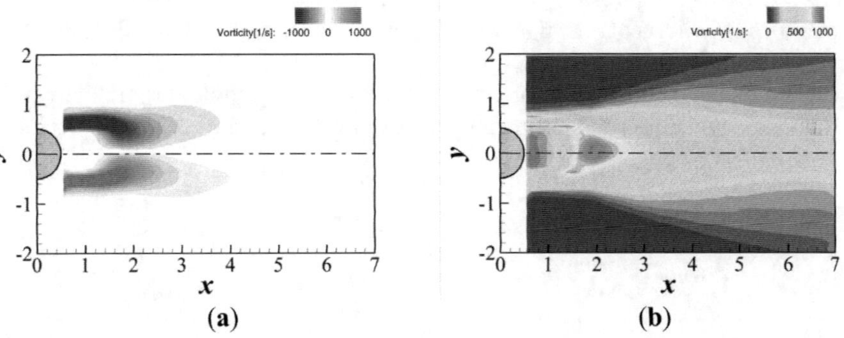

Fig. 2. Distributions of mean vorticity field (**a**) and standard deviations (**b**), respectively.

The two regions of concentrated vorticity in the near wake up-to distance 4 cylinder's diameters downstream are detected. The vorticity distribution is symmetrical around the axis x, although the vorticity sign is negative in the upper half-plane and positive in the lower half-plane. The standard deviation distribution locates well the wake border, the maximum is on the axis x in the position 2 D. the topology is symmetrical around the x-axis.

3.2 Dynamics

Dynamical analysis of the vorticity field was carried out using the OPD method. The velocity field analysis was presented in [7].

In Fig. 3 the spectrum of the dynamical analysis of velocity (a) and vorticity (b) fields are shown in Sr-p plain. The Sr_1, Sr_2 and Sr_3 represent the 1st, 2nd and 3rd harmonics of the Strouhal frequency, respectively.

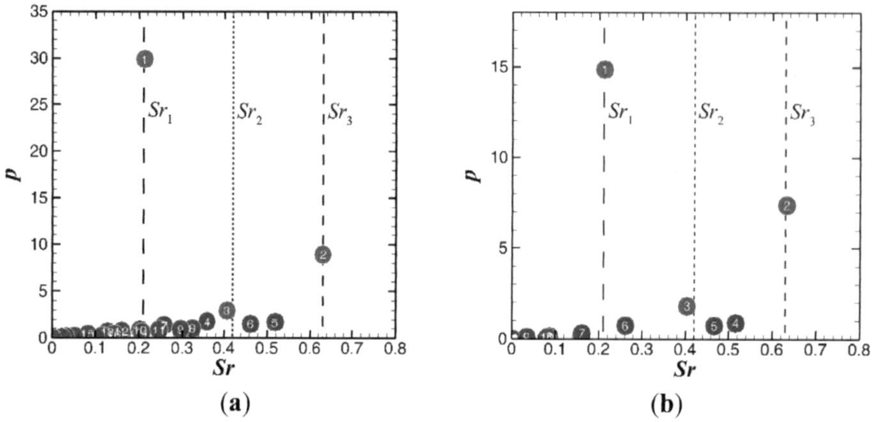

Fig. 3. OPD spectrum, velocity (**a**), vorticity (**b**).

The spectra are almost identical, however the velocity spectrum contains more high-order modes and the values of the periodicity is approximately twice as high comparing to the vorticity case. This means that the structures looks more distinct in the velocity field than in vorticity field. However in both cases we recognize the most relevant 3 OPD modes.

The most relevant OPD mode 1 is characterized by the Strouhal number $Sr_1 = 0.21$, as expected. The second mode OPD 2 corresponds to the Strouhal number $Sr_2 = 0.42$ and the second mode OPD 2 corresponds to the Strouhal number $Sr_3 = 0.63$. The OPD 3 periodicity is well above 1, thus periodical, the higher modes (4, …) have too small p indicating decaying character.

In Fig. 4 the real and imaginary parts of the OPD 1 mode topologies are shown. Please note that the vorticity values of OPD modes are normalized and thus dimensionless.

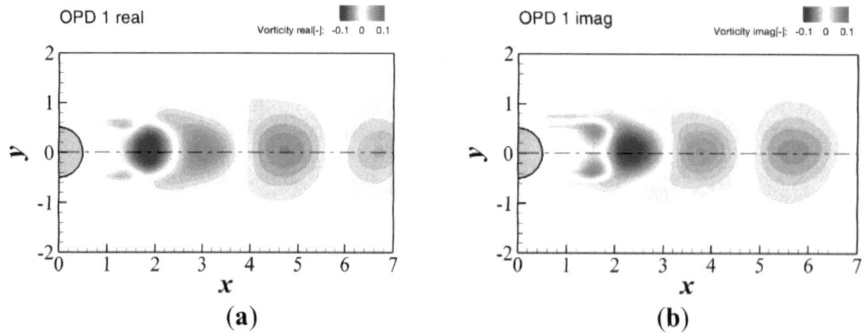

Fig. 4. OPD mode 1 topology, real (**a**) and imaginary (**b**) parts, respectively.

The real part represents 0 phase angle situation, imaginary part is the situation in $\pi/2$ phase angle (or ¼ of period, if you like), follows negative real part in π phase and negative imaginary part for phase $3\pi/2$. This cycle is repeated, the phase 2π is identical to 0 phase angle.

We could identify the vortices row with centres on the x-axis and with alternative orientations, negative and positive. In particular in the real mode topology we could recognize the negative vorticity (i.e. vortices) located in positions $x = 1.86$ and 4.71 and positive vorticity in positions 3.02 and 6.67. In imaginary part, the vortices locations are negative in positions 2.32 and 5.66 and positive in position 3.75.

The OPD mode 1 represents the wave of vortices with alternative orientations moving downstream with velocity about $0.7U$, accelerating along the axis x and getting weaker. This is exactly expected as topology of the Kármán-Bénárd vortex street.

The OPD modes 2 and 3 are again periodical with a row of vortices, real and imaginary parts are shifted in space in the same way as for the OPD 1. We will present the real parts of those OPD modes. Topologies of the OPD 2 and 3 are in Fig. 5.

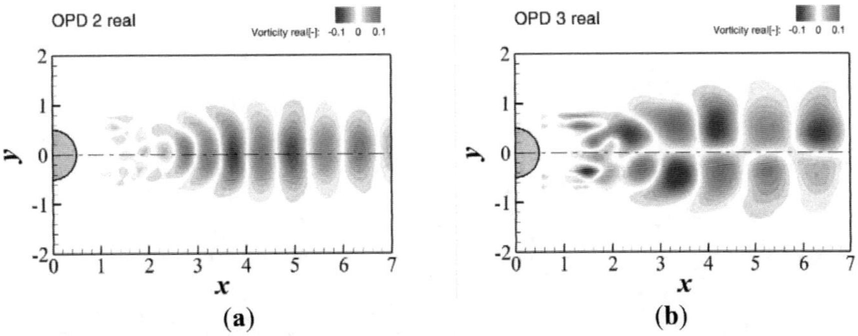

Fig. 5. OPD 2 (**a**) and 3 (**b**) topologies, real parts.

The OPD mode 2 topology in Fig. 5a is similar to the OPD 1, however the vortices spacing in x direction is about 1/3. The vortices are squeezed, as spanwise dimension remains more or less the same.

The topology of the mode OPD 3 in Fig. 5b is rather different, it is formed by two-roads of vortices in chequered pattern. The spacing is about ½ of that for the OPD 1. We remind, that this mode corresponds to the second harmonics Sr_2.

As the spacing of vortices is inversely proportional to the corresponding frequency, we could deduce that the convection velocity of the vortical patterns is more or less the same for all 3 OPD modes, around 70% of incoming velocity accelerating slowly downstream.

The higher OPD modes are characterized by relatively small periodicity p, this indicates that the appearance and disappearance of such modes is more random and short-time process. Theirs topologies are less regular, more random. We will show a few examples. The OPD 5 with relatively high frequency, $Sr = 0.52$, then OPD 6 with lower $Sr = 0.26$, even lower frequency $Sr = 0.16$ for OPD 7 and in the end the OPD 10 with $Sr = 0.08$. The real parts topologies of the above mentioned OPD modes are shown in Fig. 6a-d.

https://doi.org/10.1051/matecconf/202236905001

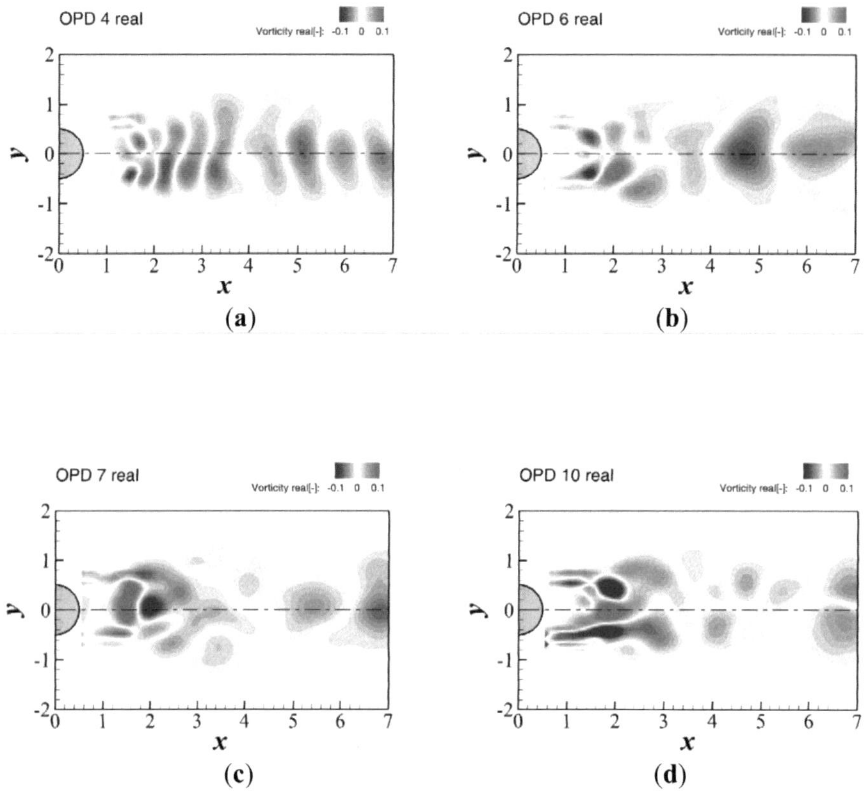

Fig. 6. OPD 4 (**a**), 6 (**b**),7 (**c**) and 10 (**d**) topologies, real parts.

The topologies of OPD modes in Fig. 6 are more pronounced in the near wake just behind the cylinder and they are linked to free shear layers in this region. The nature of dynamics is more pulsatile than wavy.

4 Conclusions

The circular cylinder turbulent wake is studied experimentally using Particle Image Velocimetry (PIV) method. The vorticity component parallel to the cylinder axis is evaluated from measured velocity field. Dynamics of the vorticity field is analysed using the Oscillation Pattern Decomposition (OPD) method.

The 3 dominant modes corresponding to the 1st, 2nd and 3rd harmonics of Strouhal frequency are detected and presented. The corresponding topologies are characterized by waves of vortical systems. The 1st and 3rd harmonic vortex waves are located on the x-axis in a single row, while the 2nd harmonic mode is represented by vortices in two-row chequered configuration. All vortex rows are convected in the streamwise direction, convection velocity is about 70 % of the incoming flow velocity in the near wake region accelerating gradually in the streamwise direction. Convection is more-or-less the same for all 3 modes.

The results obtained by the vorticity field analysis are similar to the dynamics of velocity field presented in [7], however the procedure is much easier to implement, as only half number of degrees of freedom are active.

This publication was supported by the project of University of West Bohemia in Pilsen SGS-2022-023.

References

1. von Kármán, T. (1954). Aerodynamics. *Columbus*: McGraw-Hill.
2. Roshko, A. (1955). On the development of turbulent wakes from vortex streets. *NACA Rep. 1191,* pp.801-825.
3. Williamson, C.H.K, (1996). Vortex Dynamics in the Cylinder Wake, *Annu. Rev. Fluid. Mech.* 28, pp. 477-539.
4. Uruba, V. (2016). On 3D instability of wake behind a cylinder, *AIP Conference Proceedings*, 1745, art. no. 020062.
5. Yanovych, V., Duda, D., Uruba, V. (2021). Structure turbulent flow behind a square cylinder with an angle of incidence, *European Journal of Mechanics, B/Fluids*, 85, pp. 110-123,
6. Duda, D., Uruba, V. (2019). Spatial Spectrum from Particle Image Velocimetry Data. *Journal of Nuclear Engineering and Radiation Science*, 5 (3), Article Number 030912
7. Uruba, V., and Procházka, P. (2020). On Harmonic Contents of the Wake behind a Circular Cylinder, *26th International Conference on Engineering Mechanics (IM), Engineering Mechanics 2020*, pp.500-503,
8. Uruba, V., Procházka, P., Skála, V. (2020). On the 3D Dynamics of the Wake Behind a Circular Cylinder, In *Proceedings Topical Problems of Fluid Mechanics 2020*, Prague, Edited by D. Šimurda and T. Bodnár, pp. 240-248,
9. Uruba, V. (2012). Decomposition methods in turbulent research, *EPJ Web of Conferences*, 25 01095.
10. Uruba, V. (2015). Near Wake Dynamics around a Vibrating Airfoil by Means of PIV and Oscillation Pattern Decomposition at Reynolds Number of 65 000, *Journal of Fluids and Structures*, 55, pp. 372-383.

Drop Hunter Isokinetic Probe for Cooling Tower Drift Measurement

Pavel Procházka[1,*], *Veronika* Barraclough[2], *Václav* Uruba[1,3], and *Vladislav* Skála[1]

[1]Institute of Thermomechanics, ASCR, v.v.i., Dolejškova 5, Praha 8, CR
[2]4JTech s.r.o., Ringhofferova 115/1, Praha 5, CR
[3]UWB, Faculty of Mechanical Engineering, Department of Power System Engineering, Universitní 8, Plzeň, CR

Abstract. This paper is devoted to the development of a new probe to measure a drift from cooling towers. The probe was named DPIK (Drop Hunter Isokinetic) and its development consists of a design of the aerodynamic and microelectronic part of the probe, verification methods, design and testing, verification tests in laboratory conditions and measurements on a model of the cooling tower. The paper deals mainly with the verification tools, as these methods are very closely related to the aerodynamic part of the probe design. Two options were chosen for verifying the probe principle. The first of these was the Interferometric Particle Imaging (IPI) method, the second was another optical method using similar experimental equipment Particle Image Velocimetry (PIV).

1 Introduction

Drift is the spectrum of water droplets released from cooling tower, created by the aerodynamic forces acting on droplets and films within the cooling tower and discharged into the environment. Typical drift droplet diameters range from 20 microns to 2000 microns. The cooling tower drift measurement is necessary, because the drift droplets contain both the chemicals and some biology species as the water circulating in the tower and this content could be potentially harmful and pathogenic (i.e. heavy metals and bacterium Legionella Pneumophila). Also, the drift causes considerable water loss. This loss can be 0,7-1,4 litres/m^2/hour, which represents a lower hundreds of cubic meters per one cooling tower per day (i.e. for an average nuclear power plant cooling tower).

Determining a drift rate from real cooling towers in operation becomes an important issue nowadays. Many states worldwide work on or already accepted legislation, which sets up required minimal drift eliminator (part of cooling tower capturing the drift) performance rate, e.g. for France 0.01 %, Spain 0.001 %. The drift eliminator efficiency is one of the two crucial characteristic parameters of it. The drift emission from a cooling tower is usually expressed as a percentage of the total circulating water flow rate. It can also be expressed as a mass per unit time related to water or air flow (as defined in [1]).

* Corresponding author: prochap@it.cas.cz

There are various methods used for eliminators drift measurement (real world conditions methods, laboratory conditions methods): methods employing chemically sensitive materials (sensitive papers-based methods, sensitive surfaces), isokinetic extraction-based methods (with subsequent chemical analysis), holographic method, methods based on light diffusion, interference method and laser-induced fluorescence method. More on these methods could be find in [2], [3] and [4].

In general, nowadays the drift from the real towers can only be measured using relatively complex methods, where the so-called tracers (e.g. Lithium, Sodium, Magnesium) are mixed into the whole circuit. The measurements are then disproportionately expensive and also time-consuming as the sampling is followed up by subsequent external chemical analysis. That is why a new method developed with respect to the expense and testing time would be very useful for the whole industry.

Fig. 1. The model of DPIK probe used for electronic parts development.

2 Experiment description

The paper mainly deals with the development of a new probe. The probe was named DPIK (Drop Hunter Isokinetic) and is based on calorimetric principle. It operates as follows: a liquid phase (droplets with a size of 5 μm to approx. 150 μm) is separated from the flowing air for a certain period of time and it is captured. Then it is subsequently evaporated using a specially designed heated element located inside this isokinetic probe. Heating takes place by means of power regulation, with the temperature rise being monitored. The whole development consists of a design of the aerodynamic and microelectronic part of the probe (also consideration of the heating elements correct location inside the probe belongs here), verification methods, design and testing, verification tests in laboratory conditions and measurements on the model cooling tower. The paper deals mainly with the verification tools, as these methods are very closely related to the aerodynamic part of the probe design. Two options were chosen for verifying the probe principle. The first of these was the Interferometric Particle Imaging (IPI) method, the second was another optical method using similar experimental equipment Particle Image Velocimetry (PIV).

IPI is a non-invasive optical method providing mainly information on the size of individual droplets in the inlet channel and anywhere in the probe. The results obtained may include the integral value of the cooling tower drift as well as the efficiency of the eliminator, or the probe channel efficiency in the case of the probe shape development, according to particle size, correlation between particle size and even speed, etc. The great advantage of this method is that the minimum measurable droplet sizes are almost order of magnitude smaller compared to other methods.

Similarly, the other optical method mentioned above - PIV - is a non-invasive method using the same laboratory equipment. In principle, PIV means taking two images of the same measured area within a very short time interval (in the order of hundreds of microseconds depending on the expected velocity in the measured area), the setting of which is the basic input parameter for obtaining relevant data. From these frames, the flow field in the probe can then be evaluated. The probe efficiency can be obtained from the subsequent purely mathematical processing after calculating the behavior of the two-phase flow.

The flow field was acquired using fast CMOS camera Phantom V611 with resolution 1280 x 800 pixels. Safex particles were illuminated using coherent light sheet of wavelength 527 nm from Laser New Wave Pegasus. The acquisition frequency of the recording was 100 Hz both for PIV and for IPI measurements. Only statistical evaluation of all quantities will be presented in this article.

The test stand (Fig. 2) is actually a model of a small cooling tower. It consists of an inlet part with an integrated fan, a straight part, an elbow and an output. There is a honeycomb in the outlet part to homogenize the flow entering DPIK probe. By means of the ventilator, desired speed of flow in the test stand can be set. The safex particles or water droplets are injected in front of the test stand inlet.

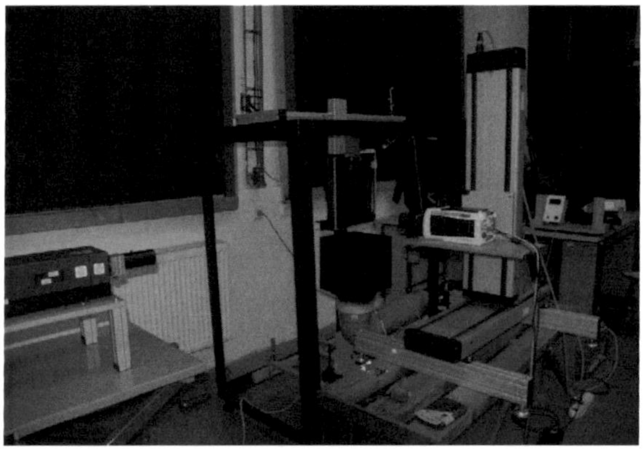

Fig. 2. Test stand in laboratory; PIV system, the DPIK probe and a cooling tower model.

Three individual models of DPIK probe were designed. There is a base case (A) in the figure 3a. The inner channel is branched into two symmetrical sections. The flow is divided just behind the inlet and it is mixed in front of the outlet section. The other two geometries utilize only one branch which significantly reduces the total size of the probe. The last variant (C) has short dead-end channel. The design of all three variants uses the same principle. The base plate is made using 3D printing (a puzzle from 4 pieces). The shape of the inner channel is defined using a transparent thin foil. The precise bending of the foil is very important since laser light is shining through it. The front panel (which is not visible in the figure 3) is used to cover and to seal the probe. It is also transparent to allow optical access for CMOS camera. The outlet section (red part) is a transition from rectangular cross-section 20 x 20 mm to circular one with second ventilator. The flow speed inside the probe is set according to the isokinetic condition.

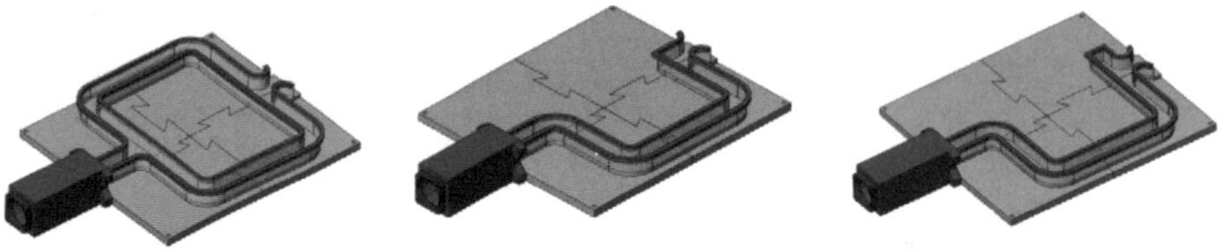

Fig. 3a, b, c. Three variants of inner channel geometry; branched channel (A) and two simple variant without (B) and with dead-end channel (C).

Dynamic similarity should be preserved to verify the efficiency of the probe. The model used for PIV and IPI measurement is twice bigger than the designed DPIK probe for real towers. It is very essential to keep the Reynolds and Stokes numbers, more details can be seen in [5]. For this model designed in scale 2:1, the flow rate is twice smaller than in reality and, on the other hand, the size of droplet should be twice bigger. Finally, we set the experiment so that the volume velocity is about 2 m/s and mean droplets size is about 30 microns.

3 Results

First, all three variant were measured by means of Particle Image Velocimetry to study the flow field topology. The overall view of statistical quantities is composed of 12 individual fields of view. Then measurement of details especially for particle mean size detection (IPI) followed.

3.1 Overall view

There is a distribution of velocity modulus for all geometries in the figure 4. The maximal velocity is approximately 3 m/s (red color). The origin of the coordinate system is placed in the middle of the entrance section. The flow enters into the probe, hits the ceiling and it is divided into two symmetrical parts in the case of the very first model. Considering the flow branching, the flow velocity is half the value with respect to inlet or outlet parts. Two elbows follow and finally both streams are mixed before entering the output fan section. This mixing phenomena significantly increases the turbulence intensity (Fig. 8a), which was, together with demand for smaller build-in dimensions, the main reason to design two more variants (B, C).

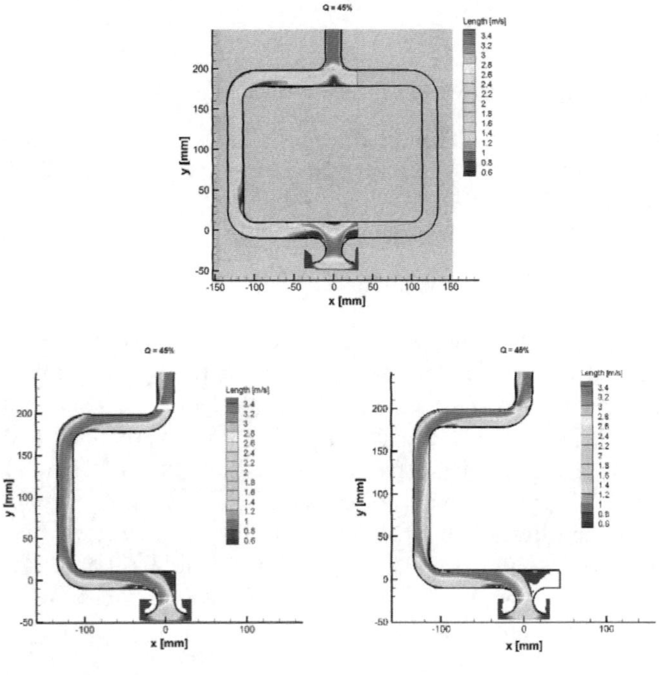

Fig. 4a, b, c. Mean flow fields of all variants; velocity distribution.

The flow enters the probe again from wide surroundings using two hemispheres. The velocity distribution is more regular – the volume flow rate is constant throughout the length of the probe. The flow separation can be seen behind each elbow. There is no mixing effect in front of the outlet part, however the velocity profile is not symmetrical across the output channel width compared to variant A.

The velocity vector lines can be seen in figure 5a, b, c. The presence of the dead-end channel seems to have no significant effect (Fig. 5c). The flow does not hit the ceiling but it is just turned to the left side in its entire volume.

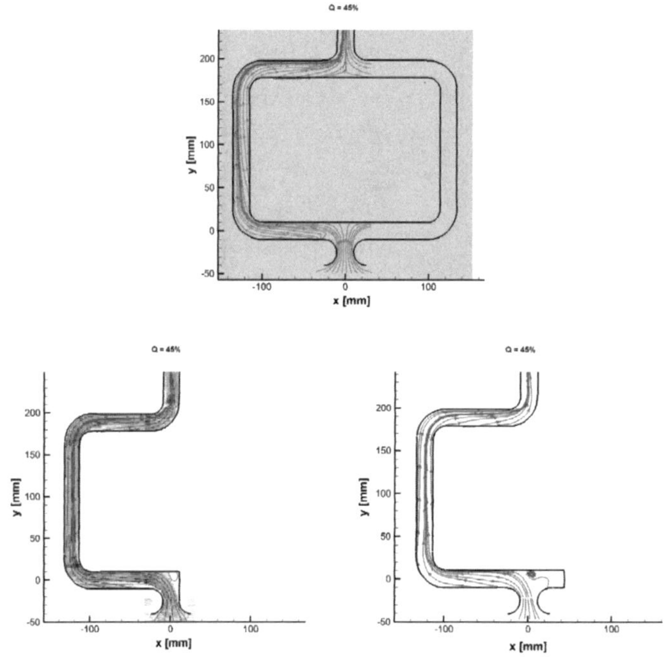

Fig. 5a, b, c. Mean flow fields of all variants; velocity vector lines distribution.

The overall velocity distribution in the probes suggests that most droplets will be captured in probe A on the wall just immediately opposite the inlet. Probes B and C will rather capture the droplets rather further in the channel.

3.2 The details - inlet and outlet parts

The most important part is the entrance section with the ceiling where we expect the droplet capturing and heated elements should be mounted (white arrows in the figure 1). There are velocity vectors for all three variants in the figure 6. The color denotes the value of velocity modulus, respectively distribution of velocity variations in the figure 7. The entrance is made by using two hemispheres so the flow enters this region regularly. The separation bubble(s) is present in the direct part of the channel (dark blue color). The separation is more significant for branched geometry. The vectors are oriented directly against the ceiling for the base case variant A which should be more suitable for capturing of the water droplets. Variants without junction do not capture so much water. Unfortunately, the effect of the dead-end on the flow field is almost negligible (compare B and C).

Variant A contains more fluctuation activity in the straight part behind the entrance section (green color in the figure 7a). Two variants without junction have rather regular flow with warp shear layers.

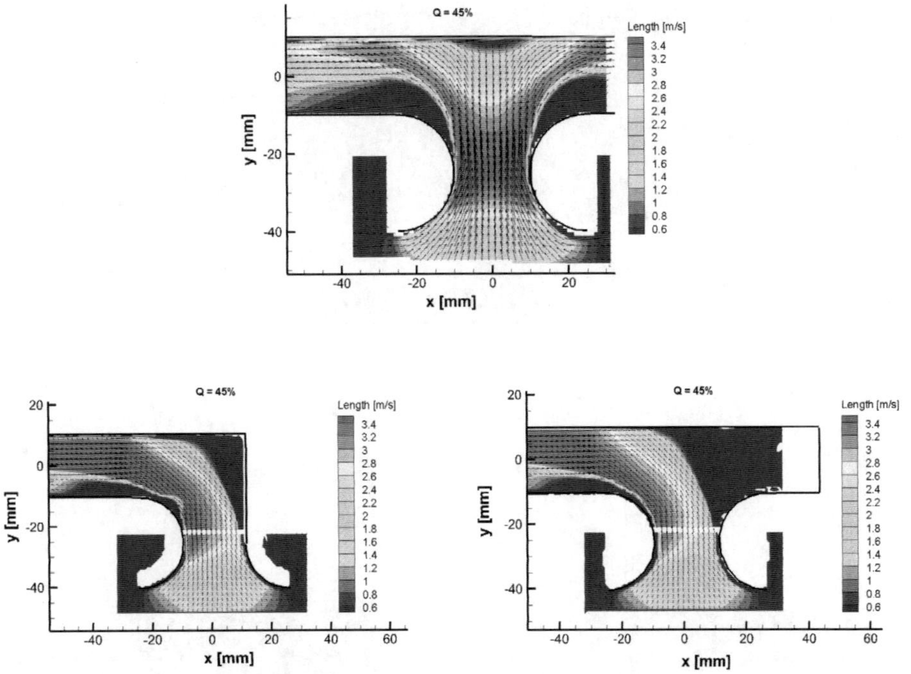

Fig. 6a, b, c. The inlet section; velocity distribution.

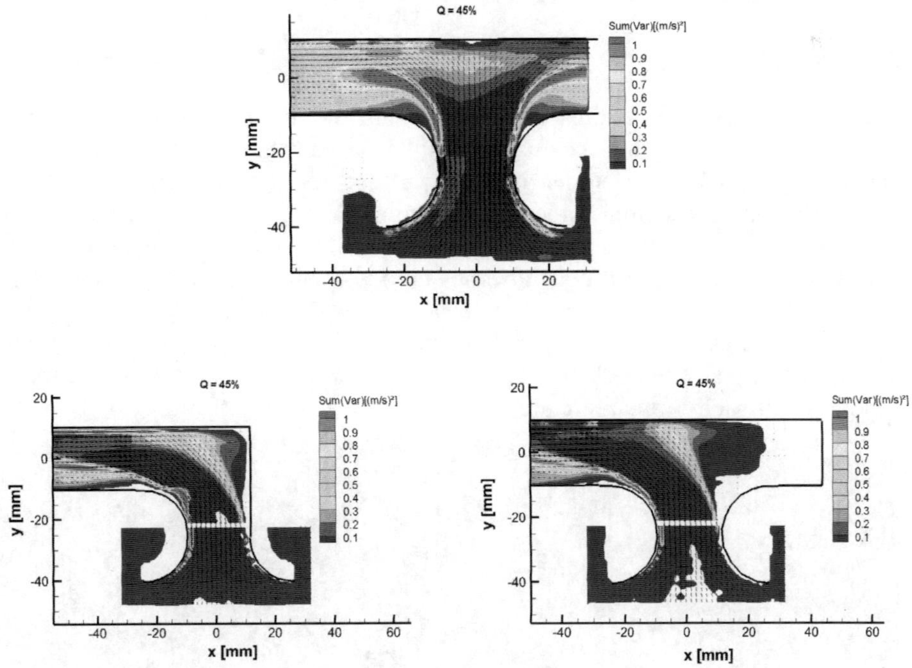

Fig. 7a, b, c. The inlet section; velocity variations (TKE) distribution.

Both streams are mixed in the T-junction of the first variant which results in high fluctuation activity (Fig. 8a). There is only small shear layer connected with tiny separation for the second and third variant equipped just with the last elbow. Future IPI measurement will show if this region also will have effect on droplets sticking.

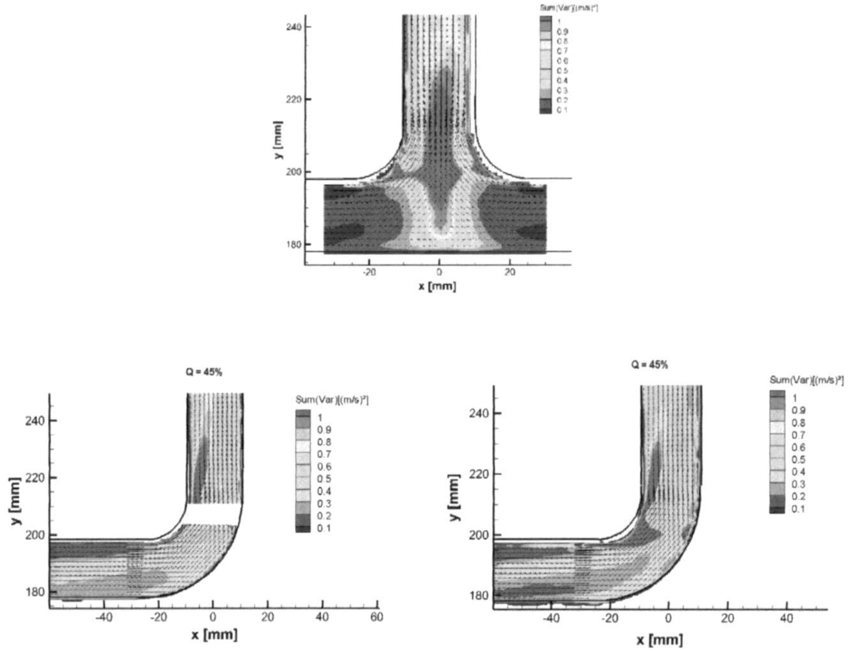

Fig. 8a, b, c. The outlet section; velocity variations distribution.

3.3 Particle detection

To evaluate the size and amount of water droplets, the IPIDET software was used. This algorithm is written in Python language as in-house software. The main purpose is to calculate the number of interferometric fringes of each defocused image of the droplets. There are IPI images acquired for entrance section (Fig. 9a) and inside the first elbow (Fig. 9b) taken for the base case variant A. As the size of the particle image is the same due to defocusing, the number of fringes gives us the information about the droplet size. The droplets are entrained at a velocity that is not very different from that of the volume velocity. The droplets in the inlet section contain more fringes (3-9) which means that droplets of different size enter the probe. Many droplets were captured as inside the elbow only droplets with 3 or 4 fringes are present.

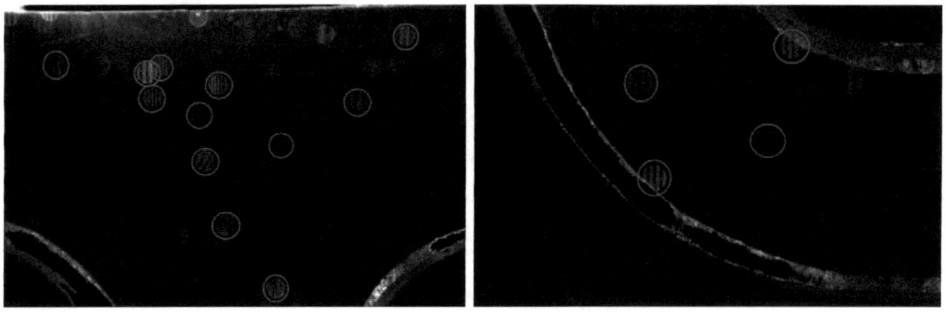

Fig. 9a, b. IPI images for inlet section and first left elbow.

Two histograms are plotted in figure 10. These histograms are calculated from 500 snapshots of IPI images. It is total amount of particles which were present in the flow for distinct number of fringes. The efficiency of used probe can be established using comparison of both histograms. Since droplet feeding is not constant in time, this information has rather qualitative character.

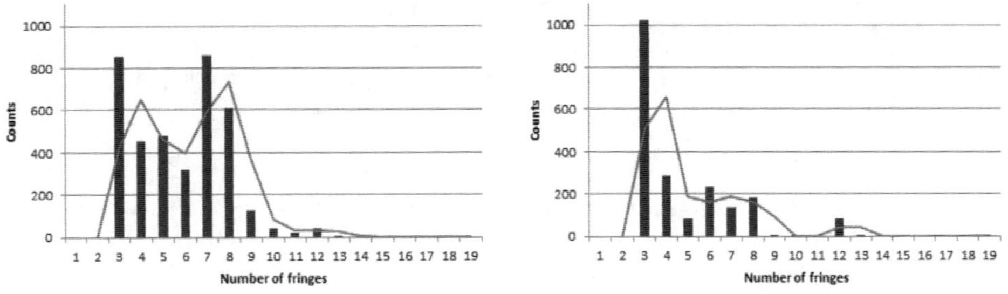

Fig. 10a, b. Calculated histograms of particles incidence for inlet section and first elbow.

4 Conclusions

Time-resolved PIV and IPI (both optical non-intrusive methods) were utilized to compare the design of three distinct DPIK probes.

Unfortunately, the geometry without branching channel is not able to capture sufficient amount of particles because the velocity vector are not oriented perpendicularly to the ceiling where heated element should be placed.

As we are not able to fulfil the repeatability of all measurements (feeding of droplets are not constant in time), during future experiments, we plan to make an acquisition using two fast cameras in the same instants.

The variant A is the most effective according to water droplets capturing. This geometry captured almost 50 % of the total number of particles present in the inlet (evaluated to the first elbow). However, the amount of captured water is even greater because especially large droplets were taken.

This work was supported by the research project of the Technology Agency of the Czech Republic, No. TK03020057 and by institutional support RVO61388998.

References

1. CTI Bulletin BUL-109: Nomenclature for Industrial Water-Colling Towers, (2019).
2. J. Čížek, L. Nováková, J. Nožička, Reducing Liquid Phase Drift from Cooling Towers, Gradient, Prague, (2009).
3. J. Čížek, L. Nováková, New methods for drift eliminators performance evaluation, CTI, TP11-11, (2011).
4. A. Roffman, L. D. Van Vleck, The State-of-the-Art of Measuring and Predicting Cooling Tower Drift and its Deposition, Journal of the Air Poll. Contr. Assoc. 24:9, 855-859, (1974).
5. P. Procházka, V. Barraclough, V. Skála, V. Uruba, The design and aerodynamic testing of DPIK probe, Engineering Mechanics 2022, Vol. 27/28, (2022).

Multi-Exposure PIV Measurements of Velocity Fields in Sprays.

Marek Mlkvik[1,*], *Robert Olšiak*[1], and *Branislav Knížat*[1]

[1]Slovak University of Technology, Faculty of Mechanical Engineering, Námestie Slobody 17, 81231, Bratislava, Slovakia

Abstract. This paper presents an approach to the use of the PIV method in the diagnosis of sprays generated by an effervescent atomiser. Due to the different density of the liquid phase depending on the distance from the nozzle, problems arise with the correct exposure of images for PIV analysis. The aim of the authors of this paper is to outline the possibility of solving this problem by composing a velocity field from partial measurements. To meet the objectives of the paper, in-house PIV equipment (hardware and software) was used rather than a commercial setup. This allowed for easier handling of the measured data and more sophisticated post-processing than offered by commercial products. It is clear from the results presented that, despite the fundamental differences in the optical properties of the spray particles, it is possible to obtain a velocity field from the discharge zone to the spray region with fine droplets. Moreover, it is possible to combine velocity measurements in the spray cone with measurements in the surrounding environment.

Research background: Spray is an environment with an abundance of tracers for PIV analysis (droplets), but their density, size and shape vary dramatically with distance from the nozzle. The use of PIV can therefore be challenging due to the demands of this method for correct image exposure.
Purpose of the article: Introduction of the application of the PIV method for environments with variable density and size of tracer particles
Methods: PIV, image processing.
Findings & Value added: By taking an appropriate approach to acquiring the source PIV images, it is possible to obtain information about the velocities throughout the spray cone as well as in the surrounding environment. The application of the proposed method requires a sufficiently large source data set (images) and sophisticated postprocessing. However, as a result, it is possible to obtain an overall view of the velocity field in the spray cone starting from the area behind the nozzle to the fine droplet region.

Keywords: *effervescent atomizer, liquid breakup, spray, particle image velocimetry, image acquisition.*

[*] Corresponding author: marek.mlkvik@stuba.sk

1 Introduction

Optical methods for measuring velocities in liquids are very popular nowadays and their importance is increasing with the decreasing investment costs of their implementation. The 1-D measurement methods, such as laser doppler anemometry (LDA) or phase doppler anemometry (PDA), play an indispensable role in spray diagnostics. These methods are the only ones that allow non-invasive measurements in the dense spray region with high temporal resolution. These methods allow information to be obtained about the processes involved in the atomisation of liquids, whether they are fuels [1, 2, 3, 4] or other substances [5, 6]. The disadvantage of 1-D methods is, in turn, the inability to obtain spatial resolution without a time-consuming traversing of the measured volume across the region of interest. For spatially resolved measurements, the Particle Image Velocimetry (PIV) method is appropriate. This works on the principle of recording the image of particles passing through the light plane at two subsequent time instants. This method is very popular in basic research in fluid mechanics [7, 8] or in applied research in the development of water and wind turbines [9], HVAC systems [10, 11] as well as aerodynamics [12]. Spray, as a medium consisting of a liquid phase dispersed in a gaseous environment, is a natural candidate for PIV diagnostics. It has been used to verify the function of new types of atomiser [13], or to monitor and describe the liquid breakup for different types of atomisers [14, 15]. This method can be used to track the motion of liquid particles in sprays [16] as the well as velocity fields in the surrounding gas [18].

This study will focus on the application of the PIV method in measuring the velocity field in a spray generated by an effervescent atomizer. This atomizer generates a two-phase mixture in an internal mixing chamber. The gas in the mixture, upon entering the free space, expands and breaks down the liquid into fibres (ligaments), which are further broken down. The reflective surface area of these structures is large compared to the small droplets in the final spray. This causes problems in the application of PIV if there is a need to measure the velocity field both in the region of the primary breakup and at greater distances from the nozzle.

In this study, the authors outline possible solutions to this problem by reconstructing the velocity field from partial measurements at different image exposures

2 Experiment

The test rig is depicted in Fig. 1. The tap water was used as the working fluid, the flow rate of which was regulated by a needle valve. Due to the very low water flows, the flow rate measurements were done by weighing the flowing amount of liquid in the collection tank over a defined time. Pressurized air, used as atomization gas, was prepared by a compressor. The air pressure value in the mixing chamber was kept constant by a pressure reducing valve. The air flow rate was measured using a laminar flow meter (OMEGA, FLR1600).

The spray was generated by an effervescent atomizer, the characteristics of which are depicted in Fig. 2. The operating modes of the atomizer were defined by the overpressure in the mixing chamber and the GLR parameter, defining the gas-liquid mass flow rate ratio. The overpressure in the mixing chamber was 100 kPa for all experiments presented.

The PIV method was implemented in a double-pulse/double-exposure configuration. The apparatus, depicted in Fig. 3, was assembled at our facility; no commercial PIV system was used. Illumination was provided by a dual pulse laser (Quantel Evergreen) with a maximum flash energy of 70 mJ. The laser was fitted with a cylindrical and convex lens to achieve optimal laser plane parameters at a distance of 50 cm from the laser head. The flash rate was 15 Hz. The particle image was captured with a high-speed PIV camera (IDT, Y3) equipped with a 50 mm Nikon lens and a 532 nm bandpass filter. The field of view was adjusted with a 5 mm spacer ring to a dimension of 110 x 110 mm. The synchronisation of individual components was provided by an 8-channel timing hub (IDT) via TTL signals.

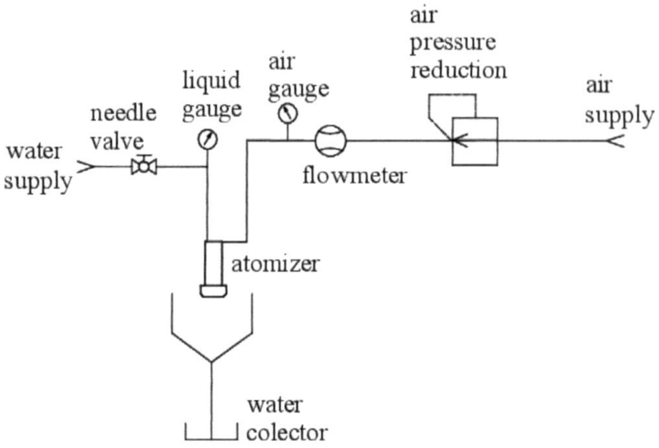

Fig. 1. Schematic drawing of the test rig.

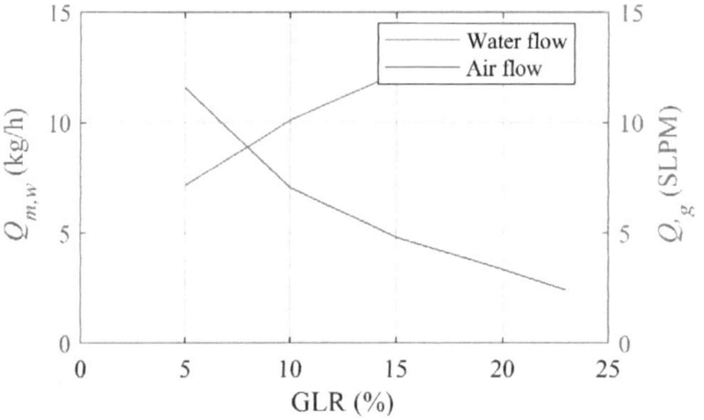

Fig. 2. Measured characteristics of the tested atomizer working with internal overpressure pressure 100 kPa.

The basic prerequisite for PIV is the presence of tracer particles in the fluid stream. Therefore, the spray appears to be a natural fit for the application of this method. However, practical experience suggests that measuring velocities in spray using PIV is a rather challenging task. The difficulty in using PIV in spray diagnostics stems from the fact that the area under investigation (spray cone) is inhomogeneous in terms of particle size and shape. Several regions can be identified in the spray generated by the effervescent atomizer (Fig. 4). In close proximity to the exit orifice, a rapid expansion of the atomized gas occurs, which is accompanied by an initial breakup of the liquid. In this region, the liquid concentration is very high due to the small diameter of the spray cone, comparable to the exit orifice. In this region the spray is very dense. As the spray cone expands, deformation of the liquid occurs due to inertial forces on the ligaments, the filaments, which further break up into large droplets ([1]). The processes described above are called primary breakup. The products of the primary breakup-the large droplets-are further deformed and undergo secondary breakup into smaller, usually spherical, droplets. After the secondary breakup of the liquid is complete, the diameter of the spray cone is much wider than in the area downstream of the exit orifice and hence the liquid concentration is also lower.

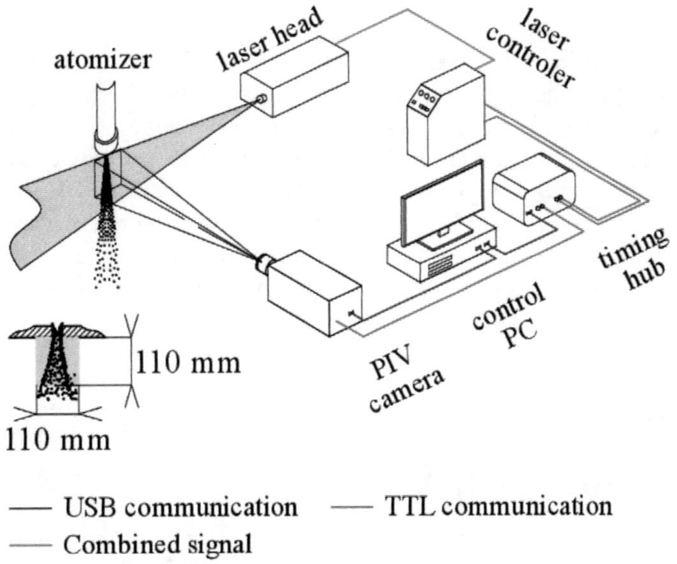

Fig. 3. Schematic drawing of the PIV setup.

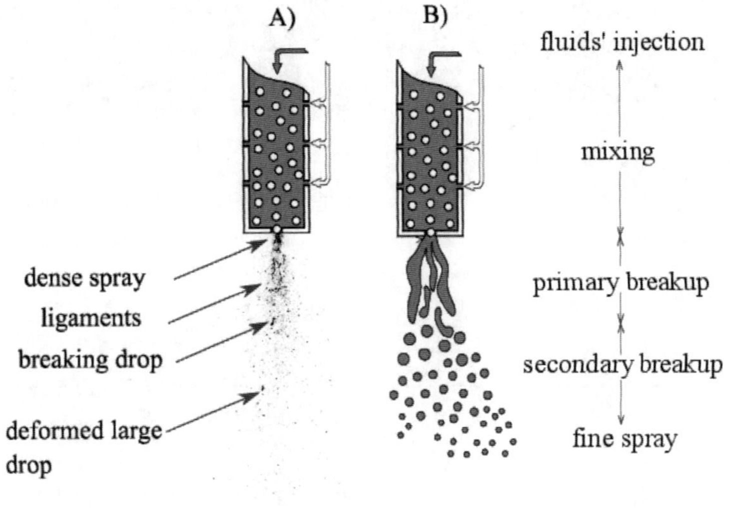

Fig. 4. Spray formation shown in inverted PIV image (A) and in schematic drawing(B).

As can be seen in Fig. 4A, the regions of spray formation described above correspond to the presence of liquid formed into structures of different shapes and densities. In the region of primary breakup, the liquid concentration is large, causing strong light reflexion and overexposure of the PIV images. The expansion of the spray cone and the gradual disintegration of the liquid into ligaments and large droplets cause the particle concentration to decrease. Therefore, in the secondary breakup region, the exposure of the image is suitable for PIV analysis. Further breakup of the large droplets into fine spray causes a gradual decrease in liquid concentration and also a decrease in the reflecting area of the individual droplets. Therefore, the fine spray region is relatively poorly exposed in the images and the low density of particles recorded may make PIV analysis impossible.

3 Results

For the above reasons, it is virtually impossible to record images suitable for PIV analysis of the velocity field in the spray from the exit orifice to the fine spray region. However, it is possible to record particle images for individual spray zones. This can be achieved by adjusting the exposure of the images (Fig. 5). To evaluate the velocity field in the primary breakup region, it is necessary to decrease the illumination intensity or increase the aperture number of the camera lens to achieve an optimal exposure in the dense spray region (Fig. 5a). On the contrary, high light intensity or low lens aperture numbers are required to obtain images of particles in the peripheral parts of the spray cone or even in the wider surroundings (Fig. 5d).

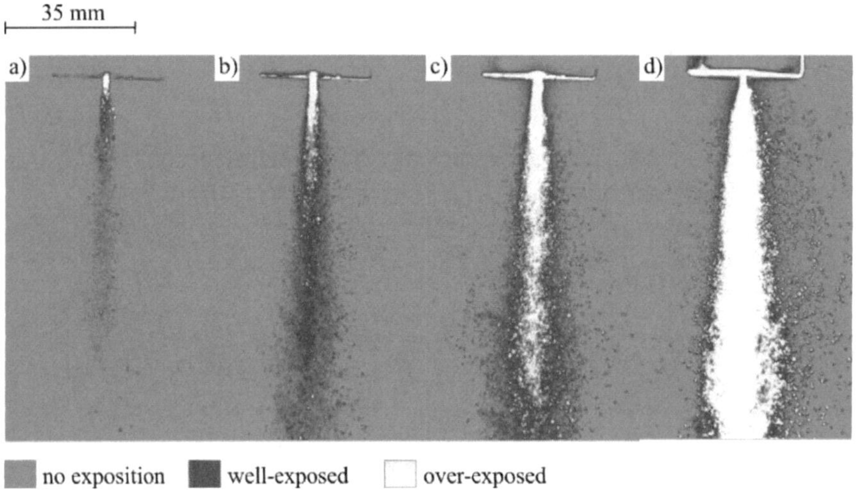

Fig. 5. Spray image obtained with different laser intensity settings (effervescent atomizer, Δp = 100 kPa, $Q_{m,w}$ = 2.85kg/h).

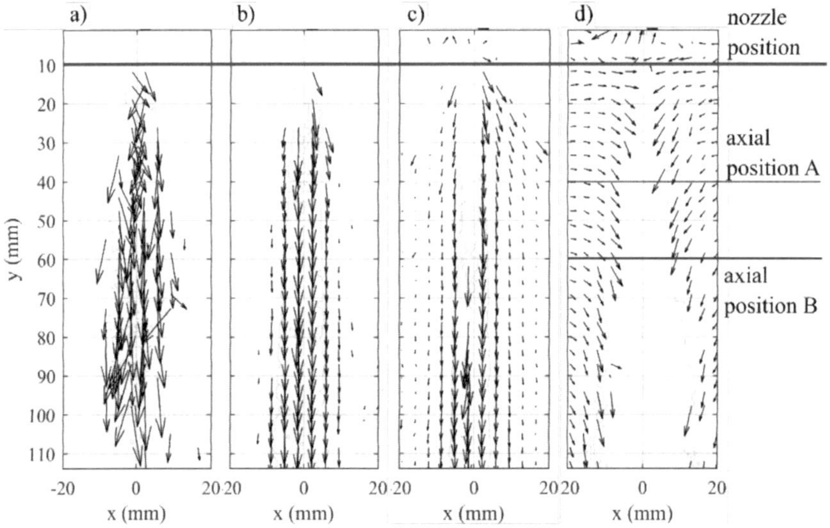

Fig. 6. Mean velocity vectors obtained with different laser intensity settings (effervescent atomizer, Δp = 100 kPa, $Q_{m,w}$ = 2.82kg/h).

The result of the time-averaged velocity field calculation for one working mode of the atomizer for each exposure setting is shown in Fig. 6. For the lowest laser intensity (Fig. 6a), all calculated vectors are concentrated near the central part of the spray, corresponding to the region with the spray cone clearly visible in Fig. 6a. From these images, in which the central, densest parts of the spray are overexposed, vectors in regions farther from the axis but not outside the spray cone are obtained (Figs. 6b, c). With sufficient illumination intensity, the entire spray cone region, except for its peripheral parts, is overexposed (Fig. 6d). However, the small droplets in the wider surroundings are sufficiently illuminated, so it is possible to calculate velocity vectors over the entire field of view of the camera. Due to the low velocities in this region compared to the velocities of the droplets in the spray cone, it is necessary to extend the time delay of the second laser flash so that the particle displacement between the two exposures is sufficient for PIV analysis. Thus, by varying the exposure and time delay of the laser flashes, partial velocity maps can be obtained for each region of the camera field of view.

The gaps in the obtained velocity field can also be seen by displaying the velocity profiles at distances of 40 mm and 60 mm behind the nozzle in Fig. 6 a,b.These distances were intentionally chosen for illustration because the velocity vectors at all image exposures presented in Fig. 5 are available here. Comparing the velocity profiles at exposures a) to c) close to the spray axis, it can be seen that the measured velocities differ. Closer to the nozzle (Fig. 7a) these differences are more noticeable. The velocity profiles at exposure a) are considerably distorted, indicating poor particle image quality in this region. This is because these locations were underexposed at low illumination intensities with a low density of visible particles. For exposure (b), the velocity profile obtained is the best quality of those presented. Further increasing the illumination intensity (exposure of figure (d)) disturbs the smoothness of the measured velocity profile due to the loss of pattern due to overexposure of the image. In Fig. 7b, the velocity profiles are depicted at a distance of 60 mm from the nozzle. At the lowest laser intensity (exposure (a)), the measured velocity profile is also of poor quality at this distance, similar to the previous case. At moderate illumination (exposures (b) and (c)), the results overlap. This indicates a good quality particle image recording at a radial distance of up to 10 mm from the center of the spray cone for both exposure regimes.

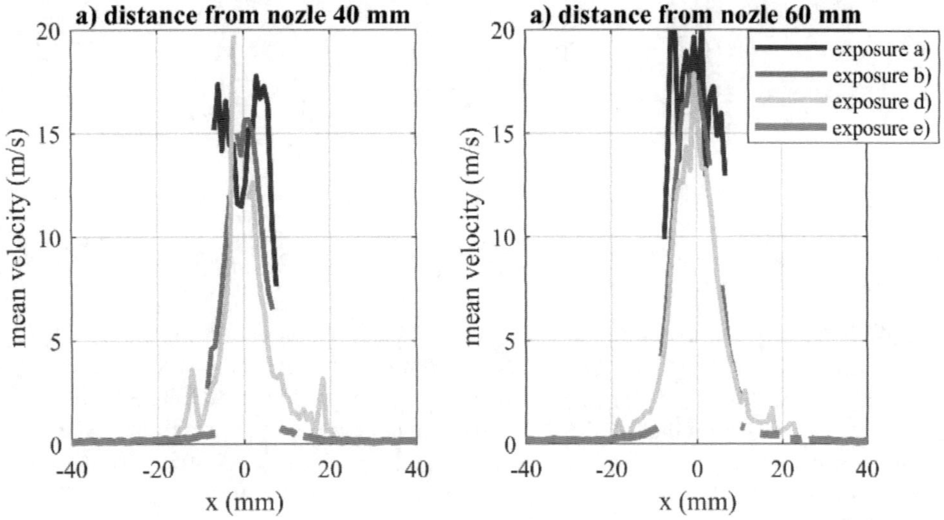

Fig. 7. Mean velocity vectors obtained with different laser intensity settings (effervescent atomizer, Δp = 100 kPa, $Q_{m,w}$ = 2.82kg/h).

By merging the partial velocity fields obtained at different exposure settings, it is possible to evaluate the velocity field over the entire field of view of the camera. The resulting vector field obtained by averaging the partial results presented in Fig. 8 is shown in Fig. a). As can be seen, the velocity field is complete from the nozzle discharge region to the regions outside the spray cone. It can be concluded that merging the vector fields by averaging them removed the influence of random errors. An exception is the region just downstream of the nozzle exit (Fig. 7a, x = 0 mm, y = 10 to 20 mm) where velocity vectors can be observed for which there is reasonable doubt as to their correctness. When the radial velocity profiles are analysed (Fig. 7b), disturbances to the smooth pattern are also evident.

The aforementioned deficiencies in the quality of the resulting velocity field arise from the poor quality of the partial data used in the averaging. Unfortunately, the presented particle image acquisition technique based on several exposure settings naturally generates regions in the images where the quality of the recorded particle image is low. This is due to the fact that the illumination intensity is adjusted with respect to the image quality only in a certain region, as demonstrated in Fig. 5. Obviously, velocity vectors obtained from regions of the image with poor exposure quality are undesirable and need to be eliminated. For this purpose, a wide range of tools are used in PIV analyses. These tools can be divided into pre-processing and post-processing tools. Pre-processing is actually image adjustment (intensity, contrast, exposure) before the actual PIV analysis. From the experience of the authors of the article, it is preferable to pay more attention to the setup of the PIV apparatus and obtain the best quality images possible. Any manipulation of the source images carries the risk of losing information or, worse, generating artificial signals.

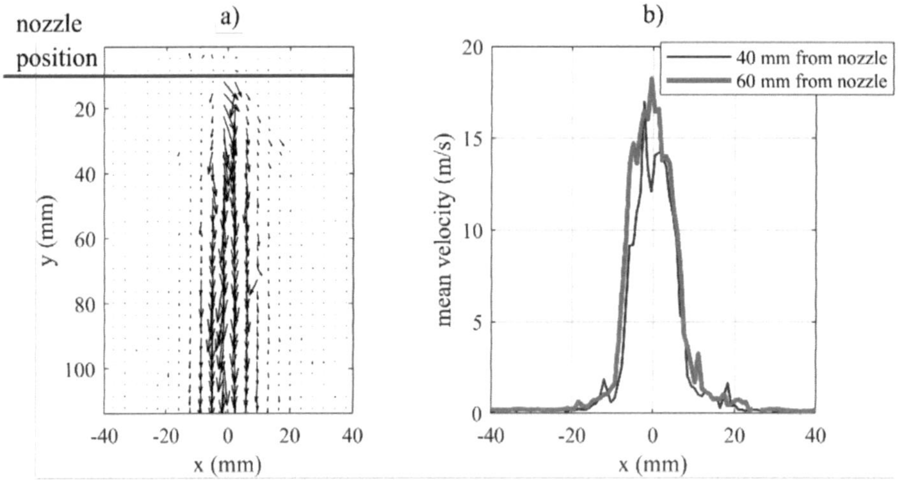

Fig. 8. Velocity field obtained by averaging of the partial data (expositions a, b, c, d) (effervescent atomizer, Δp = 100 kPa, $Q_{m,w}$ = 2.82kg/h).

Post-processing can be image-based or vector-based. When analysing velocity fields in sprays, setting up rules for identifying and discarding faulty vectors is crucial, since a large number of such vectors are generated during spray diagnostics (Fig. 9a). The vectors in this figure were obtained from the image of Fig. 5.b. The strategy in the presented velocity field measurement technique is to eliminate as many vectors as possible in the partial data. The post-processing of the data in this study is performed using four filters. No reverse flow can be expected in the central parts of the spray cone. Therefore, all vectors that have negative axial velocity can be filtered out in the first post-processing step (Fig. 9b). This operation cannot be applied for the regions in the peripheral parts of the spray, since interaction with the surroundings and the formation of vortices may occur here. A more versatile tool for vector discarding is the local median filter. This filter compares each velocity vector with the median of the vectors in its vicinity. If the analysed vector is too different, it is discarded. This filtering obviously works best in areas of high vector density. As can be seen in Fig. 9c (for y>50 mm), outside the spray centre this filter does not work optimally. To remove vectors that are incorrect, but pass the previous filtering, image-based filtering is used. In the first instance, the fault data are extracted from regions of the source images, which are too bright (not necessarily overexposed) or too dark. Here, there is a loss of drawing in the image leading to a correlation failure. The result of applying this filtering is shown in Fig. 9d. Another image defect that can lead to incorrect results is the low contrast in the analysed part of the image. By appropriately selecting the threshold of the allowed contrast, additional erroneous vectors are filtered out (Fig. 9e). The effect of applying all filters is presented in Fig. 9f. Compared to the original state (Fig. 9a), the number of vectors after filtering is considerably smaller. However, the resulting velocity field contains significantly fewer bad vectors. The smaller vector density can be compensated by a larger source data set, i.e., a larger number of frames for analysis, when creating the averaged velocity field.

In Fig. 10, the result of the evaluation of the averaged velocity fields from each of the four exposure modes is shown. Compared to the identically processed data in Fig. 8a, it can be seen that the resulting vector map is of significantly higher quality when using higher quality (filtered) partial velocity fields. The number of missing or bad vectors is minimal throughout the field of view. Similar results could not be obtained without using averaging of the partial results obtained at different exposure settings.

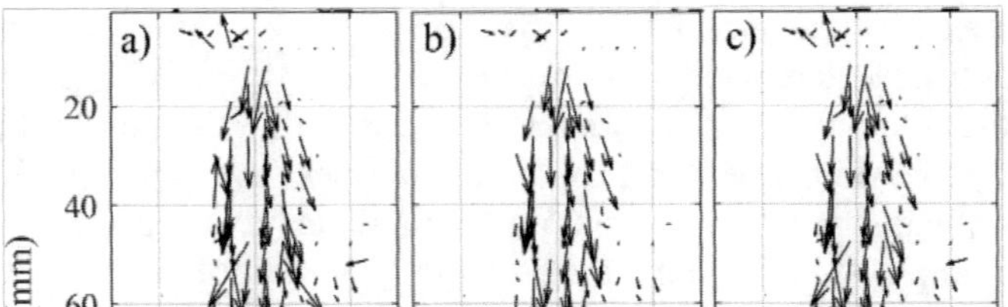

Fig. 9. Velocity field obtained by averaging of the partial data (expositions a, b, c, d) (effervescent atomizer, Δp = 100 kPa, $Q_{m,w}$ = 2.82kg/h).

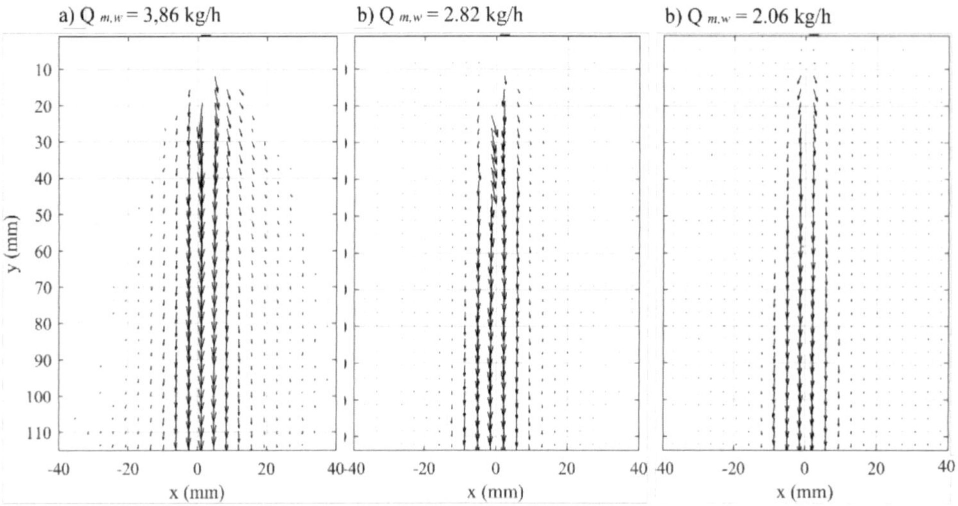

Fig. 10. Velocity field obtained by averaging of the filtered partial data (expositions a, b, c, d) (effervescent atomizer, Δp = 100 kPa).

4 Conclusion

This paper presents one possible approach to the use of the PIV method in spray diagnostics. It shows that achieving the correct exposure of droplet images is challenging. In the case where the velocity field also needs to be measured in the surrounding environment of the spray cone at the same time, this is an impossible task. In this paper, the authors present an approach to solve this problem by obtaining particle images at several exposure settings. This achieves obtaining a good quality image in the target part of the spray. If necessary, the time delay of the laser flashes can also be varied, thus increasing the range of measured velocities. This approach has the disadvantage of producing a large number of invalid vectors in areas of the image that are over- or underexposed. However, by applying appropriate filters to the sub-velocity fields, this disadvantage can be effectively removed and, as a consequence, a high-quality velocity field can be obtained from the primary liquid breakup region to the more distant surroundings of the spray cone.

This work was supported by the Slovak grant agency KEGA, project No 016STU-4/2022 and Advancing University Capacity and Competence in Research, Development and Innovation /ACCORD/ (ITMS+: 313021X329).

References

1. Kamrak, J., Kongsombut, B., Grehan, G., Saengkaew, S., Kim, K-S., Charinpanitkul, T. (2009). Mechanistic study on spraying of blended biodiesel using phase Doppler anemometry. *Biomass and Bioenergy*, 33 (10), 1452-1457.
2. Shin, D., Satija, A., Lucht, R.P. (2022). Spray characteristics of standard and alternative aviation fuels at high ambient pressure conditions. *Experimental Thermal and Fluid Science*, 130.
3. Dafsari, A. R., Yu, S., Choi, Y., Lee, J. (2021). Effect of geometrical parameters of air-induction nozzles on droplet characteristics and behaviour. *Biosystems Engineering*, 209, 14-29.

4. Zhao, W., Yan, J., Gao, S., Lee, T. H., Li, X. (2021). Effects of fuel properties and aerodynamic breakup on spray under flash boiling conditions. *International Journal of Pharmaceutics*, 387 (1-2), 82-87.
5. Liu, X., Doub,W. H., Guo,Ch. (2010). Effect of geometrical parameters of air-induction nozzles on droplet characteristics and behaviour. *Biosystems Engineering*, 209, 14-29.
6. Mlkvik, M., Stahle, P., Schuchmann, H.P., Gaukel, V., Jedelsky, J., Jicha, M. (2015). Twin-fluid atomization of viscous liquids: The effect of atomizer construction on breakup process, spray stability and droplet size. *International Journal of Multiphase flow*, 77, 19-31.
7. Habchi, Ch., Oneissi, M., Russeil, S., Bougeard, D., Lemenand, T. (2021). Comparison of eddy viscosity turbulence models and stereoscopic PIV measurements for a flow past rectangular-winglet pair vortex generator. *Chemical Engineering and Processing - Process Intensification*, 169.
8. Lee, H., Hwang, W. (2019). Error quantification of 3D homogeneous and isotropic turbulence measurements using 2D PIV, *International Journal of Heat and Fluid Flow*, 78.
9. Arpino, F., Cortellessa, G., Scungio, M., Fresilli, G., Facci, A., Frattolillo, A. (2021). PIV measurements over a double bladed Darrieus-type vertical axis wind turbine: A validation benchmark, *Flow Measurement and Instrumentation*, 82.
10. Thysen, J-H., van Hooff, T., Blocken, B., van Heijst, G. (2022). PIV measurements of opposing-jet ventilation flow in a reduced-scale simplified empty airplane cabin, *European Journal of Mechanics - B/Fluids*, 94, 212-227.
11. Rohn, M., Mathis, P., Röder, T., Müller, D. (2022), Stereoscopic particle image velocimetry study of the single side entrainment behaviour of multiple air jets with different nozzle cross-sections, *Building and Environment*, 205.
12. Shukla, S., Sinha, S.S., Saroha, S., Khan, H.H., Singh, S.N. (2022). Experimental and computational investigation of airwake aerodynamics of the generic aircraft carrier with ski-jump, *Ocean Engineering*, 249.
13. Murugan, R., Kolhe, P. S. (2021). Experimental investigation into flow blurring atomization, *Experimental Thermal and Fluid Science*, 120.
14. Zhou, G., Wang, J., Song, R., Xu, C., Wang, P. (2022). Experimental Study on Droplet Breakup and Droplet Particles Diffusion of a Pressure Nozzle Based on PIV, *Chemical Engineering Science*, (In Press).
15. Kim, D., Shin, J., Son, Y., Park. S., (2021). Characteristics of in-cylinder flow and mixture formation in a high-pressure spray-guided gasoline direct-injection optically accessible engine using PIV measurements and CFD, *Energy Conversion and Management*, 248.
16. Wang, Ch., Wang, Y., Zhu, J., Long, W., Nishida, K. (2020). Design and application of an algorithm for measuring particle velocity based on multiple characteristics of a particle, *Fuel*, 271.
17. Zhu, J., Shan, C., Nishida, K., Long, W., Dong, D. (2022). Simultaneous PIV/LIF-PIV measurements and numerical simulation of liquid flow and ambient gas flow for transient diesel spray, *Fuel*, 309.

Expanding the PIV Spectral Range and the Turbulence Generated by Grid of Prismatic Circular Cylinders

Daniel Duda[1,*], *Vitalii* Yanovych[1,2], and *Václav* Uruba[1,2]

[1]Faculty of Mechanical Engineering, University of West Bohemia, 306 14 Pilsen, Czech Republic
[2]Instituta of Thermomechanics, Czech Academy of Sciences, 180 00 Prague, Czech Republic

Abstract. The grid turbulence past a grid made of row of prismatic circular rods (rod diameter = 10 mm, rod spacing = 20 mm) perpendicular to the flow is observed by using a pair of PIV cameras. The first one has field of view larger (81 mm), the second one smaller (31 mm), which increases the dynamical range, where the spectral turbulence properties can be explored. Energy spectra displays usual behavior approximately following $k^{-5/3}$ law, the anisotropy originates at larger scales and the flatness describing strong rare events is connected with smaller scales. The spectral properties of vorticity do not collapse due to the different lengthscale of differentiation, which make questionable the previous research based on the vorticity statistics.

Research background: Grid turbulence is the best experimentally accessible prototype of ideal homogenous and isotropic turbulence, although it is known, it is not exactly the ideal one.
Purpose of the article: This contribution explores the possibility of expanding the limited dynamical range of PIV method.
Methods: Particle Image Velocimetry is based on observing the motion of small particles carried by the flow.
Findings & Value added: Anisotropy connected with large scales, flatness with small scales; it does not work for vorticity.

Keywords: Particle Image Velocimetry; Grid turbulence; Spectrum; Anisotropy; Flatness

1 Introduction

Experimental investigation of turbulence is nowadays based on two complementary methods: Hot Wire Anemometry (HWA, or CTA – constant temperature anemometry) and the Particle Image Velocimetry (PIV) [1]. HWA is a very accurate point method with very large dynamical range spanning 5 or more orders of magnitude. PIV is a planar method observing directly the flow structures, but its dynamical range is limited by the size of field of view (FoV) on one side and by the resolution of grid of interrogation areas on the other side. This range is typically 1 : 64 in dependence on the camera resolution and velocity vector

* Corresponding author: dudad@kke.zcu.cz

detection settings. In this contribution we try to enlarge this range by observing the same flow by using two cameras with different FoVs.

2 Methods

2.1 Wind tunnel

The open low-speed wind tunnel is described in our previous work [1]. Its deformation during running are checked [2] by using fast 3D optical scanner Aramis [3]. The wind tunnel test section has dimensions 740 × 300 × 200 mm. In the inlet to the test section there is located a grid of vertical rods (note that grids for turbulence research are mostly realized as a square grids [4], although different shapes and topologies are studied as well, e.g. fractal grids [5]). Despite the prismatic geometry, the flow is expected to be fully 3D as proved by Williamson [6] in the case of a single cylinder. The rods have diameter of $d = 10$ mm and its distance is $M = 20$ mm (also referred as Mesh parameter). The tunnel is operated at two velocities 4 m/s and 40 m/s in this case. The corresponding Reynolds numbers based on mesh parameter M are

$$Re_M = \frac{u_{\mathrm{ref}} M}{\nu} = \begin{cases} 5.3 \cdot 10^3 \\ 5.3 \cdot 10^4 \end{cases} \tag{1}$$

later referred just as “lower” and “higher” Re respectively.

2.2 PIV setup

The turbulent flow is measured by using Particle Image Velocimetry (PIV) [7] apparatus, which consists of a pair of FlowSense Mk II cameras, a solid-state Nd:YAG double-pulse laser and a synchronization device. Everything is supplied by the Dantec company. The seeding particles are generated by using fog generator Safex. The Fields of View (FoV) are located within a plane perpendicular to the axes of the rods. The FoV size is different for each camera (see Fig. 1), the larger FoV has size of 81 mm, i.e. 4 × M, while the smaller FoV has size of 31 mm, i.e. 1.5 × M. The smaller FoV is located inside the larger one, both centered approximately at the distance of 235 mm, i.e. aprox. 12 × M from the mesh plane (note the ruler photographed in Fig. 1 has a 10 mm starting length plus the radius of the rods).

(a) Calibration snapshot from 1st cam. (b) Calibration snapshot from 2nd cam. (c) Mapping both figures

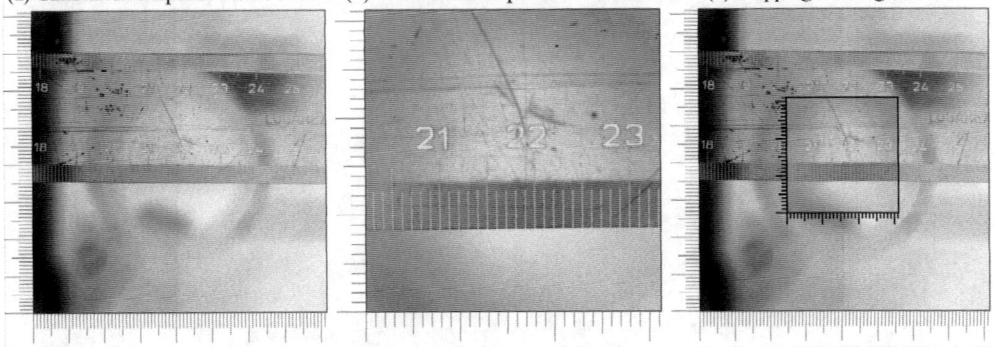

Fig. 1. Calibration snapshots of both cameras (a,b) and together (c).

The PIV method is based on correlating images taken with small time difference ΔT. If ΔT is too long, the correlation peak is too far and the algorithm does not success. On the other hand, if ΔT is small, the velocity calculation works, but the relative uncertainty is higher than it could be. Therefore, it is not useful to use the same ΔT for both FoVs – it would be too long for the smaller FoV or too short for the larger one. Despite this, we took few snapshots with the same ΔT simultaneously on both cameras in order to check the correspondence of instantaneous structures. The example of instantaneous snapshot taken by both cameras simultaneously is shown in Fig. 2 displaying a reasonable correspondence of the fluctuation velocity structures. The statistical analysis later down uses the data captured separately with different values of ΔT suitable for each FoV.

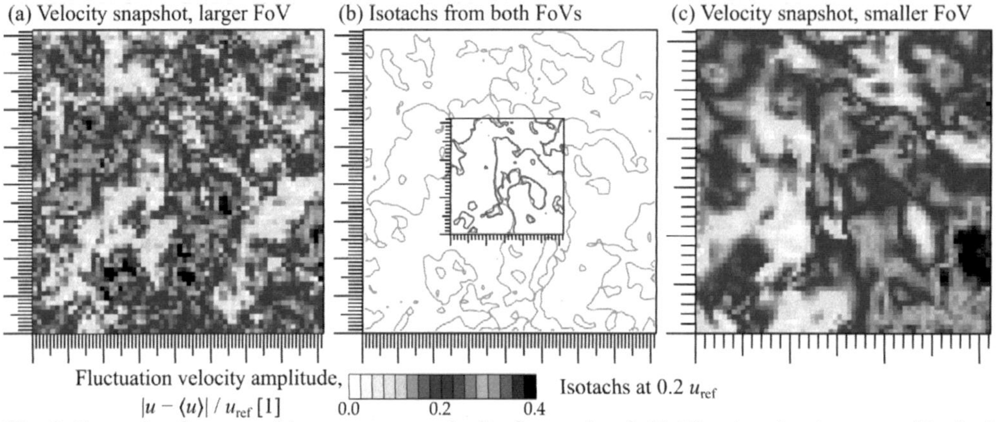

Fig. 2. Example of measured instantaneous velocity fluctuation field. The situation is captured by both cameras simultaneously by applying the same time delay 80 μs, which is too low for the larger FoV and on the edge for the smaller FoV.

3 Results – spectrum, anisotropy and flatness

The spatial spectrum of turbulent kinetic energy is shown in Fig. 3. It is calculated by using our algorithm [8] based on measuring energy of different lengthscale band of Agrawal decomposition [9] [10] [11].

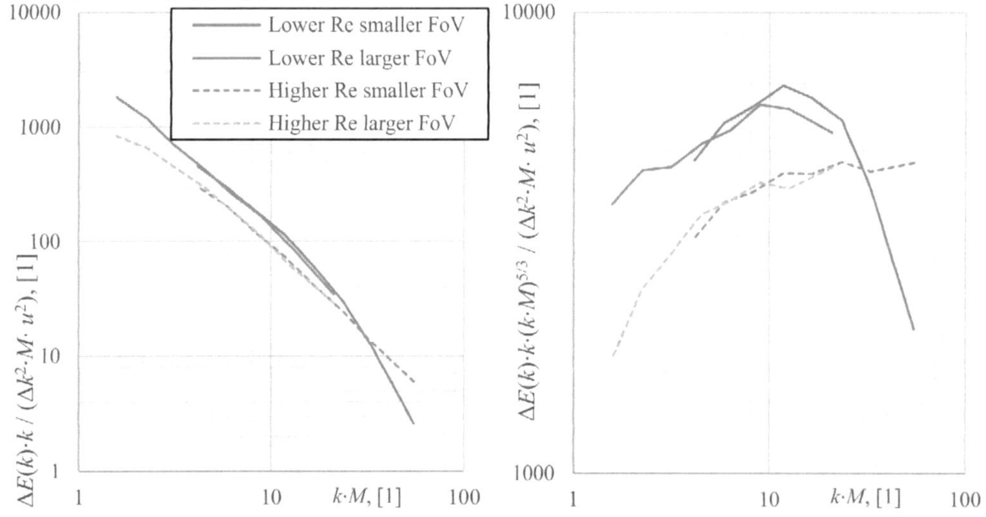

Fig. 3. Left: spatial turbulent kinetic energy spectra as calculated by our algorithm [8]; the wavenumber k is normalized by mesh parameter M and the energy in spatial band by band width Δk, M and velocity u. Right: the same data as in left normalized by the Kolmogorov scaling [12] of inertial range, which would correspond to just a horizontal line in this plot.

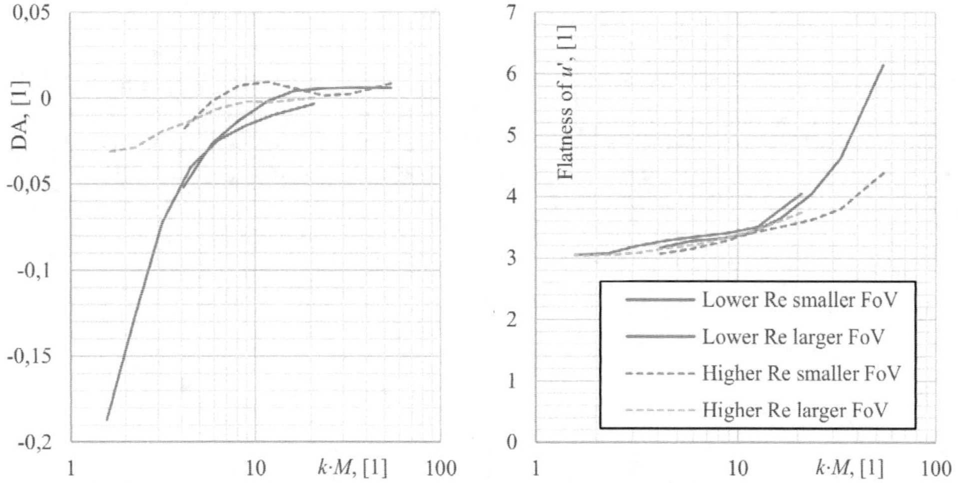

Fig. 4. Left: Degree of anisotropy as a function of wavenumber. Right: Flatness (i.e. fourth statistical moment) of stream-wise velocity fluctuations, note the Gaussian distribution has flatness equal to 3.

It is generally said, that grid turbulence is the best experimental case of Kolmogorov homogenous and isotropic turbulence. However, it is known, that this is not true – the grid turbulence is not homogenous (it evolves with time, i.e. with downstream distance), neither isotropic – the fluctuations are stronger in stream-wise direction. To check the isotropy spectral dependence, we calculate the degree of anisotropy [14] [15] in each lengthscale band as

$$DA = \frac{\sigma_u^2 - \sigma_v^2}{\sigma_u^2 + \sigma_v^2} \tag{2}$$

where σ_u and σ_v are the standard deviations in the stream-wise and perpendicular direction respectively. Fig. 4a shows, that the anisotropy originates at large scales (low k), that the perpendicular fluctuations are stronger than the stream-wise, and that its magnitude is higher at smaller Re.

The flatness or fourth statistical moment of velocity fluctuations describes the relative importance of strong rare events in the flow. It is calculated as

$$F = \frac{\langle u'^4 \rangle}{\sigma_u^4} \tag{3}$$

where u' is the fluctuation part of stream-wise velocity and $\langle\cdot\rangle$ represents ensemble averaging. The "neutral" value of F is 3, because it is flatness of Gaussian distribution. The intermittent effects in turbulence increase this value due to the presence of long-living coherent structures, which are typically of smaller sizes, as it is proven by Fig. 4b. Much more extreme is this effect in quantum turbulence, where the turbulent part of the flow is constructed of quantized vortices of very small size and thus the flatness diverges there [16].

4 Discussion – the vorticity

Vorticity and enstrophy are very interesting quantities in turbulent research [13] [17]. Unfortunately, experimental estimation of vorticity is based on numerical differentiation of the noisy measured velocity field on a grid sampled by the PIV algorithm:

$$\omega_{\mathrm{PIV}}(x,y) = \frac{u(x,y+\Delta y) - u(x,y)}{\Delta y} - \frac{v(x+\Delta x,y) - v(x,y)}{\Delta x} \tag{4}$$

It is only single component perpendicular to the measuring plane; from definition, it cannot catch structures of size of single grid point. Figure 5 shows example of single instantaneous vorticity snapshot captured by both cameras simultaneously. There is a lot of smaller structures visible at the smaller FoV, which are missing in the larger FoV. The limited resolution not only hides small structures, but decreases the amplitude as well.

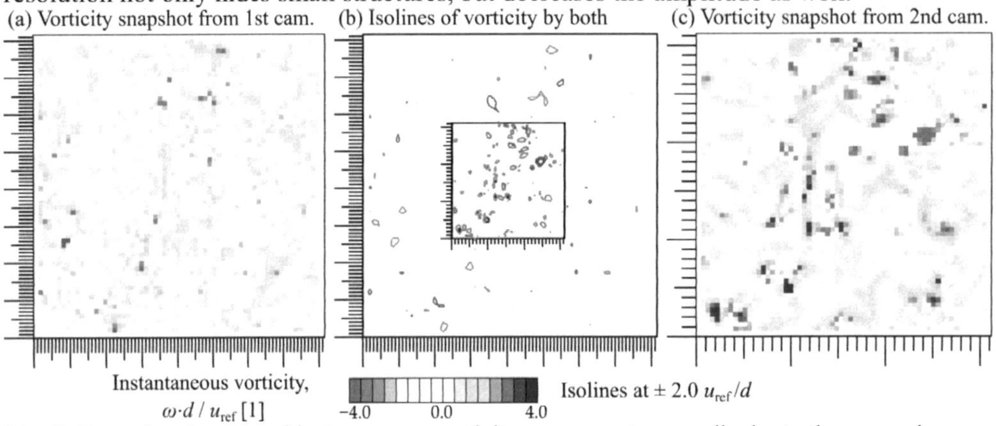

Fig. 5. Example of measured instantaneous vorticity component perpendicular to the measuring area. Note the serious lack of small-scale events in the larger FoV.

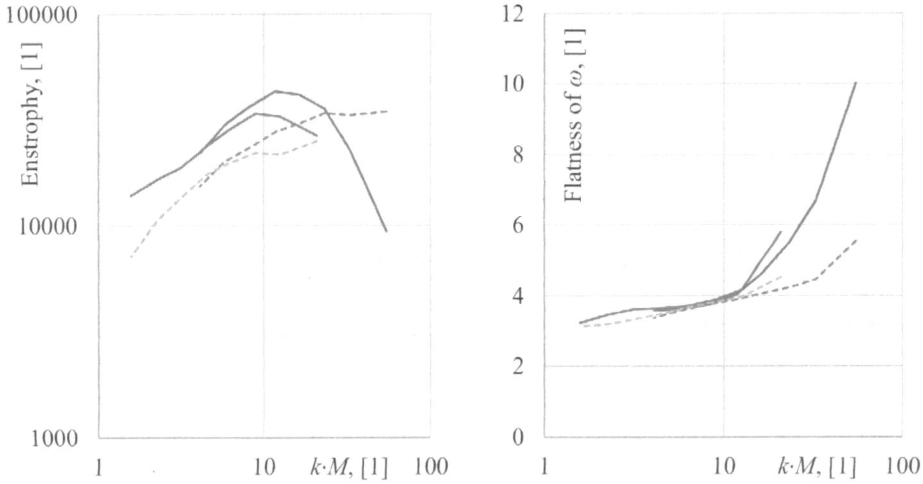

Fig. 6. Left: Estimation of enstrophy spectrum. Note the presented quantity is only rough estimation of enstrophy as it lacks 2 of 3 vorticity components. Additionally, the vorticity is calculated by numerical differentiation on single grid point (different for each FoV), which is probably responsible for the misalignment of the data for smaller and larger FoV. Right: Flatness of the vorticity. The legend is same as in Fig. 3.

Therefore, the enstrophy† estimation is heavily underestimated as well. Note that the presented value is based on one component only. Enstrophy is responsible for energy dissipation, therefore its value grows with wavenumber [13], which is not reproduced by our algorithm [8], additionally, the underestimation depends on the proximity to the PIV grid scale as apparent in Fig. 6. Enstrophy estimation in Fig. 6 is calculated as the variance of vorticity over the FoV and the ensemble and it is normalized as

† *Enstrophy* relates to vorticity analogically, as *turbulent kinetic energy* relates to velocity. Some authors include divergence as well (fully: the determinant of velocity gradient tensor)

$$Z_{\text{estimated}} = \langle \omega^2 - \langle \omega \rangle^2 \rangle \frac{kM}{\Delta k^2 u_{\text{ref}}^2} \quad (5)$$

where the average of vorticity is close zero. Here we hope for some useful comment from reviewer.

Vorticity flatness again shows increasing intermittency towards smaller scales, however, the effect is not trustworthy similarly as the enstrophy estimation although the values between different FoVs rely better. The vorticity flatness increases faster than the velocity flatness with increasing wavenumber. In our resolution, both flatnesses approach to 3 at largest probed scales; we are not sure, how this dependence continued towards larger scales – stay at 3 or go lower?

5 Conclusion

By using two cameras of PIV (particle image velocimetry) system, the lengthscale range has been expanded. The quantities related directly to velocity seem to overlay well, while the derivative quantities, such as vorticity, do not collapse. This decreases the trustworthiness of other results based on vorticity captured with single resolution only [18] [19] [20]. The spectral dependence of degree of anisotropy shows, that the anisotropy originates at larger lengthscales, while the intermittent (strong and rare) events are connected more with smaller lengthscales.

The work was supported from ERDF under project “Research Cooperation for Higher Efficiency and Reliability of Blade Machines (LoStr)” No. CZ.02.1.01/0.0/0.0/16_026/0008389.

References

[1] V. Yanovych, D. Duda, V. Horáček and V. Uruba, "Research of a wind tunnel parameters by means of cross-section analysis of air flow profiles," *AIP Conference Proceedings,* vol. 2189, p. 020024, 2019.

[2] V. Yanovych and D. Duda, "Using ARAMIS system for measurement of structural stability of running wind tunnel," *MATEC Web Conf.,* vol. 328, p. 01003, 2020.

[3] S. Zhang, Handbook of 3D Machine Vision: Optical Metrology and Imaging, Boca Raton: CRC Press, 2013.

[4] T. Kurian and J. H. M. Fransson, "Grid-generated turbulence revisited," *Fluid Dynamics Research,* vol. 41, no. 2, p. 021403, 2009.

[5] S. Usama, J. Tellez-Alvarez, J. Kopec, K. Kwiatkowski, J.-M. Redondo and N. Malik, "Turbulence behind 3D Multi-Scale Sparse Grids," *Journal of Physics: Conference Series,* vol. 1101, p. 012048, 2018.

[6] C. Williamson, "Three-dimensional wake transition," *Journal of fluid mechanics,* vol. 328, pp. 345-407, 1996.

[7] C. Tropea, A. Yarin and J. Foss, Springer Handbook of Experimental Fluid Mechanics, Berlin: Springer, 2007.

[8] D. Duda and V. Uruba, "Spatial spectrum from particle image velocimetry data," *Journal of Nuclear Engineering and Radiation Science,* vol. 5, no. 3, pp. 030912-1-030912-7, 2019.

[9] A. Agrawal and A. Prasad, "Properties of vortices in the self-similar turbulent jet," *Experiments in Fluids,* vol. 33, p. 565–577, 2002.

[10] A. Agrawal, "Measurement of spectrum with particle image velocimetry," *Experiments in Fluids,* vol. 39, p. 836–840, 2005.

[11] D. Duda and V. Uruba, "PIV of air flow over a step and discussion of fluctuation decompositions," *AIP Conference Proceedings,* vol. 2000, p. 020005, 2018.

[12] А. Н. Колмого́ров, "Локальная структура турбулентности в несжимаемой вязкой жидкости при очень больших числах Рейнольдса," *Воспроизводится по ДАН СССР,* vol. 30, p. 299, 1941.

[13] A. Alexakis and L. Biferale, "Cascades and transitions in turbulent flows," *Physics Reports,* Vols. 767-769, no. 12, pp. 1-101, 2018.

[14] I. Solís-Gallego, A. Meana-Fernández, J. Fernández Oro, K. Argüelles Díaz and S. Velarde-Suárez, "Turbulence Structure around an Asymmetric High-," *Journal of Applied Fluid Mechanics,* vol. 10, no. 4, pp. 1013-1027, 2017.

[15] L. Porreca, M. Hollenstein, A. Kalfas and R. Abhari, "Turbulence measurements and analysis in a multistage axial turbine," *Journal of Propulsion and Power,* vol. 23, pp. 227-234, 2007.

[16] M. La Mantia, P. Švančara, D. Duda and L. Skrbek, "Small-scale universality of particle dynamics in quantum turbulence," *Physical Review B,* vol. 94, p. 184512, 2016.

[17] J.-Z. Wu, H.-Y. Ma and M.-D. Zhou, Vorticity and Vortex dynamics, Berlin Heidelberg New York: Springer, 2006.

[18] D. Duda, V. Yanovych and V. Uruba, "An experimental study of turbulent mixing in channel flow past a grid," *Processes,* vol. 8, no. 11, pp. 1-17, 2020.

[19] D. Duda, J. Bém, V. Yanovych, P. Pavlíček and V. Uruba, "Secondary flow of second kind in a short channel observed by PIV," *European Journal of Mechanics, B/Fluids,* vol. 79, pp. 444-453, 2020.

[20] D. Duda, V. Yanovych, V. Tsymbalyuk and V. Uruba, "Effect of Manufacturing Inaccuracies on the Wake Past Asymmetric Airfoil by PIV," *Energies,* vol. 15, no. 3, p. 1227, 2022.

[21] D. Duda, V. Uruba and V. Yanovych, "Wake Width: Discussion of Several Methods How to Estimate It by Using Measured Experimental Data," *Energies,* vol. 14, no. 15, p. 4712, 2021.

Optimization of Air Flow in the Gasification Chamber of the Gasification Boiler Using the PIV Method

Marek Patsch[1]*, and *Peter* Pilát[1]

[1]University of Žilina, Faculty of mechanical engineering, Department of power engineering, Univerzitná 8215/1, Žilina, SK

Abstract. The article deals with the optimization of biomass combustion in a small heat source using the optimal distribution of combustion air. Uneven distribution of combustion air was observed during certification tests and in real operation of the used heat source and has an impact on uneven combustion of biomass in the gasification chamber, on increasing emissions and combustion losses. In the first phase of the research, optimization was carried out using CFD simulations, then a transparent model of a real heat source was created, on which the real distribution of combustion air in the gasification chamber was observed using the Particle Image Velocimetry (PIV) method. The results of CFD simulations and the PIV method led to the optimization of the cross-sectional profiles of the four supply channels for gasification air supply. CFD simulations and subsequent PIV measurements on the experimental device were carried out without real combustion, only the air flow in the empty gasification chamber was investigated. This approach was chosen in order to simplify calculations and experiments and on the assumption that with optimal distribution of combustion air in the empty chamber, there will be an optimal even during real combustion. The flow of primary air in the gasification chamber in real operation is influenced by the size and shape of the inserted biomass and its location in the chamber, and this influence is random and difficult to verify. After optimization, the distribution of the primary combustion air in the gasification chamber is uniform and the same amount of air flows into the chamber through the four combustion air inlets.

1 Introduction

Gasification heat sources are a modern method of energy use of biomass, in which combustion takes place with high efficiency and low emissions. During gasification, pyrolytic distillation takes place, during which all combustible parts of the fuel are gasified. Combustion takes place in a three-stage process and can be divided into three zones. In the first zone, the drying and gasification of wood takes place under the access of primary combustion air, volatile substances from the fuel are released. Gasification is basically the thermal decomposition of organic and inorganic substances in a closed chamber of the boiler

* Corresponding author: marek.patsch@fstroj.uniza.sk

under a slight overpressure, which is created by a combustion air fan. This gasification process takes place in the boiler reservoir, above the concrete nozzle. In the second zone, the released gases are mixed with preheated secondary air in the nozzle space and form a burning mixture of gases. In the third zone, the gases in the uncooled combustion chamber of the boiler are completely burned and the flue gases are discharged through the tubular heat exchanger into the chimney. This controlled method of combustion guarantees high efficiency - 80-90 %, the heat output of the boiler is continuously adjustable in the range of 40-100 % of the nominal output. [1, 2]

Test operation on a real device pointed to incomplete combustion of the fuel in its gasification chamber, unburned fuel remained in the gasification chamber, which indicated an uneven distribution of operating conditions, such as an uneven distribution of the primary combustion air. Increased emissions and a decrease of nominal power were also recorded. It was assumed that the cause is an eccentrically located combustion air fan and unequal local and longitudinal losses in the channels leading to the combustion air outlets in the gasification chamber. Therefore, according to the construction documentation, a computational CFD model and a model of a real gasification boiler were created, on which the distribution of combustion air in the gasification chamber was experimentally verified using the PIV method. The used Vigas 16 gasification boiler is shown in Fig. 1.

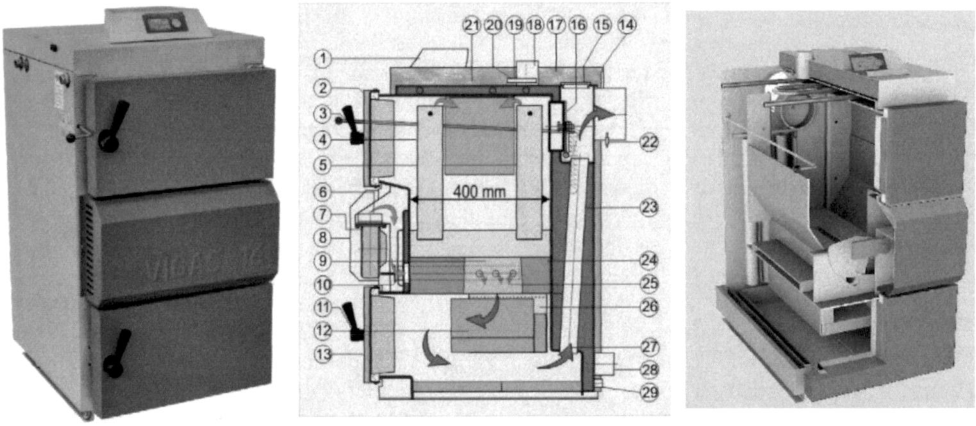

a) b) c)

Fig. 1. a) Vigas 16 gasification boiler, **b)** gasification boiler section : 1. AK 4000 regulation, 2. upper door, 3. chimney damper rod, 4. gasification chamber, 5. primary air line, 6. damper actuator, 7 . combustion air fan, 8. fan cover, 9. hot concrete block, 10. secondary air flap, 11. door cap, 12. fireclay lining, 13. lower door, 14. chimney neck, 15. heat exchanger cover, 16. heating flap . 26. combustion chamber, 27. flue gas direction, 28. return water neck, 29. filling neck, **c)** 3D cross-section of the Vigas 16 gasification boiler. [8].

Important factors that influence the formation of emissions and ash are the dividing of combustion air into primary and secondary, its distribution into the inputs channels (the heat source has four primary and six secondary air inputs) and the speed of the combustion air [9]. The optimal flow of primary combustion air in the gasification chamber and secondary combustion air in the nozzle at the entrance to the combustion chamber has a fundamental influence on the process of gasification and subsequent combustion of biomass. The supplied amount of combustion air is determined by the speed of the fan, part of the total amount of air is taken as secondary air and the rest is primary combustion air. The amount of secondary air is determined by the setting of the secondary air flap [7].

2 Methods

To visualize the flow of combustion air in the gasification chamber of the boiler we use the 2D PIV method, the following components were used:

- Litron Nd-YAG pulsed laser with two cavities, maximum light output 1200 mJ, wavelength 532 nm or 1064 nm.
- Flowsense EO high-speed camera, CCD sensor, resolution 2448px x 2050px, working in double-frame mode.
- FT 700 CE oil particle generator, particle size 2-5 μm.
- Dantec DynamicStudio software. [3, 4, 5, 6]

3 Experimental setup

For the experimental verification of the combustion air distribution, a Vigas 16 biomass gasification boiler was used, which was modified for the needs of using the 2D PIV method. The modification consisted in the removal of the flue gas-water exchanger, which was located in the rear part of the boiler, the covers and thermal insulation were removed from the sides of the boiler, peepholes were made in the side and rear walls of the gasification chamber - for laser illumination and recording with a high-speed camera, on the internal surface of the gasification chamber an anti-reflective black coating was applied. The modified boiler is shown in Fig. 2. The measurement took place with an empty gasification chamber, without combustion. The original exit of the flue gas into the heat exchanger was led through a pipe to the intake of the combustion air fan, oil droplets were fed into this pipe, the movement of which was then monitored using the PIV method. Air flow through the combustion air fan was measured with a propeller anemometer.

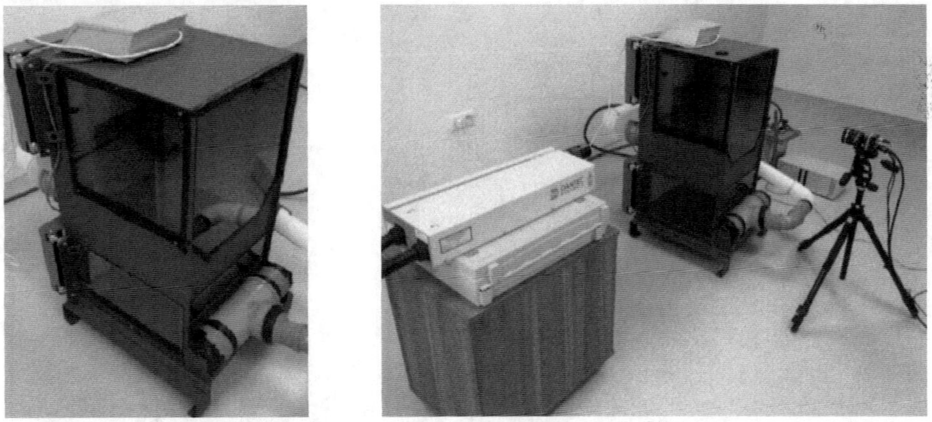

a) b)

Fig. 2. a) Modified gasification boiler Vigas 16, **b)** Assembly of the measuring system.

The entire studied volume of the gasification chamber was divided into several evaluated parts. Looking from the side at the evaluated areas in the middle (area II.) and two near the rear (area I.) and front air intake (area III.), when looking from the back at three evaluated areas near the right combustion air intake (area A), in the center (area B) and near the left air intake (area C), the marking of the parts is shown in Fig. 3. The exact designation of the evaluated area is then marked with a combination of Roman numerals and letters (Tab. 1).

Fig. 3. Marking of evaluated areas when viewed from behind of gasification boiler.

Table 1. Marking of evaluated areas and description.

I-A	rear right air intake
I-B	rear wall in the middle
I-C	left rear air intake - furthest from the combustion air fan
II-B	the center of the gasification chamber
III-A	front right air intake - closest to the combustion air fan
III-B	at the door of the gasification chamber in the middle
III-C	left front air intake

4 Results and discussion

Measurement and visualization took place in the power range of 50 to 100 % of the nominal power of the gasification boiler, which corresponded to the setting of the combustion air fan speed. The results obtained at the nominal performance were used for the evaluation. Fig. 4 shows the progress of the experimental measurement and visualization of the combustion air flow for the evaluated area I-A.

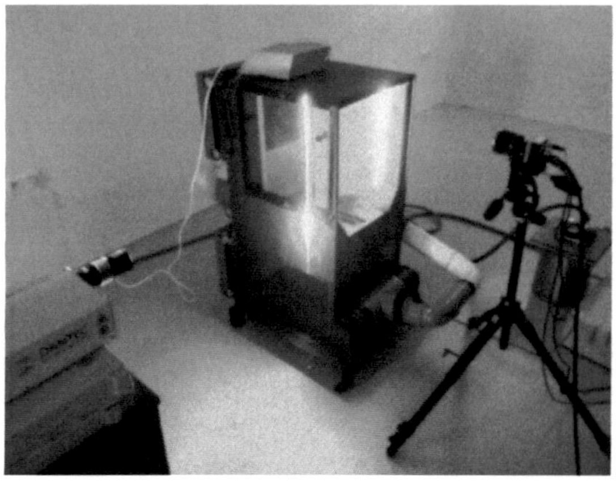

Fig. 4. The experiment.

The result of visualization of the flow in the gasification chamber are velocity profiles, which are further used to calculate the distribution of the amount of combustion air entering the gasification chamber. The measurement was carried out for each area separately at the same setting of the combustion air fan speed. An example of the measured values of the flow velocity in the I-C area is shown in Fig. 5. The calculated redistribution of combustion air in the gasification chamber is in Tab. 2. Most of the combustion air entered the gasification chamber through the air inlet that was closest to the combustion air fan – III-A, the smallest portion through the air inlet that was farthest from the fan – I-C.

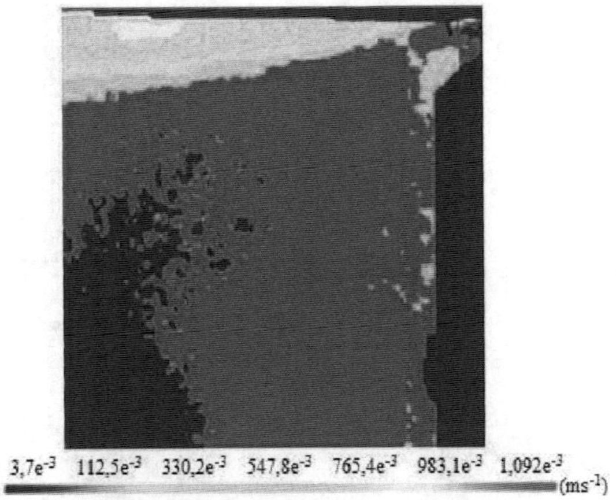

Fig. 5. The velocity scalar map in the evaluated area I-C.

Table 2. Distribution of combustion air into air inputs.

Inlet	Proportion of combustion air [%]
I-A – rear right air intake	20.4
I-C – rear left air intake (furthest from the combustion air fan)	17.6
III-A – front right air intake (closest to the combustion air fan)	33.8
III-C – left front air intake	28.2

Two variants were used to optimize the combustion air distribution. The first optimization variant was a change in the cross-section of the supply channels for the air supply. The air intakes are structurally formed by standardized U-shaped steel profiles and it is not economically advantageous to manufacture each channel by bending separately in order to achieve a change in the cross-sectional area. Therefore, a damper was placed in the outlet of each channel, which reduced the cross-section of the outlet of each channel to achieve the same flow of combustion air through each inlet channel. The second optimization variant was the placement of a low longitudinal partition in the middle of the upper wall of the gasification chamber. The purpose of the partition was to ensure that the streams of combustion air meet exactly in the center of the gasification chamber. By comparing the velocity fields in the center of the gasification chamber (I-B, II-B and III-B) before and after optimization, it can be seen that the air flow before optimization is deviated from the geometric center of the gasification chamber and does not direct axially to the combustion nozzle located at the bottom of the chamber. After optimization, the combustion air is directed exactly along the axis of the gasification chamber.

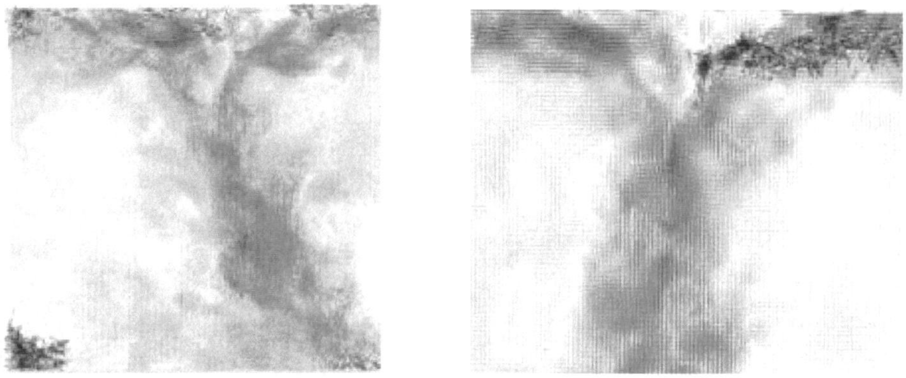

a) b)

Fig. 6. The combustion air flow in the gasification chamber - vectors, **a)** before optimization, **b)** after optimization.

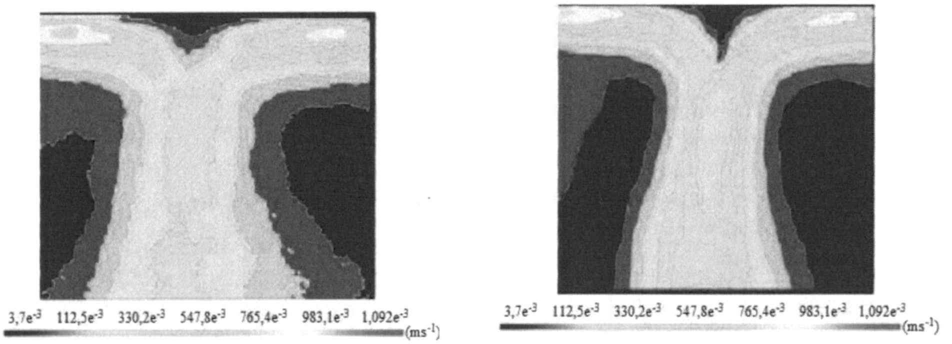

a) b)

Fig. 7. The combustion air flow in the gasification chamber – scalar velocity maps, **a)** before optimization, **b)** after optimization.

The result of the optimization steps was the equalization of the air supply to the gasification chamber, the results of the optimization of the cross-sectional areas of the combustion air inlet channels are shown in table 2.

Table 3. Distribution of combustion air into air inputs.

Inlet	**Proportion of combustion air [%]**
I-A – rear right air intake	24.1
I-C – rear left air intake (furthest from the combustion air fan)	23.8
III-A – front right air intake (closest to the combustion air fan)	27.1
III-C – left front air intake	25.0

5 Conclusions

Real operation of the biomass gasification boiler showed problems with redistribution of combustion air in the gasification chamber. This assumption was confirmed by experimental measurement and visualization of the flow using the PIV method. The uneven distribution of the combustion air was manifested by the unequal speed in the mouths of the air inlets and by entrainment of the air stream after mixing in the center of the chamber away from the axis of the gasification chamber. By optimizing the cross-sectional areas of the combustion air supply channels and using a low partition in the middle of the gasification chamber, a more even flow in the chamber was achieved. Although the operation of the gasification boiler is largely influenced by the operator - the way wood is placed in the gasification chamber, we assume that the optimal distribution of the combustion air will result in the improvement of the operating parameters of the heat source.

This article was created within the work on the task solved by the UNIVNET association no. 0201/0004/20 and it was also supported from the funds of the Ministry of Education, Youth and Sports of the Slovak Republic.

References

1. M. Holubčík, J. Jandačka, R. Nosek, J. Baranski, Particulate Matter Production of Small Heat Source Depending on the Bark Content in Wood Pellets, Emission Control Science and Technology (2018).
2. J. Jandačka, M. Malcho, M. Mikulík, *Technológie pre prípravu a energetické využitie biomasy* (2007).
3. M. Kiener, T. Nussbaumer, Influence of uneven fuel distribution on a grate on gas flow conditions and combustion quality, 20th European Biomass Conference, (Milano,18.–22.6.2012).
4. T. Nussbaumer, M. Keiner, P. Horat, Fluid dynamic optimization of grate boilers with scaled model flow experiments, CFD modeling, and measurements in practice, Biomass and Bioenergy, vol. **76**, (May 2015).
5. T. Nussbaumer, M. Kiener, Moving Grate Combustion Optimisation with CFD and PIV, 21st European Biomass Conference (Copenhagen, 3.-7.6.2013).
6. P. Procházka, V. Barraclough, V. Skála, V. Uruba, The design and aerodynamic testing of DPIK probe, Engineering Mechanics 2022, Vol. 27/28, (2022).
7. K. Sulovcová, R. Nosek, J. Jandačka, M. Holubčík, Geometrical Optimization of the Flue Gas Path with Regard to the Reduction of Particulate Matter, Emission Control Science and Technology (2018).
8. Technical documentation of gasification biomass boiler Vigas 16 (2016).
9. S. Van Loo, J. Koppejan, *The Handbook of Biomass Combustion and Co-firing* (Earthscan, London, 2008).

Demonstration of Non-Uniformity of Velocity Profiles Using 2D PIV

Jaroslav Pulec[1,*], *Jan* Novosád[1], *Petra* Dančová[1], and *Tomáš* Vít[1]

[1]TU Liberec, Faculty of Mechanical Engineering, Department of Power Engineering Equipment, Studentská 2, 46117, Liberec, Czech Republic

Abstract. This paper deals with the demonstration of the non-uniformity of velocity profiles in a DN 200 air circular pipe as part of the development of a newly designed tomographic ultrasonic flowmeter. The flowmeter should measure the flow quality independently. Therefore, different aerodynamic conditions were created by placing mechanical obstacles at the entrance of the measuring chamber, and these conditions were experimentally verified. As an experimental channel substituting a section with ultrasonic sensors, a glass segment was placed at the inlet of a suction pipeline, in front of which perforated plates were inserted to break the uniformity of the velocity profile. The two-dimensional Particle Image Velocimetry (2D PIV) method was used to measure in a plane lying in the longitudinal axis of the channel. The experimental setup and PIV system parameters are described. The results are presented as velocity profiles and contour diagrams in cross-sections at different longitudinal positions in the channel. The results show that the requirement of non-uniformity is satisfied. The used disturbing plates disturb the axial symmetry of the profiles and create a backflow.

Keywords: *tomographic ultrasonic flowmeter, disturbing of flow, non-uniform profile, particle image velocimetry.*

1 Introduction

This paper builds on a previous publication by Barraclough et al. [1], which describes the development of the ultrasonic flowmeter itself. Two methods of measuring flow with ultrasonic probes are being developed to measure flow independently of flow quality. Therefore, it is necessary to set different aerodynamic conditions and to prove that the conditions are indeed other and how.

A suitable method to verify the quality of the flow is the Particle Image Velocimetry method (PIV). PIV is an optical method for velocity field measurement. The principle of this technique is the pulsed illumination of tracer particles with a laser and the recording of a light intensity signal of an illuminated plane (planar and stereoscopic PIV) or volume (3D PIV), from which velocity vectors are calculated in sub-regions called interrogation areas according to the change in particle positions between two exposures [2].

* Corresponding author: jaroslav.pulec@tul.cz

An example of similar applications of the PIV method can be found in the paper by Novosád et al. [3], which dealt with the applicability of the method in a DN 80 pipe. It described the limits of measurement in a circular pipe of a smaller cross-section. Similar study was carried out by Wangxu et al. [4] in DN 100 tube. Zheng et al. [5] were solving a problem of 3D reconstruction of velocity field based on a 2D PIV data in water. Due to the reflections of light on the solid wall, the suitable data processing must be applied, following the methodology published by Adatrao [6] and Amaral et al. [7].

This paper presents results from 2D PIV measurements in several planes lying along the longitudinal axis of the DN 200 glass air channel. The aim is to prove that the velocity profile does not have an axisymmetric paraboloid shape or a shape close to it.

2 Methods

The FlowMaster system by LaVision company [8] with all apparatus for recording 2D PIV data was used for all experiments. The system has the following specifications:

- FlowMaster system from LaVision for 2D, stereo-PIV, and tomo-PIV.
- 4 pcs sCMOS 5.5 MPx camera, resolution 2560 × 2160 pixels, 16 bit, a minimum time interval of two frames 120 ns, exposure time 15 µs to 100 ms, spectral range 370 nm to 1100 nm, maximum frame rate 50 Hz.
- Nd: YAG double-cavity laser, 340 mJ per pulse, wavelength 532 nm, maximum repetition frequency 15 Hz, pulse duration up to 10 nm, class 4 laser. The minimal delay between double pulses d*t* = 1 µs.
- One PC for recording and evaluation, one independent PC for evaluation, Windows 10, DaVis 10.2 software.
- LaVision aerosol generator producing 1 µm diameter droplets of Di-Ethyl-Hexyl-Sebacat (DEHS).

3 Experimental setup

In Figure 1, you can find a scheme of the whole experimental track. The track consists of DN 200 segments like straight sheet-metal pipeline, control valve, Y-connection, flanges etc. The air is driven by two parallel axial suction fans. The maximum flowrate is under 800 $m^3 \cdot h^{-1}$, and the control valve is used to set lower flow rates. In one-third of the 2-meters-long straight segment before the valve, an anemometric probe SVH-14 connected to the KIMOC310 datalogger is placed to determine the velocity parameters for the PIV setup like d*t*. A glass tube is placed in front of the straight section as a measurement section. A toroid-shaped confuser is placed at the channel entrance. There is a top view PIV setup scheme in Figure 2. A laser sheet is fed through the wall from above. The glass cylindrical segment is internally black matt painted on the back half from the camera view. Therefore, the laser light passes through the upper wall only. The reverse side is blackened. Various plate obstacles can be placed between the confuser and the flange of the measuring segment. A measurement area is placed in the laser sheet and starts 100 mm behind the inlet edge of the glass channel, and it is 210 mm long. A region in the distance of 4 mm from the wall is an uncalibrated area. However, this area is covered to prevent wall reflections (see Figure 2b).

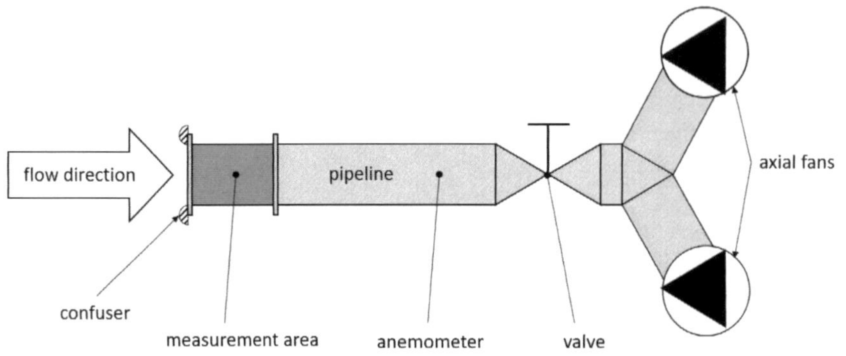

Fig. 1. Scheme of the experimental track, top view.

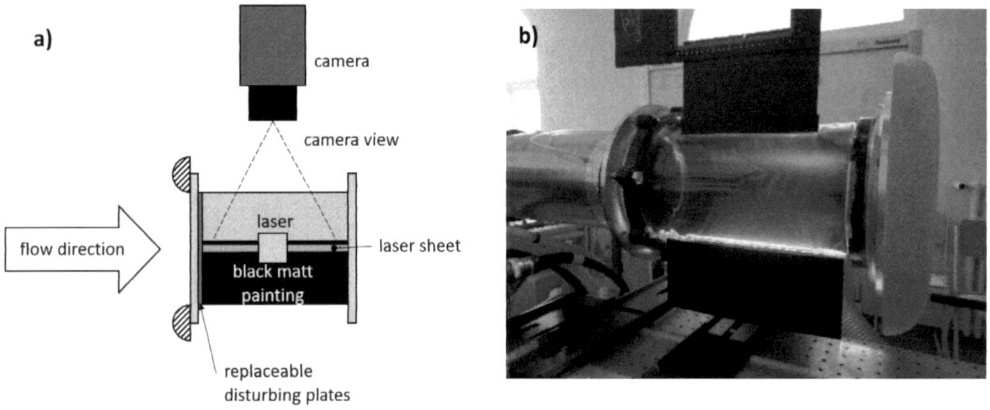

Fig. 2. PIV setup : ***a)*** scheme (top view), ***b)*** experiment.

Monitored regimes are listed in Table 1. Regime 1 symbolizes the measurement of undisturbed flow. Regimes 2 and 3 are measured with disturbing plates according to Figure 3a and Figure 3b, respectively. Different measuring planes are set by rotating the disturbing plates by 45°. All regimes were measured for five flow rates represented by the control valve position (CVP), where 0° means maximum open valve. The currently used control valve includes the mechanism with a scale to set the flap angle. The precision of ±5° fits the customer requirements. However, it was challenging to maintain the same flow rate for all angles of the plates because data for all flow rates for one plate angle were recorded firstly, and then the plate angle was changed. The data was measured in four longitudinal sections {0°, 45°, 90°, 135°} by rotating the plates around the axis. An example of the plate-sheet orientation is shown in Figure 3c.

Table 1. Monitored regimes.

Regime	Plate	Control valve position				
		0°	40°	60°	70°	80°
1	-	1a	1b	1c	1d	1e
2	I	2a	2b	2c	2d	2e
3	II	3a	3b	3c	3d	3e

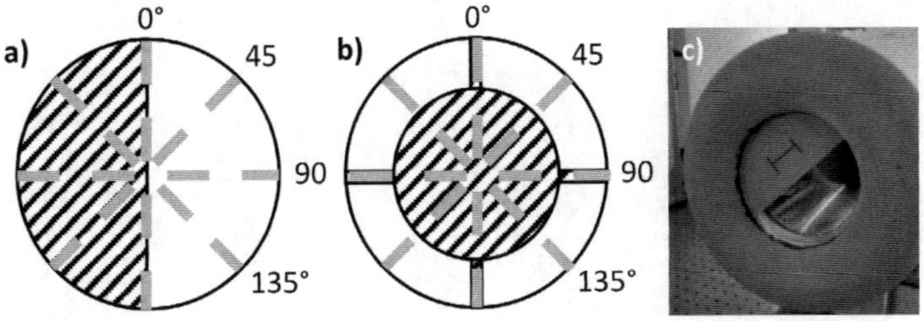

Fig. 3. Disturbing plates: ***a)*** plate I, ***b)*** plate II, ***c)*** plate I, 135°.

For every regime, 450 double frames were taken at 15 Hz. The delay time d*t* was set according to indicatively measured velocities ranging from 30 µs to 400 µs. The laser pulse A energy was 68 mJ, and pulse B was 17 mJ. The data was pre-processed by image background subtraction.

The raw data was evaluated in DaVis 10.2 software with a standard 2D PIV solver. The interrogation area was 32 pixels, overlap of 25 %. The presented data are obtained by averaging all 450 vector maps of every dataset, i.e., 30 seconds of raw data.

4 Results and discussion

The experimental results are in the form of the velocity vector maps in the laser sheet planes. A pack of measured data consists of the data for three different geometries, and each contains vector maps in 4 laser sheet planes.

4.1 Results assessment

The obtained 2D vector maps were analyzed in several cross-section lines, illustrated in Figure 4. The horizontal position *x* of the cross-sections was defined with respect to the dimensions of the ultrasonic flowmeter prototype [1], with 200 mm in length measured from the inlet. The combination of the flowmeter geometry and PIV data leads to the selection of the cross-section at the beginning of the measuring area X100, then the final position of the flowmeter X205, and finally the cross-section X152.5 in the middle.

The vector map data files were loaded and processed in the MATLAB environment. The whole process could be divided into several steps to obtain:

- Reading of the vector map.
- Cross section definitions.
- Calculation of the maximum and dimensionless velocity.
- Scatter plot of the velocity profiles in each cross-section (see Figure 5).
- Velocity profiles in X100, X152.5, and X205 according to the flow rate.
- Contour plot in each cross-section X100, X152.5, X205 for various flow rates.

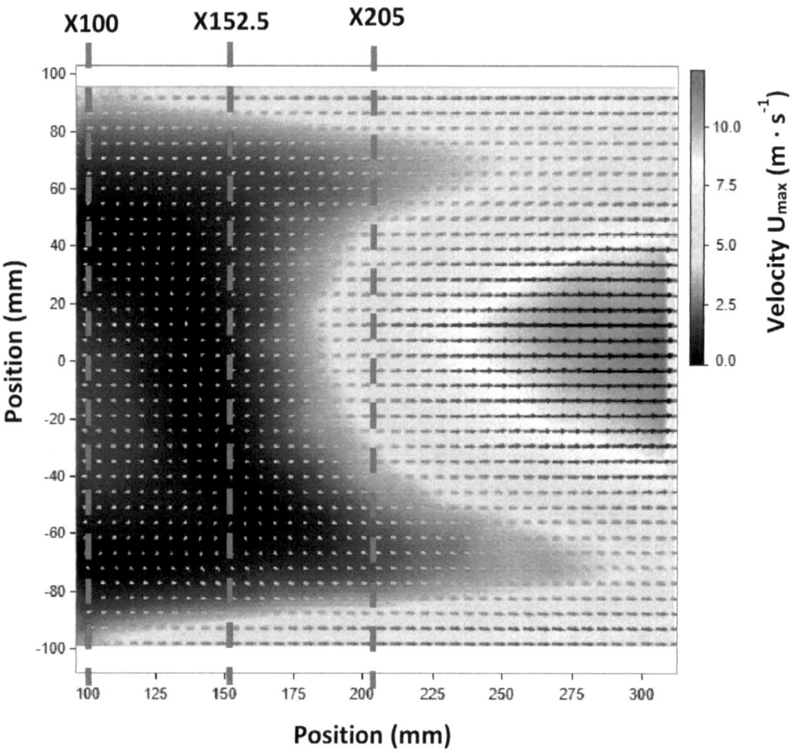

Fig. 4. Velocity vector fields – cross sections for the velocity profiles assessment.

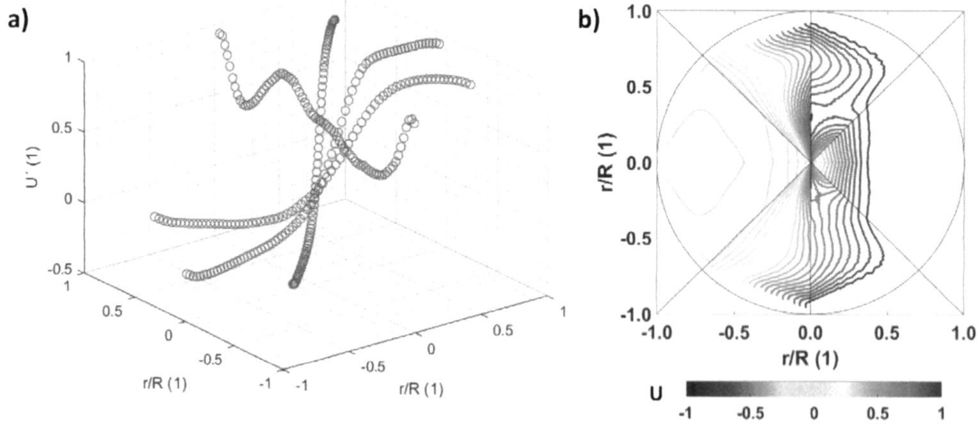

Fig. 5. Example for regime X100, CVP40°: ***a)*** velocity profiles, ***b)*** contours.

4.2 Velocity profiles

Velocity profiles have been evaluated for all cases. *X*-direction velocity profiles are presented in Figure 6 as the comparison for representative cases shows a flowrate dependence. The horizontal axis symbolizes dimensionless pipe radius r/R and the vertical coordinate is the dimensionless velocity $U'=U/U_{\max}$ based on the maximum velocity. Markers are color-coded according to the legend for each valve angle or flow rate setting. All plots are plotted at position X152.5 according to Figure 4, i.e., 152.5 mm behind the inlet edge of the experimental channel. Figure 6a represents regime 1, Figure 6b regime 2, and

Figure 6c regime 3. You can notice the strong influence of the plate. Regime 1 forms an almost flat and uniform profile where most values are close to 1. Regime 2 on the left side includes a backflow, which is an order of magnitude lower than the maximum velocity on the right side. Regime 3 forms axisymmetric profiles where backflow is more significant in the central region. Position X152.5 is an area of still evolving velocity profile. Therefore, even in its dimensionless form, there is a velocity difference of up to 20 % between the maximum and minimum flow rate curve. A small drop of velocity in the area near r/R = (-0.5) is probably due to signal noise caused by reflections on the inner wall.

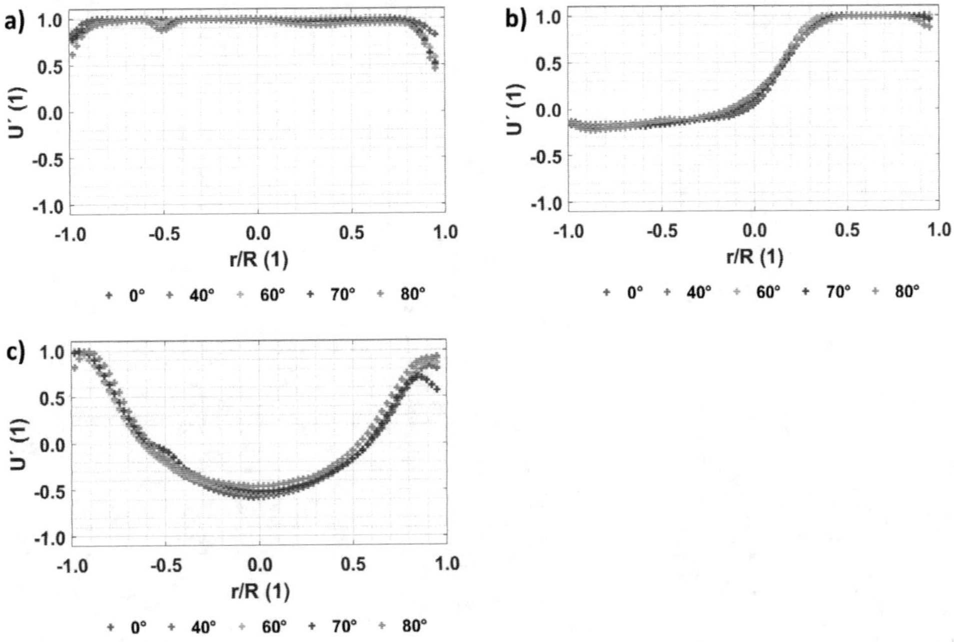

Fig. 6. Velocity profiles at X152.5 for various CVP: ***a)*** regime 1, ***b)*** regime 2, 90°, ***c)*** regime 3, 45°.

Velocity contours of *x*-velocity are shown in Figure 7, Figure 8, and Figure 9. Four-lines cross represents the measured planes. Curves between these crosses are connections of discrete values. The individual lines are graduated by 5 % of the whole velocity range. The dimensionless presentation, as in Figure 6, is maintained. The contours are plotted for three positions, X100, X152.5, and X205, according to Figure 4 to illustrate profile development. The rows are selected modes of closing the control valve in Table 1. The combination of the plotting methodology and the not quite equal flow rate while measuring the different plate angles creates non-smooth curves, i.e., unrealistic loops. The corresponding velocity can only be observed along the 45° crosses. This results presentation was chosen for better visualization.

Figure 7 shows a set of diagrams where the contours in a peripheral part of the pipe are approximately centrally symmetrical, and a higher velocity gradient is evident. Dependence on the valve closing is less noticeable.

Figure 8 shows the contours of regime 2. A half-moon plate causes the highest velocity on the right side and even backflow in the area behind the plate. From the comparison of the *x*-positions, it is evident that the flow also fills the area behind the obstacle. This can also be observed with decreasing flow rates due to lower flow momentum.

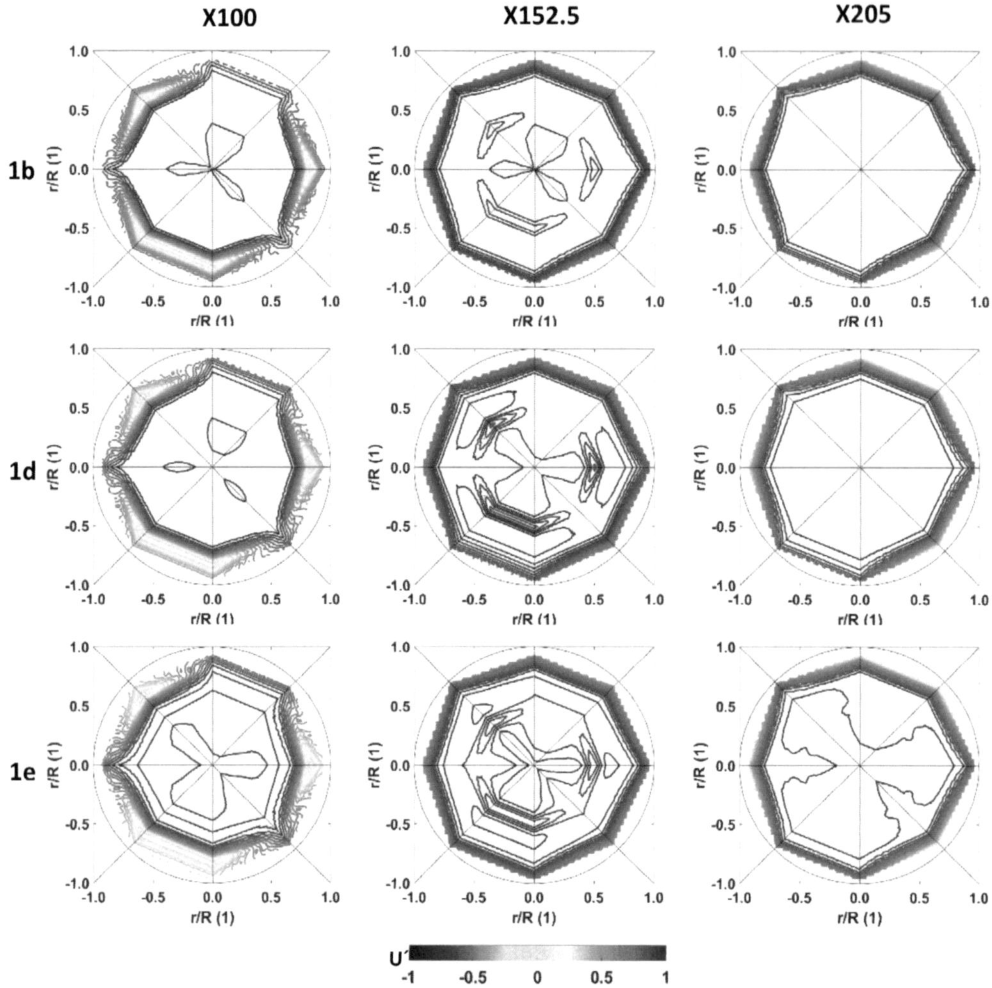

Fig. 7. Regime 1 - velocity contours at X100, X152.5, X205 for various CVP: 1b, 1d, 1e.

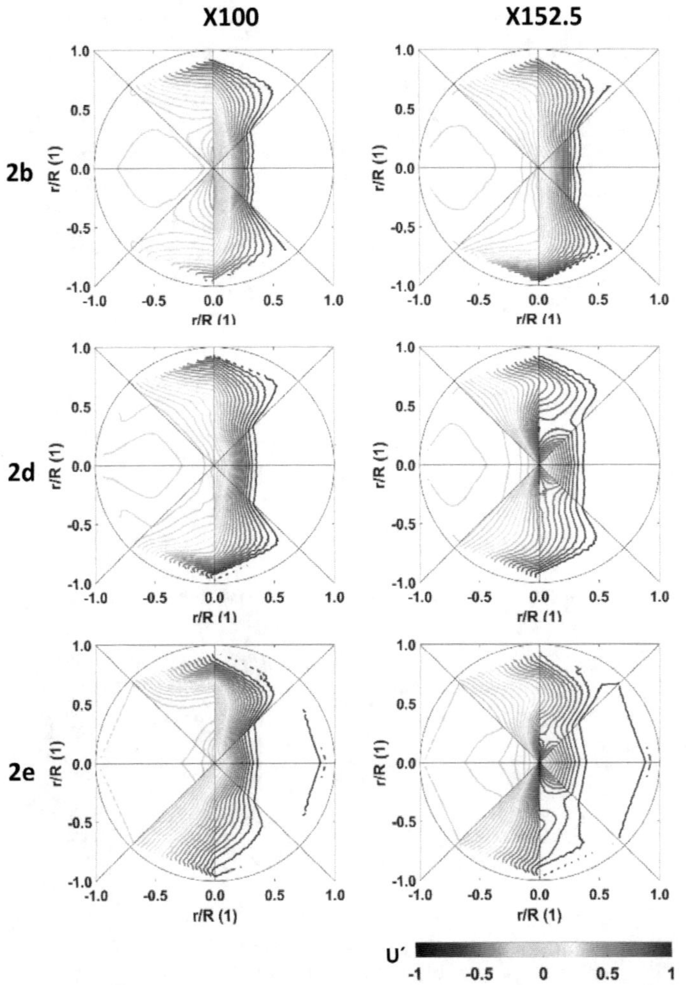

Fig. 8. Regime 2 - velocity contours at X100, X152.5, X205 for various CVP: 2b, 2d, 2e.

Figure 9 shows the contours of regime 3. This regime forms an approximately symmetrical profile. The highest velocities are in the region near 75 % of the radius. In the middle, on the other hand, the flow is backward because of the disc in the central part of the plate. In positions closer to the entrance, the high velocities contours are more densely distributed, which means a higher gradient. With lower flow rates, on the contrary, they become thinner.

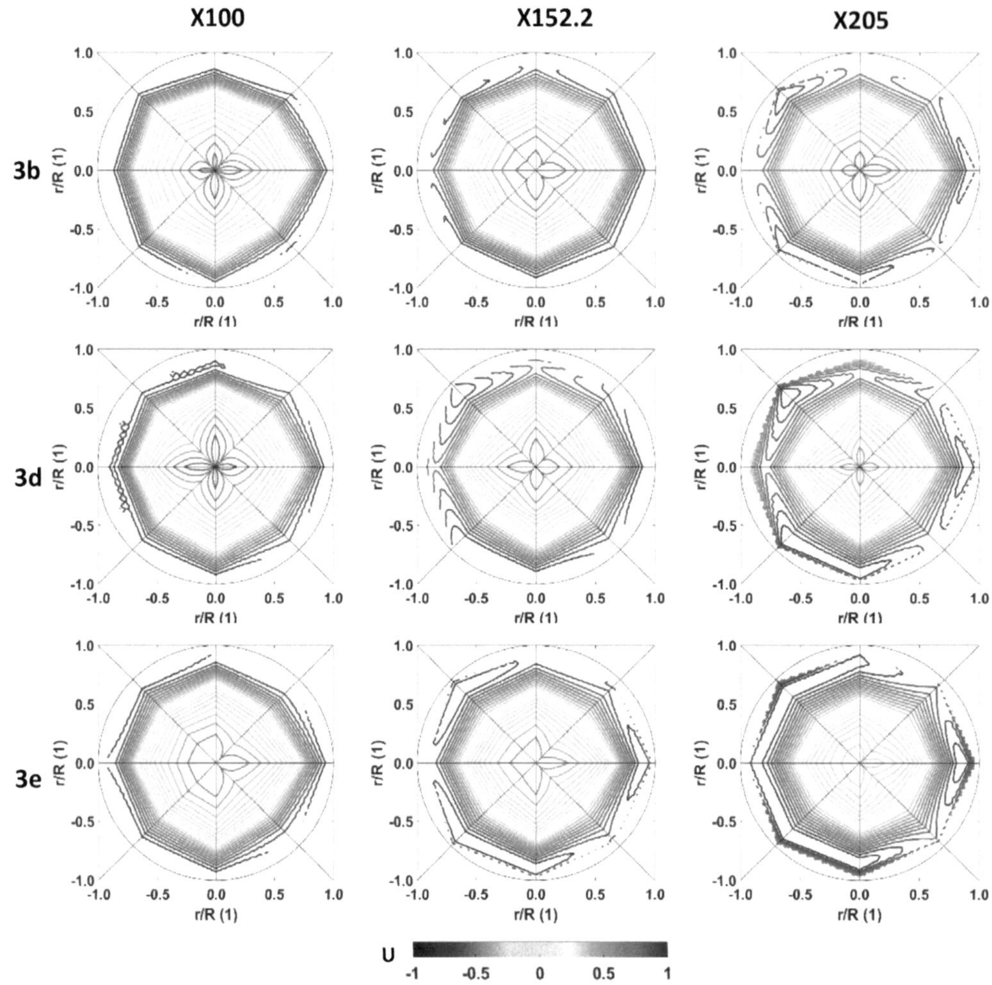

Fig. 9. Regime 3 - velocity contours at X100, X152.5, X205 for various CVP: 3b, 3d, 3e.

4.3 Results discussion

The presented results show that the velocity profile is disturbed by the plates and causes backflow in some areas. From a nearly flat profile, an axisymmetric profile is created using plate I and a centrically symmetric shape using plate II. It is also clear from Figure 6 to Figure 9 that the velocity profile is continuously evolving in the measured area. The flow gradually fills the space behind the obstacles. And with advancing distance, a uniform parabolic profile would form. It can be said that the slow plates satisfy the requirement for proving the independence of the measured values obtained by the developed flowmeter of the velocity profile in the channel.

5 Conclusion

In this paper, the experimental study related to the development of the ultrasonic flowmeter is presented. The setup of the 2D PIV experiment and data processing and post-processing are described. Circular pipe velocity profiles depending on the flow rate and disturbing plates were measured, and the approximate 3D shape of the velocity profiles was reconstructed using MATLAB environment. The data was presented in the form of 2D plots and velocity contour maps. The data showed that the disturbing plates have sufficient influence on the shape of the velocity profile. The dependence on the flow rate is also obvious. It also showed that in the part of the channel where the ultrasonic probes of the developed flowmeter will be located, there is a continuous development of the profile behind the disturbing plates. The two chosen plates are suitable for demonstrating independence of the measured flow rate on the hydrodynamic flow characteristics.

The measurements also showed deficiencies in the obtained data. A better interpolation of the obtained 2D profiles with the surface requires that the flow rate is maintained during the measurement of all positions of the disturbing plate. As a consequence of the continuous evolution of the velocity profile in the measured area, its shape is flow dependent.

The authors gratefully acknowledge financial support by Czech Technological Agency under the project TACR 02020070. This work was also partly supported by the Student Grant Competition of the Technical University of Liberec under project No. SGS-2022-5006.

Nomenclature

dt	s	laser pulse delay
r	m	radial distance; $r = \|y\|$
R	m	pipe radius
U'	1	dimensionless velocity U/U_{max}
U	$m \cdot s^{-1}$	local velocity
U_{max}	$m \cdot s^{-1}$	maximum velocity of profile
x, y	m	Cartesian coordinates

References

[1] V. Barraclough, J. Čížek, M. Strob, J. Novosád, J. Pulec a P. Dančová, „Ultrasonic methods for determining flows and velocity fields,“ *EPJ Web of Conferences,* no. 264, 2022.

[2] C. E. Willert a M. Gharib, „Digital particle image velocimetry,“ *Experiments in Fluids,* no. 10, pp. 181-193, 1991.

[3] J. Novosád, J. Pulec a P. Dančová, „Applicability of the PIV system for velocity field measurement inside the ultrasonic flowmeter,“ *MATEC Web Conf.,* no. 345, 2021.

[4] L. Wangxu, L. Zhenggui, D. Wanquan, J. Lei, Q. Yilong a C. Huiyu, „Particle image velocimetry flowmeter for natural gas applications,“ *Flow Measurement and Instrumentation,* no. 82, 2021.

[5] D. Zheng, G. Lu, J. Zhang a M. Wang, „World Congress on Intelligent Control and Automation,“ v *Experimental Method of Three-dimensional Velocity Field Measurement in Circular Pipe Based on PIV*, Guilin, 2016.

[6] S. Adatrao a A. Sciacchitano, „Elimination of unsteady background reflections in PIV images by anisotropic diffusion,“ *Measurement Science and Technology,* no. 30, 2019.

[7] R. D. Amaral, V. A. A. Bortolin, B. L. H. D. Lemos, M. Mazzeto, I. A. Cestari a J. R. Meneghini, „A novel method based on the Otsu threshold for instantaneous elimination of light reflection in PIV images,“ *Measurement Science and Technology,* č. no, 2021.

[8] „LaVision GmbH,“ [Online]. Available: www.lavision.de.

AUTHOR INDEX

AUTHOR INDEX